Deep Learning Methods for Sensing

About the Editors

Moulay A. Akhloufi

Moulay A. Akhloufi has a Bachelor of Science in Physics from University Abdelmalek Essaadi (Morocco) and a Bachelor of Engineering from Telecom Saint-Etienne (France). He received his Master's and Ph.D. in Electrical Engineering from Ecole Polytechnique of Montreal and Laval University, respectively. He also holds an MBA from Laval University. He is currently Associate Professor in Computer Science, Head of the Perception, Robotics, and Intelligent Machines (PRIME) research group, and Director of the Center for Artificial Intelligence NB Power at Université de Moncton (Canada). Before joining Université de Moncton in 2016, he worked in the computer vision industry with Matrox ES (imaging division) and in technology transfer in machine vision and robotics with CRVI inc. His research interests are in the areas of artificial intelligence, computer vision and intelligent robotic systems. Professor Akhloufi is a senior member of the Institute of Electrical and Electronics Engineers (IEEE) and member of the Society of Photo-Optical Instrumentation Engineers (SPIE).

Mozhdeh Shahbazi

Mozhdeh Shahbazi (PhD, PEng) holds a BSc degree in survey engineering, MSc in photogrammetry engineering, and PhD in remote sensing. Her expertise is in the areas of computer vision, deep learning, photogrammetry, sensor integration and self-calibration, and robot vision. She joined the Department of Geomatics Engineering at the University of Calgary as a professor in 2016. Since 2018, she has been an adjunct professor at York University. In 2019, she took on the role of the lead scientist at the Centre de Géomatique du Québec. During this period of the academic profession, she secured over CAD 5 million R&D funding as the principal investigator and project lead of competitive grants using which she established two research laboratories and a technology access center and trained over 30 highly qualified personnel. She became a machine learning specialist at Genetec Inc in 2021. In this role, she designed algorithms and software related to intelligent surveillance. In 2022, she joined the government of Canada as a computer vision and photogrammetry scientist where she is developing automated techniques for geo-referencing and processing historical images to extract various statistics related to the natural resources of the country. As part of her voluntary activities, she has been a secretary of the International Society of Photogrammetry and Remote Sensing, vice president of the Canadian Remote Sensing Society, associate editor of Canadian Journal of Remote Sensing, and co-editor in chief of Drone Systems and Applications. She has been an active participant and organizer of STEM outreach programs such as NSERC Chairs for Women in Science and Engineering, Women in Data Science, and Cyber Mentor.

Preface to "Deep Learning Methods for Remote Sensing"

The areas of machine learning and deep learning have seen impressive progress in recent years. This progress has been mainly driven by the availability of high processing performance at an affordable cost and a large quantity of data. Most state-of-the-art techniques today are based on deep neural networks. This progress has sparked innovations in technologies, algorithms, and approaches and led to results that were unachievable until recently. Among the various research areas that have been significantly impacted by this progress is remote sensing.

This book gathers cutting-edge contributions from researchers using deep learning for remote sensing. Contributions include recent work in various remote sensing areas of applications such as environmental studies, natural risks, urban analysis, and change detection. The aim of this book is to serve as a starting point for researchers, scientists, and engineers interested in learning about the use of deep learning for remote sensing.

Moulay A. Akhloufi and Mozhdeh Shahbazi
Editors

sensors

MDPI

Article

Deep Learning and Transformer Approaches for UAV-Based Wildfire Detection and Segmentation

Rafik Ghali [1], Moulay A. Akhloufi [1,*] and Wided Souidene Mseddi [2]

[1] Perception, Robotics and Intelligent Machines Research Group (PRIME), Department of Computer Science, Université de Moncton, Moncton, NB E1A 3E9, Canada; rafik.ghali@ept.rnu.tn

[2] SERCOM Laboratory, Ecole Polytechnique de Tunisie, Université de Carthage, BP 743, La Marsa 2078, Tunisia; wided.souidene@ept.rnu.tn

* Correspondence: moulay.akhloufi@umoncton.ca

Abstract: Wildfires are a worldwide natural disaster causing important economic damages and loss of lives. Experts predict that wildfires will increase in the coming years mainly due to climate change. Early detection and prediction of fire spread can help reduce affected areas and improve firefighting. Numerous systems were developed to detect fire. Recently, Unmanned Aerial Vehicles were employed to tackle this problem due to their high flexibility, their low-cost, and their ability to cover wide areas during the day or night. However, they are still limited by challenging problems such as small fire size, background complexity, and image degradation. To deal with the aforementioned limitations, we adapted and optimized Deep Learning methods to detect wildfire at an early stage. A novel deep ensemble learning method, which combines EfficientNet-B5 and DenseNet-201 models, is proposed to identify and classify wildfire using aerial images. In addition, two vision transformers (TransUNet and TransFire) and a deep convolutional model (EfficientSeg) were employed to segment wildfire regions and determine the precise fire regions. The obtained results are promising and show the efficiency of using Deep Learning and vision transformers for wildfire classification and segmentation. The proposed model for wildfire classification obtained an accuracy of 85.12% and outperformed many state-of-the-art works. It proved its ability in classifying wildfire even small fire areas. The best semantic segmentation models achieved an F1-score of 99.9% for TransUNet architecture and 99.82% for TransFire architecture superior to recent published models. More specifically, we demonstrated the ability of these models to extract the finer details of wildfire using aerial images. They can further overcome current model limitations, such as background complexity and small wildfire areas.

Keywords: wildfire detection; fire classification; fire segmentation; vision transformers; UAV; aerial images

Citation: Ghali, R.; Akhloufi, M.A.; Mseddi, W.S. Deep Learning and Transformers Approaches for UAV Based Wildfire Detection and Segmentation. *Sensors* **2022**, *22*, 1977. https://doi.org/10.3390/s22051977

Academic Editor: Chiman Kwan

Received: 13 February 2022
Accepted: 1 March 2022
Published: 3 March 2022

1. Introduction

Forest fire accidents are one of the most dangerous risks due to their frightening loss statistics. The fires cause human, financial, and environmental losses, including the death of animals and the destruction of wood, houses, and million acres of land worldwide. In 2021, forest fires have occurred in several countries such as the European Union countries, the US (United States), central and southern Africa, the Arabian Gulf, and South and North America [1]. They affect 350 million to 450 million hectares every year [2]. In the western US alone, the frequency of wildfires and the total area burned increased by 400% and 600%, respectively, in the last decade [3]. In addition, approximately 8000 wildfires affected 2.5 million hectares each year in Canada [4].

Generally, wildfires are detected using various sensors such as gas, smoke, temperature, and flame detectors. Nevertheless, these detectors have a variety of limitations such as delayed response and small coverage areas [5]. Fortunately, the advancement of computer vision techniques has made it possible to detect fire using visual features collected with cameras. However, traditional fire detection tools are being replaced by vision-based models that have many advantages such as accuracy, large coverage areas, small probability of errors, and most importantly the ability to work with existing camera surveillance systems. Through the years, researchers have proposed many innovative techniques based on computer vision in order to build accurate fire detection systems [6–9].

In recent years, Unmanned Aerial Vehicles (UAV) or drone systems were deployed in various tasks such as traffic monitoring [10], precision agriculture [11], disaster monitoring [12], smart cities [13], cover mapping [14], and object detection [15]. They are also very practical and well developed for wildfire fighting and detection. UAV-based systems help with precise fire management and provide real-time information to limit damage from fires thanks to their low cost and ability to cover large areas whether during the day or night for a long duration [16,17]. The integration of UAVs with visual and/or infrared sensors help in finding potential fires at daytime and nighttime [18]. Furthermore, fire detection and segmentation showed impressive progress thanks to the use of deep learning (DL) techniques. DL-based fire detection methods are used to detect the color of wildfire and its geometrical features such as angle, shape, height, and width. Their results are used as inputs to the fire propagation models. Thanks to the promising performances of DL approaches in wildfire classification and segmentation [19], researchers are increasingly investigating this family of methods. The existing methods use input images captured by traditional visual sensors to localize wildfire and to detect the precise shape of fire; they achieved high results [20–22]. However, it is not yet clear that these methods will perform well in detecting and segmenting forest fire using UAV images, especially in the presence of various challenges such as small object size, background complexity, and image degradation.

To address these problems, we present in this paper a novel deep ensemble learning method to detect and classify wildfire using aerial images. This method employs EfficientNet-B5 [23] and DenseNet-201 [24] models as a backbone for extracting forest fire features. In addition, we employed a deep model (EfficientSeg [25]) and two vision transformers (TransUNet [26] and TransFire) in segmenting wildfire pixels and detecting the precise shape of fire on aerial images. Then, the proposed wildfire classification method was compared to deep convolutional models (MobileNetV3-Large -Small [27], DenseNet-169 [24], EfficientNet-B1-5 [23], Xception [28,29], and InceptionV3 [29]), which showed excellent results for object classification. TransUNet, TransFire, and EfficientSeg are also evaluated with U-Net [28].

More specifically, three main contributions were reported in this paper. First, a novel DL method was proposed to detect and classify wildfire using aerial images in order to improve detection and segmentation of wildland fires. Second, vision transformers were adopted for UAV wildfire segmentation to segment fire pixels and identify the precise shape of the fire. Third, the efficiencies of CNN methods and vision transformers are demonstrated in extracting the finer details of fire using aerial images and overcoming the problems of background complexity and small fire areas.

The remainder of the paper is organized as follows: Section 2 presents recent DL approaches for UAV-based fire detection and segmentation. Section 3 describes the methods and materials used in this paper. In Section 4, experimental results and discussion are presented. Finally, Section 5 concludes the paper.

2. Related Works

DL approaches are employed for fire detection and segmentation using aerial images. They proved their ability to detect and segment wildfires [6,20]. They can be grouped into three categories: DL approaches for UAV-based fire classification, DL approaches for UAV-based fire detection, and DL approaches for UAV-based fire segmentation.

2.1. Fire Classification Using Deep Learning Approaches for UAV Images

Convolutional Neural Networks (CNNs) are the most popular AI models for images classification tasks. They extract feature maps from input images and then predict their correct classes (two classes in our case: Fire and Non-Fire). Three main types of layers, which are convolutional layers, pooling layers, and fully connected layers, are employed to build a classical CNN architecture:

- Convolution layers are a set of filters designed to extract basic and complex features such as edges, corners, texture, colors, shapes, and objects from the input images. Then, activation functions are used to add the non-linearity transformation. It helps CNN to learn complex features in the input data. Various activation functions were employed, such as Rectified Linear Unit (ReLU) function [30], Leaky ReLU (LReLU) function [31], parametric ReLU (PReLU) function [32], etc.
- Pooling layers reduce the size of each feature map resulting from the convolutional layers. The most used pooling methods are average pooling and max pooling.
- The fully connected layer is fed by the final flattened pooling or convolutional layers' output, and the class scores for the objects present in the input image are computed.

CNNs showed good results for object classification and recognition [33]. Motivated by their great success, researchers presented numerous CNN-based contributions for fire detection and classification using aerial images in the literature, and these are summarized in Table 1.

Table 1. Deep learning methods for UAV-based fire classification.

Ref.	Methodology	Smoke/Flame	Dataset	Accuracy (%)
[34]	CNN-17	Flame/Smoke	Private dataset: 2100 images	86.00
[35]	AlexNet	Flame	Private dataset: 23,053 images	94.80
	GoogLeNet			99.00
	Modified GoogLeNet			96.90
	VGG13			86.20
	Modified VGG13			96.20
[28]	Xception	Flame	FLAME dataset: 48,010 images	76.23
[36]	Fire_Net	Flame	UAV_Fire dataset: 1540 images	98.00
	AlexNet			97.10
[29]	VGG16	Flame	FLAME dataset: 8617 images	80.76
	VGG19			83.43
	ResNet50			88.01
	InceptionV3			87.21
	Xception			81.30
[37]	Fog computing and simple CNN	Flame	Private dataset: 2964 images	95.07
[38]	Fire_Net	Flame/Smoke	Private dataset: 2096 images	97.50
	AlexNet			95.00
	MobileNetv2			99.30

Chen et al. [34,39] proposed two CNNs to detect wildfire in aerial images. The first CNN contains nine layers [39]. It consists of a convolutional layer with Sigmoid function, max-pooling layer, ReLU activations, Fully connected layer, and Softmax classifier. Using 950 images collected with a six-rotor drone (DJI S900) equipped with a SONYA7 camera, the experimental results showed improvements in accuracy compared to other detection methods [39]. The second includes two CNNs for detecting fire and smoke in aerial images [34]. Each CNN contains 17 layers. The first CNN classifies Fire and Non-Fire, and the second detects the presence of smoke in the input images. Using 2100 aerial images, great performance (accuracy of 86%) was achieved, outperforming the first method and the classical method, which combines LBP (Local Binary Patterns) and SVM [34]. Lee et al. [35] employed five deep CNNs, which included AlexNet [40], GoogLeNet [41], VGG13 [42], a modified GoogLeNet, and a modified VGG13 to detect forest fires in aerial images:

- AlexNet includes eleven layers: five convolutional layers with ReLU activation function, three max-pooling layers, and three fully connected layers;
- VGG13 is a CNN with 13 convolutional layers;

- GoogLeNet contains 22 inception layers, which employ, simultaneously and in parallel, multiple convolutions with various filters and pooling layers;
- Modified VGG13 is a VGG13 model with a number of channels of each convolutional layer and fully connected layers equal to half of that of the original VGG13;
- Modified GoogLeNet is a GoogLeNet model with a number of channels of each convolutional layer and fully connected layer equal to half of that of the original GoogLeNet.

GoogLeNet and the modified GoogLeNet achieved high accuracies thanks to data augmentation techniques (cropping, vertical, and horizontal flip). They showed their ability in detecting wildfires in aerial images [35]. Shamsoshoara et al. [28] proposed a novel method based on the Xception model [43] for wildfire classification. Xception architecture is an extension of the Inceptionv3 model [44] with the modified depth-wise separable convolution, which contains 1×1 convolution followed by a $n \times n$ convolution and no intermediate ReLU activations. Using 48,010 images of the FLAME dataset [45] and data augmentation techniques (horizontal flip and rotation), this method achieved an accuracy of 76.23%. Treneska et al. [29] also adopted four deep CNNs, namely InceptionV3, VGG16, VGG19, and ResNet50 [46], to detect wildfire in aerial images. ResNet50 achieved the best accuracy with 88.01%. It outperformed InceptionV3, VGG16, and VGG19 and the recent state-of-the-art model, Xception, using transfer learning techniques and the FLAME dataset as learning data. Srinivas et al. [37] also proposed a novel method, which integrates CNN and Fog computing to detect forest fire using aerial images at an early stage. The proposed CNN consists of six convolutional layers followed by the ReLU activation function and max-pooling layers, three fully connected layers, and a sigmoid classifier that determines the output as Fire or Non-Fire. This method showed a great performance (accuracy of 95.07% and faster response time) and proved its efficiency to detect forest fires. Zhao et al. [36] presented a novel model called "Fire_Net" to extract fire features and classified them as Fire and Non-Fire. Fire_Net is a deep CNN with 15 layers. It consists of eight convolutional layers with ReLU activation functions, four max-pooling layers, three fully connected layers, and a softmax classifier. Using the UAV_Fire dataset [36], Fire_Net achieved an accuracy of 98% and outperformed previous methods. Wu et al. [38] used a pretrained MobileNetv2 [47] model to detect both smoke and fire. MobileNetv2 is an extended version of MobileNetv1 [48], which is a lightweight CNN with depth-wise separable convolutions. It requires small data and reduces the number of parameters of the model and its computational complexity. It employs inverted residuals and linear bottlenecks to improve the performance of MobileNetv1. Using transfer learning and data augmentation strategies, this method achieved an accuracy of 99.3%. It outperformed published state-of-the-art methods such as Fire_Net and AlexNet and proved its suitability in detecting forest fire on aerial monitoring systems [38].

2.2. Fire Detection Using Deep Learning Approaches for UAV

Region-based CNNs are used to detect, identify, and localize objects in an image. They determine the detected objects' locations in the input image using bounding boxes. These techniques are divided into two categories: one-stage detectors and two-stage detectors [49]. One-stage detectors detect and localize objects as a simple regression task in an input image. Two-stage detectors generate the ROI (Region of Interest) in the first step using the region proposal network. Then, the generated region is classified and its bounding box is determined. Region-based CNNs showed excellent accuracy for object detection problems. They are also employed to reveal the best performance in detecting fires on aerial images.

Table 2 presents deep learning methods for UAV-based fire detection. Jiao et al. [50] exploited the one-stage detector, YOLOv3 [51], to detect forest fires. YOLOv3 is the third version of YOLO deep object detectors. It was proposed to improve the detection performance of older versions by making detections at three different scales and using the Darknet-53 model, which contains 53 convolutional layers as a backbone [51]. Testing results revealed great performances and high speed [50]. Jiao et al. [52] also proposed the UAV-FFD (UAV forest fire detection) platform, which employs YOLOv3 to detect smoke and fire by using UAV-acquired images. YOLOv3 showed high performance with reduced computational time (F1-score of 81% at a frame rate of 30 frames per second). It proved its potential in detecting smoke/fire with high precision in real-time UAV applications [52]. Alexandrov et al. [53] adopted two one-stage detectors (SSD [54] and YOLOv2 [55]) and a two-stage detector (Faster R-CNN [56]) to detect wildfires. Using large data of real and simulated images, YOLOv2 showed the best performance compared to Faster R-CNN, SSD, and hand-crafted classical methods. It proved its reliability in detecting smoke at an early stage [53]. Tang et al. [57] also presented a novel application to detect wildfire using 4K images, which have a high resolution of 3840 × 2160 pixels collected by UAS (Unmanned Aerial Systems). A coarse-to-fine strategy was proposed to detect fires that are sparse, small, and irregularly shaped. At first, an ARSB (Adaptive sub-Region Select Block) was employed to select subregions, which contain the objects of interest in 4K input images. Next, these subregions were zoomed to maintain the area bounding box's size. Then, YOLOv3 was used to detect the objects. Finally, the bounding boxes in the subregions were combined. Using 1400 4K aerial images, this method obtained a mean average precision (mAP) of 67% at an average speed of 7.44 frames per second.

Table 2. Fire detection using Deep learning methods for UAVs.

Ref.	Methodology	Smoke/Flame	Dataset	Results (%)
[50]	YOLOv3	Flame	Private dataset: 3,840,000 images	F1-score = 81.0
[53]	YOLOv2	Smoke	Private dataset: 12,000 images	Accuracy = 98.3
	Faster R-CNN			Accuracy = 95.9
	SSD			Accuracy = 81.1
[52]	YOLOv3	Flame/Smoke	Private dataset: 3,684,000 images	F1-score = 81.0
[57]	YOLOv3 and ARSB method	Flame	Private dataset: 1400 K images	mAP = 67.0

2.3. Fire Segmentation Using Deep Learning Approaches for UAV

Image segmentation is very important in computer vision. It determines the exact shape of the objects in the images. With the progress of deep learning models, numerous problems were tackled and a variety of solutions was proposed with good results.

Deep learning models are also used to segment fire pixels and detect the precise shape of smoke and/or flame using aerial images. Table 3 shows deep learning methods for UAV-based fire segmentation. For example, Barmpoutis et al. [58] proposed a 360-degree remote sensing system to segment both fire and smoke using RGB 360-degree images, which were collected from UAV. Two DeepLab V3+ [59] models that are encoder–decoder detectors with ASPP (Atrous Spatial Pyramid Pooling) were applied to identify smoke and flame regions. Then, an adaptive post-validation scheme was employed to reject smoke/flame false-positive regions, especially regions with similar characteristics with smoke and flame. Using 150 360-degree images of urban and forest areas, experiments achieved an F1-score of 94.6% and outperformed recent state-of-the-art methods such as DeepLabV3+. These results showed the robustness of the proposed method in segmenting smoke/fire and reducing the false-positive rate [58]. Similarly to wildfire classification, Shamsoshoara et al. [28] proposed a method based on the encoder–decoder U-Net [60] for wildfire segmentation. Using a dropout strategy and the FLAME dataset, U-Net obtained an F1-score of 87.75% and proved its ability to segment wildfire and identify the precise shapes of flames [28]. Frizzi et al. [61] also proposed a method based on VGG16 to segment both smoke and fire. This method showed good results (accuracy of 93.4% and segmentation time per image of 21.1 s) using data augmentation techniques such as rotation, flip, changing brightness/contrast, crop, and adding noises. It outperformed previous published models and proved its efficiency in detecting and classifying fire/smoke pixels [61].

Table 3. Fire segmentation using deep learning methods for UAVs.

Ref.	Methodology	Smoke/Flame	Dataset	Results (%)
[58]	DeepLabV3+ DeepLabV3+ + validation approach	Flame/Smoke	Fire detection 360-degree dataset: 150 360-degree images	F1-score = 81.4 F1-score = 94.6
[60]	U-Net	Flame	FLAME dataset: 5137 images	F1-score = 87.7
[61]	U-Net CNN based on VGG16	Flame/Smoke	Private dataset: 366 images	Accuracy = 90.2 Accuracy = 93.4

3. Materials and Methods

In this section, we first introduced our proposed methods for wildfire classification and segmentation. Then, we describe the dataset used in training and testing. Finally, we present the evaluation metrics employed in this work.

3.1. Proposed Method for Wildfire Classification

To detect and classify fire, we propose a novel method based on deep ensemble learning using EfficientNet-B5 [23] and DenseNet-201 [24] models. EfficientNet models proved their efficiency to reduce the parameters and Floating-Point Operations Per Second using an effective scaling method that employs a compound coefficient to uniformly scale model depth, resolution, and width. EfficientNet-B5 showed excellent accuracy and outperformed state-of-the-art models such as Xception [43], AmoebaNet-A [62], PNASNet [63], ResNeXt-101 [64], InceptionV3 [44], and InceptionV4 [65]. DenseNet (Dense Convolutional Network) connects each layer to all preceding layers to create very diversified feature maps. It has several advantages, including feature reuse, elimination of the vanishing-gradient problem, improved feature propagation, and a reduction in the number of parameters. Using extracted features of all complexity levels, DenseNet shows interesting results in various competitive object recognition benchmark tasks such as ImageNet, SVHN (Street View House Numbers), CIFAR-10, and CIFAR-100 [24].

Figure 1 presents the architecture of the proposed method. First, this method is fed with RGB aerial images. EfficientNet-B5 and DenseNet-201 models were employed as a backbone to extract two feature maps. Next, the feature maps of the two models are concatenated. The concatenated map was then fed an average pooling layer. Then, a dropout of 0.2 was employed to avoid overfitting. Finally, a Sigmoid function was applied to classify the input image into Fire or Non-Fire classes.

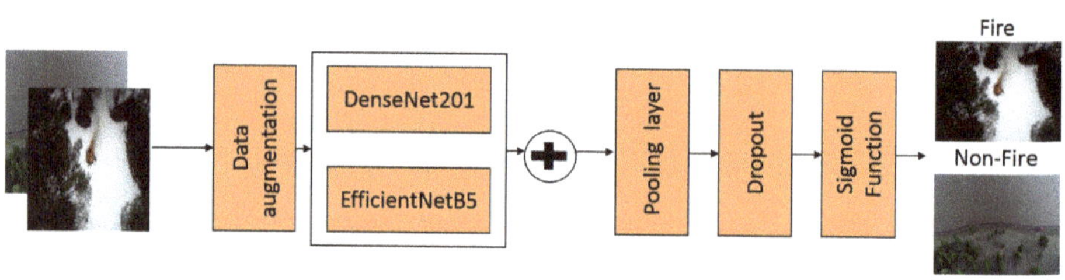

Figure 1. The proposed architecture for wildfire classification.

3.2. Proposed Methods for Wildfire Segmentation

To segment wildfires, we used a CNN model, EfficientSeg [25], and two vision transformers, which are TransUNet [26] and TransFire.

3.2.1. TransUNet

TransUNet [26] is a vision transformer based on U-Net architecture. It employs global dependencies between inputs and outputs using self-attention methods. It is an encoder–decoder. The encoder uses a hybrid CNN-transformer architecture consisting of ResNet-50 and pretrained ViT (Vision Transformer) to extract feature maps. It contains MLP (Multi-Layer Perceptron) and MSA (Multihead Self-Attention) blocks. The decoder employs CUP (cascaded up-sampler) blocks to decode the extracted features and outputs the binary segmentation mask. Each CUP includes a 3 × 3 convolutional layer, ReLU activation function, and two upsampling operators. Figure 2 depicts the architecture of TransUNet.

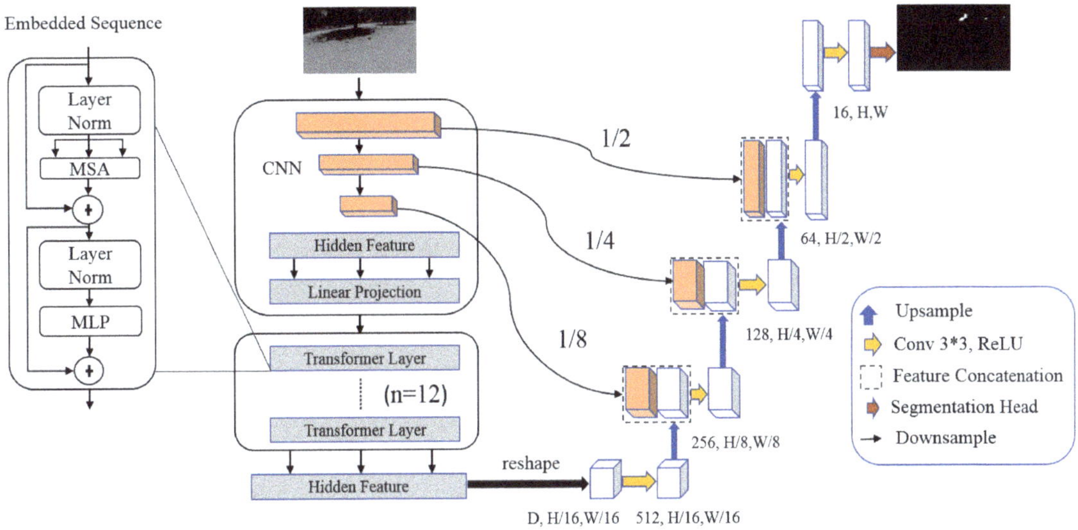

Figure 2. The proposed TransUNet architecture.

3.2.2. TransFire

TransFire is based on MedT (Medical Transformer) architecture. MedT [66] was proposed in order to segment medical images with no requirement of a large dataset for training. Two concepts, gated position-sensitive axial attention and LoGo (Local-Global) training methodology, were employed to improve segmentation performance. Gated position-sensitive axial attention was used to determine long-range interactions between the input features with high computational efficiency. LoGo training methodology used two branches, which are global branch and local branch, to extract feature maps. The first branch works on the image's original resolution. It consists of 2 encoders and 2 decoders. The second operates on image patches. It contains 5 encoders and 5 decoders. The input to both of these branches is the feature extracted using a convolutional block, which includes 3 convolutional layers with ReLU activation function and batch normalization.

TransFire is a modified MedT architecture. It includes one encoder and one decoder in the global branch. It also employs a dropout strategy in the local branch (after the fourth first encoders and the last decoder), in the global branch (after the decoder), and in each input of both of these branches. TransFire was developed to overcome the memory problem of MedT and to prevent overfitting. Figure 3 illustrates the architecture of TransFire.

3.2.3. EfficientSeg

EfficientSeg [25] is a semantic segmentation method, which is based on a U-Net structure and uses MobileNetV3 [27] blocks. It showed impressive results and outperformed U-Net in some medical image segmentation tasks [25].

Figure 4 depicts the architecture of EfficientSeg. It is an encoder–decoder with 4 concatenation shortcuts. It includes five types of blocks, which are MobileNetV3 blocks (Inverted Residual blocks), Downsampling operator, Upsampling operator, and 1 × 1 and 3 × 3 convolutional blocks with ReLU activation function and batch normalization layer.

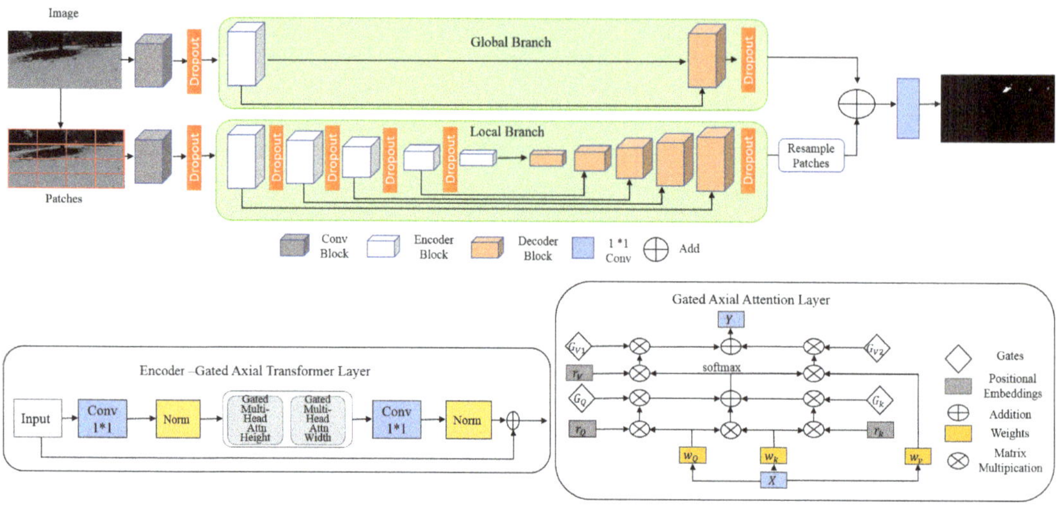

Figure 3. The proposed TransFire architecture.

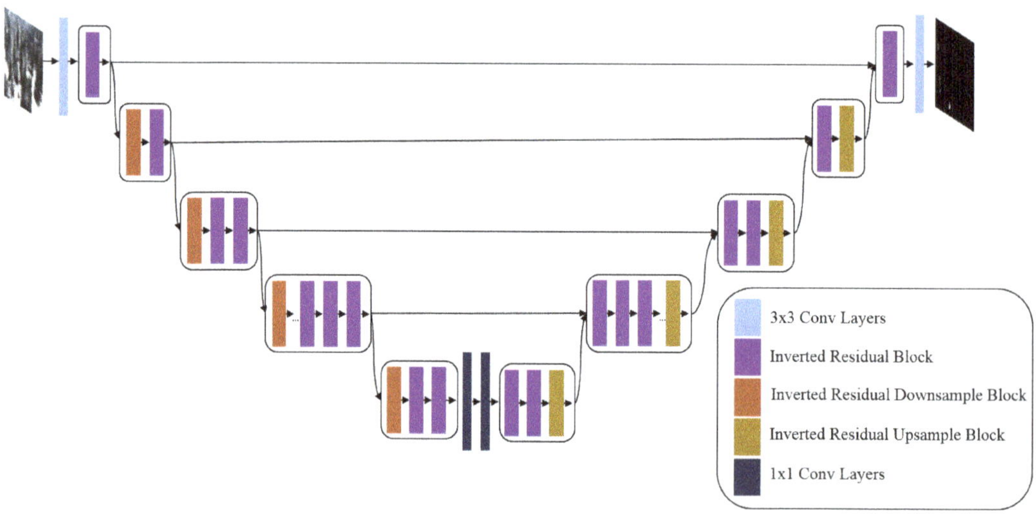

Figure 4. The proposed EfficientSeg architecture.

3.3. Dataset

In the area of deep learning, many large datasets are available for researchers to train their models and perform benchmarking by making comparisons with other methods. However, until recently, there was a lack of a UAV dataset for fire detection and segmentation. In this work, we use a public database called FLAME dataset (Fire Luminosity Airborne-based Machine learning Evaluation) [45] to train and evaluate our proposed methods. The FLAME dataset contains aerial images and raw heat-map footage captured by visible spectrum and thermal cameras onboard a drone. It consists of four types of videos, which are a normal spectrum, white-hot, fusion, and green-hot palettes.

In this paper, we focus on RGB aerial images. We used 48,010 RGB images, which are split into 30,155 Fire images and 17,855 Non-Fire images for wildfire classification task. Figure 5 presents some samples of the FLAME dataset for fire classification. On the other hand, we used 2003 RGB images and their corresponding masks for fire segmentation task. Figure 6 illustrates some examples of RGB aerial images and their corresponding binary masks.

Figure 5. Examples from the FLAME dataset. Top line: Fire images and bottom line: Non-Fire images.

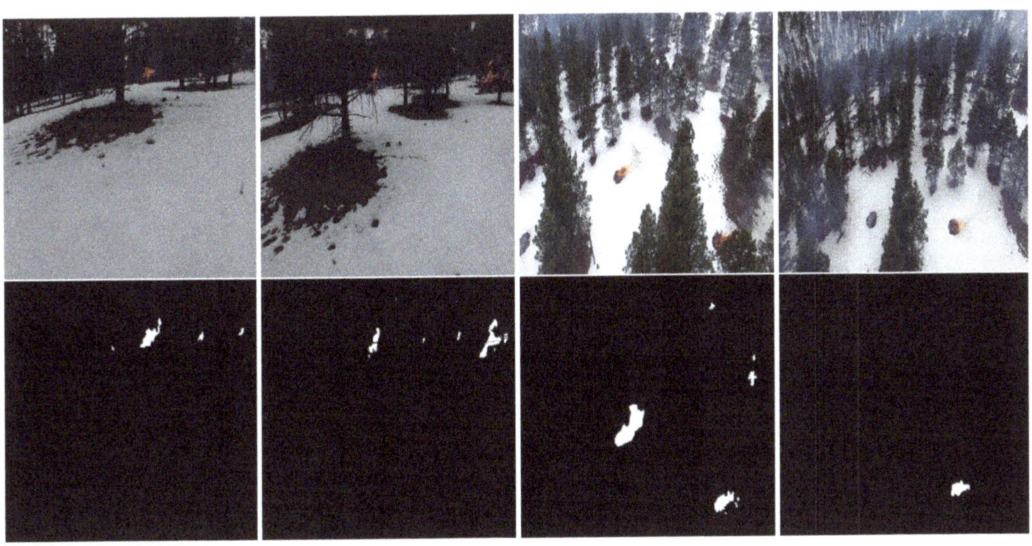

Figure 6. Examples from the FLAME dataset. Top line: RGB images; bottom line: their corresponding binary masks.

3.4. Evaluation Metrics

We used F1-score, accuracy, and inference time to evaluate our proposed approaches for fire classification and segmentation:

- F1-score combines precision and recall metrics to determine the ability of the model in detecting wildfire pixels (as shown by Equation (1)):

$$\text{F1-score} = \frac{2 \times Precision \times Recall}{Precision + Recall} \tag{1}$$

$$Precision = \frac{TP}{TP + FP} \tag{2}$$

$$Recall = \frac{TP}{TP + FN} \tag{3}$$

where TP is the true positive rate, FP is the false positive rate, and FN is the false negative rate.

- Accuracy is the proportion of correct predictions over the number of total ones, achieved per the proposed model (as given by Equation (4)):

$$\text{Accuracy} = \frac{TP + TN}{TP + FP + TN + FN} \tag{4}$$

where TN is the true negative rate, FN is the false negative rate, TP is the true positive rate, and FP is the false positive rate.

- Inference time is the average time of segmentation or classification using our testing images.

4. Results and Discussion

For wildfire classification, we used TensorFlow [67] and trained the proposed models on a machine with NVIDIA Geforce RTX 2080Ti GPU. The learning data were split as follows: 31,515 images for training, 7878 images for validation, and 8617 images for testing as presented in Table 4.

Table 4. Dataset subsets for classification.

Dataset	Fire Images	Non-Fire Images
Training set	20,015	11,500
Validation set	5003	2875
Testing set	5137	3480

We employed categorical cross-entropy loss (CE) [68], which measures the probability of the presence of a wildfire in the input image (as shown in Equation (5)):

$$CE = -\sum_{c=1}^{M} z_{b,c} \log\left(p_{b,c}\right) \tag{5}$$

where M is the number of classes (in our case two classes (Fire and Non-Fire)), p is the predicted probability, and z is the binary indicator.

For our experiments, we used input RGB images with 254×254 resolution, a batch size of 16, and Adam as an optimizer. We also employed the following data augmentation techniques: rotation, shear, zoom, and shift with random values.

For wildfire segmentation, we developed the proposed methods using Pytorch [69] on an Nvidia V100l GPU. Learning data were divided into three sets: 1401 images for training, 201 images for validation, and 401 images for testing. We employed dice loss [70] to measure the difference between the predicted binary mask and the corresponding input mask (as given by Equation (6)). We also used two data augmentation methods, which are a horizontal flip and a rotation of 15 degrees:

$$DC = 1 - \frac{2|Z \cap W|}{|Z| + |W|} \tag{6}$$

where Z is the input aerial image, W is the predicted image, and $\cap$ is the intersection of the input and the predicted images.

The input data are RGB aerial images with a 512×512 resolution and their corresponding binary mask. The TransFire Transformer was trained from scratch (no pretraining) using a hybrid CNN-Transformer as a backbone, patch sizes of 16, and a learning rate of 10^{-3}. TransUNet is evaluated using a learning rate of 10^{-3}, patch size of 16, and two backbones that include a pretrained ViT and a hybrid backbone, which includes ResNet50

(R-50) and pretrained ViT. EfficientSeg also was tested from scratch using a learning rate of 10^{-1}.

We analyzed the proposed methods' performance (accuracy and F1-score) as well as their speed (inference time). In addition, we compared our novel wildfire classification method to state-of-the-art models (Xception [28,29] and InceptionV3 [29]) and deep CNNs (MobileNetV3-Large [27], MobileNetV3-Small [27], DenseNet-169 [24], and EfficientNet-B1-5 [23]), which already showed excellent results for object classification. We also compared the proposed wildfire segmentation methods, including TransUNet, TransFire, and EfficientSeg, to U-Net [28].

4.1. Wildfire Classification Results

We trained wildfire classification methods on aerial images collected using the Matrice 200 drone with a Zenmuse X4S camera. Testing data are collected using the Phantom drone with a Phantom camera.

Table 5 reports a comparative analysis of our proposed method and deep CNN methods using the test data. We can observe that our proposed method achieved the best performance (accuracy of 85.12% and F1-score of 84.77%) thanks to scaled and diversified feature maps extracted by EfficientNet-B5 and DenseNet-201 models. It outperformed recent models for object classification (MobileNetV3-Large, MobileNetV3-Small, DensNet-169, and EfficientNet models (EfficientNet-B2, -B3, -B4, and -B5)) and inception models (Xception and InceptionV3). It proved its good ability to detect and classify forest fires on aerial images. However, it needed a high inference time with 0.018 s.

Figure 7 presents the confusion matrix on test data. We can see that the rate of true positives (classifying Fire as Fire) and the rate of true negatives (classifying No-Fire as No-Fire) are higher than the rate of the false positives (classifying Fire as No-Fire) and the rate of false negatives (classifying No-Fire as Fire), respectively. Our proposed method showed interesting results in detecting and classifying fires, even for very small fire areas. It proved its efficiency to overcome challenging problems such as uneven object intensity and background complexity.

To conclude, our proposed method revealed the best result based on the trade-off between performance and inference time. It showed an excellent capacity to classify forest fires in aerial images and managed to overcome the problems of small fire areas and background complexity.

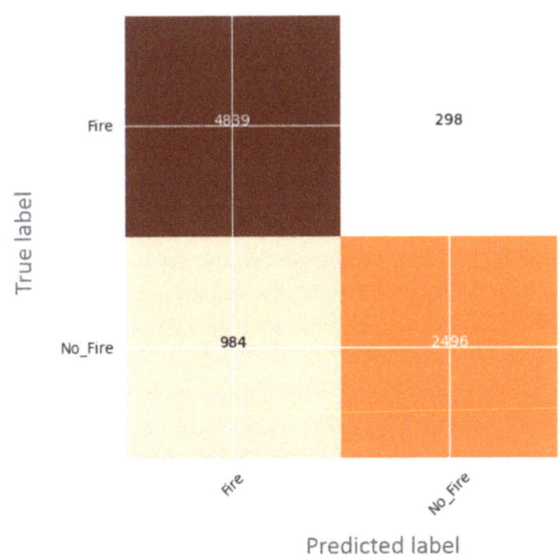

Figure 7. Confusion matrix for fire classification.

Table 5. Performance evaluation of wildfire classification models.

Models	Accuracy (%)	F1-Score (%)	Inference Time (s)
Xception	78.41	78.12	0.002
Xception [28]	76.23	—	—
EfficientNet-B5	75.82	73.90	0.010
EfficientNet-B4	69.93	65.51	0.008
EfficientNet-B3	65.81	64.02	0.004
EfficientNet-B2	66.04	60.71	0.002
InceptionV3	80.88	79.53	0.002
DenseNet169	80.62	79.40	0.003
MobileNetV3-Small	51.64	44.97	**0.001**
MobileNetV3-Large	65.10	60.91	**0.001**
Proposed ensemble model	**85.12**	**84.77**	0.018

4.2. *Wildfire Segmentation Results*

Table 6 illustrates the quantitative results of fire segmentation using the FLAME dataset. We can see that TransUNet, TransFire, and EfficientSeg obtained excellent results and outperformed U-Net used as a baseline model.

Vision Transformers (TransUNet and TransFire) obtained higher performances compared to deep CNN models (EfficientSeg and U-Net) due to their ability to determine long-range interactions within input features and extract the finer details of the input images. TransUNet-R50-ViT achieved the best performance with an accuracy of 99.9% and an F1-score of 99.9% thanks to local and global features extracted using a hybrid backbone, which includes a CNN, R-50, and pretrained ViT Transformer.

Figure 8 depicts examples of the segmentation of TransUNet-R50-ViT. We can see that this model accurately detected the finer details of fire and distinguished between wildfire and background. In addition, TransUNet-R50-ViT showed its efficiency in localizing and detecting the precise shape of wildfire, especially with respect to small fire areas on aerial images.

TransUNet-ViT also showed excellent performances (accuracy of 99.86% and F1-score of 99.86%) and high speeds (inference time of 0.4 s) compared to TransFire and EfficientSeg. We can see in Figure 8 that TransUNet with ViT transformer accurately segmented wildfire pixels and detected wildfire regions even for small fire areas.

TransUNet models proved their ability in segmenting wildfire, in detecting the exact shape of fire areas, and in overcoming challenging problems such as small fire areas and background complexity. However, they still depend on a pretrained vision transformer (ViT) on a large dataset.

TransFire also showed a higher accuracy with 99.83% and an F1-score of 99.82% due to high-level information and finer features extracted in the global branch and local branch, respectively. It outperformed EfficientSeg and U-Net. It proved its excellent capacity in segmenting wildfire pixels and detecting the exact fire areas, especially small fire areas as shown in Figure 8. It also segmented forest fire pixels under the presence of smoke.

EfficientSeg also obtained a high accuracy with 99.63% and an F1-score of 99.66% thanks to its extracted finer details. It outperformed U-Net. It showed its efficiency in segmenting fire pixels and in detecting the precise shape of fire areas as depicted in Figure 8. However, It had a higher inference time with 1.38 s compared to vision transformers.

To conclude, TransUNet, TransFire, and EfficientSeg showed excellent performances. They proved an impressive potential in segmenting wildfire pixels and determining the precise shape of fire. Based on the F1-score, TransFire showed great performance and outperformed deep convolutional models (EfficientSeg and U-Net) and was very close to the performance of vision transformer (TransUNet). In addition, it demonstrated its reliability in detecting and segmenting wildland fires; in particular, it was the best performing in detecting small fire areas under the presence of smoke, as observed in Figure 9.

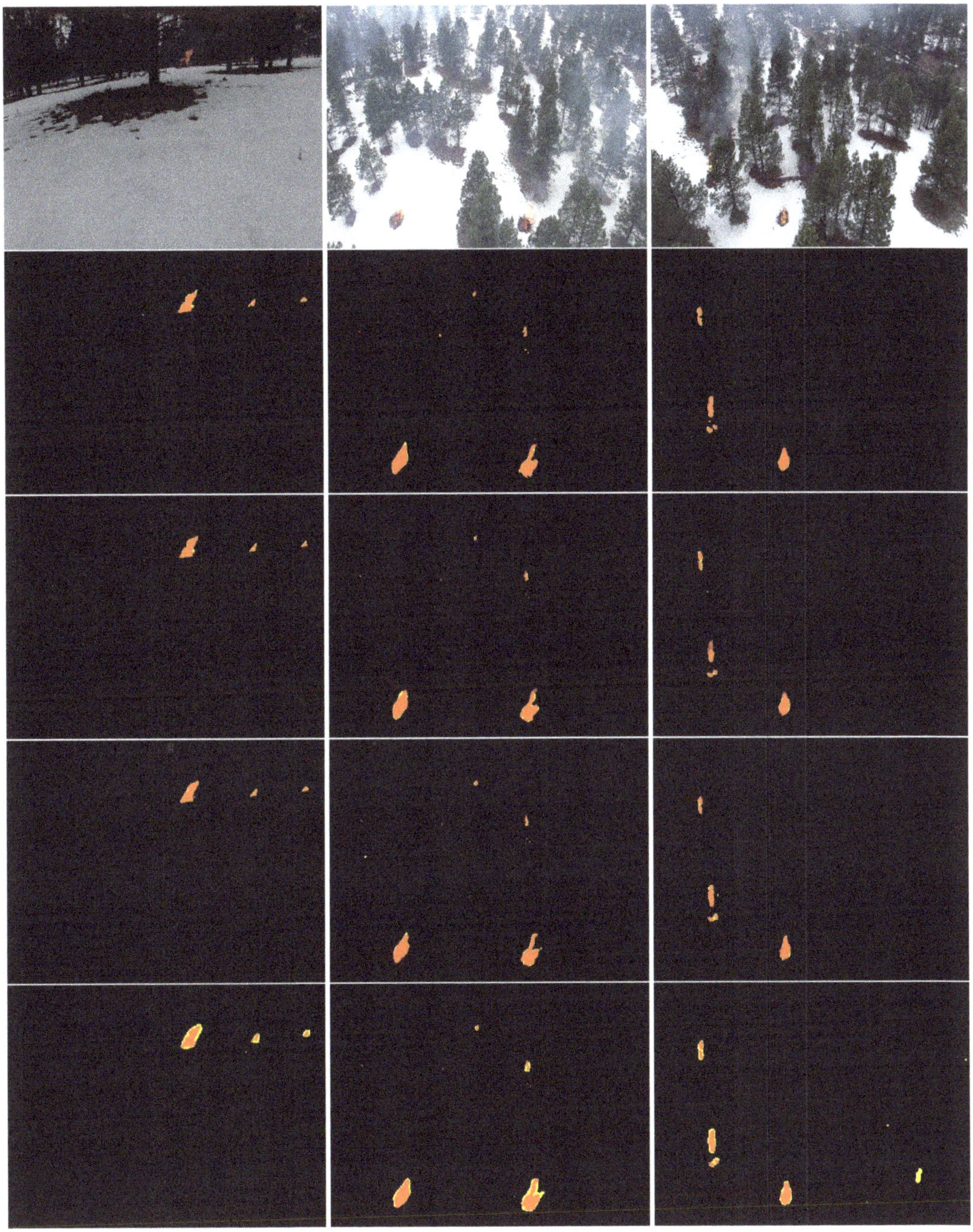

Figure 8. Segmentation results of the proposed models. From top to bottom: RGB aerial images and the results of TransUNet-R50-ViT, TransUNet-ViT, TransFire, and EfficientSeg. Orange represents *TP* (true positives), yellow depicts *FP* (false positives), and red shows *FN* (false negatives).

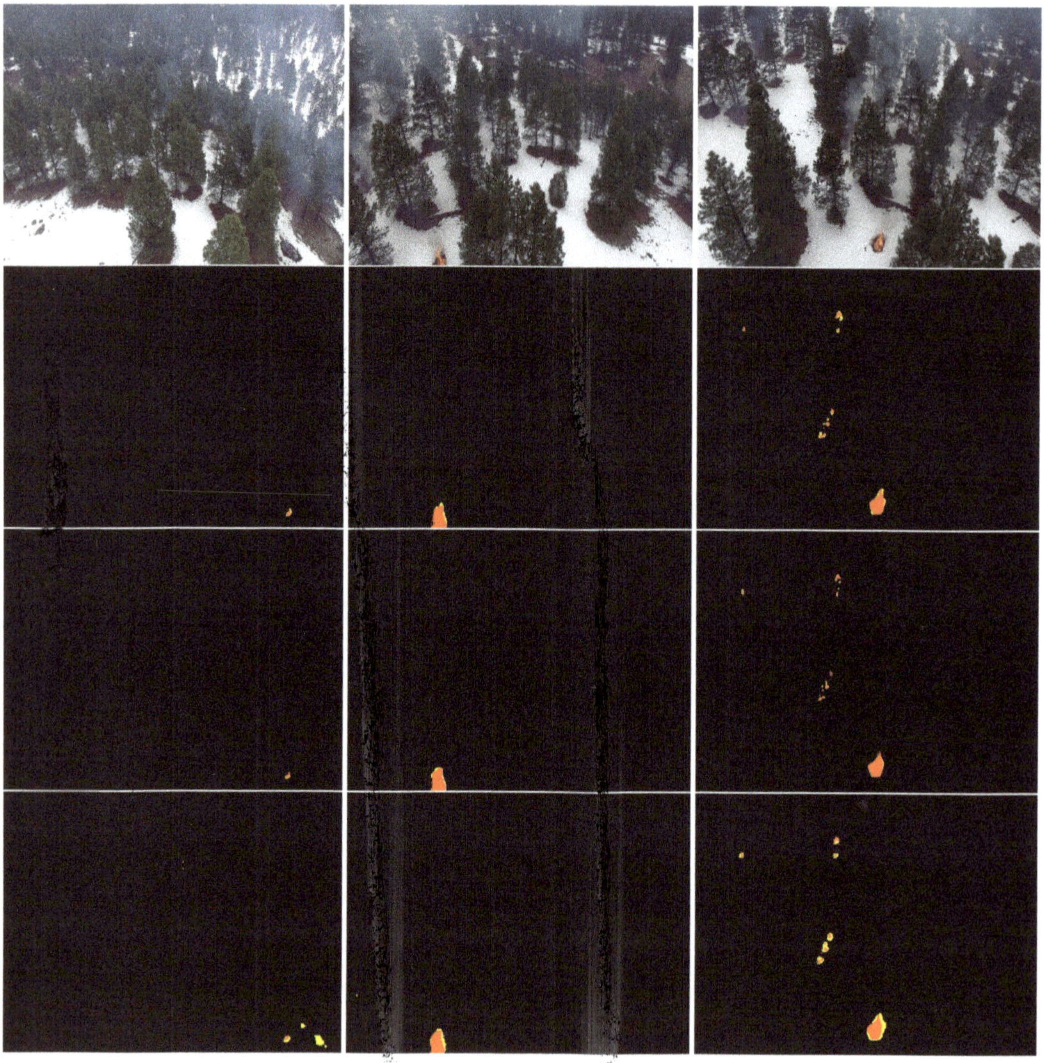

Figure 9. Results of TransFire, TransUNet-R50-ViT, and EfficientSeg. From top to bottom: RGB aerial images and the results of TransFire, TransUNet-R50-ViT, and EfficientSeg. Orange represents *TP*, yellow depicts *FP*, and red shows *FN*. We can see the interesting results of TransFire in determining the precise size of small wildfire areas under the presence of smoke compared to TransUNet and EfficientSeg models.

Table 6. Performance evaluation of wildfire segmentation models.

Models	Accuracy (%)	F1-Score (%)	Inference Time (s)
TransUNet-R50-ViT	**99.90**	**99.90**	0.51
TransUNet-ViT	99.86	99.86	0.40
TransFire	99.83	99.82	1.00
EfficientSeg	99.63	99.66	1.38
U-Net	99.00	99.00	**0.29**

5. Conclusions

In this paper, we address the problem of wildfire classification and segmentation on aerial images using deep learning models. A novel ensemble learning method, which combines EfficientNet-B5 and DenseNet-201 models, was developed to detect and classify wildfires. Using the FLAME dataset, experimental results showed that our proposed method was the most reliable in wildfire classification tasks, presenting a higher performance than recent state-of-the-art models. Furthermore, two vision transformers (TransUNet and TransFire) and a deep CNN (EfficientSeg) are developed to segment wildfires and detect the region of fire areas on aerial images. This is the first proposed approach (in our knowledge) using Transformers for UAV wildfire image segmentation. These models showed impressive results and outperformed recent published methods. They proved their ability in segmenting wildfire pixels, detecting the precise shape of fire. Based on the F1-score, TransFire obtained great performance, outperforming deep models such as EfficientSeg and U-Net. It also showed its excellent potential in detecting and segmenting forest fires and in overcoming challenging problems such as small fire areas and background complexity.

Author Contributions: Conceptualization, M.A.A. and R.G.; methodology, R.G. and M.A.A.; software, R.G.; validation, R.G. and M.A.A.; formal analysis, R.G., M.A.A. and W.S.M.; writing—original draft preparation, R.G.; writing—review and editing, M.A.A. and W.S.M.; funding acquisition, M.A.A. All authors have read and agreed to the published version of the manuscript.

Funding: This research was enabled in part by support provided by the Natural Sciences and Engineering Research Council of Canada (NSERC), funding reference number RGPIN-2018-06233.

Institutional Review Board Statement: Not applicable.

Informed Consent Statement: Not applicable.

Data Availability Statement: This work uses a publicly FLAME dataset, which is available on IEEE-Dataport [45]. More details about the data are available under Section 3.3.

Acknowledgments: The authors would like to thank the support of WestGrid (www.westgrid.ca/) and Compute Canada (www.computecanada.ca).

Conflicts of Interest: The authors declare no conflict of interest.

References

1. Aytekin, E. Wildfires Ravaging Forestlands in Many Parts of Globe. 2021. Available online: https://www.aa.com.tr/en/world/wildfires-ravaging-forestlands-in-many-parts-of-globe/2322512 (accessed on 20 November 2021).
2. Dimitropoulos, S. Fighting fire with science. *Nature* **2019**, *576*, 328–329. [CrossRef]
3. Westerling, A.L.; Hidalgo, H.G.; Cayan, D.R.; Swetnam, T.W. Warming and Earlier Spring Increase Western U.S. Forest Wildfire Activity. *Science* **2006**, *313*, 940–943. [CrossRef] [PubMed]
4. Canadian Wildland Fire Information System. Canada Wildfire Facts. 2021. Available online: https://www.getprepared.gc.ca/cnt/hzd/wldfrs-en.aspx (accessed on 20 November 2021).
5. Gaur, A.; Singh, A.; Kumar, A.; Kulkarni, K.S.; Lala, S.; Kapoor, K.; Srivastava, V.; Kumar, A.; Mukhopadhyay, S.C. Fire Sensing Technologies: A Review. *IEEE Sens. J.* **2019**, *19*, 3191–3202. [CrossRef]
6. Ghali, R.; Jmal, M.; Souidene Mseddi, W.; Attia, R. Recent Advances in Fire Detection and Monitoring Systems: A Review. In Proceedings of the 18th International Conference on Sciences of Electronics, Technologies of Information and Telecommunications (SETIT'18), Genoa, Italy, 18–20 December 2018; Springer International Publishing: Berlin/Heidelberg, Germany, 2018; Volume 1, pp. 332–340.
7. Gaur, A.; Singh, A.; Kumar, A.; Kumar, A.; Kapoor, K. Video flame and smoke based fire detection algorithms: A literature review. *Fire Technol.* **2020**, *56*, 1943–1980. [CrossRef]
8. Dao, M.; Kwan, C.; Ayhan, B.; Tran, T.D. Burn scar detection using cloudy MODIS images via low-rank and sparsity-based models. In Proceedings of the IEEE Global Conference on Signal and Information Processing GlobalSIP), Washington, DC, USA, 7–9 December 2016; pp. 177–181.
9. Töreyin, B.U.; Dedeoğlu, Y.; Güdükbay, U.; Çetin, A.E. Computer vision based method for real-time fire and flame detection. *Pattern Recognit. Lett.* **2006**, *27*, 49–58. [CrossRef]
10. Zhang, J.S.; Cao, J.; Mao, B. Application of deep learning and unmanned aerial vehicle technology in traffic flow monitoring. In Proceedings of the International Conference on Machine Learning and Cybernetics (ICMLC), Ningbo, China, 9–12 July 2017; Volume 1, pp. 189–194.

11. Chen, C.J.; Huang, Y.Y.; Li, Y.S.; Chang, C.Y.; Huang, Y.M. An AIoT Based Smart Agricultural System for Pests Detection. *IEEE Access* **2020**, *8*, 180750–180761. [CrossRef]
12. Geraldes, R.; Gonçalves, A.; Lai, T.; Villerabel, M.; Deng, W.; Salta, A.; Nakayama, K.; Matsuo, Y.; Prendinger, H. UAV-Based Situational Awareness System Using Deep Learning. *IEEE Access* **2019**, *7*, 122583–122594. [CrossRef]
13. Lee, H.; Jung, S.; Kim, J. Distributed and Autonomous Aerial Data Collection in Smart City Surveillance Applications. In Proceedings of the IEEE VTS 17th Asia Pacific Wireless Communications Symposium (APWCS), Osaka, Japan, 30–31 August 2021; pp. 1–3.
14. Giang, T.L.; Dang, K.B.; Toan Le, Q.; Nguyen, V.G.; Tong, S.S.; Pham, V.M. U-Net Convolutional Networks for Mining Land Cover Classification Based on High-Resolution UAV Imagery. *IEEE Access* **2020**, *8*, 186257–186273. [CrossRef]
15. Aposporis, P. Object Detection Methods for Improving UAV Autonomy and Remote Sensing Applications. In Proceedings of the IEEE/ACM International Conference on Advances in Social Networks Analysis and Mining (ASONAM), The Hague, The Netherlands, 7–10 December 2020; pp. 845–853.
16. Akhloufi, M.A.; Castro, N.A.; Couturier, A. UAVs for wildland fires. In *Autonomous Systems: Sensors, Vehicles, Security, and the Internet of Everything*; International Society for Optics and Photonics: Orlando, FL, USA, 3 May 2018; pp. 134–147.
17. Khennou, F.; Ghaoui, J.; Akhloufi, M.A. Forest fire spread prediction using deep learning. In *Geospatial Informatics XI*; Palaniappan, K., Seetharaman, G., Harguess, J.D., Eds.; International Society for Optics and Photonics: Bellingham, WA, USA, 2021; pp. 106–117.
18. Akhloufi, M.A.; Couturier, A.; Castro, N.A. Unmanned Aerial Vehicles for Wildland Fires: Sensing, Perception, Cooperation and Assistance. *Drones* **2021**, *5*, 15. [CrossRef]
19. Ghali, R.; Akhloufi, M.A.; Jmal, M.; Souidene Mseddi, W.; Attia, R. Wildfire Segmentation Using Deep Vision Transformers. *Remote Sens.* **2021**, *13*, 3527. [CrossRef]
20. Yuan, C.; Zhang, Y.; Liu, Z. A survey on technologies for automatic forest fire monitoring, detection, and fighting using unmanned aerial vehicles and remote sensing techniques. *Can. J. For. Res.* **2015**, *45*, 783–792. [CrossRef]
21. Mseddi, W.S.; Ghali, R.; Jmal, M.; Attia, R. Fire Detection and Segmentation using YOLOv5 and U-NET. In Proceedings of the 29th European Signal Processing Conference (EUSIPCO), Dublin, Ireland, 23–27 August 2021; pp. 741–745.
22. Ghali, R.; Akhloufi, M.A.; Jmal, M.; Mseddi, W.S.; Attia, R. Forest Fires Segmentation using Deep Convolutional Neural Networks. In Proceedings of the IEEE International Conference on Systems, Man, and Cybernetics (SMC), Melbourne, Australia, 17–20 October 2021; pp. 2109–2114.
23. Tan, M.; Le, Q. EfficientNet: Rethinking Model Scaling for Convolutional Neural Networks. In Proceedings of the 36th International Conference on Machine Learning, Long Beach, CA, USA, 9–15 June 2019; pp. 6105–6114.
24. Huang, G.; Liu, Z.; van der Maaten, L.; Weinberger, K.Q. Densely Connected Convolutional Networks. In Proceedings of the IEEE Conference on Computer Vision and Pattern Recognition (CVPR), Honolulu, HI, USA, 21–26 July 2017; pp. 4700–4708.
25. Yesilkaynak, V.B.; Sahin, Y.H.; Unal, G.B. EfficientSeg: An Efficient Semantic Segmentation Network. *arXiv* **2020**, arXiv:2009.06469.
26. Chen, J.; Lu, Y.; Yu, Q.; Luo, X.; Adeli, E.; Wang, Y.; Lu, L.; Yuille, A.L.; Zhou, Y. TransUNet: Transformers Make Strong Encoders for Medical Image Segmentation. *arXiv* **2021**, arXiv:2102.04306.
27. Howard, A.; Sandler, M.; Chu, G.; Chen, L.C.; Chen, B.; Tan, M.; Wang, W.; Zhu, Y.; Pang, R.; Vasudevan, V.; et al. Searching for MobileNetV3. In Proceedings of the IEEE/CVF International Conference on Computer Vision (ICCV), Seoul, Korea, 27 October–2 November 2019; pp. 1314–1324.
28. Shamsoshoara, A.; Afghah, F.; Razi, A.; Zheng, L.; Fulé, P.Z.; Blasch, E. Aerial imagery pile burn detection using deep learning: The FLAME dataset. *Comput. Netw.* **2021**, *193*, 108001. [CrossRef]
29. Treneska, S.; Stojkoska, B.R. Wildfire detection from UAV collected images using transfer learning. In Proceedings of the 18th International Conference on Informatics and Information Technologies, Skopje, North Macedonia, 6–7 May 2021.
30. Glorot, X.; Bordes, A.; Bengio, Y. Deep Sparse Rectifier Neural Networks. In Proceedings of the Fourteenth International Conference on Artificial Intelligence and Statistics, Fort Lauderdale, FL, USA, 11–13 April 2011; Volume 15, pp. 315–323.
31. Maas, A.L.; Hannun, A.Y.; Ng, A.Y. Rectifier nonlinearities improve neural network acoustic models. In Proceedings of the ICML, Atlanta, GA, USA, 16–21 June 2013; p. 3.
32. Jin, X.; Xu, C.; Feng, J.; Wei, Y.; Xiong, J.; Yan, S. Deep Learning with S-shaped Rectified Linear Activation Units. *arXiv* **2015**, arXiv:1512.07030.
33. Zhao, B.; Feng, J.; Wu, X.; Yan, S. A survey on deep learning-based fine-grained object classification and semantic segmentation. *Int. J. Autom. Comput.* **2017**, *14*, 119–135. [CrossRef]
34. Chen, Y.; Zhang, Y.; Xin, J.; Wang, G.; Mu, L.; Yi, Y.; Liu, H.; Liu, D. UAV Image-based Forest Fire Detection Approach Using Convolutional Neural Network. In Proceedings of the 14th IEEE Conference on Industrial Electronics and Applications (ICIEA), Xi'an, China, 19–21 June 2019; pp. 2118–2123.
35. Lee, W.; Kim, S.; Lee, Y.T.; Lee, H.W.; Choi, M. Deep neural networks for wild fire detection with unmanned aerial vehicle. In Proceedings of the IEEE International Conference on Consumer Electronics (ICCE), Taipei, Taiwan, 12–14 June 2017; pp. 252–253.
36. Zhao, Y.; Ma, J.; Li, X.; Zhang, J. Saliency Detection and Deep Learning-Based Wildfire Identification in UAV Imagery. *Sensors* **2018**, *18*, 712. [CrossRef]

37. Srinivas, K.; Dua, M. Fog Computing and Deep CNN Based Efficient Approach to Early Forest Fire Detection with Unmanned Aerial Vehicles. In Proceedings of the International Conference on Inventive Computation Technologies, Coimbatore, India, 26–28 February 2020; pp. 646–652.
38. Wu, H.; Li, H.; Shamsoshoara, A.; Razi, A.; Afghah, F. Transfer Learning for Wildfire Identification in UAV Imagery. In Proceedings of the 54th Annual Conference on Information Sciences and Systems (CISS), Princeton, NJ, USA, 18–20 March 2020; pp. 1–6.
39. Chen, Y.; Zhang, Y.; Xin, J.; Yi, Y.; Liu, D.; Liu, H. A UAV-based Forest Fire Detection Algorithm Using Convolutional Neural Network. In Proceedings of the 37th Chinese Control Conference (CCC), Wuhan, China, 25–27 July 2018; pp. 10305–10310.
40. Krizhevsky, A.; Sutskever, I.; Hinton, G.E. ImageNet classification with deep convolutional neural networks. *Adv. Neural Inf. Process. Syst. (NIPS)* **2012**, *25*, 1097–1105. [CrossRef]
41. Szegedy, C.; Liu, W.; Jia, Y.; Sermanet, P.; Reed, S.; Anguelov, D.; Erhan, D.; Vanhoucke, V.; Rabinovich, A. Going Deeper with Convolutions. In Proceedings of the IEEE Conference on Computer Vision and Pattern Recognition (CVPR), Boston, MA, USA, 7–12 June 2015; pp. 1–9.
42. Simonyan, K.; Zisserman, A. Very deep convolutional networks for large-scale image recognition. *arXiv* **2014**, arXiv:1409.1556.
43. Chollet, F. Xception: Deep Learning With Depthwise Separable Convolutions. In Proceedings of the IEEE Conference on Computer Vision and Pattern Recognition (CVPR), Honolulu, HI, USA, 21–26 July 2017; pp. 1251–1258.
44. Szegedy, C.; Vanhoucke, V.; Ioffe, S.; Shlens, J.; Wojna, Z. Rethinking the Inception Architecture for Computer Vision. In Proceedings of the IEEE Conference on Computer Vision and Pattern Recognition (CVPR), Las Vegas, NV, USA, 27–30 June 2016; pp. 2818–2826.
45. Shamsoshoara, A.; Afghah, F.; Razi, A.; Zheng, L.; Fulé, P.; Blasch, E. *The FLAME Dataset: Aerial Imagery Pile Burn Detection Using Drones (UAVs)*; IEEE DataPort: New York, NY, USA, 2020. [CrossRef]
46. He, K.; Zhang, X.; Ren, S.; Sun, J. Deep Residual Learning for Image Recognition. In Proceedings of the IEEE Conference on Computer Vision and Pattern Recognition (CVPR), Las Vegas, NV, USA, 27–30 June 2016; pp. 770–778.
47. Sandler, M.; Howard, A.; Zhu, M.; Zhmoginov, A.; Chen, L.C. MobileNetV2: Inverted Residuals and Linear Bottlenecks. In Proceedings of the IEEE/CVF Conference on Computer Vision and Pattern Recognition, Salt Lake City, UT, USA, 18–23 June 2018; pp. 4510–4520.
48. Howard, A.G.; Zhu, M.; Chen, B.; Kalenichenko, D.; Wang, W.; Weyand, T.; Andreetto, M.; Adam, H. MobileNets: Efficient Convolutional Neural Networks for Mobile Vision Applications. *arXiv* **2017**, arXiv:1704.04861.
49. Liu, L.; Ouyang, W.; Wang, X.; Fieguth, P.; Chen, J.; Liu, X.; Pietikäinen, M. Deep learning for generic object detection: A survey. *Int. J. Comput. Vis.* **2020**, *128*, 261–318. [CrossRef]
50. Jiao, Z.; Zhang, Y.; Xin, J.; Mu, L.; Yi, Y.; Liu, H.; Liu, D. A Deep Learning Based Forest Fire Detection Approach Using UAV and YOLOv3. In Proceedings of the 1st International Conference on Industrial Artificial Intelligence (IAI), Shenyang, China, 23–27 July 2019; pp. 1–5.
51. Redmon, J.; Farhadi, A. YOLOv3: An Incremental Improvement. *arXiv* **2018**, arXiv:1804.02767.
52. Jiao, Z.; Zhang, Y.; Mu, L.; Xin, J.; Jiao, S.; Liu, H.; Liu, D. A YOLOv3-based Learning Strategy for Real-time UAV-based Forest Fire Detection. In Proceedings of the Chinese Control And Decision Conference (CCDC), Hefei, China, 22–24 August 2020; pp. 4963–4967.
53. Alexandrov, D.; Pertseva, E.; Berman, I.; Pantiukhin, I.; Kapitonov, A. Analysis of Machine Learning Methods for Wildfire Security Monitoring with an Unmanned Aerial Vehicles. In Proceedings of the 24th Conference of Open Innovations Association (FRUCT), Moscow, Russia, 8–12 April 2019; pp. 3–9.
54. Liu, W.; Anguelov, D.; Erhan, D.; Szegedy, C.; Reed, S.; Fu, C.Y.; Berg, A.C. Ssd: Single shot multibox detector. In Proceedings of the European Conference on Computer Vision (ECCV), Amsterdam, The Netherlands, 8–16 October2016; pp. 21–37.
55. Redmon, J.; Farhadi, A. YOLO9000: Better, Faster, Stronger. In Proceedings of the IEEE Conference on Computer Vision and Pattern Recognition (CVPR), Honolulu, HI, USA, 21–26 July 2017; pp. 7263–7271.
56. Ren, S.; He, K.; Girshick, R.B.; Sun, J. Faster R-CNN: Towards Real-Time Object Detection with Region Proposal Networks. *arXiv* **2015**, arXiv:1506.01497.
57. Tang, Z.; Liu, X.; Chen, H.; Hupy, J.; Yang, B. Deep Learning Based Wildfire Event Object Detection from 4K Aerial Images Acquired by UAS. *AI* **2020**, *1*, 10. [CrossRef]
58. Barmpoutis, P.; Stathaki, T.; Dimitropoulos, K.; Grammalidis, N. Early Fire Detection Based on Aerial 360-Degree Sensors, Deep Convolution Neural Networks and Exploitation of Fire Dynamic Textures. *Remote Sens.* **2020**, *12*, 3177. [CrossRef]
59. Chen, L.C.; Zhu, Y.; Papandreou, G.; Schroff, F.; Adam, H. Encoder-Decoder with Atrous Separable Convolution for Semantic Image Segmentation. In Proceedings of the European Conference on Computer Vision (ECCV), Munich, Germany, 8–14 September 2018; pp. 801–818.
60. Ronneberger, O.; Fischer, P.; Brox, T. U-net: Convolutional networks for biomedical image segmentation. In Proceedings of the International Conference on Medical Image Computing and Computer-Assisted Intervention, Munich, Germany, 5–9 October 2015; Springer International Publishing: Cham, Switzerland, 2015; pp. 234–241.
61. Frizzi, S.; Bouchouicha, M.; Ginoux, J.M.; Moreau, E.; Sayadi, M. Convolutional neural network for smoke and fire semantic segmentation. *IET Image Process.* **2021**, *15*, 634–647. [CrossRef]

62. Real, E.; Aggarwal, A.; Huang, Y.; Le, Q.V. Regularized Evolution for Image Classifier Architecture Search. In Proceedings of the AAAI Conference on Artificial Intelligence, Honolulu, HI, USA, 27 January–1 February 2019; Volume 33, pp. 4780–4789. [CrossRef]
63. Liu, C.; Zoph, B.; Neumann, M.; Shlens, J.; Hua, W.; Li, L.J.; Fei-Fei, L.; Yuille, A.; Huang, J.; Murphy, K. Progressive Neural Architecture Search. In Proceedings of the European Conference on Computer Vision (ECCV), Munich, Germany, 8–14 September 2018; pp. 19–34.
64. Xie, S.; Girshick, R.; Dollar, P.; Tu, Z.; He, K. Aggregated Residual Transformations for Deep Neural Networks. In Proceedings of the IEEE Conference on Computer Vision and Pattern Recognition (CVPR), Honolulu, HI, USA, 21–26 July 2017; pp. 1492–1500.
65. Szegedy, C.; Ioffe, S.; Vanhoucke, V. Inception-v4, Inception-ResNet and the Impact of Residual Connections on Learning. In Proceedings of the Thirty-First AAAI Conference on Artificial Intelligence, San Francisco, CA, USA, 4–9 February 2017; pp. 4278–4284.
66. Valanarasu, J.M.J.; Oza, P.; Hacihaliloglu, I.; Patel, V.M. Medical Transformer: Gated Axial-Attention for Medical Image Segmentation. *arXiv* **2021**, arXiv:2102.10662.
67. Dillon, J.V.; Langmore, I.; Tran, D.; Brevdo, E.; Vasudevan, S.; Moore, D.; Patton, B.; Alemi, A.; Hoffman, M.D.; Saurous, R.A. TensorFlow Distributions. *arXiv* **2017**, arXiv:1711.10604.
68. Ma, Y.; Liu, Q.; Qian, Z. Automated image segmentation using improved PCNN model based on cross-entropy. In Proceedings of the International Symposium on Intelligent Multimedia, Video and Speech Processing, Hong Kong, China, 20–22 October 2004; pp. 743–746.
69. Paszke, A.; Gross, S.; Massa, F.; Lerer, A.; Bradbury, J.; Chanan, G.; Killeen, T.; Lin, Z.; Gimelshein, N.; Antiga, L.; et al. PyTorch: An Imperative Style, High-Performance Deep Learning Library. *Adv. Neural Inf. Process. Syst.* **2019**, *32*, 8026–8037.
70. Sudre, C.H.; Li, W.; Vercauteren, T.; Ourselin, S.; Jorge Cardoso, M. Generalised Dice Overlap as a Deep Learning Loss Function for Highly Unbalanced Segmentations. In *Deep Learning in Medical Image Analysis and Multimodal Learning for Clinical Decision Support*; Springer: Cham, Switzerland, 2017; pp. 240–248.

sensors

MDPI

Article

High-Resolution U-Net: Preserving Image Details for Cultivated Land Extraction

Wenna Xu [1,2,†], Xinping Deng [1,3], Shanxin Guo [1,3,†], Jinsong Chen [1,3,*], Luyi Sun [1,3], Xiaorou Zheng [1,2], Yingfei Xiong [1,2], Yuan Shen [1,2] and Xiaoqin Wang [4]

1 Center for Geo-Spatial Information, Shenzhen Institutes of Advanced Technology, Chinese Academy of Sciences, Shenzhen 518055, China; wn.xu@siat.ac.cn (W.X.); xp.deng1@siat.ac.cn (X.D.); sx.guo@siat.ac.cn (S.G.); ly.sun@siat.ac.cn (L.S.); xiaorou.zheng@siat.ac.cn (X.Z.); yf.xiong@siat.ac.cn (Y.X.); yuan.shen@siat.ac.cn (Y.S.)

2 University of Chinese Academy of Sciences, Beijing 101407, China

3 Shenzhen Engineering Laboratory of Ocean Environmental Big Data Analysis and Application, Shenzhen 518055, China

4 Key Laboratory of Spatial Data Mining & Information Sharing of Ministry of Education, National & Local Joint Engineering Research Center of Satellite Geospatial Information Technology, Fuzhou University, Fuzhou 350000, China; wangxq@fzu.edu.cn

* Correspondence: js.chen@siat.ac.cn; Tel.: +86-755-86392331

† These authors contributed equally to this work.

Received: 29 May 2020; Accepted: 17 July 2020; Published: 22 July 2020

Abstract: Accurate and efficient extraction of cultivated land data is of great significance for agricultural resource monitoring and national food security. Deep-learning-based classification of remote-sensing images overcomes the two difficulties of traditional learning methods (e.g., support vector machine (SVM), K-nearest neighbors (KNN), and random forest (RF)) when extracting the cultivated land: (1) the limited performance when extracting the same land-cover type with the high intra-class spectral variation, such as cultivated land with both vegetation and non-vegetation cover, and (2) the limited generalization ability for handling a large dataset to apply the model to different locations. However, the "pooling" process in most deep convolutional networks, which attempts to enlarge the sensing field of the kernel by involving the upscale process, leads to significant detail loss in the output, including the edges, gradients, and image texture details. To solve this problem, in this study we proposed a new end-to-end extraction algorithm, a high-resolution U-Net (HRU-Net), to preserve the image details by improving the skip connection structure and the loss function of the original U-Net. The proposed HRU-Net was tested in Xinjiang Province, China to extract the cultivated land from Landsat Thematic Mapper (TM) images. The result showed that the HRU-Net achieved better performance (Acc: 92.81%; kappa: 0.81; F1-score: 0.90) than the U-Net++ (Acc: 91.74%; kappa: 0.79; F1-score: 0.89), the original U-Net (Acc: 89.83%; kappa: 0.74; F1-score: 0.86), and the Random Forest model (Acc: 76.13%; kappa: 0.48; F1-score: 0.69). The robustness of the proposed model for the intra-class spectral variation and the accuracy of the edge details were also compared, and this showed that the HRU-Net obtained more accurate edge details and had less influence from the intra-class spectral variation. The model proposed in this study can be further applied to other land cover types that have more spectral diversity and require more details of extraction.

Keywords: full convolutional network; U-Net; cultivated land extraction; deep learning; remote sensing

1. Introduction

Accurate area and change of cultivated land is one of the fundamental types of data for precision agriculture, food security analysis, yields forecasting, and land-use/land-cover research [1]. In the arid and semi-arid regions, this information is particularly important as it is related to the regional water balance and the local ecosystem health [2]. Currently, the increasing free remote sensing data (such as U.S. Geological Survey (USGS) Landsat and European Space Agency (ESA) Sentinel) provides sufficient data sources and the opportunity to extract and monitor the dynamic change of the cultivated land [3–5].

However, the cultivated land, as a man-made concept, usually shows different spectral characteristics due to the varying types of crops, different irrigation methods, and different soil types, as well as fallow land plots. As a result, for classification, the intra-class variation increases and the inter-class separability decreases [6,7]. The frequently used traditional pixel-based classifiers, such as support vector machine (SVM), K-nearest neighbors (KNN), and random forest (RF) [8,9], and the object-based farmland extraction models, such as the stratified object-based farmland extraction [6], the superpixels and supervised machine-learning model [10], and the time-series-based methods [11], usually require the prior knowledge to model the high intra-class variation of the spatial or spectral features. Due to this, the features learned by these methods are often limited to the specific datasets, time, and locations, which is known as limited model generalization ability. The re-training process is usually required when applying these models to different datasets, time, and locations.

With the rapid development of deep learning [12], convolutional neural networks (CNNs) have gained state-of-the-art performance in land cover classification, which overcomes the abovementioned difficulties [13]. Possible reasons for its success include: (1) the capacity of learning from a large dataset; (2) the tolerance for larger intra-class variation of the object features; and (3) the high generalization ability. Benefitting from the large training dataset, the feature variation of the target object across different locations or time can be modeled. CNNs have shown great advantage in urban land use and land cover mapping [14–18], scene classification [19–21], and object extraction [22–25]. Among the popular CNNs, the U-Net is reported to achieve state-of-the-art performance on several benchmark datasets even with limited training data [26,27]. It was widely used in many fields as a result.

However, the "pooling" process in most deep convolutional networks, which (1) provides the invariance (translation, rotation, and scale-invariance) capacity for the model to capture the major feature of the target; (2) reduces the number of parameters for multi-scale training; and (3) increases the receptive field by involving the down-sampling process (converting images from the high to low spatial resolution) with certain calculation (maximum, average, etc.), leads to significant detail loss from the image, including edges, gradients, and image texture details [28,29]. This problem could decrease the accuracy of extraction of land cover type even more when dealing with remote sensing images considering the high intra-class spectral variation [30]. Ideas to solve this problem can currently be organized in two categories: (1) learning and recovering high-resolution details from low-resolution feature maps or (2) maintaining high-resolution details throughout the network [31].

In the first category, the detailed texture is recovered from the low-resolution feature maps after the pooling process by applying certain up-sampling methods (such as bilinear interpolation or deconvolution) [32–34] of the representative model in this category, such as the fully convolutional network (FCN) [35], In SegNet [36], DeconvNet [37], RefineNet [38] et al.

For remotely sensed images, this idea was widely used. For instance, the FCN-based network achieved an overall accuracy of 89.1% on the International Society for Photogrammetry and Remote Sensing (ISPRS) Vaihingen Dataset without a down-sampling layer to obviate deconvolution in the latter part of the structure [39]. Marmanis, et al. (2016) designed a segmentation network at the pixel-level that synthesized the FCN and deconvolution layers and refined the results using fully connected conditional random fields (CRF) [40]. ASPP-Unet and ResASPP-Unet recovered the spatial texture by adding the Atrous Spatial Pyramid Pooling (ASPP) technique in network to increase the effective field-of-view in convolution and capture the features in multiple scales [41]. MultiResoLCC provides a two-branch CNN architecture to improve the image details by jointly using panchromatic (PAN) and multi spectral (MS) imagery [42]. For hyperspectral image classification tasks, the CNN structure can also improve the accuracy by extracting the hierarchical features [43] and creating the low-dimensional feature space to increase the separability [44].

This type of method recovers the high-resolution details by learning from low-resolution feature maps. Although various skip connection methods have been used to optimize the obtained high-resolution details, the effect is limited since the lost details are usually recovered only from low-spatial resolution features. This often causes the recovering procedural to be ill-posed as the number of pixels of the output is always bigger than that of the input.

In the second category, high-resolution details are first extracted and maintained through the whole process, typically by a network that is formed by connecting multi-level convolutions with repeated information exchange across parallel convolutions. Under this idea, the skip connection is usually redesigned between the pooling nodes and the up-sampling nodes. For instance, (1) adding more skip connections to link different convolution nodes at the same scale, and (2) adding more skip connections to link the convolution nodes at the different scales. Representative models include convolutional neural fabrics [45], interlinked CNNs [46], and high-resolution networks (HRNet) [47]. This kind of method avoids the ill-posed problem; however, the time consumed in the training process can dramatically increase. More free parameters in the model require more data to train.

In this paper, we propose a new end-to-end cultivated land extraction algorithm, high-resolution U-Net (HRU-Net), to extract cultivated land from Landsat TM images. The new network is based on the U-Net structure, in which the skip connections are redesigned following the ideas of the second category mentioned above to obtain more details. Inspired by HRNet, the loss function of the original U-Net is also improved to take into account features extracted at both shallow and deep levels. The proposed HRU-Net was tested in Xinjiang Province, China for the cultivated land extraction based on three years' worth of Landsat TM images, and was compared with the original U-Net, U-Net++, and the RF method. The major contributions of this study can be summarized as: (1) we redesigned the skip connection structure of the U-Net to keep the high-resolution details for remote sensing image classification; (2) we modified the original U-Net loss function to achieve a higher extraction accuracy for the target with a high intra-class variation; (3) we proposed a new end-to-end cultivated land extraction algorithm, the high-resolution U-Net (HRU-Net), which demonstrated good performance in extracting the target with high edge details and high intra-class spectral variation.

2. Related Work

2.1. Learning and Recovering High-Resolution Details from Low-Resolution Feature Maps

The representative model in this category is the fully convolutional network (FCN) [35]. In each stage of an FCN, an up-sampling subnetwork, like a decoder, was used as the up-sampling procedure, which attempts to recover the fine-spatial resolution details from the coarse-spatial resolution feature maps [33,34,48]. In SegNet [36], the up-sampling strategy is a mirrored symmetric version from the pooling subnetwork by grabbing the indices directly for the pooling subnetwork. The up-sampling strategy can be combined with the deconvolution process, such as in DeconvNet [37], where the locations and values of the highest gradience are kept by the up-sampling strategy, and the sparseness of the up-sampling output is repaired by the deconvolution layers. In RefineNet [38], instead of using only one feature map from one pooling layer, the long-range residual connections were used to combine all information along with all pooling layers to refine the high-resolution details. Other asymmetric structures, such as the light up-sampling process [49], light pooling, heavy up-sample processes [50], and re-combinator networks [51], were all reported have good performance for object detection.

2.2. Maintaining High-Resolution Details throughout the Network

Representative models include convolutional neural fabrics [45], interlinked CNNs [46], and high-resolution networks (HRNet) [47]. In an HRNet, a high-resolution subnetwork was first established as the first stage, then the high-to-low resolution subnetworks were added consecutively to form more low-level stages. This structure maintains the high-resolution details through the whole process and has achieved state-of-the-art performance in the field of human pose estimation [47]. Fu et al. (2019) and Wu et al. (2018) also improved skip connections by stacking multiple DeconvNets/UNets/Hourglasses with dense connections [52,53].

3. Study Area and Datasets

In this paper, the intra-class spectral variation of cultivated land can be reflected in three perspectives: (1) intra-class spectral variation over different time, (2) intra-class spectral variation over different geo-locations, (3) intra-class spectral variation over different crop types. These three variation factors can be represented with multiple times (both winter and summer) and different locations within a large area. The study area is located in the Urumqi and Bosten farmlands in Xinjiang, China (Figure 1), which mainly grow cash crops, such as cotton and pears. The crops are planted in large areas with high yield and require a huge amount of water supply every year. Extracting cultivated land of these two regions is of great significance to the agricultural and water resource monitoring to ensure the national food security of Xinjiang and China.

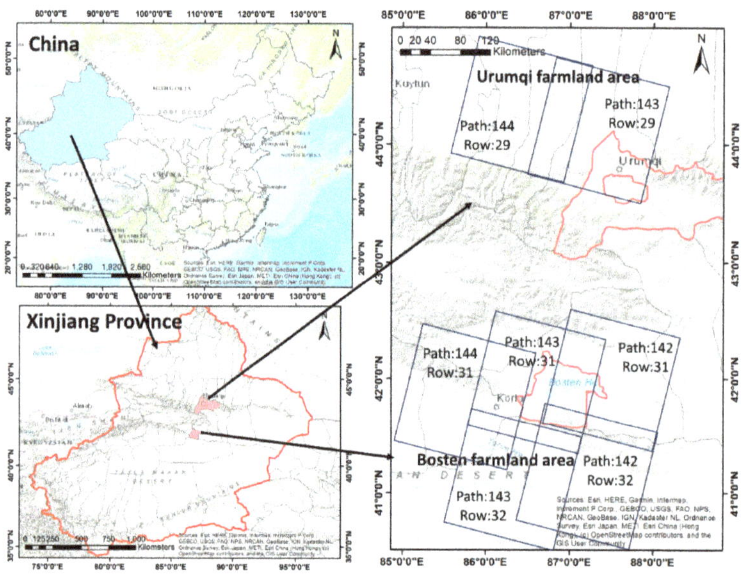

Figure 1. The study area.

Landsat5 thematic mapper (TM) top of atmosphere (TOA) reflectance (the USGS Earth Explorer: https://earthexplorer.usgs.gov/) from 2009 to 2011 was collected as the dataset in this study. The TM sensor has seven spectral bands (Table 1), but we only selected six bands with a resolution of 30 m: B1 (blue), B2 (green), B3 (red), B4 (near-infrared, NIR), B5 (short-wave-infrared, SWIR 1), and B7 (short-wave-infrared, SWIR 2). The thermal band was not used in this study as it could vary during the different observation dates, which was caused by the different local environmental factors, such as the radiative energy the land received or the wind speed. Only cloud free images were chosen in this study. The image details are shown in Table 2.

Table 1. Parameters of Landsat4—5 thematic mapper (TM).

Sensor	Bands	Wavelength/μm	Resolution/m
Landsat4—5 TM	B1-Blue	0.45–0.52	30
	B2-Green	0.52–0.60	30
	B3-Red	0.63–0.69	30
	B4-NIR	0.76–0.90	30
	B5-SWIR1	1.55–1.75	30
	B6-TIR	10.40–12.5	120
	B7-SWIR2	2.08–2.35	30

Table 2. Data list of Landsat 5 top of atmosphere (TOA) products used in this study.

Date	Area	Image File Names	Path/Row
10 Jun 2010	Bosten farmland	LT51420312010161IKR00.tar	142/31
13 Aug 2010		LT51420312010225IKR00.tar	
2 Sep 2010		LT51430312010248IKR00.tar	143/31
26 Aug 2009		LT51420322009238IKR00.tar	142/32
27 Sep 2009		LT51420322009270KHC00.tar	
15 Jul 2011		LT51420322011196IKR00.tar	
3 Oct 2011		LT51420322011276KHC01.tar	
29 Aug 2010		LT51420322010241IKR00.tar	
30 Sep 2010		LT51420322010273IKR00.tar	
20 Aug 2010		LT51430322010232IKR00.tar	143/32
4 Aug 2010		LT51430322010216IKR00.tar	
21 Sep 2010		LT51430322010264IKR00.tar	
11 Aug 2010		LT51440312010223IKR01.tar	144/31
15 Nov 2010		LT51440312010319KHC00.tar	
11 Aug 2010	Urumqi farmland	LT51440292010223IKR01.tar	144/29
15 Nov 2010		LT51440292010319KHC00.tar	
11 Aug 2010		LT51430292010232IKR00.tar	143/29
21 Sep 2010		LT51430292010264IKR00.tar	
4 Jun 2011		LT51430292011155IKR00.tar	
7 Aug 2011		LT51430292011219KHC01.tar	
23 Aug 2011		LT51430292011235KHC01.tar	

We used the historical landcover map in the 2010 version from the local government to extract the ground truth manually based on the Landsat 5 image at 30 m scale. The changes in the land cover types were considered to be consistent from 2009 to 2011 and were neglected in this study. The original historical landcover map contained five land cover types (the urban area, cultivated land, forest, water, and desert). We classified the historical landcover map by only two types (cultivated land and other). The historical landcover map was then transformed from the original polygon to the raster format with the same spatial resolution of the Landsat data. For convenience, we added the ground truth data (the historical landcover map) to the Landsat dataset as the seventh band. After that, the TM images and corresponding ground truth were split into 256 × 256-pixel tiles to keep the memory consumption low during the training and validation. These tiles were adjacent and non-overlapping.

To evaluate different combinations of spectral bands on the performance of cultivated land extraction, we defined three datasets, namely, TM-NRG, TM-RGB, TM-All, with a varying number of spectral bands. An overview of each dataset is provided in Table 3. To avoid overfitting during training, we selected 4050 tiles (approximately 70%) randomly for training, 867 tiles (approximately 15%) as validation data for adjusting the model hyperparameters during training, and the remaining 868 tiles (approximately 15%) for independent testing. The methods we used for comparation (RF, U-Net, and U-Net++) were all based and tested on the same datasets.

Table 3. Three different TM datasets used in this study.

Dataset	Bands	Resolution/m	Training Sample	Validating Sample in Training	Testing Sample
TM-NRG	NIR, Red, Green	30	4050 (70%)	867 (15%)	868 (15%)
TM-RGB	Red, Green, Blue	30			
TM-All	Blue, Green, Red, NIR, SWIR1, SWIR2	30			

4. Methodology

In this paper, a new end-to-end cultivated land extraction algorithm, high-resolution U-Net (HRU-Net), was proposed, with the aim to extract the same land-cover type with different spectra accurately and preserve the image details by improving the skip connection structure and loss function of the original U-Net. Figure 2 shows an overview of the workflow of this study.

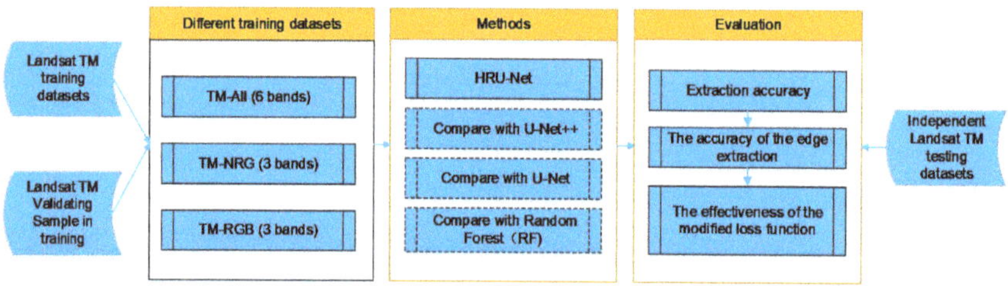

Figure 2. Overview of the performance evaluation framework. High-resolution U-Net (HRU-Net).

4.1. The Original U-Net and U-Net++

Initially, the U-Net was developed for biomedical image segmentation. We chose it as the base network to extract cultivated land as it achieves state-of-the-art performance on benchmark datasets even with limited training data [27,28]. Figure 3a shows the structure of the original U-Net network. It contains two main pathways: the contracting pathway on the left side and the expansive pathway on the right side.

In the contracting path, the input image was first sent to the feature detection by operating a 2-dimensional convolution by the typical architecture of a convolutional network, which repeated the block of two 3×3 convolutions, a rectified linear unit, and a 2×2 max-pooling operation, iteratively. To enlarge the "sense field" of the convolution kernel and give the network more ability for a global view of the features of the object, the "pooling operation" was added to contract the feature map into the lower level. Meanwhile, a skip connection structure attempted to reduce the loss of image details in the "pooling operation" in the contraction path by adding a feature vector to the expansive path at the same level, as indicated by the gray arrow in Figure 3a.

In the expansive path, the central idea was to combine the low-level feature maps to expand the image size. First, the low-level feature map was up-sampled by a 2×2 transpose convolution. Secondly, the output was combined with the corresponding feature map from the skip connection at the same level. Thirdly, two 3×3 convolutions and the rectified linear unit (ReLU) activation function were applied for further feature detection.

At the final layer, to match the number of channels to the number of classes in the final output, a 1×1 convolution with the Softmax activation function was used. The output of this network was the predicted probabilities of each class $p(x)$. The final class labels were calculated by selecting the highest probability class in the vector $p(x)$. In this structure, the skip connection was the only path to restore the high-resolution details in every convolution level.

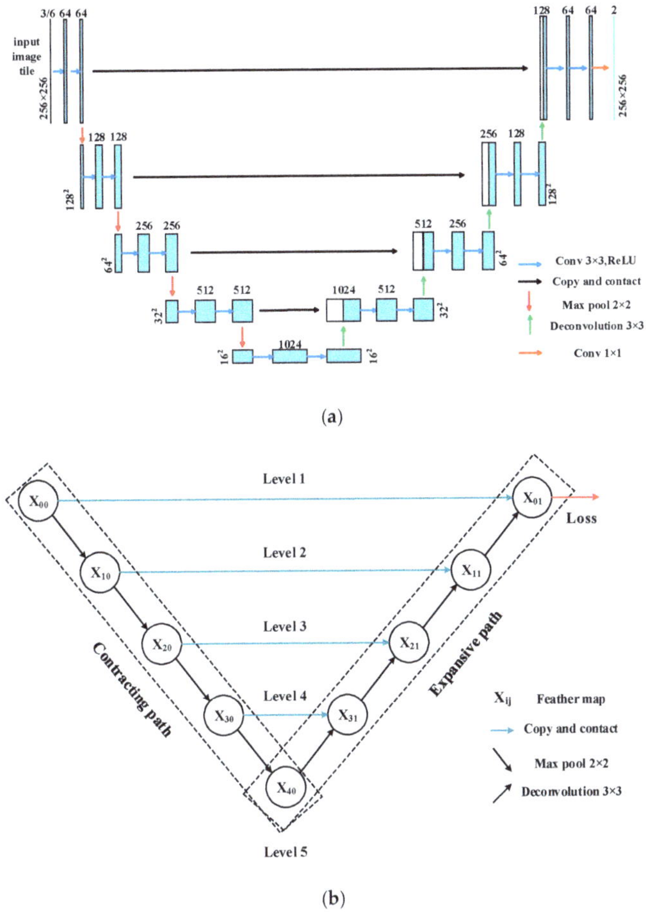

Figure 3. (**a**) U-Net architecture [28] and (**b**) Simplified U-Net topology diagram from (**a**).

As shown in Figure 3b, in order to emphasize the skip connections between the feature maps at the different levels, the structure of the U-Net (Figure 3a) was simplified by replacing the convolution process in Figure 3a with the symbol X_{ij}, where i is the level index and j is the convolution node index at the same level. For example, the X_{10} represented the first convolution module at the second level.

The other benefit of the U-Net is that the number of the trainable parameters is relatively small. Other networks, such as FCN and DeconvNet, are more complicated with more trainable parameters, and require a bigger training set and a longer time to train [35,37]. Usually, to reduce the training time of networks, a pre-trained network can be used to retrain the top layer on a new dataset. However, the pre-trained network is usually trained on natural pictures with RGB bands. As we hope to take full advantage of the multi-band data of remote-sensing images instead of only RGB channels, this strategy cannot work well when the channel difference happens between the pre-trained and the new datasets. For this reason, the U-Net network in this study was trained from scratch.

Under the hypothesis that the feature maps from contracting path (encoder networks) can enrich the prior for the expansive path (decoder networks), UNet++ was proposed to increase the segmentation accuracy for medical images [54]. In UNet++, a small down-triangle structure was designed as the basic unit. With this unit, UNet++ can be easily extended to different levels depending on the accuracy and performance required for the different tasks. The intuitive purpose of the UNet++ is to reach the high overall accuracy of segmentation in medical images for improving disease diagnosis. In this paper, we focused on the application of a deep learning model for satellite images, specifically to recover the edge details of the land cover types which were lost during the "pooling" process. More details of the HRU-Net will be described in the next section.

4.2. The High-Resolution U-Net

Giving the network the ability to learn the high-resolution details of the image is the key to solving the problems of insufficient accuracy of cultivated land extraction due to a loss of image details. The idea of the U-Net network is to learn and recover high-resolution details directly from a low-resolution feature map by simply combining the feature maps from the skip connection at the same level. In the first step, learning and recovering high-resolution information from the lower level feature map is extremely difficult as it requires the recovery of non-existent details. In the second step, simply adding the feature map from the skip connection to a low-level feature map could disturb the concise features learned from the low level. The image details from the skip connection are limited as it has already suffered the "pooling process" in the previous feature detection.

Considering the multi-level structure of the U-Net and the higher level that the convolutional nodes locate, a smaller number of the "pooling process" were applied to these nodes. As a result, more texture details remained in these feature maps. The key to solving this problem was to find a proper strategy to enrich the feature map details by involving information from the higher level and reducing the noise amplifying effect at the same time. The new structure we proposed in this study, the HRU-Net, used the idea of maintaining high-resolution details during the whole process to ensure that the multi-resolution descriptions of the image were always present (Figure 4).

In this structure, the image details not only came from the same level but were also enriched from the higher level. To reduce the noise from the higher level and produce more deep sematic features, several convolutional nodes were added in the skip connection path. The new convolutional nodes increased the number of overall parameters, so in this study, to learn the network parameters more efficiently, the idea of deep supervision was adopted to re-design the loss function. The network architecture is illustrated in Figure 4a. Compared to the original U-Net architecture, the HRU-Net kept the same structure in the contracting and expansive path. More skip connections were added between the contracting and expansive path. The simplified topology diagram of the HRU-Net is shown in Figure 4b, simplified from Figure 4a by replacing the convolution process with the symbol X_{ij} to make clearer the structure of the skip connection in the HRU-Net.

In the following part, we will further discuss from the two perspectives: (1) how to improve the skip connection structure and (2) how to use the idea of deep supervision to design the loss function.

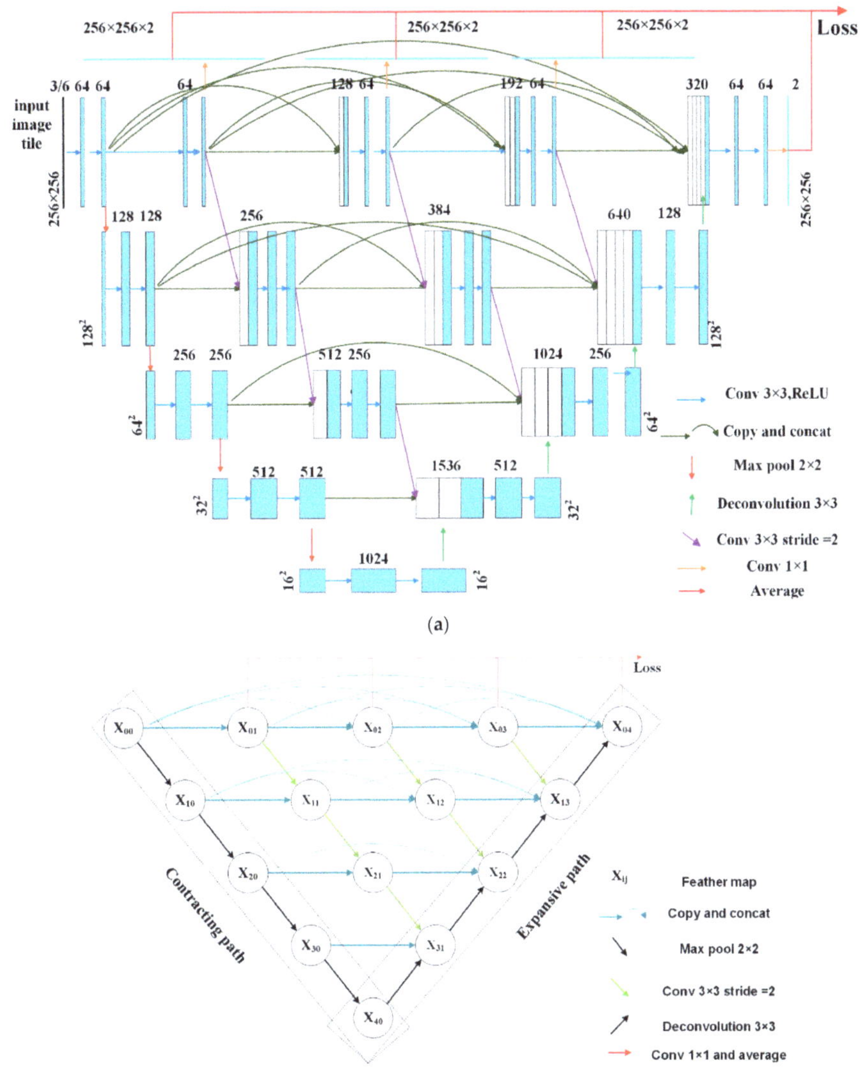

(a)

(b)

Figure 4. (**a**) The HRU-Net architecture and (**b**) the simplified topology diagram of the HRU-Net.

4.2.1. Improving the Skip Connection Structure

The skip connections were first introduced in the FCN [37]. Starting from the FCN, this structure has been widely introduced in many models to retain the high-resolution details across the different levels. In the U-Net, the feature maps in the contracting path are directly sent to the expansive path by skip connections. To simply copy the feature map from the contracting path and merge to the expansive path with the feature map from the lower level does not always work as the details have already been lost before the skip connections. The basic idea to solve this problem is to borrow the image details from a higher level to minimize the effect of the "pooling" (the green-sampling arrow in Figure 4b). Followed by this idea, in the HRU-Net the skip connection was improved in the following two aspects:

(1) Maintained resolution details at the same level

First, the HRU-Net maintained feature maps at the same layer by applying a repeated convolution module (shown in blue arrows in Figure 4b). Each module consisted of two 3 × 3 convolutions and a rectified linear unit. Then, it incorporated shallow features into deep features at each layer by a skip connection at the same level to retain details (shown in blue curved arrows in Figure 4).

(2) Fused multi-scale details cross different levels

The HRU-Net converted the high-resolution feature map into the same size and the same number of channels as the lower-level required by applying a 3 × 3 convolution with a stride of 2 (shown in green arrows in Figure 4b); then, the HRU-Net combined this high-level feature map with the feature map from the previous node by a convolution operation and a concatenation operation; at last, two 3 × 3 convolutions and a rectified linear unit were applied for further feature detection (shown in blue arrows in Figure 4a,b).

The HRU-Net can be formulated as follows:

$$X_{ij} = \begin{cases} c\left(d\left(X_{(i-1)j}\right)\right) & j = 0 \\ c\left(\left[X_{ik}\right]_{k=0}^{j-1}\right) & i = 0,\ j = 1,\ 2,\ 3 \\ c\left(\left[\left[X_{ik}\right]_{k=0}^{j-1}, u\left(X_{(i+1)(j-1)}\right)\right]\right) & i = 0\ and\ j = 4 \\ c\left(\left[\left[X_{ik}\right]_{k=0}^{j-1}, d\left(X_{(i-1)j}\right)\right]\right) & j > 0, i > 0\ and\ i + j < 4 \\ c\left(\left[\left[X_{ik}\right]_{k=0}^{j-1}, d\left(X_{(i-1)j}\right), u\left(X_{(i+1)(j-1)}\right)\right]\right) & j > 0, i > 0\ and\ i + j = 4 \end{cases} \tag{1}$$

where X_{ij} is the output feature map of the node (i, j), where i is the level index and j is the convolution node index at the same level. Function $c(\cdot)$ represents the convolution operation, $u(\cdot)$ denotes an up-sampling operation, $d(\cdot)$ is a pooling or down-sampling operation, and $[\cdot]$ is the concatenation operation. The overall structure can be described as follows:

- The nodes at level $j = 0$, X_{i0} can be gained by only one input $X_{(i-1)0}$, which is from the previous layer in the contracting path. The max pooling and convolution operation are applied in nodes $X_{(i-1)0}$.
- The nodes at level $i = 0$ and $j < 4$ receive the j feature maps of the previous nodes at the same level. For example, X_{03} can be gained by X_{00}, X_{01}, and X_{02}. The inputs are concatenated by concatenation operation, then the convolution operation is performed.
- The nodes at level $i = 0$ and $j = 4$ receive the j feature maps from the previous nodes at the same level and the up-sampled feature maps from the lower level. In particular, X_{04} can be gained by X_{00}, X_{01}, X_{02}, X_{03} and X_{13}. The up-sampled X_{13} is concatenated with X_{00}, X_{01}, X_{02}, X_{03} nodes.
- The nodes at the middle of the network, where $j > 0, i > 0$ and $i + j < 4$, receive $j + 1$ inputs (j inputs from are the j feature maps form previous nodes at the same level, one input is the down-sampled output from the higher level).
- The nodes at the end of each layer, where $j > 0, i > 0$ and $i + j = 4$, receive $j + 2$ inputs, (j inputs are from the j feature maps form previous nodes at the same level, one input is the down-sampled output from the higher-level, and one input is up-sampled output from the lower-level).

4.2.2. Using the Idea of Deep Supervision to Modify the Loss Function

When designing the input of the loss function, the U-Net only obtains the classification probabilities from X_{04}. Compared to the U-Net, the HRU-Net generated full-resolution feature maps from multiple levels, $\left\{X_{0j},\ j \in (1,2,3,4)\right\}$, which can be used to apply deep supervision. We first obtained the classification probabilities at different semantic levels, from $\left\{X_{0j},\ j \in (1,2,3,4)\right\}$, through 1 × 1 convolutions with the Softmax activation function (as marked by red arrows in Figure 4), and then obtained the predicted class probabilities $P(x)$ by averaging all probabilities,

$$P(x) = \left[P_0(x), P_1(x)\right]^T \tag{2}$$

where $P_i(x)$ is the predicted probability of x belonging to class i ($i = 0$ for cultivated land, and $i = 1$ for non-cultivated land). The class label y of a given image can be calculated by obtaining the label from the maximized probability in $P(x)$:

$$y = argmax(P(x)). \tag{3}$$

The loss function of HRU-net is defined as

$$H\left(Y, \overline{Y}\right) = -\frac{1}{N}\sum_{i} Y_i log\left(\overline{Y}_i\right) \tag{4}$$

with $\overline{Y}_i$ and Y_i denoting the predicted and the actual probability of class i, respectively, and N being the batch size.

4.2.3. Assessment

The accuracy evaluation metrics in this paper include (1) the overall accuracy, (2) Cohen's kappa coefficient, and (3) the F1-score. The overall accuracy is defined as the number of correctly classified pixels over the total number of pixels. It is simple and intuitive but may fail to assess the performance thoroughly when the number of samples for different classes varies significantly. Cohen's kappa coefficient is more robust, as it takes into consideration the possibility of agreements occurring randomly. Let p_0 be the percentage of pixels correctly classified, and p_e be the expected probability of agreement when the classifier assigns class labels by chance, Cohen's kappa coefficient is defined as:

$$K = \frac{p_0 - p_e}{1 - p_e}. \tag{5}$$

Usually, we characterize $K < 0$ as no agreement, $[0, 0.20]$ as poor agreement, $[0.20, 0.40]$ as fair agreement, $[0.40, 0.60]$ as moderate agreement, $[0.60, 0.80]$ as good agreement, and $[0.80, 1]$ as almost perfect agreement. The F1-score is defined as the harmonic mean of the precision rate and recall rate:

$$\mathrm{F1} = \frac{2 \times P \times R}{P + R} \tag{6}$$

where P is the number of positive classes predicted correctly (TP) divided by the number of all positive results (including both true positive TP and false positive FP), and R is the number of true positive results (TP) divided by the number of all relevant samples (true positive plus false negative FN):

$$P = \frac{TP}{TP + FP} \tag{7}$$

$$R = \frac{TP}{TP + FN} \tag{8}$$

An F1 score reaches its best value at 1 (perfect precision and recall) and its worst at 0.

5. Results and Discussion

5.1. The Learning Process of the HRU-Net

In this study, we hoped to make full use of the advantage of the multi-band data of remote-sensing images instead of only RGB images. Thus, we decided to train the all network (HRU-Net, U-Net++, the original U-Net, and RF) from scratch. To compare the performance of the different numbers of bands, three datasets were prepared (Table 2). The performance of the near-infrared (NIR) band can be analyzed when comparing the results of the TM-NRG with those of the TM-RGB. Similarly, comparing the results from the TM-All to the TM-NRG datasets, the improvement of the shortwave-infrared (SWIR) can be investigated.

The HRU-Net, U-Net++, U-Net, and RF were trained and tested on the three datasets (Table 2) separately. In each dataset, all samples were randomly split into three: the training set, the validation set, and the testing set. The training set was used for model training. The validation set was used to calibrate the hyperparameters of the deep learning model, and the testing set was used to apply the independent assessment for the different models.

All experiments of the HRU-Net, the U-Net++, and U-Net were carried out on four TITAN X GPUs. We used PyTorch backend as the deep-learning framework (https://pytorch.org/). To maximize the GPU memory usage, we set a different batch size for each network (HRU-Net and U-Net++:24, U-Net:48), and each network model was trained by starting with a different initial learning rate (HRU-Net:0.0015, U-Net++:0.002, U-Net:0.0002). For three networks, the gradient descent optimization (SGD) optimizer with a momentum of 0.95 and a weight decay of 10^{-4} was adopted, and the learning rate was decreased every iteration by a factor of $0.5 \times (1 + cos(\pi \frac{iter}{max\ iters}))$. The batch-norm parameters were learned with a decay rate of 0.9, and the input crop size for each training image was set to 256×256. Figure 5 shows the training history of the HRU-Net, U-Net++, and U-Net. Considering the popularity and the success of the RF in the classification of remote-sensing images, we also trained the traditional RF classifier on the same datasets as a comparison. The Scikit-learn (http://scikit-learn.org, 2018) implementation was adopted for RF in our experiments, which employed several optimized *C4.5* decision trees to improve the prediction accuracy while controlling the over-fitting at the same time [55]. The detailed parameters of the random forest are shown in Table 4.

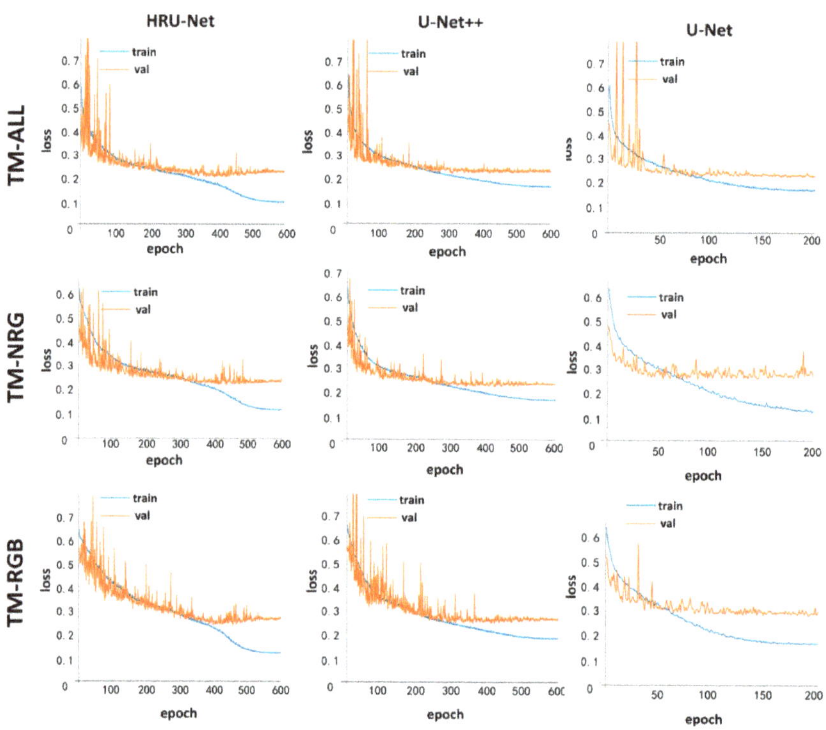

Figure 5. Visualizations of the training history for the HRU-Net, U-Net, and U-Net++ models.

Table 4. The parameters used in the random forest algorithm.

Parameter	Description	Value
n_estimators	Max number of the decision trees	160
criterion	The principle function used to separate a branch	Gini
max_features	Max number of the features considering when separating a branch	All
max_depth	Max depth of the tree	No limit
min_samples split	The minimum number of samples at least remains in one node that can be split.	10
min_samples leaf	The minimum number of samples at least remains in leaf nodes.	1

The visualizations of the training history for the HRU-Net, U-Net, and U-Net++ models are shown in Figure 5. The blue line represents the loss calculated by the training set at each epoch. The orange line represents the loss calculated by the validation set at each epoch. Both values of the loss are high at the beginning of the training process. As the model developed by each epoch, both loss values decrease. The main purpose of Figure 5 is to avoid overfitting during the training. As shown in Figure 5, all orange lines converge to a certain value, indicating that there is no overfitting that happens during the training process. In other words, all three models were sufficiently trained and can be compared fairly with each other.

5.2. *Comparation of the HRU-Net with U-Net, U-Net++, and RF*

We tested the results of the HRU-Net, U-Net, U-Net++, and RF from three aspects: (1) the overall accuracy, (2) the accuracy of the edge details, and (3) the robustness for the intra-class variation.

5.2.1. The Overall Extraction Accuracy

Table 5 and Figure 6 show the assessment of each method on the independent testing datasets. Over the three datasets, the HRU-Net outperformed the other three models concerning the overall accuracy (Acc), Cohen's kappa coefficient (K), and F1 score (F1).

Table 5. Extraction accuracy of HRU-Net, U-Net++, U-Net, and RF.

	TM-All Dataset			TM-NRG Dataset			TM-RGB Dataset		
	Acc.	K	F1	Acc.	K	F1	Acc.	K	F1
HRU-Net	**92.81**	**0.81**	**0.90**	**92.01**	**0.79**	**0.89**	**91.05**	**0.75**	**0.88**
U-Net++	91.74	0.79	0.89	91.31	0.78	0.88	90.31	0.74	0.86
U-Net	89.83	0.74	0.86	88.47	0.72	0.85	86.33	0.66	0.82
RF	76.13	0.48	0.69	75.22	0.36	0.66	71.87	0.25	0.61

First, the results in Table 5 indicate that the NIR and SWIR bands could significantly improve the overall accuracy by 1–4%. The TM-All dataset achieved the highest accuracy compared to the results from the TM-NRG and TM-RGB datasets. The highest improvement appeared when adding NIR to the RF model (3.35%). This may be related to the model capacity to capture the higher-scale features (such as the possible nonlinear band combinations). As the deep learning model can do better at this perspective, less improvement appears when adding the new bands for training.

Secondly, the HRU-Net achieved the highest extraction accuracy in all three datasets. Especially on the TM-All dataset, the HRU-Net achieved an overall accuracy of 92.81%, improved by 1.07% compared with U-Net++, 2.98% to U-Net, and more than 16% compared with RF. The HRU-Net had the best kappa coefficient of 0.75–0.81, increased by 0.01–0.02 compared with U-Net++, 0.07–0.09 compared with U-Net and 0.33–0.50 compared with RF. A similar result can be found in the F1 score.

Thirdly, as we can see from the Table 5, the NIR band and the SWIR band can provide some useful features to help to distinguish the cultivated land and others, but the improvement was bigger in the RF model (1–4% improvement in Acc) rather than in deep learning models (0.4–1% improvement in Acc). One possible reason could be that the deep learning models have more learning capacity which can extract deeper level features such as the shape and gradients. The other reason could be that under the high intra-class spectral variation, the benefit of the NIR and SWIR band to separate the vegetation and non-vegetation pixels is less effective to distinguish the cultivated land and non-cultivated land since cultivated land can be covered by vegetation or not during the different times.

Figure 6 shows the confusion matrix for the three models over the TM-All dataset. The results indicated the HRU-Net achieved the highest recall and precision. The type 1 and type 2 error in the HRU-Net also remained the lowest compared to the U-Net++, U-Net, and RF.

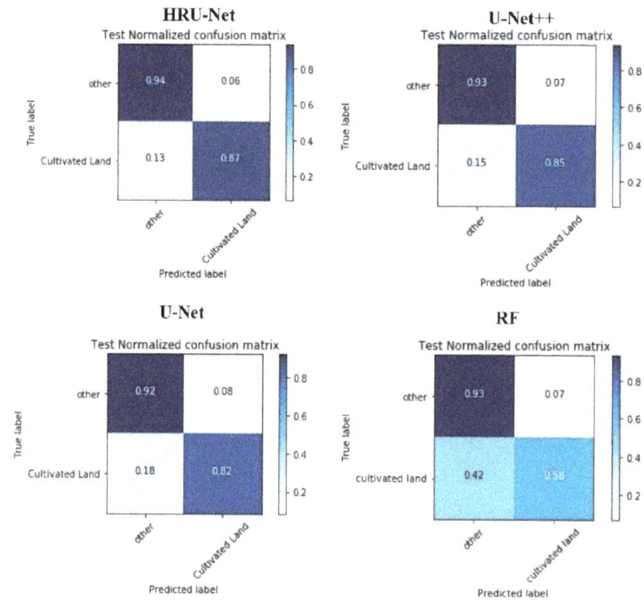

Figure 6. Confusion matrix for the HRU-Net, U-Net++, U-Net, and RF models over the independent test dataset for the Landsat TM-All dataset.

Table 6 shows the overall accuracy of the HRU-Net under 50%, 60%, and 70% training sets. As we expected, the smaller training set, the lower the accuracy will be, but as we can see, even with the 50% training samples, the accuracy decreases slowly in HRU-Net.

Table 7 shows the time consumption during the training of the HRU-Net, U-Net++, and U-Net. The RF is excluded as it was trained by CPU rather than the GPU; thus, it is not comparable to the other three GPU-based algorithms. Compare to the original U-Net, the training time increased approximately 2.6 times as more model parameters were involved by adding more complex skip connections. The time consumption of the HRU-Net was similar to the U-Net++ as these two networks had a similar number of parameters when the level was the same.

Table 6. The overall accuracy of the HRU-Net under 50%, 60% and 70% training sets.

	Percentage of the Training Data	TM-All Dataset			TM-NRG Dataset			TM-RGB Dataset		
		Acc.	K	F1	Acc.	K	F1	Acc.	K	F1
HRU-Net	50%	89.16	0.75	0.88	88.65	0.74	0.87	86.62	0.70	0.85
	60%	89.70	0.76	0.88	88.99	0.73	0.87	86.68	0.67	0.84
	70%	92.81	0.81	0.90	92.01	0.79	0.89	91.05	0.75	0.88

Table 7. The time consumption and network complexity of the training of the HRU-Net, U-Net++, and U-Net.

	Network Complexity	TM-All Dataset		TM-NRG Dataset		TM-RGB Dataset	
	Number of Free Parameters	s/epoch	epoch	s/epoch	epoch	s/epoch	epoch
HRU-Net	3.85×10^7	102.08	600	98.84	600	98.47	600
U-Net++	3.62×10^7	107.44	600	103.09	600	103.47	600
U-Net	1.34×10^7	39.23	200	36.79	200	36.92	200

5.2.2. The Accuracy of the Edge Details

As shown in Figure 7, the accuracy of the edge details was evaluated by visual interpretation. The results of the HRU-Net had clearer edges and richer details than those of the U-Net++ and U-Net. Specifically, comparing with the U-Net++, the more detailed edge remained in the output. The edge of the output from the HRU-Net was much more accurate than the edge of the original U-Net, as the loss of details could not be recovered from the lower nodes in the U-Net. In the output of the RF, the edge was sharp. However, the farmland without the crop covering was not detected correctly as it suffered from intra-class variation.

The robustness for the intra-class variation for the different models can be seen in Figure 8. In Figure 8, the overall accuracy of each tile in the testing dataset was plotted. The different tiles were randomly located and captured the main spectral variation of the cultivated land. The variation of the overall accuracy can be seen as the performance of the model handling the intra-class variation. As shown in Figure 8a, the RF model had the highest variation as indicated by its limited generalization ability to cross different spectra. Figure 8b shows a clearer comparison among the HRU-Net, U-Net++, and U-Net by removing RF from Figure 8a. In Figure 8b the variation of HRU-Net is similar to the U-Net++, however, it achieves higher accuracy in all three datasets. This indicates the effectiveness of the HRU-Net for solving the intra-class variation problem for accurate classification.

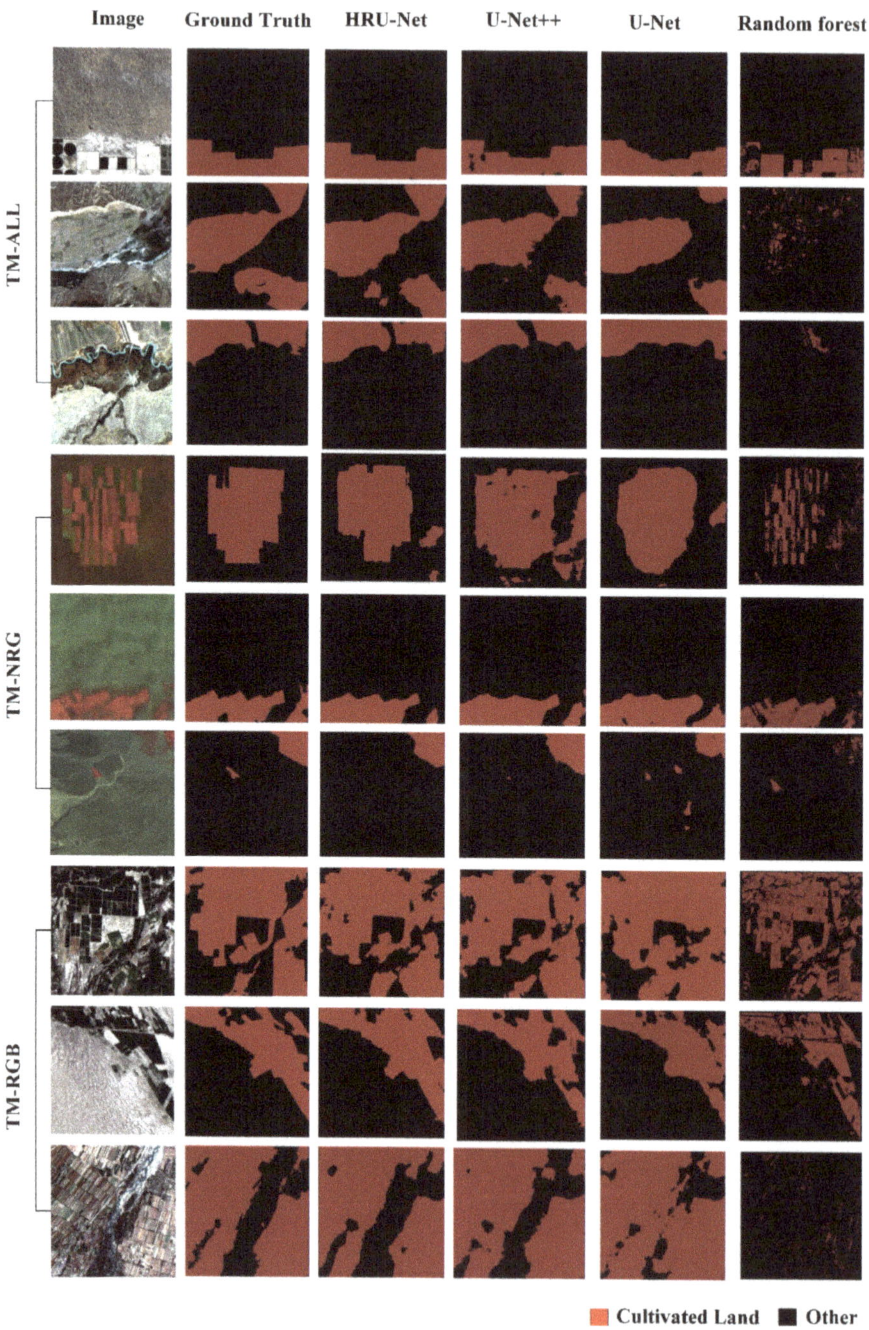

Figure 7. Selected results for the best RFU-Net, U-Net, and random forest models on all three datasets.

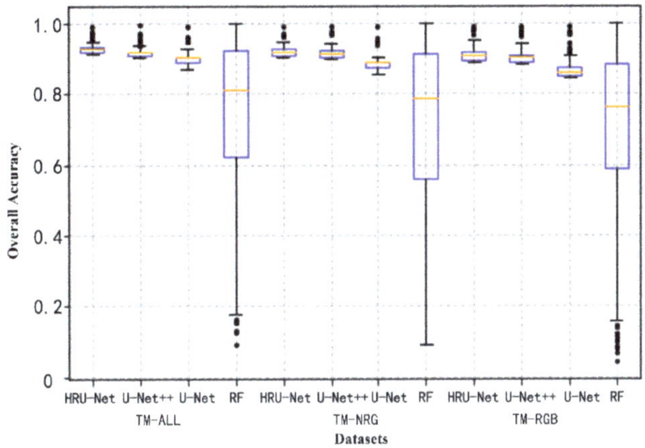

(a)

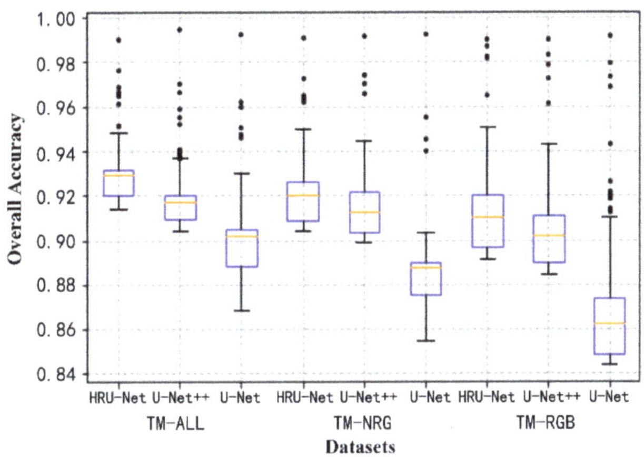

(b)

Figure 8. Boxplot of the overall accuracy distribution over the test dataset (868 tiles). (**a**) The comparison between RF and deep learning algorithms. (**b**) The comparison among HRU-Net, U-Net++, and U-Net.

5.2.3. The Effectiveness of the Modified Loss Function in the HRU-Net

To clarify the effectiveness of the modified loss function, we compared the HRU-Net with the modified loss function and the HRU-Net with the original loss function designed by U-Net. Figure 9 shows the difference in the training history of these two models. The HRU-Net with the modified loss function can be trained with more epochs; the slight overfitting happened after 500 epochs compared to after 125 epochs with the original U-Net loss function. In all three datasets, the HRU-Net with the original U-Net loss function (right column) appeared to have quicker overfitting, which was expected when training with a smaller number of bands (such as the TM-NRG and TM-RGB datasets).

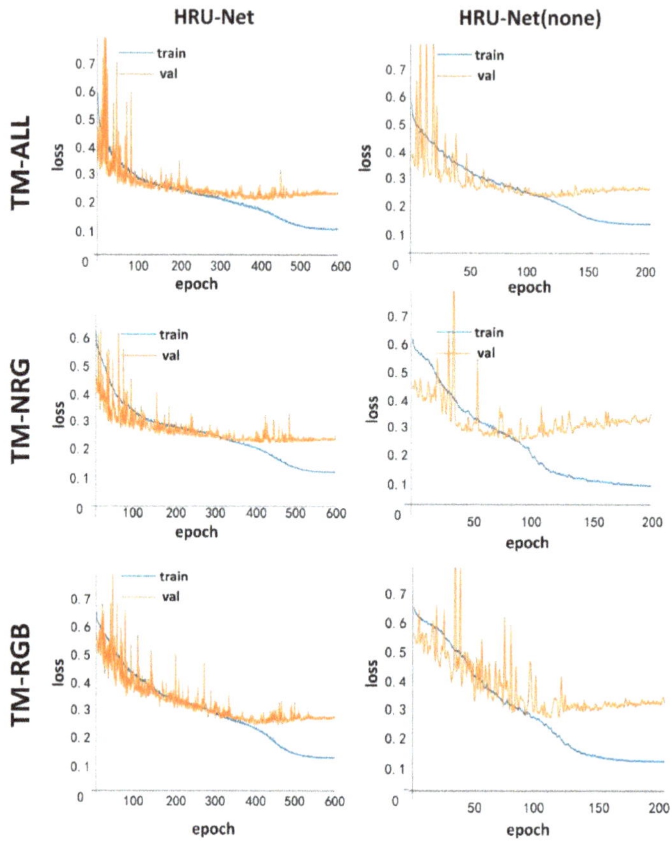

Figure 9. Visualization of the training history for the HRU-Net. HRU-Net (none) represents the HRU-Net without the modification of the loss function.

Table 8 shows the overall accuracy compared with or without the modified loss function. The results indicated that the modified loss function contributed nearly 4–5%, 5–16%, and 2–8% improvement of the overall accuracy, kappa, and F1 score over the three datasets. This indicates the modified loss function in the HRU-Net can help the model learn the spectral features of cultivated land more effectively from any perspective.

Table 8. Comparison of the HRU-Net with or without the modified loss function.

	TM-All Dataset			TM-NRG Dataset			TM-RGB Dataset		
	Acc.	K	F1	Acc.	K	F1	Acc.	K	F1
HRU-Net with the new loss function	**92.81**	**0.81**	**0.90**	**92.01**	**0.79**	**0.89**	**91.05**	**0.75**	**0.88**
HRU-Net with original U-Net loss function	90.63	0.76	0.88	87.80	0.63	0.81	87.30	0.68	0.84

5.3. Discussion

How to fix the smooth effectiveness of the pooling process to maintain or recover the image details for the deep learning model has become a topic of concern in recent years. The model we presented in this study followed the idea of maintaining and enhancing the image details during all convolution processes. The structure of the HRU-Net was similar to the 5-level U-Net++; however, the initial purposes were different. As mentioned before, the HRU-Net aimed to maintain and transfer the image details from shallow to deep levels. However, the purpose of the U-Net++ was to balance the speed and accuracy by redefining the original U-Net structure with the combination of the basic down-triangle units to achieve a more flexible structure for the different sizes of the network. The difference is that in U-Net++ more feature maps from lower levels were merged rather than higher-levels feature maps being combined in HRU-Net.

At a 30 m scale, the spectral mixing pixel is one of the sources of the classification uncertainty. The model, such as endmember extraction or mixed pixel decomposition, could help this situation. Fortunately, for the study area of this study, Xinjiang China, cultivated land is located in the huge flat area near the river or lake. The farmland is adjacent rather than separated, so the influence of the mixing pixels relatively low. This problem could be more serious when applying this model in more broken farmland, such as the southeast province of China.

The experiment in this study basically is a binary decision which mainly classify the cultivated land versus other (everything else). One of the questions is whether all the other vegetated areas (but non-cultivated) like grass fields or forest plots are well separated and classified as non-cultivated. To answer this question, we further evaluate the classification accuracy of the HRU-Net under the vegetated area. As we can see from Figure 10, all vegetated areas (grassland and forest) are correctly classed to the "others" category. This indicated the deep features from the spectral, texture, and time series may help the deep learning model like HRU-Net to better distinguish the cultivated land with other vegetated land cover.

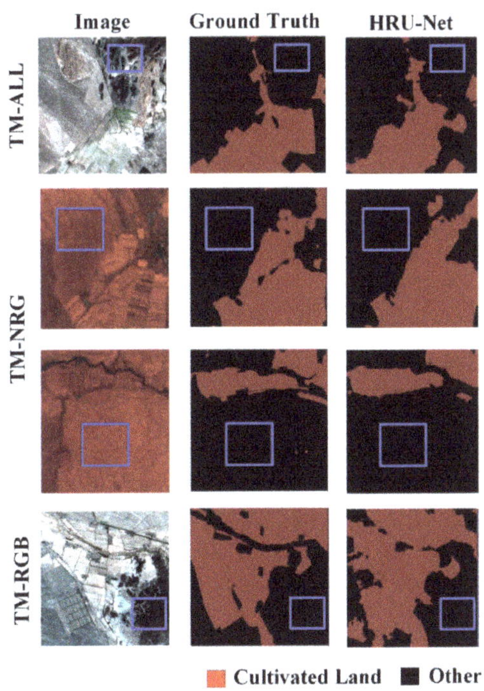

Figure 10. Visual investigation the classification accuracy of the HRU-Net under the vegetated area.

In this study, we used three years of data to capture the spectral variation of the cultivated land under different conditions. The labels of the training and testing data were obtained from historical landcover maps and manual interpretation of the corresponding satellite images. They may contain errors as the accuracy depended on the performance of a human analyst. In particular, regarding the accuracy of the edge and cultivated land extraction with different spectra, interpretation and delineation of cultivated land could be partially subjective.

More accurate extraction could be achieved by involving more prior knowledge, such as the time-series features of the cultivated land or by enhancing the spectral features of the soil or crops by adding the vegetation index as auxiliary channels.

6. Conclusions

In this study, we proposed a new end-to-end cultivated land extraction algorithm, the high-resolution U-Net (HRU-Net). Compared with the original U-Net, the HRU-Net had two improvements: (1) it improved the skip connection structure, and (2) it used the idea of deep supervision to modify the loss function. We tested the proposed method and compared it with the U-Net++, U-Net, and the RF on three Landsat TM datasets with different spectral band combinations and drew the following conclusions:

(1) The NIR and SWIR band improved the extraction accuracy of the cultivated land extraction. This follows the commonsense idea that more independent features can better help with class separation.
(2) Due to the high intra-class variation of the cultivated land, the traditional machine learning RF model had a high variation in the classification accuracy. This may be related to the Hughes phenomenon when more divergent features are involved in the model.
(3) The edge details were improved by the new structure of the HRU-Net. The HRU-Net model achieved the best results in all three Landsat TM images datasets with the lowest accuracy variation for the difference spectra of the cultivated land.

The HRU-Net model presented in this study demonstrated good performance in extracting the target with high edge details and high intra-class spectral variation. This model can be further used to extract the target within these characteristics. The model introduced in this study can be extended or combined to more other high spatial resolution satellite data, such as Sentinel-2, GF1, and GF2.

Author Contributions: Data curation, W.X. and X.W.; formal analysis, W.X.; investigation, L.S.; methodology, W.X. and S.G.; project administration, J.C.; software, X.Z., Y.X. and Y.S.; supervision, X.D., S.G. and L.S.; writing—original draft, W.X.; writing—review & editing, S.G. All authors have read and agreed to the published version of the manuscript.

Funding: This research was supported by the National Key Research and Development Program of China (Project No. 2017YFB0504203), the Natural science foundation of China project (41601212, 41801358, 41801360, 41771403), and the Fundamental Research Foundation of Shenzhen Technology and Innovation Council (JCYJ20170818155853672).

Acknowledgments: The authors thank C. Ling from SIAT for discussion and suggestions.

Conflicts of Interest: The authors declare no conflict of interest.

References

1. Atzberger, C. Advances in remote sensing of agriculture: Context description, existing operational monitoring systems and major information needs. *Remote Sens.* **2013**, *5*, 949–981. [CrossRef]
2. Thenkabail, P.S. Global croplands and their importance for water and food security in the twenty-first century: Towards an ever green revolution that combines a second green revolution with a blue revolution. *Remote Sens.* **2010**, *2*, 2305–2312. [CrossRef]
3. Xiao, X.; Boles, S.; Frolking, S.; Li, C.; Babu, J.Y.; Salas, W.; Moore, B. Mapping paddy rice agriculture in South and Southeast Asia using multi-temporal MODIS images. *Remote Sens. Environ.* **2006**, *100*, 95–113. [CrossRef]
4. Alcantara, C.; Kuemmerle, T.; Prishchepov, A.V.; Radeloff, V.C. Mapping abandoned agriculture with multi-temporal MODIS satellite data. *Remote Sens. Environ.* **2012**, *124*, 334–347. [CrossRef]

5. Massey, R.; Sankey, T.T.; Yadav, K.; Congalton, R.G.; Tilton, J.C. Integrating cloud-based workflows in continental-scale cropland extent classification. *Remote Sens. Environ.* **2018**, *219*, 162–179. [CrossRef]
6. Xu, L.; Ming, D.; Zhou, W.; Bao, H.; Chen, Y.; Ling, X. Farmland extraction from high spatial resolution remote sensing images based on stratified scale pre-estimation. *Remote Sens.* **2019**, *11*, 108. [CrossRef]
7. Zhao, W.; Du, S. Learning multiscale and deep representations for classifying remotely sensed imagery. *ISPRS J. Photogramm. Remote Sens.* **2016**, *113*, 155–165. [CrossRef]
8. Thanh Noi, P.; Kappas, M. Comparison of Random Forest, k-Nearest Neighbor, and Support Vector Machine Classifiers for Land Cover Classification Using Sentinel-2 Imagery. *Sensors* **2017**, *18*, 18. [CrossRef]
9. Mountrakis, G.; Im, J.; Ogole, C. Support vector machines in remote sensing: A review. *ISPRS J. Photogramm. Remote Sens.* **2011**, *66*, 247–259. [CrossRef]
10. García-Pedrero, A.; Gonzalo-Martín, C.; Lillo-Saavedra, M. A machine learning approach for agricultural parcel delineation through agglomerative segmentation. *Int. J. Remote Sens.* **2017**, *38*, 1809–1819. [CrossRef]
11. Li, Q.; Wang, C.; Zhang, B.; Lu, L. Object-based crop classification with Landsat-MODIS enhanced time-series data. *Remote Sens.* **2015**, *7*, 16091–16107. [CrossRef]
12. Schmidhuber, J. Deep Learning in neural networks: An overview. *Neural Netw.* **2015**, *61*, 85–117. [CrossRef]
13. Zhu, X.X.; Tuia, D.; Mou, L.; Xia, G.S.; Zhang, L.; Xu, F.; Fraundorfer, F. Deep Learning in Remote Sensing: A Comprehensive Review and List of Resources. *IEEE Geosci. Remote Sens. Mag.* **2017**, *5*, 8–36. [CrossRef]
14. Zhang, C.; Wei, S.; Ji, S.; Lu, M. Detecting large-scale urban land cover changes from very high resolution remote sensing images using CNN-based classification. *ISPRS Int. J. Geo-Inf.* **2019**, *8*, 189. [CrossRef]
15. Zhang, C.; Sargent, I.; Pan, X.; Li, H.; Gardiner, A.; Hare, J.; Atkinson, P.M. Joint Deep Learning for land cover and land use classification. *Remote Sens. Environ.* **2019**, *221*, 173–187. [CrossRef]
16. Luus, F.P.S.; Salmon, B.P.; Van Den Bergh, F.; Maharaj, B.T.J. Multiview Deep Learning for Land-Use Classification. *IEEE Geosci. Remote Sens. Lett.* **2015**, *12*, 2448–2452. [CrossRef]
17. Kussul, N.; Lavreniuk, M.; Skakun, S.; Shelestov, A. Deep Learning Classification of Land Cover and Crop Types Using Remote Sensing Data. *IEEE Geosci. Remote Sens. Lett.* **2017**, *14*, 778–782. [CrossRef]
18. Zhao, W.; Du, S. Spectral-Spatial Feature Extraction for Hyperspectral Image Classification: A Dimension Reduction and Deep Learning Approach. *IEEE Trans. Geosci. Remote Sens.* **2016**, *54*, 4544–4554. [CrossRef]
19. Qin, Z.; Lihao, N.; Tong, Z.; Qian, W. Deep Learning Based Feature Selection for Remote Sensing Scene Classification. *IEEE Geosci. Remote Sens. Lett.* **2015**, *12*, 2321–2325.
20. Zhang, F.; Du, B.; Zhang, L. Scene classification via a gradient boosting random convolutional network framework. *IEEE Trans. Geosci. Remote Sens.* **2016**, *54*, 1793–1802. [CrossRef]
21. Nogueira, K.; Penatti, O.A.B.; dos Santos, J.A. Towards better exploiting convolutional neural networks for remote sensing scene classification. *Pattern Recognit.* **2017**, *61*, 539–556. [CrossRef]
22. Hu, F.; Xia, G.S.; Hu, J.; Zhang, L. Transferring deep convolutional neural networks for the scene classification of high-resolution remote sensing imagery. *Remote Sens.* **2015**, *7*, 14680–14707. [CrossRef]
23. Lin, T.-Y.; Dollár, P.; Girshick, R.; He, K.; Hariharan, B.; Belongie, S. Feature Pyramid Networks for Object Detection. In Proceedings of the IEEE Conference on Computer Vision and Pattern Recognition (CVPR), Honolulu, HI, USA, 21–26 July 2017.
24. Ren, S.; He, K.; Girshick, R.; Sun, J. Faster R-CNN: Towards Real-Time Object Detection with Region Proposal Networks. *IEEE Trans. Pattern Anal. Mach. Intell.* **2017**, *39*, 1137–1149. [CrossRef] [PubMed]
25. Cai, Z.; Vasconcelos, N. Cascade R-CNN: Delving into High Quality Object Detection. In Proceedings of the 2018 IEEE/CVF Conference on Computer Vision and Pattern Recognition, Salt Lake City, UT, USA, 18–22 June 2018; pp. 6154–6162.
26. Redmon, J.; Farhadi, A. YOLO9000: Better, faster, stronger. In Proceedings of the 2017 IEEE Conference on Computer Vision and Pattern Recognition, Honolulu, HI, USA, 21–26 July 2017; pp. 6517–6525.
27. Wieland, M.; Li, Y.; Martinis, S. Multi-sensor cloud and cloud shadow segmentation with a convolutional neural network. *Remote Sens. Environ.* **2019**, *230*, 111203. [CrossRef]
28. Ronneberger, O.; Fischer, P.; Brox, T. U-Net: Convolutional Networks for Biomedical Image Segmentation. *arXiv* **2015**, arXiv:1505.04597.
29. Wu, X.; Irie, G.; Hiramatsu, K.; Kashino, K. Weighted Generalized Mean Pooling for Deep Image Retrieval. In Proceedings of the 2018 IEEE International Conference on Image Processing, Athens, Greece, 7–10 October 2018; pp. 495–499.

30. Yildirim, O.; Baloglrf, U.B. Regp: A new pooling algorithm for deep convolutional neural networks. *Neural Netw. World* **2019**, *29*, 45–60. [CrossRef]
31. Yu, B.; Yang, L.; Chen, F. Semantic Segmentation for High Spatial Resolution Remote Sensing Images Based on Convolution Neural Network and Pyramid Pooling Module. *IEEE J. Sel. Top. Appl. Earth Obs. Remote Sens.* **2018**, *11*, 3252–3261. [CrossRef]
32. Sun, K.; Zhao, Y.; Jiang, B.; Cheng, T.; Xiao, B.; Liu, D.; Mu, Y.; Wang, X.; Liu, W.; Wang, J. High-Resolution Representations for Labeling Pixels and Regions. *arXiv* **2019**, arXiv:1904.04514.
33. He, K.; Zhang, X.; Ren, S.; Sun, J. Deep residual learning for image recognition. In Proceedings of the IEEE Computer Society Conference on Computer Vision and Pattern Recognition, Las Vegas, NV, USA, 17–30 June 2016; Volume 2016, pp. 770–778.
34. Huang, G.; Liu, Z.; Maaten, L.; van der Weinberger, K.Q. Densely Connected Convolutional Networks. *arXiv* **2018**, arXiv:1608.06993.
35. Long, J.; Shelhamer, E.; Darrell, T. Fully Convolutional Networks for Semantic Segmentation. *IEEE Trans. Pattern Anal. Mach. Intell.* **2014**, *39*, 640–651.
36. Badrinarayanan, V.; Kendall, A.; Cipolla, R. SegNet: A Deep Convolutional Encoder-Decoder Architecture for Image Segmentation. *IEEE Trans. Pattern Anal. Mach. Intell.* **2017**, *39*, 2481–2495. [CrossRef] [PubMed]
37. Noh, H.; Hong, S.; Han, B. Learning deconvolution network for semantic segmentation. In Proceedings of the IEEE International Conference on Computer Vision, Santiago, Chile, 7–13 December 2015; pp. 1520–1528.
38. Lin, G.; Milan, A.; Shen, C.; Reid, I. RefineNet: Multi-path refinement networks for high-resolution semantic segmentation. In Proceedings of the 2017 IEEE Conference on Computer Vision and Pattern Recognition, Honolulu, HI, USA, 21–26 July 2017; pp. 5168–5177.
39. Sherrah, J. Fully Convolutional Networks for Dense Semantic Labelling of High-Resolution Aerial Imagery. *arXiv* **2016**, arXiv:1606.02585.
40. Marmanis, D.; Wegner, J.D.; Galliani, S.; Schindler, K.; Datcu, M.; Stilla, U. Semantic Segmentation of Aerial Images With an Ensemble of Cnns. *ISPRS Ann. Photogramm. Remote Sens. Spat. Inf. Sci.* **2016**, *III-3*, 473–480. [CrossRef]
41. Zhang, P.; Ke, Y.; Zhang, Z.; Wang, M.; Li, P.; Zhang, S. Urban land use and land cover classification using novel deep learning models based on high spatial resolution satellite imagery. *Sensors* **2018**, *18*, 3717. [CrossRef]
42. Gaetano, R.; Ienco, D.; Ose, K.; Cresson, R. A two-branch CNN architecture for land cover classification of PAN and MS imagery. *Remote Sens.* **2018**, *10*, 1746. [CrossRef]
43. Cheng, G.; Li, Z.; Han, J.; Yao, X.; Guo, L. Exploring Hierarchical Convolutional Features for Hyperspectral Image Classification. *IEEE Trans. Geosci. Remote Sens.* **2018**, *56*, 6712–6722. [CrossRef]
44. Zhou, P.; Han, J.; Cheng, G.; Zhang, B. Learning Compact and Discriminative Stacked Autoencoder for Hyperspectral Image Classification. *IEEE Trans. Geosci. Remote Sens.* **2019**, *57*, 4823–4833. [CrossRef]
45. Saxena, S.; Verbeek, J. Convolutional neural fabrics. *Adv. Neural Inf. Process. Syst.* **2016**, 4060–4068.
46. Zhou, Y.; Hu, X.; Zhang, B. Interlinked convolutional neural networks for face parsing. *arXiv* **2018**, arXiv:1806.02479.
47. Sun, K.; Xiao, B.; Liu, D.; Wang, J. Deep high-resolution representation learning for human pose estimation. In Proceedings of the IEEE Conference on Computer Vision and Pattern Recognition, Long Beach, CA, USA, 16–20 June 2019; pp. 5686–5696.
48. Drozdzal, M.; Vorontsov, E.; Chartrand, G.; Kadoury, S.; Pal, C. The Importance of Skip Connections in Biomedical Image Segmentation. *arXiv* **2016**, arXiv:1608.04117.
49. Bulat, A.; Tzimiropoulos, G. Human Pose Estimation via Convolutional Part Heatmap Regression. *arXiv* **2016**, arXiv:1609.01743.
50. Valle, R.; Buenaposada, J.M.; Valdés, A.; Baumela, L. A deeply-initialized coarse-to-fine ensemble of regression trees for face alignment. In *Computer Vision—ECCV 2018 Lecture Notes in Computer Science*; Ferrari, V., Hebert, M., Sminchisescu, C., Weiss, Y., Eds.; Springer: Cham, Swizerland, 2018; pp. 609–624.
51. Honari, S.; Yosinski, J.; Vincent, P.; Pal, C. Recombinator networks: Learning coarse-to-fine feature aggregation. In Proceedings of the 2016 IEEE Conference on Computer Vision and Pattern Recognition (CVPR 2016), Las Vegas, NV, USA, 27–30 June 2016; pp. 5743–5752.
52. Fu, J.; Liu, J.; Wang, Y.; Zhou, J.; Wang, C.; Lu, H. Stacked Deconvolutional Network for Semantic Segmentation. *IEEE Trans. Image Process.* **2019**. [CrossRef] [PubMed]

53. Wu, W.; Qian, C.; Yang, S.; Wang, Q.; Cai, Y.; Zhou, Q. Look at Boundary: A Boundary-Aware Face Alignment Algorithm. In Proceedings of the IEEE Conference on Computer Vision and Pattern Recognition 2018, Salt Lake City, UT, USA, 18–23 June 2018; pp. 2129–2138.
54. Zhou, Z.; Siddiquee, M.M.R.; Tajbakhsh, N.; Liang, J. UNet++: A Nested U-Net Architecture for Medical Image Segmentation. In *Deep Learning in Medical Image Analysis and Multimodal Learning for Clinical Decision Support*; Stoyanov, D., Taylor, Z., Carneiro, G., Syeda-Mahmood, T., Martel, A., Maier-Hein, L., Tavares, J.M.R.S., Bradley, A., Papa, J.P., Belagiannis, V., et al., Eds.; Springer: Cham, Switzerland, 2018; pp. 3–11.
55. Mitchell, M.T. *Machine Learning*; McGraw-Hill: Singapore, 1997.

sensors

MDPI

Article

Mask R-CNN and OBIA Fusion Improves the Segmentation of Scattered Vegetation in Very High-Resolution Optical Sensors

Emilio Guirado [1,2,*], Javier Blanco-Sacristán [3], Emilio Rodríguez-Caballero [4,5], Siham Tabik [6], Domingo Alcaraz-Segura [7,8], Jaime Martínez-Valderrama [1] and Javier Cabello [2,9]

1 Multidisciplinary Institute for Environment Studies "Ramon Margalef" University of Alicante, Edificio Nuevos Institutos, Carretera de San Vicente del Raspeig s/n San Vicente del Raspeig, 03690 Alicante, Spain; jaime.mv@ua.es
2 Andalusian Center for Assessment and monitoring of global change (CAESCG), University of Almeria, 04120 Almeria, Spain; jcabello@ual.es
3 College of Engineering, Mathematics and Physical Sciences, University of Exeter, Penryn Campus, Cornwall TR10 9EZ, UK; jb1230@exeter.ac.uk
4 Agronomy Department, University of Almeria, 04120 Almeria, Spain; e.rodriguez-caballer@mpic.de
5 Centro de Investigación de Colecciones Científicas de la Universidad de Almería (CECOUAL), 04120 Almeria, Spain
6 Department of Computer Science and Artificial Intelligence, University of Granada, 18071 Granada, Spain; siham@ugr.es
7 Department of Botany, Faculty of Science, University of Granada, 18071 Granada, Spain; dalcaraz@ugr.es
8 iEcolab, Inter-University Institute for Earth System Research, University of Granada, 18006 Granada, Spain
9 Department of Biology and Geology, University of Almeria, 04120 Almeria, Spain
* Correspondence: e.guirado@ual.es

Citation: Guirado, E.; Blanco-Sacristán, J.; Rodríguez-Caballero, E.; Tabik, S.; Alcaraz-Segura, D.; Martínez-Valderrama, J.; Cabello, J. Mask R-CNN and OBIA Fusion Improves the Segmentation of Scattered Vegetation in Very High-Resolution Optical Sensors. *Sensors* **2021**, *21*, 320. https://doi.org/10.3390/s21010320

Received: 17 December 2020
Accepted: 1 January 2021
Published: 5 January 2021

Publisher's Note: MDPI stays neutral with regard to jurisdictional claims in published maps and institutional affiliations.

Abstract: Vegetation generally appears scattered in drylands. Its structure, composition and spatial patterns are key controls of biotic interactions, water, and nutrient cycles. Applying segmentation methods to very high-resolution images for monitoring changes in vegetation cover can provide relevant information for dryland conservation ecology. For this reason, improving segmentation methods and understanding the effect of spatial resolution on segmentation results is key to improve dryland vegetation monitoring. We explored and analyzed the accuracy of Object-Based Image Analysis (OBIA) and Mask Region-based Convolutional Neural Networks (Mask R-CNN) and the fusion of both methods in the segmentation of scattered vegetation in a dryland ecosystem. As a case study, we mapped *Ziziphus lotus*, the dominant shrub of a habitat of conservation priority in one of the driest areas of Europe. Our results show for the first time that the fusion of the results from OBIA and Mask R-CNN increases the accuracy of the segmentation of scattered shrubs up to 25% compared to both methods separately. Hence, by fusing OBIA and Mask R-CNNs on very high-resolution images, the improved segmentation accuracy of vegetation mapping would lead to more precise and sensitive monitoring of changes in biodiversity and ecosystem services in drylands.

Keywords: deep-learning; fusion; mask R-CNN; object-based; optical sensors; scattered vegetation; very high-resolution

1. Introduction

Dryland biomes cover ~47% of the Earth's surface [1]. In these environments, vegetation appears scattered [2] and its structure, composition and spatial patterns are key indicators of biotic interactions [3], regulation of water, and nutrient cycles at landscape level [4]. Changes in the cover and spatial patterns of dryland vegetation occur in response to land degradation processes [5]. Hence, methods to identify and characterize vegetation patches and their structural characteristics can improve our ability to understand dryland functioning and to assess desertification risk [5–8]. Progress has been made using remote sensing tools in this regard (e.g., quantification of dryland vegetation structure at landscape scale [9], monitoring vegetation trends [10], spatial patterns identifying ecosystem multifunctionality [11], characterizing flood dynamics [12], among many others). However, the improvement in the accuracy of vegetation cover measurement is still being studied to obtain maximum performance from data and technology. Estimating and monitoring changes in vegetation cover through remote sensing is key for dryland ecology and conservation [6]. Both historical temporal and spatial data are the base for remote sensing studies to identify the functioning and structure of vegetation [13,14].

The analysis of very high-resolution images to detect and measure vegetation cover and its spatial arrangement across the landscape starts typically by segmenting the objects to be identified in the images [7]. Object-Based Image Analysis (OBIA) [15] and Mask Region-based Convolutional Neural Networks (Mask R-CNN) [16] are among the most used and state-of-the-art segmentation methods. Though they provide a similar product, both methods rely on very different approaches. OBIA combines spectral information from each pixel with its spatial context [17,18]. Similar pixels are then grouped in homogenous objects that are used as the basis for further classification. Mask R-CNN, on the other hand, a type of artificial intelligence whose functioning is inspired by the human brain provides transferable models between zones and semantic segmentation with unprecedented accuracy [19,20]. Besides, fusion has recently been used to improve spectral, spatial, and temporal resolution from remote sensing images [21–23]. However, the fusion of methods for vegetation mapping has not been evaluated.

Remote sensing studies based on very high-resolution images have increased in the last years (e.g., [24–27]), partly because of the availability of Google Earth images worldwide [28–30] and the popularization of unmanned aerial vehicles (UAV). Although these images have shown a high potential for vegetation mapping and monitoring [31–33], two main problems arise when they are used. First, higher spatial resolution increases the spectral heterogeneity among and within vegetation types, resulting in a salt and pepper effect in their identification that does not correctly characterize the actual surface [34]. Second, the processing time of very high-resolution images and the computational power required is larger than in the case of low-resolution images [35]. Under these conditions, traditional pixel-based analysis has proved to be less accurate than OBIA or Mask R-CNN for scattered vegetation mapping [15,36]. There are many applications for OBIA [37–39] and deep learning segmentation methods [40,41]. For example, mapping greenhouses [42], monitoring disturbances affecting vegetation cover [5], or counting scattered trees in Sahel and Sahara [43]. These methods have been compared with excellent results in both segmenting and detecting tree cover and scattered vegetation [7,44,45]. However, greater precision is always advisable in problems of very high sensitivity [46]. Despite methodological advances, selecting the appropriate image source is key to produce accurate segmentations of objects, like in vegetation maps [47,48], and there is no answer to the question of which image or method to choose for segmenting objects. Understanding how the spatial resolution of the imagery used affects these segmentation methods or the fusing of both is key for their correct application to obtain better accuracy in object segmentation in vegetation mapping in drylands.

To evaluate which is the most accurate method between OBIA and Mask R-CNN to segment scattered vegetation in drylands and to understand the effect of the spatial resolution of the images used in this process, we assessed the accuracy of these two methods in the segmentation of scattered dryland shrubs and compared how final accuracy varies as does spatial resolution. We also check the accuracy of the fusion of both methods.

This work is organized as follows. Section 2 describes the study area, the dataset used, and the methodologies tested. Section 3 describes the experiments addressed to assess the accuracies of the methods used. The experimental results and discussion are presented in Section 4, and conclusions are given in Section 5.

2. Materials and Methods

2.1. Study Area

We focused on the community of *Ziziphus lotus* shrubs, an ecosystem of priority conservation interest at European level (habitat 5220* of Directive 92/43/EEC), located in Cabo de Gata-Níjar Natural Park (36°49′43″ N, 2°17′30″ W, SE Spain), one of the driest areas of continental Europe. This type of vegetation is scarce and patchy, which appears surrounded by a matrix of bare soil and small shrubs (e.g., *Launea arborescens*, *Lygeum spartum* and *Thymus hyemalis*). *Z. lotus* is a facultative phreatophyte [49] and forms large hemispherical canopies (1–3 m tall) that constitute fertility islands where many other species of plants and animals live [50]. These shrubs are long-lived species contributing to the formation of geomorphological structures, called nebkhas [51], that protect from the intense wind erosion activity that characterizes the area, thereby retaining soil, nutrients, and moisture.

2.2. Dataset

The data set consisted of two plots (Plot 1 and Plot 2) with 3 images of different spatial resolution in each one. The plots had an area of 250 × 250 m with scattered *Z. lotus* shrubs. The images were obtained from optical remote sensors in the visible spectral range, Red, Green and Blue bands (RGB) and spatial resolutions of < 1 m/pixel:

- A 0.5 × 0.5 m spatial resolution RGB image obtained from Google Earth [52].
- A 0.1 × 0.1 m spatial resolution image acquired using an RGB camera sensor of 50 megapixels (Hasselblad H4D) equipped with a 50 mm lens and charge-coupled device (CCD) sensor of 8176 pixels × 6132 pixels mounted on a helicopter with a flight height of 550 m.
- A 0.03 × 0.03 m spatial resolution image acquired using a 4K pixels resolution RGB camera sensor on a professional UAV Phantom 4 UAV (DJI, Shenzhen, China) and with a flight height of 40 m.

2.3. OBIA

OBIA-based segmentation is a method of image analysis that divides the image into homogeneous objects of interest (i.e., groups of pixels also called segments) based on similarities of shape, spectral information, and contextual information [17]. It identifies homogeneous and discrete image objects by setting an optimal combination of values for three parameters (i.e., Scale, Shape, and Compactness) related to their spectral and spatial variability. There are no unique values for any of these parameters, and their final combination always depends on the object of interest, so finding this optimal combination represents a challenge due to the vast number of possible combinations. First, it is necessary to establish an appropriate Scale level depending on the size of the object studied in the image [43]; for example, low Scale values for small shrubs and high Scale values for large shrubs [44,45]. Recent advances have been oriented in developing techniques (e.g., [53–59]) and algorithms (e.g., [60–63]) to automatically find the optimal value of the Scale parameter [64], which is the most important for determining the size of the segmented objects [65,66]. The Shape and the Compactness parameters must be configured too. While high values of the Shape parameter prioritize the shape over the colour, high values of the Compactness parameter prioritize compactness of the objects over the smoothness of their edges [67].

2.4. Mask R-CNN

In this problem of locating and delimiting the edges of dispersed shrubs, we used a computer vision technique named instance segmentation [68]. Such technique infers a label for each pixel considering other nearby objects, thus including the boundaries of the object. We used Mask R-CNN segmentation model [16], which extends Faster R-CNN detection model [16] and provides three outputs for each object: (i) a class label, (ii) a bounding box that delimits the object and (iii) a mask which delimits the pixels that constitute each object.

In the binary problem addressed in this work, Mask R-CNN generates for each predicted object instance a binary mask (values of 0 and 1), where values of 1 indicate a *Z. lotus* pixel and 0 indicates a bare soil pixel.

Mask R-CNN relies on a classification model for the task of feature extraction. In this work, we used ResNet 101 [69] to extract increasingly higher-level characteristics from the lowest to the deepest layer levels.

The learning process of Mask R-CNN is influenced by the number of epochs, which is the number of times the network goes through the training phase, and by other optimizations such as transfer-learning or data-augmentation (see Section 3.2). Finally, the 1024 × 1024 × 3 band image input is converted to 32 × 32 × 2048 to represent objects at different scales via the characteristic network pyramid.

2.5. Segmentation Accuracy Assessment

The accuracy of the segmentation task in this work was assessed with respect to ground truth by using the Euclidean Distance v.2 (ED2; [70]), which evaluates the geometric and arithmetic discrepancy between reference polygons and the segments obtained during the segmentation process. Both types of discrepancy need to be assessed. As reference polygons, we used the perimeter of 60 *Z. lotus* shrubs measured with photo-interpretation in all images by a technical expert. We estimated the geometric discrepancy by the "Potential Segmentation Error" (PSE; Equation (1)), defined as the ratio of the total area of each segment obtained in the segmentation that falls outside the reference segment and the total area of reference polygons as:

$$\mathrm{PSE} = \frac{\Sigma|\mathrm{s_i} - \mathrm{r_k}|}{\Sigma|\mathrm{r_k}|} \tag{1}$$

where PSE is the "Potential Segmentation Error", $\mathrm{r_k}$ is the area of the reference polygon and si is the overestimated area of the segment obtained during the segmentation. A value of 0 indicates that segments obtained from the segmentation fit well into the reference polygons. Conversely, larger values indicate a discrepancy between reference polygons and the segments.

Although the geometric relation is necessary, it is not enough to describe the discrepancies between the segments obtained during the segmentation process and the corresponding reference polygons. To solve such problem the ED2 index includes an additional factor, the "Number-of-Segmentation Ratio" (NSR), that evaluates the arithmetic discrepancy between the reference polygons and the generated segments (Equation (2)):

$$\mathrm{NSR} = \frac{\mathrm{abs}(\mathrm{m} - \mathrm{v})}{\mathrm{m}} \tag{2}$$

where NSR is the arithmetic discrepancy between the polygons of the resulting segmentation and the reference polygons and abs is the absolute value of the difference of the number of reference polygons, m, and the number of segments obtained, v.

Thus, the ED2 can be defined as the joint effect of geometric and arithmetic differences (Equation (3)), estimated from PSE and NSR, respectively, as:

$$\mathrm{ED2} = \sqrt{(\mathrm{PSE})^2 + (\mathrm{NSR})^2} \tag{3}$$

where ED2 is Euclidean Distance v.2, PSE is Potential Segmentation Error, and NSR is Number-of-Segmentation Ratio. According to Liu et al. [70], values of ED2 close to 0 indicate good arithmetic and geometric coincidence, while high values indicate a mismatch between them.

3. Experiments

We set several experiments to assess the accuracy of the two different OBIA and Mask R-CNN segmenting scattered vegetation in drylands. We used the images of Plot 1 to test the OBIA and Mask R-CNN segmentation methods. The images of Plot 2 were used for the training phase in Mask R-CNN experiments exclusively (Figure 1). In Section 3.1, we describe OBIA experiments, focused on detecting the best parameters (i.e., Scale, Shape and Compactness) of a popularly used "multi-resolution" segmentation algorithm [71]. In Section 3.2. we described the Mask R-CNN experiments, in which we first evaluated the precision in the detection of shrubs (capture or notice the presence of shrubs) and second how accurate is the segmentation of those shrubs. Finally, in Section 3.3. we described the fusion of both methods and compared all the accuracies between them in Section 4.3.

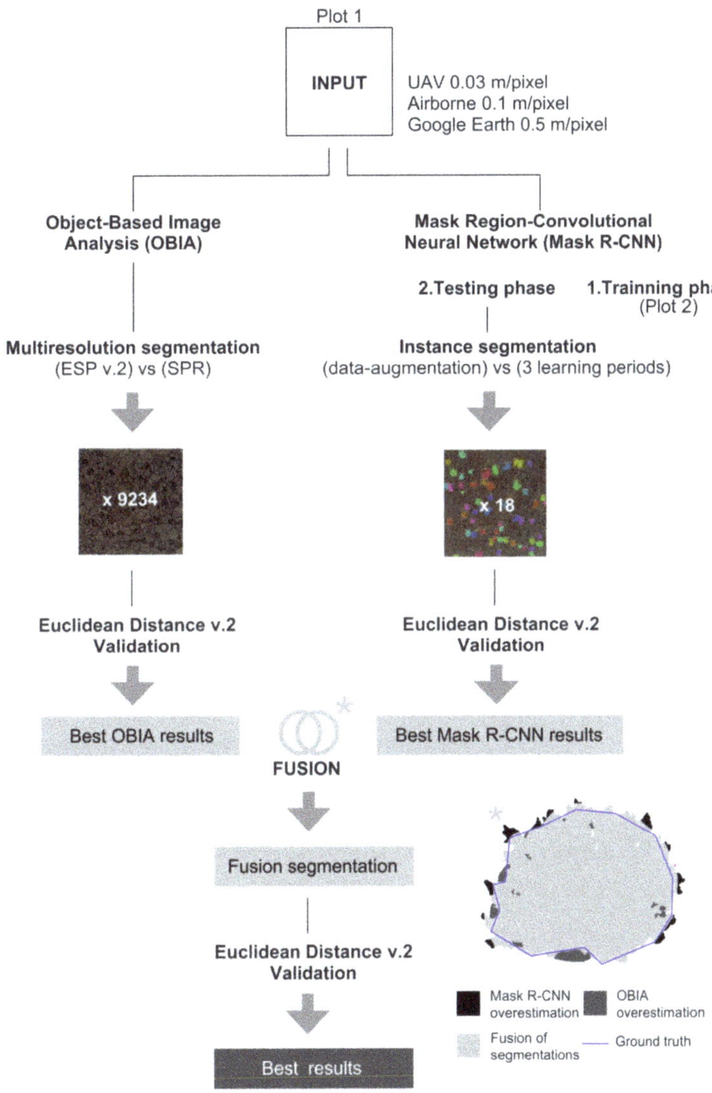

Figure 1. Workflow with the main processes carried out in this work. Asterisk shows an example of the result of the fusion of the segmentation results from OBIA and Mask R-CNN. OBIA: Object-Based Image Analysis; Mask R-CNN: Mask Region-based Convolutional Neural Networks; ESP v.2: Estimation of Scale Parameter v.2; SPR: Segmentation Parameters Range.

3.1. OBIA Experiments

To obtain the optimal value of each parameter of the OBIA segmentation, we use two approaches:

(i) A ruleset called Segmentation Parameters Range (SPR) in eCognition v8.9 (Definiens, Munich, Germany) with the "multi-resolution" algorithm that segmented the images of Plot 1 by systematically increasing the Scale parameter in steps of 5 and the Shape and Compactness parameters in steps of 0.1. The Scale parameter ranged from 80 to 430, and the Shape and the Compactness from 0.1 to 0.9. We generated a total of 9234 results with possible segmentations of *Z. lotus* shrubs. The Scale parameter ranges were evaluated considering the minimum cover size (12 m^2) and maximum cover size (311 m^2) of the shrubs measured in the plot and the pixel size.

(ii) We also performed the semi-automatic method Estimation of Scale Parameter v.2 (ESP2; [70]) to select the best scale parameter. This tool performs semi-automatic segmentation of multiband images within a range of increasing Scale values (Levels), while the user previously defines the values of the Compactness and Shape parameters. Three options available in the ESP2 tool were tested: a) the hierarchical analysis Top-down (HT), starting from the highest level and segmenting these objects for lower levels; b) the hierarchical analysis Bottom-up (HB), which starts from the lower level and combines objects to get larger levels; and c) analysis without hierarchy (NH), where each scale parameter is generated independently, based only on the level of the pixel [64].

3.2. Mask R-CNN Experiments

Mask R-CNN segmentation is divided in two phases: i) Training and ii) Testing phases. In the training phase, we selected 100 training polygons representing 100 shrub individuals with different sizes. The sampling was done using VGG Image Annotator [72] to generate a JSON file, which includes the coordinates of all the vertices of each segment, equivalent to the perimeter of each shrub. To increase the number of samples and reduce overfitting of the model, we applied data-augmentation and transfer-learning:

- Data augmentation aims to artificially increase the size of the dataset by slightly modifying the original images. We applied the filters of vertical and horizontal flip; Scale decrease and increase in the horizontal and vertical axis between 0.8 to 1.2; Rotation of 0 to 365 degrees; Shearing factor between −8 to 8; Contrast normalization with values of 0.75 and 1.5 per channel; Emboss with alpha 0, 0.1; Strength with 0 to 2.0; Multiply 0.5 and 1.5, per channel to change the brightness of the image (50–150% of the original value).
- Transfer-learning consists in using knowledge learnt from one problem to another related one [73], and we used it to improve the neural network. Since the first layers of a neural network extract low-level characteristics, such as colour and edges, they do not change significantly and can be used for other visual recognition works. As our new dataset was small, we applied fine adjustment to the last part of the network by updating the penultimate weights, so that the model was not overfitting, as mainly occurs between the first layers of the network. We specifically used transfer-learning on ResNet 101 [69] and used Region-based CNN with the pre-trained weights of the same architectures on COCO dataset (around 1.28 million images over 1000 generic object classes) [74].

We tested three different learning periods (100 steps per epoch) per model:

(A) 40 epochs with transfer-learning in heads,
(B) 80 epochs with 4 fist layers transfer-learning,
(C) 160 epochs with all layers transfer-learning.

We trained the algorithm based on the ResNet architecture with a depth of 101 layers with each of the three proposed spatial resolutions. We then evaluated the trained models in all possible combinations between the resolutions. We evaluated the use of data-augmentation and transfer-learning from more superficial layers to the whole architecture with different stages in the training process. Particularly:

(1.1) Trained with UAV images.
(1.2) Trained with UAV images and data-augmentation.
(2.1) Trained with airborne images.
(2.2) Trained with airborne images and with data-augmentation.
(3.1) Trained with Google Earth images.
(3.2) Trained with Google Earth images and data-augmentation.

We did the test phase using Plot 1. To identify the most accurate experiments, we evaluated the detection of the CNN-based models, and determined their Precision, Recall, and F1-measure [75] as:

$$\text{Precision} = \frac{\text{True Positives}}{\text{True Positives} + \text{False Positives}}, \quad (4)$$

$$\text{Recall} = \frac{\text{True Positives}}{\text{True positives} + \text{False Negatives}}, \quad (5)$$

$$\text{F1} - \text{measure} = 2 \times \frac{\text{Precision} \times \text{Recall}}{\text{Precision} + \text{Recall}} \quad (6)$$

3.3. *Fusion of OBIA and Mask R-CNN*

We combined the most accurate segmentations obtained using OBIA and Mask R-CNN, according to ED2 values (Figure 1). We let o_i denote the i-th OBIA polygon within the OBIA segmentation, O, and m_j denote the j-th Mask R-CNN polygon within the Mask R-CNN segmentation, C. Then we have $O = \{o_i: i = 1, 2, ..., m\}$ and $C = \{c_j: j = 1, 2, ..., n\}$. Here, the subscripts i and j are sequential numbers for the polygons of the OBIA and Mask R-CNN segmentations, respectively. m and n indicate the total numbers of the objects segmented with OBIA and Mask R-CNN, respectively. m and n must be equal. Finally, the corresponding segment data sets extracted (Equation (7)) by the fusion are considered a consensus among the initially segmented objects as:

$$OC_{ij} = \text{area}O_i \cap \text{area}C_j \quad (7)$$

where OC_{ij} is the intersected area between the segments of the OBIA segmentation (O_i) and the area of the segments of the Mask R-CNN segmentation (C_j).

Finally, we estimate ED2 values of the final segmentation using validation shrubs from Plot 1, and we compared it with segmentation accuracy obtained by the different methods.

4. Results and Discussion

4.1. *OBIA Segmentation*

In total, 9234 segmentations were performed by SPR, 3078 for each image type (e.g., Google Earth, airborne and UAV). OBIA segmentation accuracy using the SPR presented large variability (Table 1), with values of ED2 ranging between 0.05 and 0.28. Segmentation accuracy increased with image spatial resolution. Thus, the higher the spatial resolution, the higher the Scale values and more accurate the segmentation was. This result was represented by a decrease in ED2 values of 0.14, 0.10 and 0.05 for Google Earth, airborne and UAV images, respectively. The best combinations of segmentation parameters along the different images were (Figure 2): (i) for the Google Earth image, Scale values ranging from 105 to 110, low Shape values of 0.3 and high Compactness values from 0.8 to 0.9; (ii) for the orthoimage from the airborne sensor, Scale values between 125 and 155, Shape of 0.6 and Compactness of 0.9; and (iii) for the UAV image, the optimal segmentation showed the highest Scale values, ranging from 360 to 420, whereas Shape and Compactness values were similar to the values of the Google Earth image.

Table 1. Segmentation accuracies of Object-Based Image Analysis (OBIA) among the three spatial resolutions evaluated. For each segmentation type, only the most accurate combination of Scale, Shape, and Compactness is shown. ESP2/HB: Estimate Scale Parameter v.2 (ESP2) with Bottom-up Hierarchy; ESP2/HT: ESP2 with Top-down Hierarchy; ESP2/NH: ESP2 Non-Hierarchical; SPR: Segmentation with Parameters Range. Closer values to 0 indicate accurate segmentations. In bold the most accurate results.

			Segmentation Parameters			Segmentation Quality	
Image Source	**Resolution (m/Pixel)**	**Segmentation Method**	**Scale**	**Shape**	**Compactness**	**ED2**	**Average Time (s)**
Google Earth	0.5	ESP2/HB	100	0.6	0.9	0.25	365
		ESP2/HT	105	0.7	0.5	0.26	414
		ESP2/NH	105	0.5	0.1	0.28	2057
		SPR	**90**	**0.3**	**0.8**	**0.2**	**18**
Airborne	0.1	ESP2/HB	170	0.5	0.9	0.14	416
		ESP2/HT	160	0.5	0.9	0.15	650
		ESP2/NH	160	0.5	0.5	0.14	3125
		SPR	**155**	**0.6**	**0.9**	**0.1**	**24**
UAV	0.03	ESP2/HB	355	0.3	0.7	0.12	5537
		ESP2/HT	370	0.5	0.7	0.11	8365
		ESP2/NH	350	0.5	0.7	0.1	40,735
		SPR	**420**	**0.1**	**0.8**	**0.05**	**298**

When we applied the semi-automatic method ESP2 to estimate the optimum value of the Scale parameter, we observed a similar pattern to that described for the SPR, with an increase in accuracy when increasing spatial resolution. The highest value of ED2 was for the Google Earth image segmentation results (ED2 = 0.25), decreasing for the orthoimage from the airborne sensor (ED2 = 0.15) and reaching the minimum value (best) in the UAV image (ED2 = 0.12). However, the results obtained by ESP2 were worse than the results obtained by the SPR method in all the images analysed (Table 1) with the largest differences in the image with the lowest spatial resolution (Google Earth). In the Google Earth images, the best method of analysis of the three options presented by the ESP2 tool was the hierarchical bottom level, with acceptable ED2 values, lower than 0.14 (Table 1). For the airborne images, the results were equal to Google Earth images (hierarchical bottom level). Conversely, the segmentation of the UAV image produced the best ED2 values when applying the ESP2 without hierarchical level. The computational time for the segmentation of the images was higher in ESP2 than SPR approach. In addition, the computation time of the analysis was also influenced by the number of pixels to analyse, it increased in higher spatial resolution images in computer with a Core i7-4790K, 4 GHz and 32G of RAM memory (Intel, Santa Clara, CA, USA) (Table 1).

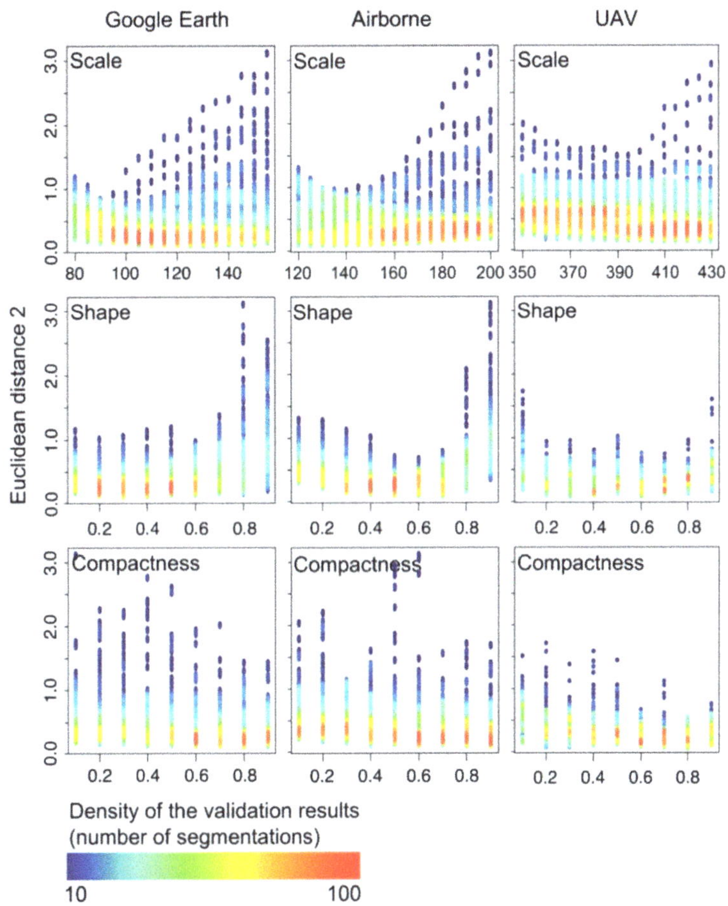

Figure 2. Relationship between Scale, Shape and Compactness parameters (X axis) evaluated using Euclidean distance v.2 (ED2; Y axis) in 9234 Object-based image analysis (OBIA) segmentations from Google Earth, Airborne and unmanned aerial vehicle (UAV) images. The rainbow palette shows the density of validation results. In red high density and in blue low density.

4.2. *Mask R-CNN Segmentation*

4.2.1. Detection of Scattered Shrubs

We obtained the best detection results for the models trained and evaluated with UAV images (F1-measure = 0.91) and the models trained with the highest number of epochs and data-augmentation activated (Table 2). The best transfer from a UAV trained model to a test with another resolution was to the image from the airborne sensor. Nevertheless, the Google Earth test image produced a similar result of F1-measure = 0.90. We consider that a model trained with data-augmentation and very high spatial resolution images (0.03 m/pixel) can generalize well to less accurate images such as those from Google Earth (0.5 m/pixel). Furthermore, when we trained the models with Google Earth images, we observed that it also generalised well to more precise resolutions (F1-measure = 0.90). For this reason, the detection of *Z. lotus* shrubs might be generalizable from any resolution less than 1 m/pixel.

Table 2. Test results of Mask Region-based Convolutional Neural Networks (Mask R-CNN) experiments in three different spatial resolutions images. TP: True Positive; FP: False Negative; FN: False Negative. Precision, Recall, and F1-measure were used for detection results. In bold the most accurate results.

Experiments/Image		TP	FP	FN	Precision	Recall	F1
	UAV	55	5	10	0.92	0.85	0.88
1.1.A	Airborne	56	4	9	0.93	0.86	0.90
	GE	50	1	15	0.98	0.77	0.86
	UAV	59	6	6	0.91	0.91	0.91
1.1.B	Airborne	60	7	5	0.90	0.92	0.91
	GE	55	2	10	0.96	0.85	0.90
	UAV	**55**	**1**	**10**	**0.98**	**0.85**	**0.91**
1.1.C	Airborne	52	3	13	0.94	0.80	0.87
	GE	53	0	12	1	0.81	0.89
	UAV	53	1	12	0.98	0.82	0.89
1.2.A	Airborne	54	1	11	0.98	0.83	0.90
	GE	42	3	23	0.93	0.65	0.76
	UAV	55	1	10	0.98	0.85	0.91
1.2.B	Airborne	50	2	15	0.96	0.77	0.85
	GE	50	2	15	0.96	0.77	0.85
	UAV	**56**	**3**	**8**	**0.95**	**0.87**	**0.91**
1.2.C	Airborne	52	3	13	0.94	0.80	0.87
	GE	54	1	12	0.98	0.81	0.89
	UAV	41	0	24	1	0.63	0.77
2.1.A	Airborne	38	0	27	1	0.58	0.74
	GE	34	1	31	0.97	0.52	0.68
	UAV	47	0	18	1	0.72	0.84
2.1.B	**Airborne**	**55**	**3**	**10**	**0.95**	**0.85**	**0.89**
	GE	50	1	16	0.98	0.76	0.85
	UAV	52	1	13	0.98	0.80	0.88
2.1.C	Airborne	58	3	7	0.95	0.88	0.91
	GE	54	1	12	0.98	0.82	0.89
	UAV	31	0	34	1	0.48	0.65
2.2.A	Airborne	48	1	17	0.98	0.74	0.84
	GE	38	1	27	0.97	0.58	0.73
	UAV	38	1	27	0.97	0.58	0.73
2.2.B	Airborne	46	1	19	0.98	0.71	0.82
	GE	47	3	18	0.94	0.72	0.82
	UAV	46	1	19	0.98	0.70	0.82
2.2.C	**Airborne**	**51**	**2**	**14**	**0.96**	**0.78**	**0.86**
	GE	50	2	15	0.96	0.77	0.85
	UAV	37	0	28	1	0.57	0.73
3.1.A	Airborne	43	0	22	1	0.66	0.80
	GE	41	1	24	0.98	0.63	0.77
	UAV	48	1	17	0.98	0.74	0.84
3.1.B	Airborne	51	1	14	0.98	0.78	0.87
	GE	**54**	**1**	**11**	**0.98**	**0.83**	**0.90**
	UAV	52	1	13	0.98	0.80	0.88
3.1.C	Airborne	52	1	13	0.98	0.80	0.88
	GE	54	2	11	0.96	0.83	0.89
	UAV	54	1	11	0.98	0.83	0.90
3.2.A	Airborne	56	4	9	0.93	0.86	0.90
	GE	53	2	12	0.96	0.82	0.88
	UAV	**56**	**3**	**9**	**0.95**	**0.86**	**0.90**
3.2.B	Airborne	54	5	11	0.92	0.83	0.87
	GE	53	3	12	0.95	0.82	0.88
	UAV	54	3	11	0.95	0.83	0.89
3.2.C	Airborne	52	3	13	0.95	0.80	0.87
	GE	52	3	13	0.95	0.80	0.87

4.2.2. Segmentation Accuracy for Detected Shrubs

The best segmentation accuracy was obtained with the models trained and tested with the same source of images, reaching values of ED2 = 0.07 in Google Earth ones. However, when the model trained with Google Earth images was tested in a UAV image, the ED2 resulted in 0.08. Moreover, the effect of data-augmentation was counterproductive in models trained with airborne images and only lowered ED2 (best results) in models trained with the UAV image. In general, data-augmentation helped to generalise between images but did not obtain a considerable increase in precision in models trained and tested with the same image resolution (Table 3 and Figure 3).

Table 3. Segmentation accuracies of Mask Region-based Convolutional Neural Networks (Mask R-CNN). PSE: Potential Segmentation Error; NSR: Number Segmentation Ratio; ED2: Euclidean Distance v.2. In bold the most accurate results.

Best Experiment	Image Train	Image Test	PSE	NSR	ED2
1.1.C	UAV	UAV	0.0532	0.1290	0.1396
1.2.C	UAV	UAV	0.0512	0.0967	0.1095
2.1.C	**Airborne**	**Airborne**	**0.0408**	**0.0645**	**0.0763**
2.2.C	Airborne	Airborne	0.0589	0.0645	0.0873
3.1.B	**GE**	**GE**	**0.0414**	**0.0645**	**0.0767**
3.2.B	**GE**	**UAV**	**0.0501**	**0.0645**	**0.0816**

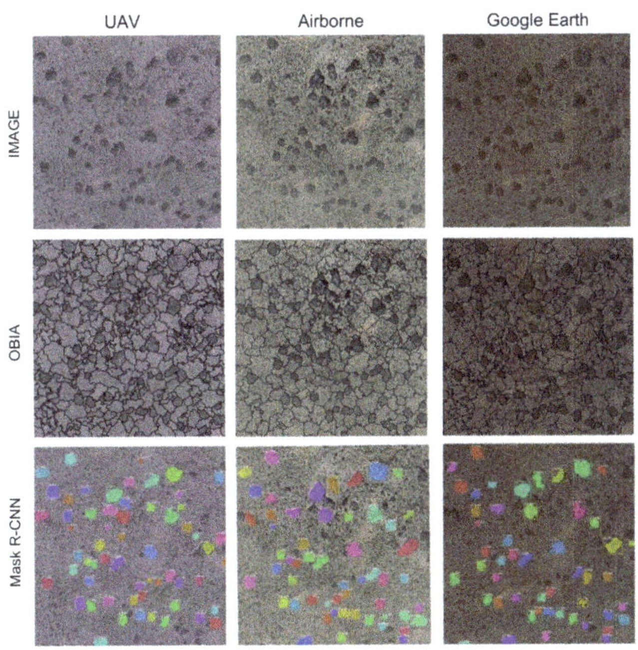

Figure 3. Examples of segmentation of images from Plot 1 using Object-based Image Analysis (OBIA; **Top**) and Mask Region-based Convolutional Neural Networks (Mask R-CNN; **Down**) on Google Earth, Airborne and Unmanned Aerial Vehicle (UAV) images. The different colours in the Mask R-CNN approach are to differentiate the shrubs individually.

4.3. Fusion of OBIA and Mask R-CNN

Our results showed that the fusion between OBIA and Mask R-CNN methods in very high-resolution RGB images is a powerful tool for mapping scattered shrubs in drylands.

We found that the individual segmentations by using OBIA and Mask R-CNN independently were worse than the fusion of both. The accuracy of the fusion of OBIA and Mask R-CNN was higher than the accuracies of the separate segmentations (Table 4), being the most accurate segmentation of all the experiments tested in this work, with an ED2 = 0.038. However, the fusion between results on Google Earth images only improved the ED2 by 0.02. Therefore, the fusion of both segmentation methods provided the best segmentation over the previous methods (OBIA (ED2 = 0.05) and Mask R-CNN (ED2 = 0.07)), in very high-resolution images to segment scattered vegetation in drylands. Moreover, by merging the results of both methodologies (OBIA ∩ Mask R-CNN), the accuracy increases with an ED2 = 0.03.

Table 4. Segmentation accuracies of the fusion of Object-Based Image Analysis (OBIA) and Mask Region-based Convolutional Neural Networks (Mask R-CNN). PSE: Potential Segmentation Error; NSR: Number Segmentation Ratio; ED2: Euclidean Distance v.2. In bold the most accurate results.

Best Experiment	Best OBIA (ED2)	Best Mask R-CNN (ED2)	PSE	NSR	ED2
1.1.C	**0.05**	**0.13**	**0.02**	**0.03**	**0.0386**
1.2.C	0.05	0.10	0.02	0.03	0.0417
2.1.C	0.10	0.07	0.02	0.03	0.0388
2.2.C	0.10	0.08	0.05	0.06	0.0395
3.1.B	0.20	0.07	0.00	0.06	0.0645
3.2.B	0.20	0.08	0.00	0.06	0.0645

To our knowledge, the effect of mixing these two methodologies has not been studied until the date, and it might be vital to improving future segmentation methods. As can be seen in the conceptual framework (Figure 1), it is reasonable to think that the higher the resolution and, therefore, the higher the detail at the edges of vegetation represented in the images, the fusion will improve the final precision of the segmentation. Nevertheless, in images with lower resolution, the fusion improved but to a minor degree.

The spatial resolution of the images affected the accuracy of the segmentation, providing outstanding results in all segmentation methods and spatial resolutions. However, according to [57], we observed that the spatial resolution and Scale parameter played a key role during the segmentation process and controlled the accuracy of the final segmentations. In non-fusion segmentation methods (OBIA or Mask R-CNN) the segmentation accuracy was higher in the spatial resolution image from UAV and OBIA up to ED2 = 0.05. However, when the object to be segmented is larger than the pixel size of the image, the spatial resolution of the image is of secondary importance [37,57,76,77]. For this reason, as the scattered vegetation in this area presents a mean size of 100 m^2 [5], corresponding to 400 pixels of Google Earth image, only slight increases in segmentation accuracy were observed as the spatial resolution increased. Moreover, the overestimation of the area of each shrub was not significant as the images spatial resolution increased. Therefore, Google Earth images could be used to map scattered vegetation in drylands, if the plants to be mapped are larger than the pixel size. This result opens a wide range of new opportunities for vegetation mapping in remote areas where UAV or airborne image acquisition is difficult or acquiring commercial imagery of very high-resolution is very expensive. These results are promising and highlight the usefulness of free available Google Earth images for big shrubs mapping with only a negligible decrease in segmentation accuracy when compared with commercial UAV or airborne images. However, the segmentation of vegetation could be better if we use the near infrared NIR band since vegetation highlights in this range of the spectrum (e.g., 750 to 2500 nm) or used in vegetation indices such as the normalized difference vegetation index (NDVI) or Enhanced vegetation index (EVI). Finally, very high spatial resolution UAV images need much more computational time and are expensive and not always possible to obtain at larger scales in remote areas, hampering their use.

5. Conclusions

Our results showed that both OBIA and Mask R-CNN methods are powerful tools for mapping scattered vegetation in drylands. However, both methods were affected by the spatial resolution of the orthoimages utilized. We have shown for the first time that the fusion of the results from these methods increases, even more, the precision of the segmentation. This methodology should be tested on other types of vegetation or objects in order to prove to be fully effective. We propose an approach that offers a new way of fusing these methodologies to increase accuracy in the segmentation of scattered shrubs and should be evaluated on other objects in very high-resolution and hyperspectral images.

Using images with very high spatial resolution could provide the required precision to further develop methodologies to evaluate the spatial distribution of shrubs and dynamics of plant populations in global drylands, especially when utilizing free-to-use images, like the ones obtained from Google Earth. Such evaluation is of particular importance in drylands of developing countries, which are particularly sensitive to anthropogenic and climatic disturbances and may not have enough resources to acquire airborne or UAV imagery. For these reasons, future methodologies as the one presented in this work should focus on using freely available datasets.

In this context, the fusion of OBIA and Mask R-CNN could be extended to a larger number of classes of shrub and tree species or improved with the inclusion of more spectral and temporal information. Furthermore, this approach could improve the segmentation and monitoring of the crown of trees and arborescent shrubs in general, which are of particular importance for biodiversity conservation and for reducing uncertainties in carbon storages worldwide [78]. Recently, scattered trees have been identified as key structures for maintaining ecosystem services provision and high levels of biodiversity [43]. Global initiatives could benefit largely from CNNs, including those recently developed by FAO [79] to provide the forest extent in drylands. The uncertainties in this initiative [80,81] might be reduced implementing our approach CNN-based to segment trees. Tree and shrub segmentation methods could provide a global characterization of forest ecosystem structures and population abundances as part of the critical biodiversity variables initiative [82,83]. In long-lived shrubs where the precision of the segmentation is key for monitoring the detection of disturbances (e.g., pests, soil loss or seawater intrusion) [5]. Finally, the monitoring of persistent vegetation with minimal cover changes over decades could benefit from fusion approaches in the segmentation methods proposed.

Author Contributions: Conceptualization, E.G., J.B.-S., E.R.-C. and S.T.; methodology, E.G. and J.B.-S.; writing—original draft preparation, E.G., J.B.-S. and E.R.-C.; writing—review and editing, E.G., J.B.-S., E.R.-C., S.T., J.M.-V., D.A.-S., J.C.; funding acquisition, J.C. All authors have read and agreed to the published version of the manuscript.

Funding: This research was funded by the European Research Council (ERC Grant agreement 647038 [BIODESERT]), the European LIFE Project ADAPTAMED LIFE14 CCA/ES/000612, the RH2O-ARID (P18-RT-5130) and RESISTE (P18-RT-1927) funded by Consejería de Economía, Conocimiento, Empresas y Universidad from the Junta de Andalucía, and by projects A-TIC-458-UGR18 and DETECTOR (A-RNM-256-UGR18), with the contribution of the European Union Funds for Regional Development. E.R-C was supported by the HIPATIA-UAL fellowship, founded by the University of Almeria. S.T. is supported by the Ramón y Cajal Program of the Spanish Government (RYC-2015-18136).

Institutional Review Board Statement: Not applicable.

Informed Consent Statement: Not applicable.

Data Availability Statement: All drone and airborne orthomosaic data, shapefile and code will be made available on request to the correspondent author's email with appropriate justification.

Acknowledgments: We are very grateful to the reviewers for their valuable comments that helped to improve the paper. We are grateful to Garnata Drone SL, Andalusian Centre for the Evaluation and Monitoring of Global Change (CAESCG) for providing the data set for the experiments.

Conflicts of Interest: The authors declare no conflict of interest.

Abbreviations

The following abbreviations are used in this manuscript:

Abbreviation	**Description**
CCD	Charge-Coupled Device
ED2	Euclidean Distance v.2
ESP2	Estimation Scale Parameter v.2
ETRS	European Terrestrial Reference System
HB	Bottom-up Hierarchy
HT	Top-down Hierarchy
JSON	JavaScript Object Notation
NH	Non-Hierarchical
NSR	Number-of-Segmentation Ratio
OBIA	Object-based Image Analysis
R-CNN	Region—Convolutional Neural Networks
RGB	Red Green Blue
SPR	Segmentation Parameters Range
UAV	Unmanned aerial vehicle
UTM	Universal Transverse Mercator
VGG	Visual Geometry Group

References

1. Koutroulis, A.G. Dryland changes under different levels of global warming. *Sci. Total Environ.* **2019**, *655*, 482–511. [CrossRef] [PubMed]
2. Puigdefábregas, J. The role of vegetation patterns in structuring runoff and sediment fluxes in drylands. *Earth Surf. Process. Landf.* **2005**, *30*, 133–147. [CrossRef]
3. Ravi, S.; Breshears, D.D.; Huxman, T.E.; D'Odorico, P. Land degradation in drylands: Interactions among hydrologic–aeolian erosion and vegetation dynamics. *Geomorphology* **2010**, *116*, 236–245. [CrossRef]
4. Gao, Z.; Sun, B.; Li, Z.; Del Barrio, G.; Li, X. Desertification monitoring and assessment: A new remote sensing method. In Proceedings of the 2016 IEEE International Geoscience and Remote Sensing Symposium (IGARSS), Beijing, China, 10–15 June 2016.
5. Guirado, E.; Blanco-Sacristán, J.; Rigol-Sánchez, J.; Alcaraz-Segura, D.; Cabello, J. A Multi-Temporal Object-Based Image Analysis to Detect Long-Lived Shrub Cover Changes in Drylands. *Remote Sens.* **2019**, *11*, 2649. [CrossRef]
6. Guirado, E.; Alcaraz-Segura, D.; Rigol-Sánchez, J.P.; Gisbert, J.; Martínez-Moreno, F.J.; Galindo-Zaldívar, J.; González-Castillo, L.; Cabello, J. Remote-sensing-derived fractures and shrub patterns to identify groundwater dependence. *Ecohydrology* **2018**, *11*, e1933. [CrossRef]
7. Guirado, E.; Tabik, S.; Alcaraz-Segura, D.; Cabello, J.; Herrera, F. Deep-learning Versus OBIA for Scattered Shrub Detection with Google Earth Imagery: *Ziziphus lotus* as Case Study. *Remote Sens.* **2017**, *9*, 1220. [CrossRef]
8. Kéfi, S.; Guttal, V.; Brock, W.A.; Carpenter, S.R.; Ellison, A.M.; Livina, V.N.; Seekell, D.A.; Scheffer, M.; van Nes, E.H.; Dakos, V. Early warning signals of ecological transitions: Methods for spatial patterns. *PLoS ONE* **2014**, *9*, e92097. [CrossRef]
9. Cunliffe, A.M.; Brazier, R.E.; Anderson, K. Ultra-fine grain landscape-scale quantification of dryland vegetation structure with drone-acquired structure-from-motion photogrammetry. *Remote Sens. Environ.* **2016**, *183*, 129–143. [CrossRef]
10. Brandt, M.; Hiernaux, P.; Rasmussen, K.; Mbow, C.; Kergoat, L.; Tagesson, T.; Ibrahim, Y.Z.; Wélé, A.; Tucker, C.J.; Fensholt, R. Assessing woody vegetation trends in Sahelian drylands using MODIS based seasonal metrics. *Remote Sens. Environ.* **2016**, *183*, 215–225. [CrossRef]
11. Berdugo, M.; Kéfi, S.; Soliveres, S.; Maestre, F.T. Author Correction: Plant spatial patterns identify alternative ecosystem multifunctionality states in global drylands. *Nat. Ecol. Evol.* **2018**, *2*, 574–576. [CrossRef]
12. Mohammadi, A.; Costelloe, J.F.; Ryu, D. Application of time series of remotely sensed normalized difference water, vegetation and moisture indices in characterizing flood dynamics of large-scale arid zone floodplains. *Remote Sens. Environ.* **2017**, *190*, 70–82. [CrossRef]
13. Tian, F.; Brandt, M.; Liu, Y.Y.; Verger, A.; Tagesson, T.; Diouf, A.A.; Rasmussen, K.; Mbow, C.; Wang, Y.; Fensholt, R. Remote sensing of vegetation dynamics in drylands: Evaluating vegetation optical depth (VOD) using AVHRR NDVI and in situ green biomass data over West African Sahel. *Remote Sens. Environ.* **2016**, *177*, 265–276. [CrossRef]
14. Taddeo, S.; Dronova, I.; Depsky, N. Spectral vegetation indices of wetland greenness: Responses to vegetation structure, composition, and spatial distribution. *Remote Sens. Environ.* **2019**, *234*, 111467. [CrossRef]
15. Blaschke, T. Object based image analysis for remote sensing. *ISPRS J. Photogramm. Remote Sens.* **2010**, *65*, 2–16. [CrossRef]
16. He, K.; Gkioxari, G.; Dollar, P.; Girshick, R. Mask R-CNN. In Proceedings of the 2017 IEEE International Conference on Computer Vision (ICCV), Venice, Italy, 22–29 October 2017.

17. Zhang, J.; Jia, L. A comparison of pixel-based and object-based land cover classification methods in an arid/semi-arid environment of Northwestern China. In Proceedings of the 2014 Third International Workshop on Earth Observation and Remote Sensing Applications (EORSA), Changsha, China, 11–14 June 2014.
18. Amitrano, D.; Guida, R.; Iervolino, P. High Level Semantic Land Cover Classification of Multitemporal Sar Images Using Synergic Pixel-Based and Object-Based Methods. In Proceedings of the IGARSS 2019—2019 IEEE International Geoscience and Remote Sensing Symposium, Yokohama, Japan, 28 July–2 August 2019.
19. Li, S.; Yan, M.; Xu, J. Garbage object recognition and classification based on Mask Scoring RCNN. In Proceedings of the 2020 International Conference on Culture-oriented Science & Technology (ICCST), Beijing, China, 28–31 October 2020.
20. Zhang, Q.; Chang, X.; Bian, S.B. Vehicle-Damage-Detection Segmentation Algorithm Based on Improved Mask RCNN. *IEEE Access* **2020**, *8*, 6997–7004. [CrossRef]
21. Ghassemian, H. A review of remote sensing image fusion methods. *Inf. Fusion* **2016**, *32*, 75–89. [CrossRef]
22. Belgiu, M.; Stein, A. Spatiotemporal Image Fusion in Remote Sensing. *Remote Sens.* **2019**, *11*, 818. [CrossRef]
23. Moreno-Martínez, Á.; Izquierdo-Verdiguier, E.; Maneta, M.P.; Camps-Valls, G.; Robinson, N.; Muñoz-Marí, J.; Sedano, F.; Clinton, N.; Running, S.W. Multispectral high resolution sensor fusion for smoothing and gap-filling in the cloud. *Remote Sens. Environ.* **2020**, *247*, 111901. [CrossRef]
24. Alphan, H.; Çelik, N. Monitoring changes in landscape pattern: Use of Ikonos and Quickbird images. *Environ. Monit. Assess.* **2016**, *188*, 81. [CrossRef]
25. Mahdianpari, M.; Granger, J.E.; Mohammadimanesh, F.; Warren, S.; Puestow, T.; Salehi, B.; Brisco, B. Smart solutions for smart cities: Urban wetland mapping using very-high resolution satellite imagery and airborne LiDAR data in the City of St. John's, NL, Canada. *J. Environ. Manag.* **2020**, 111676, In press. [CrossRef]
26. Mahdavi Saeidi, A.; Babaie Kafaky, S.; Mataji, A. Detecting the development stages of natural forests in northern Iran with different algorithms and high-resolution data from GeoEye-1. *Environ. Monit. Assess.* **2020**, *192*, 653. [CrossRef] [PubMed]
27. Fawcett, D.; Bennie, J.; Anderson, K. Monitoring spring phenology of individual tree crowns using drone—Acquired NDVI data. *Remote Sens. Ecol. Conserv.* **2020**. [CrossRef]
28. Hu, Q.; Wu, W.; Xia, T.; Yu, Q.; Yang, P.; Li, Z.; Song, Q. Exploring the Use of Google Earth Imagery and Object-Based Methods in Land Use/Cover Mapping. *Remote Sens.* **2013**, *5*, 6026–6042. [CrossRef]
29. Venkatappa, M.; Sasaki, N.; Shrestha, R.P.; Tripathi, N.K.; Ma, H.O. Determination of Vegetation Thresholds for Assessing Land Use and Land Use Changes in Cambodia using the Google Earth Engine Cloud-Computing Platform. *Remote Sens.* **2019**, *11*, 1514. [CrossRef]
30. Sowmya, D.R.; Deepa Shenoy, P.; Venugopal, K.R. Feature-based Land Use/Land Cover Classification of Google Earth Imagery. In Proceedings of the 2019 IEEE 5th International Conference for Convergence in Technology (I2CT), Bombay, India, 29–31 March 2019.
31. Li, W.; Buitenwerf, R.; Munk, M.; Bøcher, P.K.; Svenning, J.-C. Deep-learning based high-resolution mapping shows woody vegetation densification in greater Maasai Mara ecosystem. *Remote Sens. Environ.* **2020**, *247*, 111953. [CrossRef]
32. Uyeda, K.A.; Stow, D.A.; Richart, C.H. Assessment of volunteered geographic information for vegetation mapping. *Environ. Monit. Assess.* **2020**, *192*, 1–14. [CrossRef]
33. Ancin-Murguzur, F.J.; Munoz, L.; Monz, C.; Hausner, V.H. Drones as a tool to monitor human impacts and vegetation changes in parks and protected areas. *Remote Sens. Ecol. Conserv.* **2020**, *6*, 105–113. [CrossRef]
34. Yu, Q.; Gong, P.; Clinton, N.; Biging, G.; Kelly, M.; Schirokauer, D. Object-based Detailed Vegetation Classification with Airborne High Spatial Resolution Remote Sensing Imagery. *Photogramm. Eng. Remote Sens.* **2006**, *72*, 799–811. [CrossRef]
35. Laliberte, A.S.; Herrick, J.E.; Rango, A.; Winters, C. Acquisition, Orthorectification, and Object-based Classification of Unmanned Aerial Vehicle (UAV) Imagery for Rangeland Monitoring. *Photogramm. Eng. Remote Sens.* **2010**, *76*, 661–672. [CrossRef]
36. Whiteside, T.G.; Boggs, G.S.; Maier, S.W. Comparing object-based and pixel-based classifications for mapping savannas. *Int. J. Appl. Earth Obs. Geoinf.* **2011**, *13*, 884–893. [CrossRef]
37. Hossain, M.D.; Chen, D. Segmentation for Object-Based Image Analysis (OBIA): A review of algorithms and challenges from remote sensing perspective. *ISPRS J. Photogramm. Remote Sens.* **2019**, *150*, 115–134. [CrossRef]
38. Arvor, D.; Durieux, L.; Andrés, S.; Laporte, M.-A. Advances in Geographic Object-Based Image Analysis with ontologies: A review of main contributions and limitations from a remote sensing perspective. *ISPRS J. Photogramm. Remote Sens.* **2013**, *82*, 125–137. [CrossRef]
39. Johnson, B.A.; Ma, L. Image Segmentation and Object-Based Image Analysis for Environmental Monitoring: Recent Areas of Interest, Researchers' Views on the Future Priorities. *Remote Sens.* **2020**, *12*, 1772. [CrossRef]
40. Guo, Y.; Liu, Y.; Georgiou, T.; Lew, M.S. A review of semantic segmentation using deep neural networks. *Int. J. Multimed. Inf. Retr.* **2018**, *7*, 87–93. [CrossRef]
41. Singh, R.; Rani, R. Semantic Segmentation using Deep Convolutional Neural Network: A Review. *SSRN Electron. J.* **2020**. [CrossRef]
42. Aguilar, M.A.; Aguilar, F.J.; García Lorca, A.; Guirado, E.; Betlej, M.; Cichon, P.; Nemmaoui, A.; Vallario, A.; Parente, C. Assessment of multiresolution segmentation for extracting greenhouses from worldview-2 imagery. *ISPRS-Int. Arch. Photogramm. Remote Sens. Spat. Inf. Sci.* **2016**, *XLI-B7*, 145–152. [CrossRef]

43. Brandt, M.; Tucker, C.J.; Kariryaa, A.; Rasmussen, K.; Abel, C.; Small, J.; Chave, J.; Rasmussen, L.V.; Hiernaux, P.; Diouf, A.A.; et al. An unexpectedly large count of trees in the West African Sahara and Sahel. *Nature* **2020**, *587*, 78–82. [CrossRef]
44. Guirado, E.; Tabik, S.; Rivas, M.L.; Alcaraz-Segura, D.; Herrera, F. Whale counting in satellite and aerial images with deep learning. *Sci. Rep.* **2019**, *9*, 14259. [CrossRef]
45. Guirado, E.; Alcaraz-Segura, D.; Cabello, J.; Puertas-Ruíz, S.; Herrera, F.; Tabik, S. Tree Cover Estimation in Global Drylands from Space Using Deep Learning. *Remote Sens.* **2020**, *12*, 343. [CrossRef]
46. Zhou, Y.; Zhang, R.; Wang, S.; Wang, F. Feature Selection Method Based on High-Resolution Remote Sensing Images and the Effect of Sensitive Features on Classification Accuracy. *Sensors* **2018**, *18*, 2013. [CrossRef]
47. Foody, G.; Pal, M.; Rocchini, D.; Garzon-Lopez, C.; Bastin, L. The Sensitivity of Mapping Methods to Reference Data Quality: Training Supervised Image Classifications with Imperfect Reference Data. *ISPRS Int. J. Geo-Inf.* **2016**, *5*, 199. [CrossRef]
48. Lu, D.; Weng, Q. A survey of image classification methods and techniques for improving classification performance. *Int. J. Remote Sens.* **2007**, *28*, 823–870. [CrossRef]
49. Torres-García, M.; Salinas-Bonillo, M.J.; Gázquez-Sánchez, F.; Fernández-Cortés, A.; Querejeta, J.L.; Cabello, J. Squandering Water in Drylands: The Water Use Strategy of the Phreatophyte *Ziziphus lotus* (L.) Lam in a Groundwater Dependent Ecosystem. *Am. J. Bot.* **2021**, *108*, 2, in press.
50. Tirado, R.; Pugnaire, F.I. Shrub spatial aggregation and consequences for reproductive success. *Oecologia* **2003**, *136*, 296–301. [CrossRef] [PubMed]
51. Tengberg, A.; Chen, D. A comparative analysis of nebkhas in central Tunisia and northern Burkina Faso. *Geomorphology* **1998**, *22*, 181–192. [CrossRef]
52. Fisher, G.B.; Burch Fisher, G.; Amos, C.B.; Bookhagen, B.; Burbank, D.W.; Godard, V. Channel widths, landslides, faults, and beyond: The new world order of high-spatial resolution Google Earth imagery in the study of earth surface processes. *Google Earth Virtual Vis. Geosci. Educ. Res.* **2012**, *492*, 1–22. [CrossRef]
53. Li, M.; Ma, L.; Blaschke, T.; Cheng, L.; Tiede, D. A systematic comparison of different object-based classification techniques using high spatial resolution imagery in agricultural environments. *Int. J. Appl. Earth Obs. Geoinf.* **2016**, *49*, 87–98. [CrossRef]
54. Yu, W.; Zhou, W.; Qian, Y.; Yan, J. A new approach for land cover classification and change analysis: Integrating backdating and an object-based method. *Remote Sens. Environ.* **2016**, *177*, 37–47. [CrossRef]
55. Yan, J.; Lin, L.; Zhou, W.; Ma, K.; Pickett, S.T.A. A novel approach for quantifying particulate matter distribution on leaf surface by combining SEM and object-based image analysis. *Remote Sens. Environ.* **2016**, *173*, 156–161. [CrossRef]
56. Colkesen, I.; Kavzoglu, T. Selection of Optimal Object Features in Object-Based Image Analysis Using Filter-Based Algorithms. *J. Indian Soc. Remote Sens.* **2018**, *46*, 1233–1242. [CrossRef]
57. Lefèvre, S.; Sheeren, D.; Tasar, O. A Generic Framework for Combining Multiple Segmentations in Geographic Object-Based Image Analysis. *ISPRS Int. J. Geo-Inf.* **2019**, *8*, 70. [CrossRef]
58. Hurskainen, P.; Adhikari, H.; Siljander, M.; Pellikka, P.K.E.; Hemp, A. Auxiliary datasets improve accuracy of object-based land use/land cover classification in heterogeneous savanna landscapes. *Remote Sens. Environ.* **2019**, *233*, 111354. [CrossRef]
59. Gonçalves, J.; Pôças, I.; Marcos, B.; Mücher, C.A.; Honrado, J.P. SegOptim—A new R package for optimizing object-based image analyses of high-spatial resolution remotely-sensed data. *Int. J. Appl. Earth Obs. Geoinf.* **2019**, *76*, 218–230. [CrossRef]
60. Ma, L.; Cheng, L.; Li, M.; Liu, Y.; Ma, X. Training set size, scale, and features in Geographic Object-Based Image Analysis of very high resolution unmanned aerial vehicle imagery. *ISPRS J. Photogramm. Remote Sens.* **2015**, *102*, 14–27. [CrossRef]
61. Zhang, X.; Du, S.; Ming, D. Segmentation Scale Selection in Geographic Object-Based Image Analysis. *High Spat. Resolut. Remote Sens.* **2018**, 201–228.
62. Yang, L.; Mansaray, L.; Huang, J.; Wang, L. Optimal Segmentation Scale Parameter, Feature Subset and Classification Algorithm for Geographic Object-Based Crop Recognition Using Multisource Satellite Imagery. *Remote Sens.* **2019**, *11*, 514. [CrossRef]
63. Mao, C.; Meng, W.; Shi, C.; Wu, C.; Zhang, J. A Crop Disease Image Recognition Algorithm Based on Feature Extraction and Image Segmentation. *Traitement Signal* **2020**, *37*, 341–346. [CrossRef]
64. Drăguţ, L.; Csillik, O.; Eisank, C.; Tiede, D. Automated parameterisation for multi-scale image segmentation on multiple layers. *ISPRS J. Photogramm. Remote Sens.* **2014**, *88*, 119–127. [CrossRef]
65. Torres-Sánchez, J.; López-Granados, F.; Peña, J.M. An automatic object-based method for optimal thresholding in UAV images: Application for vegetation detection in herbaceous crops. *Comput. Electron. Agric.* **2015**, *114*, 43–52. [CrossRef]
66. Josselin, D.; Louvet, R. Impact of the Scale on Several Metrics Used in Geographical Object-Based Image Analysis: Does GEOBIA Mitigate the Modifiable Areal Unit Problem (MAUP)? *ISPRS Int. J. Geo-Inf.* **2019**, *8*, 156. [CrossRef]
67. Blaschke, T.; Lang, S.; Hay, G. *Object-Based Image Analysis: Spatial Concepts for Knowledge-Driven Remote Sensing Applications*; Springer Science & Business Media: Berlin/Heidelberg, Germany, 2008; ISBN 9783540770589.
68. Watanabe, T.; Wolf, D.F. Instance Segmentation as Image Segmentation Annotation. In Proceedings of the 2019 IEEE Intelligent Vehicles Symposium (IV), Paris, France, 9–12 June 2019.
69. Demir, A.; Yilmaz, F.; Kose, O. Early detection of skin cancer using deep learning architectures: Resnet-101 and inception-v3. In Proceedings of the 2019 Medical Technologies Congress (TIPTEKNO), Izmir, Turkey, 3–5 October 2019.
70. Liu, Y.; Bian, L.; Meng, Y.; Wang, H.; Zhang, S.; Yang, Y.; Shao, X.; Wang, B. Discrepancy measures for selecting optimal combination of parameter values in object-based image analysis. *ISPRS J. Photogramm. Remote Sens.* **2012**, *68*, 144–156. [CrossRef]

71. Nussbaum, S.; Menz, G. eCognition Image Analysis Software. In *Object-Based Image Analysis and Treaty Verification*; Springer: Berlin/Heidelberg, Germany, 2008; pp. 29–39.
72. Dutta, A.; Gupta, A.; Zissermann, A. VGG Image Annotator (VIA). Available online: http://www.robots.ox.ac.uk/~{}vgg/software/via (accessed on 11 December 2020).
73. Shin, H.-C.; Roth, H.R.; Gao, M.; Lu, L.; Xu, Z.; Nogues, I.; Yao, J.; Mollura, D.; Summers, R.M. Deep Convolutional Neural Networks for Computer-Aided Detection: CNN Architectures, Dataset Characteristics and Transfer Learning. *IEEE Trans. Med. Imaging* **2016**, *35*, 1285–1298. [CrossRef] [PubMed]
74. Caesar, H.; Uijlings, J.; Ferrari, V. COCO-Stuff: Thing and Stuff Classes in Context. In Proceedings of the 2018 IEEE/CVF Conference on Computer Vision and Pattern Recognition, Salt Lake City, UT, USA, 18–22 June 2018.
75. Wang, B.; Li, C.; Pavlu, V.; Aslam, J. A Pipeline for Optimizing F1-Measure in Multi-label Text Classification. In Proceedings of the 2018 17th IEEE International Conference on Machine Learning and Applications (ICMLA), Orlando, FL, USA, 17–20 December 2018.
76. Zhan, Q.; Molenaar, M.; Tempfli, K.; Shi, W. Quality assessment for geo–spatial objects derived from remotely sensed data. *Int. J. Remote Sens.* **2005**, *26*, 2953–2974. [CrossRef]
77. Chen, G.; Weng, Q.; Hay, G.J.; He, Y. Geographic object-based image analysis (GEOBIA): Emerging trends and future opportunities. *Gisci. Remote Sens.* **2018**, *55*, 159–182. [CrossRef]
78. Cook-Patton, S.C.; Leavitt, S.M.; Gibbs, D.; Harris, N.L.; Lister, K.; Anderson-Teixeira, K.J.; Briggs, R.D.; Chazdon, R.L.; Crowther, T.W.; Ellis, P.W.; et al. Mapping carbon accumulation potential from global natural forest regrowth. *Nature* **2020**, *585*, 545–550. [CrossRef] [PubMed]
79. Bastin, J.-F.; Berrahmouni, N.; Grainger, A.; Maniatis, D.; Mollicone, D.; Moore, R.; Patriarca, C.; Picard, N.; Sparrow, B.; Abraham, E.M.; et al. The extent of forest in dryland biomes. *Science* **2017**, *356*, 635–638. [CrossRef]
80. Schepaschenko, D.; Fritz, S.; See, L.; Bayas, J.C.L.; Lesiv, M.; Kraxner, F.; Obersteiner, M. Comment on "The extent of forest in dryland biomes". *Science* **2017**, *358*, eaao0166. [CrossRef]
81. de la Cruz, M.; Quintana-Ascencio, P.F.; Cayuela, L.; Espinosa, C.I.; Escudero, A. Comment on "The extent of forest in dryland biomes". *Science* **2017**, *358*, eaao0369. [CrossRef]
82. Fernández, N.; Ferrier, S.; Navarro, L.M.; Pereira, H.M. Essential Biodiversity Variables: Integrating In-Situ Observations and Remote Sensing Through Modeling. *Remote Sens. Plant Biodivers.* **2020**, *18*, 485–501.
83. Vihervaara, P.; Auvinen, A.-P.; Mononen, L.; Törmä, M.; Ahlroth, P.; Anttila, S.; Böttcher, K.; Forsius, M.; Heino, J.; Heliölä, J.; et al. How Essential Biodiversity Variables and remote sensing can help national biodiversity monitoring. *Glob. Ecol. Conserv.* **2017**, *10*, 43–59. [CrossRef]

sensors

MDPI

Article

MAFF-Net: Multi-Attention Guided Feature Fusion Network for Change Detection in Remote Sensing Images

Jinming Ma, Gang Shi *, Yanxiang Li and Ziyu Zhao

College of Information Science and Engineering, Xinjiang University, Urumqi 830046, China; majinming@stu.xju.edu.cn (J.M.); liyanxiang@stu.xju.edu.cn (Y.L.); 107551901060@stu.xju.edu.cn (Z.Z.)
* Correspondence: shigang@xju.edu.cn; Tel.: +86-135-7999-8016

Abstract: One of the most important tasks in remote sensing image analysis is remote sensing image Change Detection (CD), and CD is the key to helping people obtain more accurate information about changes on the Earth's surface. A Multi-Attention Guided Feature Fusion Network (MAFF-Net) for CD tasks has been designed. The network enhances feature extraction and feature fusion by building different blocks. First, a Feature Enhancement Module (FEM) is proposed. The FEM introduces Coordinate Attention (CA). The CA block embeds the position information into the channel attention to obtain the accurate position information and channel relationships of the remote sensing images. An updated feature map is obtained by using an element-wise summation of the input of the FEM and the output of the CA. The FEM enhances the feature representation in the network. Then, an attention-based Feature Fusion Module (FFM) is designed. It changes the previous idea of layer-by-layer fusion and chooses cross-layer aggregation. The FFM is to compensate for some semantic information missing as the number of layers increases. FFM plays an important role in the communication of feature maps at different scales. To further refine the feature representation, a Refinement Residual Block (RRB) is proposed. The RRB changes the number of channels of the aggregated features and uses convolutional blocks to further refine the feature representation. Compared with all compared methods, MAFF-Net improves the $F1$-Score scores by 4.9%, 3.2%, and 1.7% on three publicly available benchmark datasets, the CDD, LEVIR-CD, and WHU-CD datasets, respectively. The experimental results show that MAFF-Net achieves state-of-the-art (SOTA) CD performance on these three challenging datasets.

Keywords: remote sensing images; change detection; attention mechanism; cross-layer feature fusion

Citation: Ma, J.; Shi, G.; Li, Y.; Zhao, Z. MAFF-Net: Multi-Attention Guided Feature Fusion Network for Change Detection in Remote Sensing Images. *Sensors* **2022**, *22*, 888. https://doi.org/10.3390/s22030888

Academic Editors: Moulay A. Akhloufi and Mozhdeh Shahbazi

Received: 21 December 2021
Accepted: 22 January 2022
Published: 24 January 2022

Publisher's Note: MDPI stays neutral with regard to jurisdictional claims in published maps and institutional affiliations.

1. Introduction

Remote sensing image change detection (CD) uses two or more remote sensing images of the same area at different times to compare and analyze the atmospheric, spectral, and sensor information through artificial intelligence or mathematical statistics to obtain the change information of the area [1,2]. CD is an important research direction in the field of remote sensing and plays a great role in many fields such as land planning, urban expansion [3,4], environmental monitoring [5–7], and disaster assessment [8] as a key technology for monitoring surface conditions.

Recently, with the gradual maturity of remote sensing imaging technology, remote sensing image data with high resolution (HR) have been emerging. Compared with medium-resolution and low-resolution remote sensing images, HR remote sensing images have richer geometric and spatial information, which provide favorable conditions for humans to monitor surface changes more accurately. Therefore, the authors have paid more attention to the processing of HR remote sensing images. Effectively extracting the rich feature information of HR remote sensing images, better focusing on the change regions, avoiding the interference of other factors, and reducing the interference of pseudo-changes are the key issues of remote sensing image CD research [9].

There are many CD methods proposed, and different authors have made a more comprehensive summary classification from different aspects. In this paper, we will summarize and compare two perspectives from traditional methods and deep learning-based methods.

The traditional methods are divided into pixel-based remote sensing image CD methods and object-oriented remote sensing image CD methods according to the size of the basic unit [10]. The pixel-based remote sensing image CD method usually directly processes the input image according to the pixel-level spectral features, texture features, and other specific meaningful features (water bodies, vegetation indices). It obtains the difference image by difference or ratio. The change information is then extracted using a threshold segmentation method [11]. In the early days, methods such as image difference [12], image ratio [13], and regression analysis [14] were commonly used. However, these methods usually failed to obtain complete change information. To better utilize the spectral information of images, methods based on image transformation such as independent component analysis (ICA) [15] and multivariate alteration detection (MAD) [16,17] have emerged one after another and have achieved good results in land CD. For multispectral remote sensing images, the change vector analysis (CVA) [18] method is proposed to detect different changes in the ground. The CVA methods calculate the amplitude and phase angle and use the phase angle information to subdivide the changes. However, the performance of this type of method depends heavily on the quality of the spectral bands involved in the calculation, and the stability of the algorithm cannot be guaranteed. Therefore, improved versions of the CVA technique have been proposed during 2012–2016 to further improve the performance of CD [19–22]. With the development of HR optical remote sensing satellite technology, more and more HR remote sensing images are used for CD.

The characteristic of "different objects in the same spectrum" in HR remote sensing images easily leads to the phenomenon of "salt and pepper" in the detection results. This problem further limits the practical application of pixel-level CD methods in HR remote sensing images [23]. Object-based CD methods are commonly used in HR remote sensing image CDs. This is because it allows for a richer representation of information. Ma et al. [24] investigated the effects of semantic strategy, scale, and feature space on an unsupervised, object-based CD method in urban areas. Subsequently, Zhang et al. [25] proposed an object-based CD method for unsupervised CD by incorporating a multi-scale uncertainty analysis. Zhang et al. [26] proposed a method based on the box-whisker plot with cosine law, which outperformed the traditional CD method. For CD tasks where "from–to" change information has to be determined, Gil-Yepes et al. [27] and Qin et al. [28] utilized a post-classification comparison strategy. Although the object-based CD method can better utilize the spatial feature information of HR remote sensing images compared with the pixel-based CD method, it also relies on the traditional manual feature extraction method, which is not only complicated and low-efficiency, but also has less stable CD performance [9]. In recent years, deep learning methods have been widely used in natural language processing, speech recognition [29,30], and image processing [31–33]. Deep learning methods have excellent learning ability and do not require the manual design of feature factors to extract features. With the success of deep learning in the field of image processing, deep learning-based CD for remote sensing images has quickly attracted the interest of scholars. With the continuous development of technology, the field of remote sensing CD has also started to make some excellent research based on convolutional neural networks (CNNs) [34]. CNNs do not require feature extraction by manually designed features. In the field of remote sensing CD, ResNet [35], full convolutional networks (FCN) [36], and UNet [37] structures have been widely used for feature map extraction with certain results. With continuous research, the model of remote sensing CD has been continuously optimized and improved.

For example, the FC-EF [38] network performs a concatenation operation before feeding two images into the backbone network of the UNet structure, then processes the images separately through two branches of the network. These two branches have the same network structure and shared parameters, and, finally, the outputs of the two branches are combined using convolutional layers. The FC-Siam-conc [38] and FC-Siam-diff [38] improve the network by jump-connecting the three feature maps from the two encoder branches and the corresponding decoder layer. FC-Siam-diff improves the network by first differencing the feature maps of the two decoder branches, then finding the absolute value of the difference, finally using a skip connection strategy to connect with the corresponding decoder layer. Subsequently, the FCN-based UNet network was successfully applied to the CD task [39,40], which was trained in an end-to-end manner from scratch using only available CD datasets. Coarse-to-fine [41] proposes a detection framework based on coarse-to-fine detection to detect remote sensing change regions. It firstly uses an encoder and decoder to obtain coarse change maps of bi-temporal images, then applies the idea of residuals to obtain refined change maps. The method can effectively detect the change regions with good results. After considering the feature maps between different layers with the idea of residuals, many scholars also use the attention mechanism in the direction of remote sensing CD to extract richer and finer feature maps. ResNet is used as a backbone by STANet [42], and then a self-attention module for CD is added in the process of feature extraction, which can calculate any two pixels. The authors of this model introduced Transformer on top of ResNet, which makes the network performance further improved [43]. DASNet [44] proposes a dual-attention mechanism to generate better feature representations to enhance the performance of the network. Zhang et al. [45] first use the two Siamese network architectures as the raw images feature extraction network. To enhance the integrity of change map boundaries and internal densities, multi-level depth features are fused with image difference map features by an attention mechanism. In 2021, Hou et al. [46] proposed a novel attention mechanism for mobile networks by embedding location information into channel attention, calling it Coordinate Attention (CA). CA enhances feature representation. In addition, in 2021, HDFNet [47] uses the idea of a hierarchical fusion and dynamic convolution model to obtain a fine feature map. The network makes innovations in the fusion of features at different levels, which makes the network recognition performance superior. The above methods have achieved certain results in the field of remote sensing CD. However, the accurate extraction of effective feature representations and the adequate fusion of feature information at different scales are still research challenges in the field of remote sensing CD. For the benefit of retrieval, a summary of the above-mentioned methods is presented in Table 1.

Table 1. Summary of contemporary CD methods.

Method	Category	Example Studies
Traditional CD methods	Pixel-based CD	Wang et al. [11], Quarmby et al. [12], Howarth et al. [13], Ludeke et al. [14], Zhang et al. [15], Nielsen et al. [16], Nielsen et al. [17], Bovolo et al. [18], Bovolo et al. [19], Liu et al. [20], Liu et al. [21], Frank et al. [22]
	Object-based CD	Ma et al. [24], Zhang et al. [25], Zhang et al. [26], Gil-Yepes et al. [27], Qin et al. [28]
Deep learning CD methods		FC-EF [38], FC-Siam-conc [38], FC-Siam-diff [38], Daudt et al. [39], FCN-PP [40], BA^{2}Net [41], STANet [42], BIT-CD [43], DASNet [44], IFN [45], HDFNet [47]

In this paper, we propose a Multi-Attention Guided Feature Fusion Network (MAFF-Net) for remote sensing images to address the above problems effectively. The main contributions of this article are as follows:

1. We propose the Feature Enhancement Module (FEM), which solves the problem that the features extracted from the backbone network have much interference information

and the feature representation is not clear enough. The FEM captures not only cross-channel information but also direction-aware and location-sensitive information, which helps the model to locate the region of interest more accurately and enhance the representation of changing region features.

2. To solve the problem of inadequate feature fusion and insufficient feature communication in different layers or scales, we designed the attention-based Feature Fusion Module (FFM), which is divided into FFM_ S1 and FFM_S2 according to the input feature maps. FFM_S1 fuses the high-level feature maps with the low-level feature maps by a cross-layer approach. This cross-layer feature fusion approach is of great benefit to highlight the spatial consistency of objects. FFM_S2 fuses two feature maps of the same scale, and it should be noted that one is the feature map of T1 and one is the feature map of T2. The role of FFM_S2 is to fully fuse the feature maps of the bi-temporal image pairs to obtain a better change map.
3. We propose a Refinement Residual Block (RRB) using a residual structure, which can compensate for the shortcomings of using a single 3 × 3 convolutional kernel to refine the feature representation method.

We tested the model on three publicly available remote sensing image datasets. The experimental results validate the effectiveness of our proposed algorithm. The remainder of this article is organized as follows: Section 2 describes the proposed method in detail. In Section 3, corresponding experiments are designed to verify the effectiveness of the method in this article, and the experimental results are analyzed and discussed. Section 4 draws some conclusions about our method.

2. Methodology

In this section, a detailed description of the network proposed for the remote sensing image CD task is presented. First, the backbone of the architecture is described. Second, a detailed description of the proposed FEM is presented. Next, the attention-guided feature fusion mechanism is the focus of this section description, and these modules are described separately in this section. Then, the RRB proposed in this paper is introduced. Finally, the final prediction results are generated by applying convolutional operations [48,49] on the final fused feature maps.

2.1. Network Architecture

The overall structure of the proposed network in this paper is shown in Figure 1. The proposed network uses ResNet18 as its backbone network. Based on some previous work [42,50,51], the proposed network modifies Res-Net18 by removing the last max-pooling layer and the fully connected layer and retaining the layers in the first five convolutional blocks (Conv1 to Conv5).

First, the bi-temporal image pairs $(T1, T2)$ are input to the feature extraction network to obtain sets of feature maps, $\left(F_{T1}{}_1^0, F_{T1}{}_1^1, F_{T1}{}_1^2, F_{T1}{}_1^3, F_{T1}{}_1^4\right)$ and $\left(F_{T2}{}_2^0, F_{T2}{}_2^1, F_{T2}{}_2^2, F_{T2}{}_2^3, F_{T2}{}_2^4\right)$. For each set of feature maps, the proposed method uses only the last four feature maps. These feature maps are then fed into the Feature Enhancement Module (FEM) according to their respective scales to obtain two sets of updated feature maps, $\left(F_1^1, F_1^2, F_1^3, F_1^4\right)$ and $\left(F_2^1, F_2^2, F_2^3, F_2^4\right)$. Next, the cross-layer feature fusion strategy is employed for each of the two updated feature maps. It should be noted here that our cross-layer feature fusion strategy targets different scale features of the same image. Specifically, take image $T1$ as an example. First, bilinear up-sampling [52–54] and convolution operations are performed on high-level features $F_1^3 \in \mathbb{R}^{4C \times H/4 \times W/4}$ to obtain $F_1^3 \in \mathbb{R}^{C \times H \times W}$, where $H \times W$ is the size of the feature map $F_1^1 \in \mathbb{R}^{C \times H \times W}$ and C is the channel dimension of F_1^1. Then, the feature maps F_1^1 and F_1^3 of the $T1$ image are concatenated to obtain feature $F_1^{13} \in \mathbb{R}^{2C \times H \times W}$. F_1^{13} is input to the convolutional block attention module (CBAM) [55] and then output to $F_1^{13} \in \mathbb{R}^{C \times H \times W}$ after using 3 × 3 convolution on it. The same method is used to fuse $F_1^2 \in \mathbb{R}^{2C \times H/2 \times W/2}$ and $F_1^4 \in \mathbb{R}^{8C \times H/8 \times W/8}$ of $T1$ to obtain $F_1^{24} \in \mathbb{R}^{2C \times H/2 \times W/2}$. With the FFM module, four feature maps F_1^{13}, F_2^{13}, F_1^{24}, and F_2^{24} were obtained. Depending on the corresponding scales, the fused feature map pairs, (F_1^{13}, F_2^{13}) and (F_1^{24}, F_2^{24}), are fed into our proposed RRB to further refine the feature representation to obtain $F_{12}^{13} \in \mathbb{R}^{C\times H\times W}$ and $F_{12}^{24} \in \mathbb{R}^{2C\times H/2\times W/2}$, respectively. Finally, the two feature maps, F_{12}^{13} and F_{12}^{24}, are sent to the FFM for final fusion. The prediction map is obtained after applying a pixel classifier (equipped with the sequence 3×3 Conv, batch normalization (BN) [56], and ReLU [57]).

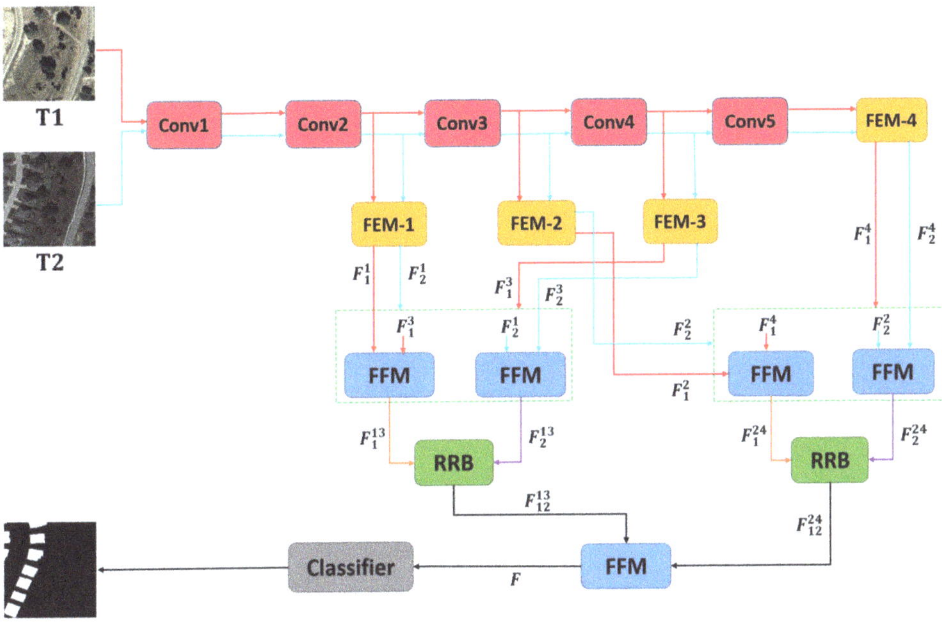

Figure 1. Architecture of the proposed MAFF-Net network. The green dotted box shows the cross-layer fusion strategy. $(F_1^1, F_1^2, F_1^3, F_1^4)$ and $(F_2^1, F_2^2, F_2^3, F_2^4)$ denote the two sets of features updated by the FEM.

2.2. Feature Enhancement Module

The existing CD methods for HR remote sensing images have received less attention to the position information and channel relationships. HR remote sensing images have rich location-spatial information. To obtain accurate position information, a Feature Enhancement Module (FEM) based on coordinate attention (CA) is proposed in this paper to obtain the accurate location information and channel relationships of HR remote sensing images. The module can consider both position information and channel information. The structure of the FEM is shown in Figure 2.

In Figure 2, first, a 3×3 convolution operation is performed on the input $F1$. Then it is fed into the CA block to obtain the weighted feature map, $F2 \in \mathbb{R}^{C\times H\times W}$. Feature maps $F1$ and $F2$ are merged into one feature map by element-wise summation, and a 3×3 convolution operation is used to obtain $F3$.

In Figure 2, the coordinate attention module encodes H and W respectively. In the HR remote sensing image, for a given position (i, j), its pixel value on channel c is $x_c(i, j)$. The H average pooling output of the c-th channel at height h is as Equation (1) [46]:

$$z_c^h(h) = \frac{1}{W} \sum_{0\le i<W} x_c(h, i) \tag{1}$$

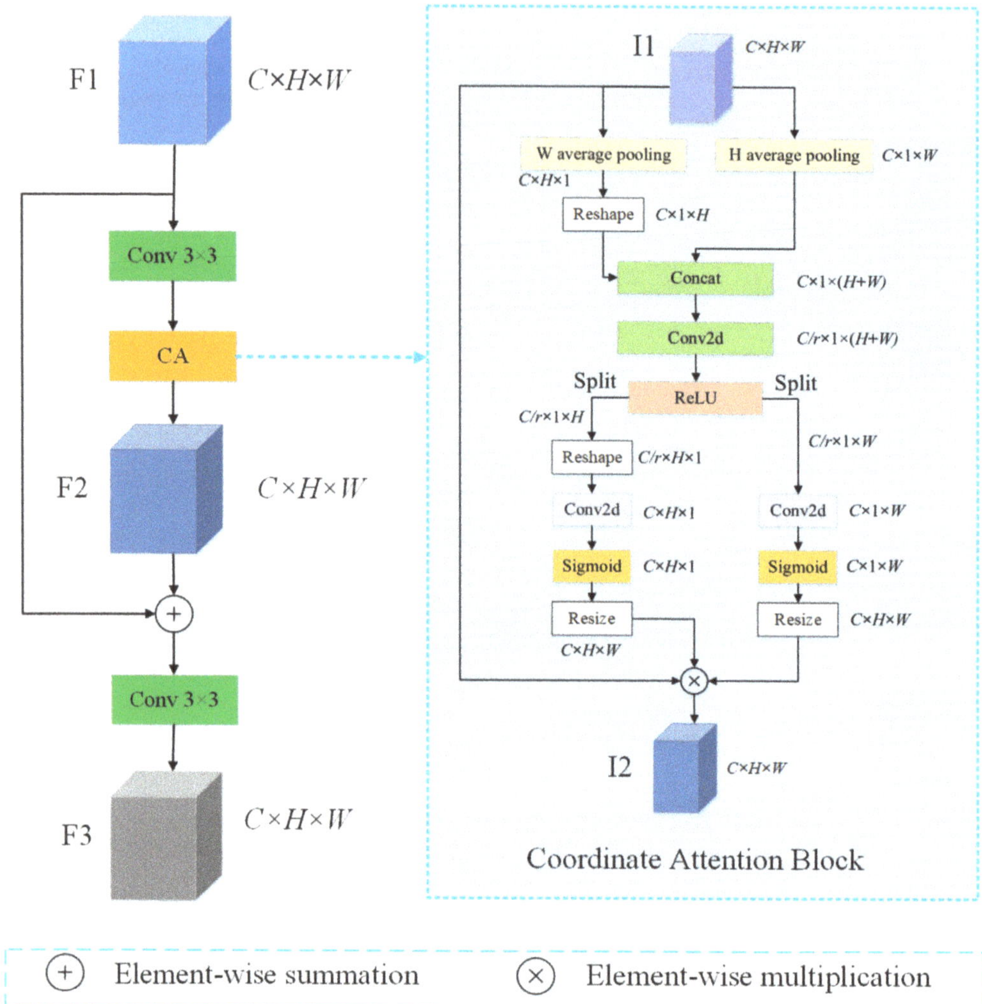

Figure 2. Feature Enhancement Module (FEM). "*W* average pooling" and "*H* average pooling" refer to 1D horizontal global average pooling and 1D vertical global average pooling, respectively. The r indicates the reduction ratio, where r is set to 16. The Reshape operation permutes the Dimension of the tensor. The Resize operation extends the tensor to the same size as the input I1.

Similarly, the *W* average pooling output of the *c*-th channel at width w is as Equation (2) [46]:

$$z_c^w(w) = \frac{1}{H} \sum_{0 \le j < H} x_c(j, w) \tag{2}$$

Then, the Reshape operation is used to permute the dimensionality of the z_c^h tensor to be the same as that of the z_c^w tensor. Next, the coordinate attention module uses concatenation, convolution, and activation function operations. The related definition is as Equation (3) [46]:

$$f = \delta\left(F_C\left(\left[z_c^h, z_c^w\right]\right)\right) \tag{3}$$

where [,] indicates a concatenation operation, F_C indicates a 1×1 convolution operation, and δ indicates the ReLU activation function. f is the output feature map of the ReLU layer.

After the split operation, f can be decomposed into $f^h \in \mathbb{R}^{C/r\times 1\times H}$ and $f^w \in \mathbb{R}^{C/r\times 1\times W}$. The reshape operation is used again to permute the dimension of the tensor f^h to obtain $f^h \in \mathbb{R}^{C/r\times H\times 1}$. Next, two 1×1 convolutional transforms F_h and F_w are used to transform f^h and f^w into tensor with the same number of channels as the input I1, respectively. Then, applying the sigmoid activation function [58] to the tensors updated by F_h and F_w, respectively, two outputs are obtained as shown in Equation (4) and Equation (5) [46]:

$$g^h = \sigma\left(F_h\left(f^h\right)\right) \tag{4}$$

$$g^w = \sigma(F_w(f^w)) \tag{5}$$

where σ indicates sigmoid activation function. The Resize operation expands the size of $g^h \in \mathbb{R}^{C\times H\times 1}$ and $g^w \in \mathbb{R}^{C\times 1\times W}$ to the same size as the input I1 $\in \mathbb{R}^{C\times H\times W}$, respectively, and the g^h and g^w, after being Resized, are used as attention weights. Finally, the output feature map I2 of the CA block is defined as Equation (6) [46]:

$$y_c(i,j) = \left(x_c(i,j) \times g_c^h(i)\right) \times g_c^w(j) \tag{6}$$

where c is the c-th channel, $g_c^h(i)$ is the weight of the i-th position in the H direction, $g_c^w(j)$ is the weight of the j-th position in the W direction, and $y_c(i,j)$ is the value of the output feature map I2.

2.3. Feature Fusion Module

With the study of deep learning-based CD, it has been found that the CD task is unsatisfactory if it relies only on simple feature extraction networks. On the one hand, this is because simple feature extraction networks cannot eliminate semantic interference such as seasonal appearance differences and cannot accurately label change regions in the presence of diverse object shapes and complex boundaries. On the other hand, it is not fully exploited to multi-scale information, and the fusion of multi-scale features to make them communicate can help our network improve its performance.

Therefore, as shown in Figure 1, an attention-based Feature Fusion Module (FFM) is introduced into the CD network. The detail of the FFM is shown in Figure 3.

The proposed FFM is slightly different at different stages. The FFM whose input features are from FEM is named FFM_S1, and the FFM whose input features are from RRB is named FFM_S2. Specifically, the difference between FFM_S1 and FFM_S2 lies in the input part. The inputs of FFM_S1 are two feature maps of different scales of one image, while the input of FFM_S2 is two feature maps of the same scale of two images.

After FEM processing, two sets of updated feature maps, $\left(F_1^1, F_1^2, F_1^3, F_1^4\right) \in T1$ and $\left(F_2^1, F_2^2, F_2^3, F_2^4\right) \in T2$, were obtained. For FFM_S1, the inputs are the feature map pairs $\left(F_1^1, F_1^3\right)$ and $\left(F_1^2, F_1^4\right)$ and $\left(F_2^1, F_2^3\right)$ and $\left(F_2^2, F_2^4\right)$, respectively. Figure 3 shows FFM_S1, and the structure of FFM_S2 is not drawn separately because the two only have different inputs. However, it should be emphasized here that FFM_S2, which has two input feature maps of the same scale, does not distinguish between high-level features and low-level features, and also does not need to up-sample high-level features such as FFM_S1.

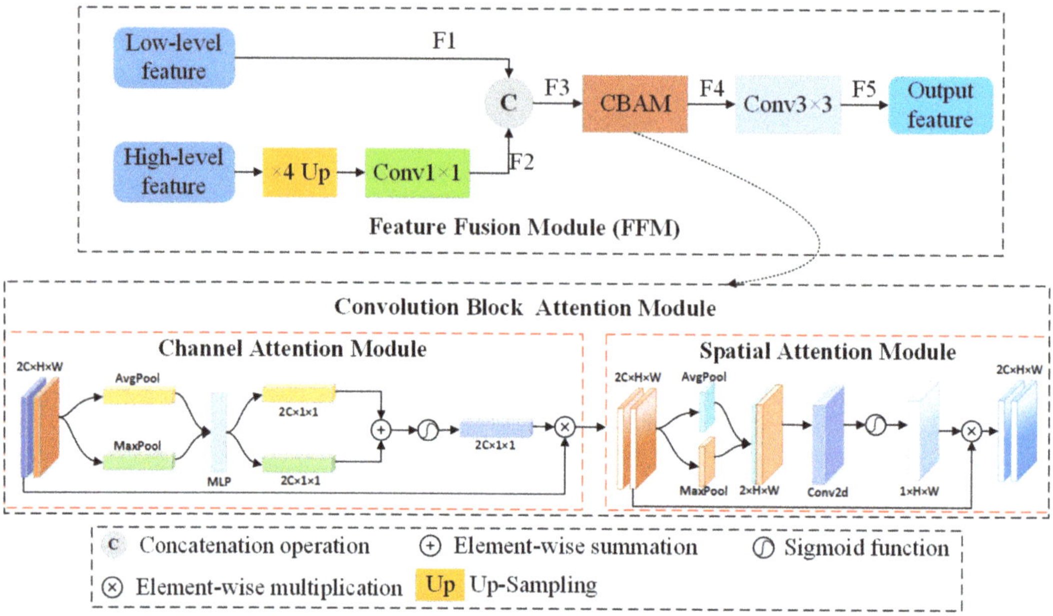

Figure 3. Feature Fusion Module (FFM). *F1–F5* represent the feature maps that are output by different blocks.

The next step is to describe FFM_S1. After experiments, it is found that the fusion of features by cross-layer is more effective. This may be because the high-level features will lose some semantic information carried by the original image or low-level features, such as some edge features, as the number of convolution layers increases, and the fusion with low-level features can compensate for this deficiency. At the same time, the semantic information carried by the feature maps between neighboring layers is not so obviously different, so the fusion method by cross-layer plays a role. For an original image $T1$, feature map pairs (F_1^1, F_1^3) and (F_1^2, F_1^4) are fed into FFM_S1, respectively. For original image $T2$, feature map pairs (F_2^1, F_2^3) and (F_2^2, F_2^4) are fed into FFM_S1, respectively. As shown in Figure 3, the high-level feature needs an up-sampling operation to make the feature map shape consistent with the low-level feature. Next, one 1×1 convolution is used to obtain the feature map $F2 \in \mathbb{R}^{C \times H \times W}$. The two inputs $F1 \in \mathbb{R}^{C \times H \times W}$ and $F2$ are concatenated to obtain the feature map $F3 \in \mathbb{R}^{2C \times H \times W}$. The resulting feature map can be viewed as a feature map with different channels. The calculation process of F3 is shown in Equation (7):

$$F3 = [Conv(Up(F1)), F2] \tag{7}$$

where $Conv$ denotes the 1×1 convolution, and [.,.] denotes the concatenation operation. Considering that this direct aggregation of features in cross-layer does not yet communicate well in the channel and spatial dimensions, feed $F3$ to the *CBAM*. *CBAM* is an attention module consisting of the channel and spatial attention. It considers both the importance of pixels in different channels and the importance of pixels in different positions in the same channel. The *CBAM* outputs the feature map $F4 \in \mathbb{R}^{2C \times H \times W}$. Then, the 3×3 convolution block is used, the main purpose of which is to recover the channels of the aggregated feature map to the number of channels of the input feature map. The above calculation process is shown in Equation (8):

$$F4 = Conv(CBAM(F3)) \tag{8}$$

where *Conv* denotes the 3×3 convolution block. Next, in two subsubsections, two parts of *CBAM*, namely the channel attention module and the spatial attention module, are described in detail.

2.3.1. Channel Attention Module

In the Channel Attention Module (CAM), the vectors described as $AF^{ca}_{avg} \in \mathbb{R}^{B\times C\times 1}$ and $F^{ca}_{max} \in \mathbb{R}^{B\times C\times 1}$ are obtained by the average-pooling and max-pooling operations, respectively. Then, each of them is input to the shared multi-layer perceptron (MLP) with one hidden layer, respectively, to get two vectors, and the two vectors are merged to one feature vector by element-wise summation. After sigmoid activation, the feature map of the CAM is finally obtained. This is shown in Equation (9) [55]:

$$M_{ca}(D) = \delta\left(FC_1\left(FC_0\left(F^{ca}_{avg}\right)\right) + FC_1(FC_0(F^{ca}_{max}))\right) \quad (9)$$

where FC_0 and FC_1 denote the convolution operation in MLP and δ denotes the sigmoid function. The CAM compresses the feature map spatial dimensions to obtain a one-dimensional vector before manipulating it. Channel attention is concerned with what is significant on this feature map. The average-pooling has feedback for every pixel point on the feature map, while max-pooling has feedback for gradients only where the response is greatest in the feature map when performing gradient backpropagation calculations.

2.3.2. Spatial Attention Module

In the Spatial Attention Module (SAM), it is the feature map output from the CAM that is used as input. First, do a max-pooling and average-pooling based on the channel to get the element-wise summation, and then a concatenation operation is performed on the two layers. Then, convolution is performed and reduced to 1 channel, and then the feature map output from the SAM is obtained by sigmoid activation. This is given by Equation (10) [55]:

$$M_{sa}(D^{ca}) = \delta\left(f^{7\times 7}\left(Cat\left(F^{sa}_{avg}, F^{sa}_{max}\right)\right)\right) \quad (10)$$

where *Cat* is the concatenation operation, $f^{7\times 7}$ represents a convolutional layer with a filter size of 7×7, and δ denotes the sigmoid function. The SAM is a channel compression mechanism that performs average-pooling and max-pooling in the channel dimension respectively. The max-pooling operation is to extract the maximum value on the channel, and the number of extractions is $H \times W$. The average-pooling operation is to extract the average value on the channel, and the number of extractions is also $H \times W$. Thus, a 2-channel feature map can be obtained.

2.4. *Refinement Residual Block*

The use of a single 3×3 convolutional kernel has some shortcomings in refining the feature representation. Inspired by Yu et al. [59], a Refinement Residual Block (RRB) is introduced to modify the channels of the aggregated feature map to be consistent with the input feature map and further refine the feature representation before the final feature fusion using FFM_S2. Its structure is shown in Figure 4.

As can be seen in Figure 4, the RRB has three inputs, one of which is the difference map of two feature maps. The three feature maps are first subjected to a concatenation operation, followed by two consecutive convolution blocks, each consisting of *Conv* 3×3, BN, and ReLU. The two convolution blocks output the feature maps $F1 \in \mathbb{R}^{C\times H\times W}$ and $F2 \in \mathbb{R}^{C\times H\times W}$, respectively. Here, it should be noted that the number of channels of each convolutional block output is different. In addition, the module adds additional residual connections with the 1×1 convolutional layers for obtaining some additional spatial information of the remote sensing images. Finally, the four feature maps are subjected to element-wise summation and the final output feature map $F4 \in \mathbb{R}^{C\times H\times W}$ is obtained.

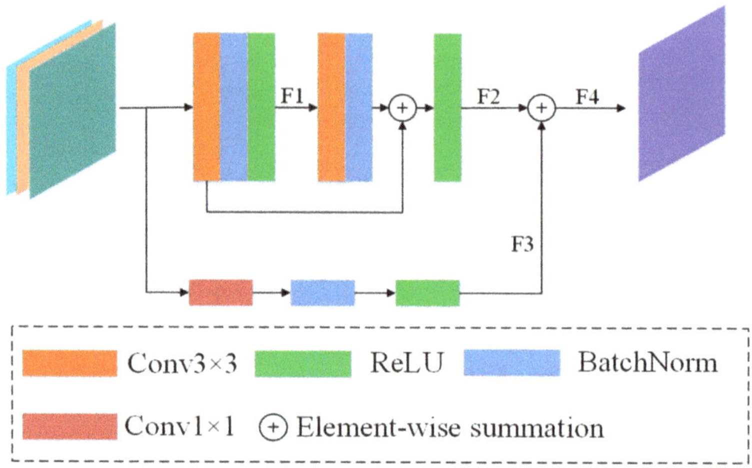

Figure 4. Refinement Residual Block (RRB). *F1–F4* represent the feature maps that are output by different blocks.

2.5. Loss Function

In the training stage, a cross-entropy loss function optimized by Chen et al. [43] is used, which minimizes the cross-entropy loss to optimize the network parameters. Formally, the loss function is defined as Equation (11) [43]:

$$L = \frac{1}{H_{0\times}W_0}\sum_{h=1,w=1}^{H,W} l(P_{hw}, Y_{hw}) \tag{11}$$

where $l(P_{hw}, y) = -log\left(P_{hwy}\right)$ is the cross-entropy loss and Y_{hw} is the label for the pixel at location (h, w) [43].

3. Experiments and Results

In this section, the proposed network MAFF-Net is evaluated on three publicly available benchmark datasets to demonstrate its effectiveness. First, the details of the three datasets, the CDD dataset [60], the LEVIR-CD dataset [42], and the WHU-CD dataset [61], are introduced. Next, the implementation details are presented, including the experimental environment and evaluation metrics. Then, seven state-of-the-art (SOTA) comparison methods are introduced. In this section, quantitative and qualitative analyses of these methods are presented on three datasets.

3.1. Datasets and Settings

The CDD dataset has three types of images, synthetic images with no relative movement of objects, synthetic images with less relative movement of objects, and real remote sensing images with seasonal changes (obtained from Google Earth). In this paper, a subset of remote sensing image data with seasonal changes is selected. This subset has 16,000 images with an image size of 256 × 256 pixels, of which 10,000 images are used as the training set, 3000 images as the validation set, and 3000 images as the test set. As shown in Figure 5, the change scenarios of this dataset include building changes, road changes, and vehicle changes. The data set was considered for different sizes of objects.

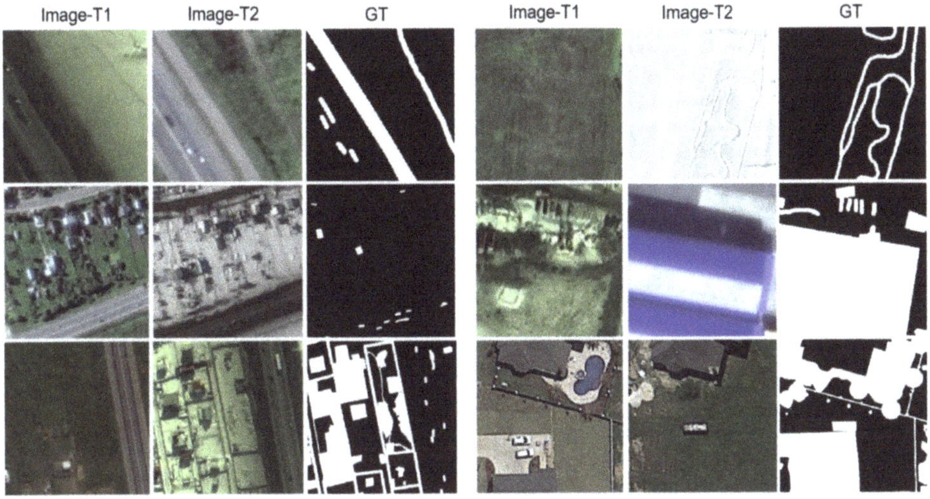

Figure 5. Illustration of samples from CDD. (Image-T1) and (Image-T2) indicate the bi-temporal image pairs. (GT) indicates the ground truth.

LEVIR-CD contains 637 very high resolution (VHR, 0.5 m/pixel) Google Earth image patch pairs, 1024 × 1024 pixels in size. These bitmap images spanning 5 to 14 years have significant land-use changes, especially building growth. LEVIR-CD covers various types of buildings such as villas, high-rise apartments, small garages, and large warehouses. The fully annotated LEVIR-CD contains a total of 31,333 individual instances of change construction. As shown in Figure 6, each sample is cropped into 16 small patches of size 256 × 256, generating 7120 image patch pairs for training, 1024 for validation, and 2048 for testing.

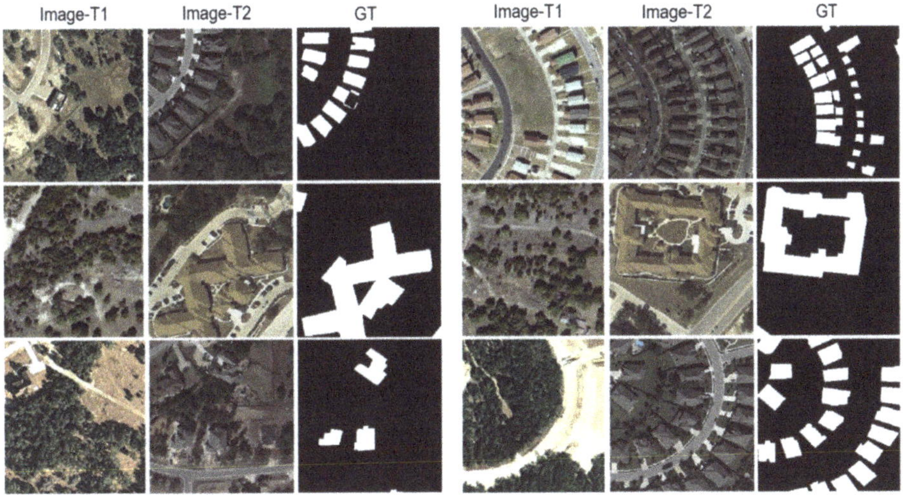

Figure 6. Illustration of samples from LEVIR-CD. (Image-T1) and (Image-T2) indicate the bi-temporal image pairs. (GT) indicates the ground truth.

The third dataset is named the WHU-CD dataset, which is a CD dataset of public buildings. The dataset covers the area where the 6.3 magnitude earthquake occurred in

February 2011 and has been reconstructed in the following years. It consists of a pair of HR (0.075 m) aerial images of size 32,507 × 15,354. Considering that the authors of the original paper did not provide a solution for data segmentation, as shown in Figure 7, the solution of cropping the image into small pieces of size 224 × 224 was finally chosen, and dividing them into three random parts: 7918/987/955 for training/validation/testing, respectively.

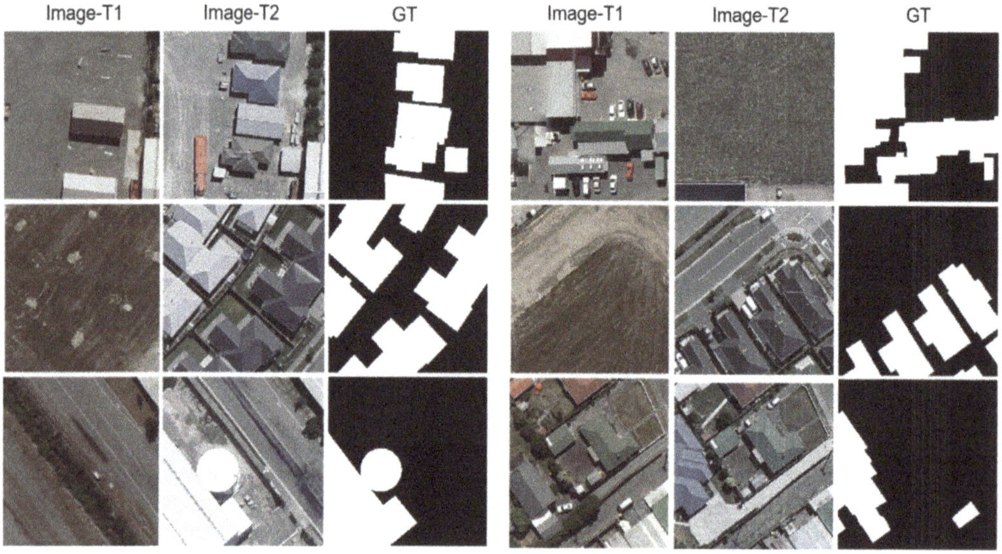

Figure 7. Illustration of samples from WHU-CD. (Image-T1) and (Image-T2) indicate the bi-temporal image pairs. (GT) indicates the ground truth.

3.2. Evaluation Metrics and Settings

For quantitative assessment, three indices, namely the $F1$-score ($F1$), *Kappa* coefficient (*Kappa*), and overall accuracy (OA) are used as the evaluation metrics. These three indices can be calculated as follows:

$$P = \frac{TP}{TP + FP} \tag{12}$$

$$R = \frac{TP}{TP + FN} \tag{13}$$

$$F1 = \frac{2}{P^{-1} + R^{-1}} \tag{14}$$

$$OA = \frac{TP + TN}{TP + FP + TN + FN} \tag{15}$$

$$PRE = \frac{(TP + FN) \times (TP + FP) + (TN + FP) \times (TN + FN)}{(TP + TN + FP + FN)^2} \tag{16}$$

$$Kappa = \frac{OA - PRE}{1 - PRE} \tag{17}$$

where OA and PRE denote the overall accuracy and expected accuracy, respectively. The TP, FP, TN, and FN are the number of true positives, false positives, true negatives, and false negatives, respectively.

We implemented our proposed method with PyTorch, supported by NVIDIA CUDA with a GeForce GTX 2080Ti GPU. In the training stage, the feature extraction backbone of the proposed MAFF-Net is initialized from ResNet18. We used the Adam ($\beta_1 = 0.5$, $\beta_2 = 0.9$) optimizer and the entire training period was set to 200 epochs. The initial learning rate is 0.001 in the first 100 epochs, in the next 100 epochs, the value of the learning rate decays linearly to 0. Considering the GPU size, we set the batch size to 8 to facilitate GPU training.

3.3. *Comparison of Experimental Results*

In this section, the performance of the different methods is compared on the three datasets CDD, LEVIR-CD, and WHU-CD, respectively. The advantages and disadvantages of each method are further described based on the results of the quantitative and qualitative analyses. In addition, an ablation study is performed on the proposed method to compare and analyze the effectiveness of each of its modules.

3.3.1. Comparison Methods

To verify the effectiveness and superiority of our methods, we selected seven methods that are represented in the CD task and compared the performance of these methods in CDD, LEVIR-CD, and WHU-CD, respectively, and a brief description of the selected methods is as follows:

1. CD-Net [62] combines the multi-sensor fusion SLAM and fast density 3D reconstruction for coarse alignment of image pairs followed by deep learning methods for pixel-level CD.
2. FC-EF [38] refers to early fusion with full convolution. It concatenates the two input images before feeding them into the network, treating them as different channels of one image. It is then fed into a standard U-Net.
3. FC-Siam-conc [38] connects three feature maps from the two encoder branches and the corresponding layer of the decoder.
4. FC-Siam-diff [38] first finds the absolute value of the difference between the feature maps of the two decoder branches and then makes a skip-connection to the corresponding layer of the decoder.
5. DASNet [44] is a CD model based on a dual-attentive fully convolutional twin neural network and proposes a weighted double-margin contrastive loss (WDMC) to be able to solve the sample imbalance problem.
6. IFN [45] first uses the two Siamese network architectures as the raw images feature extraction network. To enhance the integrity of change map boundaries and internal densities, multi-level depth features are fused with image difference map features by an attention mechanism.
7. STANet [42] proposes a new spatial-temporal attention neural network based on twin networks. The network exploits spatial-temporal dependence and designs a CD self-attentive mechanism to model spatial-temporal relations. A new HR remote sensing image dataset, LEVIR-CD, is also proposed.

3.3.2. CDD Dataset

For quantitative comparison, we calculated and summarized the evaluation metrics for CDD, LEVIR-CD, and WHU-CD, as shown in Tables 1–3, respectively. To compare the performance of each method more visually, we visualized the test results of each method on the three data sets, as shown in Figures 8–10, respectively. The white color indicates the changes that were correctly detected. Black indicates that no changes have been correctly detected. Red indicates false alarms. Blue indicates unpredicted changes.

Table 2. Comparison of CDD dataset results. The best scores are highlighted in bold.

Method	*F1* (%)	*Kappa* (%)	*OA* (%)
CDNet	81.9	79.6	95.9
FC-EF	83.0	80.8	96.0
FC-Siam-conc	84.0	81.9	96.3
FC-Siam-diff	84.8	82.8	96.4
DASNet	90.1	88.7	97.5
IFN	90.6	89.2	97.6
STANet	91.6	90.4	97.9
MAFF-Net	**96.5**	**96.0**	**99.2**

Table 3. Comparison of LEVIR-CD dataset results. The best scores are highlighted in bold.

Method	*F1* (%)	*Kappa* (%)	*OA* (%)
CDNet	78.0	76.9	97.8
FC-EF	80.7	79.7	98.0
FC-Siam-conc	82.2	81.2	98.0
FC-Siam-diff	83.7	82.8	98.3
DASNet	84.6	83.7	98.4
IFN	86.2	85.4	98.6
STANet	86.5	85.9	**98.9**
MAFF-Net	**89.7**	**89.1**	**98.9**

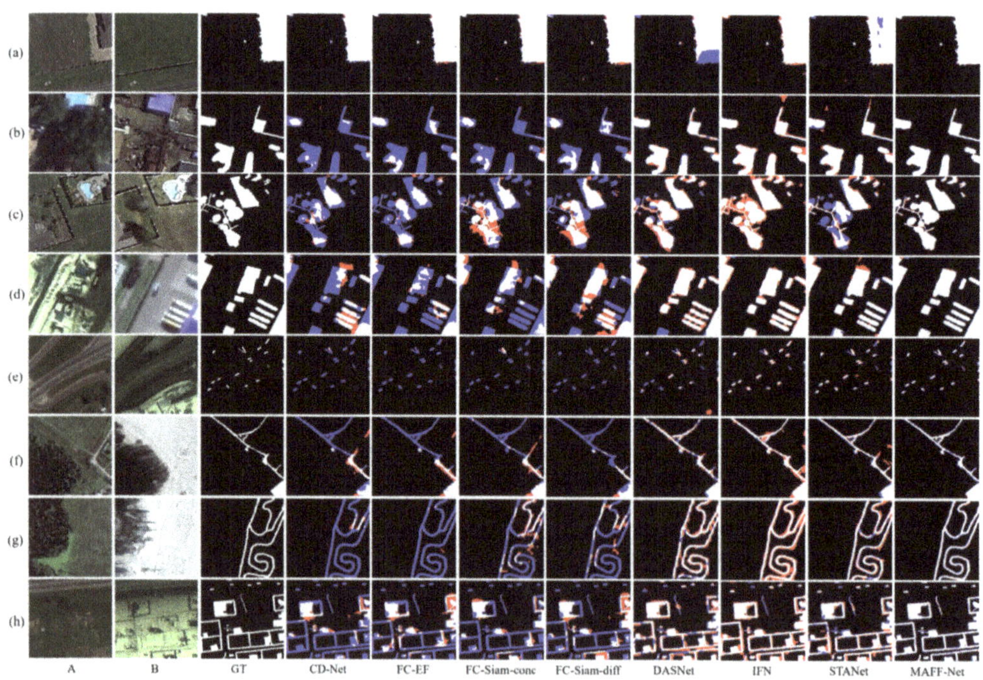

Figure 8. Illustration of a qualitative comparison on dataset CDD. (**a**–**h**) indicate samples from CDD and the change maps obtained with different methods. The white color indicates the changes that were correctly detected. Black indicates that no changes have been correctly detected. Red indicates false alarms. Blue indicates unpredicted changes.

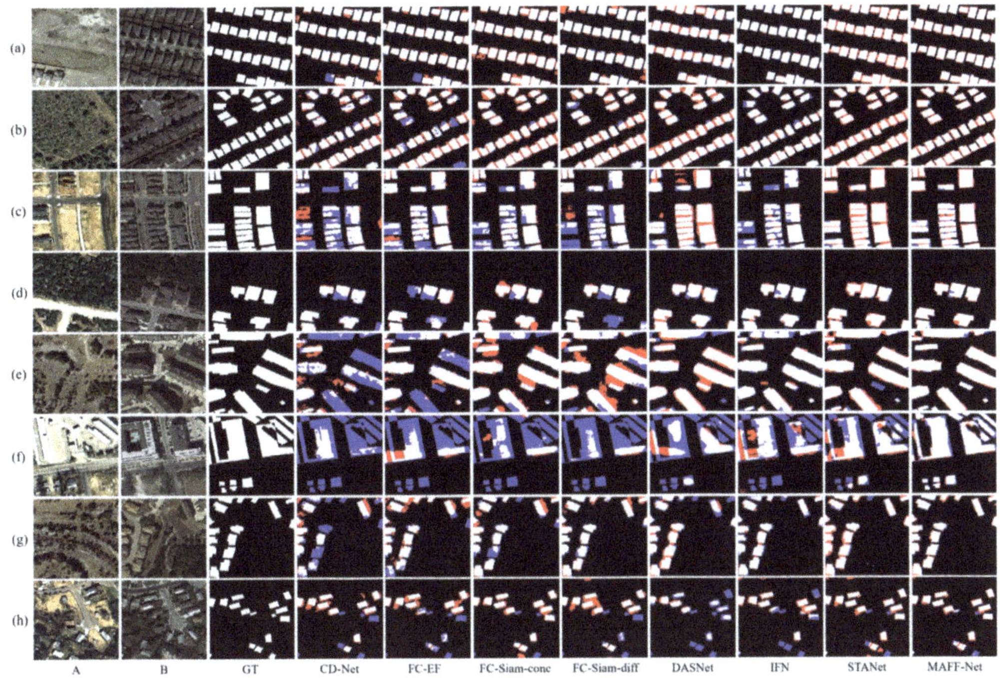

Figure 9. Illustration of a qualitative comparison on dataset LEVIR-CD. (**a–h**) indicate samples from LEV-IR-CD and the change maps obtained with different methods. The white color indicates the changes that were correctly detected. Black indicates that no changes have been correctly detected. Red indicates false alarms. Blue indicates unpredicted changes.

As can be seen from Table 2, the proposed MAFF-Net reached the first on *F*1, *Kappa*, and *OA* on the CDD dataset. This also indicates that the proposed network performs optimally on this dataset. It is also evident from Figure 8 that the proposed network can better mark the change region, while there are few cases of wrong and missing detections. Specifically, as can be seen from the data in Table 2, CD-Net, which does not pay attention to the connections and interactions between multi-scale features, performs relatively poorly in the three evaluation metrics, 14.6% lower than the proposed MAFF-Net in terms of *F*1 score. This is somewhat related to its fewer network levels and relatively simple structure. Considering early fusion and late fusion strategies separately and using skip-connected encoding-decoding, the baselines of FC-EF, FC-Siam-conc, and FC-Siam-diff achieve better performance with their compact and efficient structures. Among these three baselines, the late fusion baseline shows a clear advantage over the early fusion baseline. The fusion of feature maps using bi-temporal image pairs with their difference maps achieves better results than the fusion of feature maps using only bi-temporal image pairs. FC-Siam-Diff scores 0.8%, 0.9%, and 0.1% higher than FC-Siam-conc on *F*1, *Kappa*, and *OA*, respectively. This is because the original image coding features are preserved as much as possible while obtaining the difference maps. This helps the network to achieve better performance.

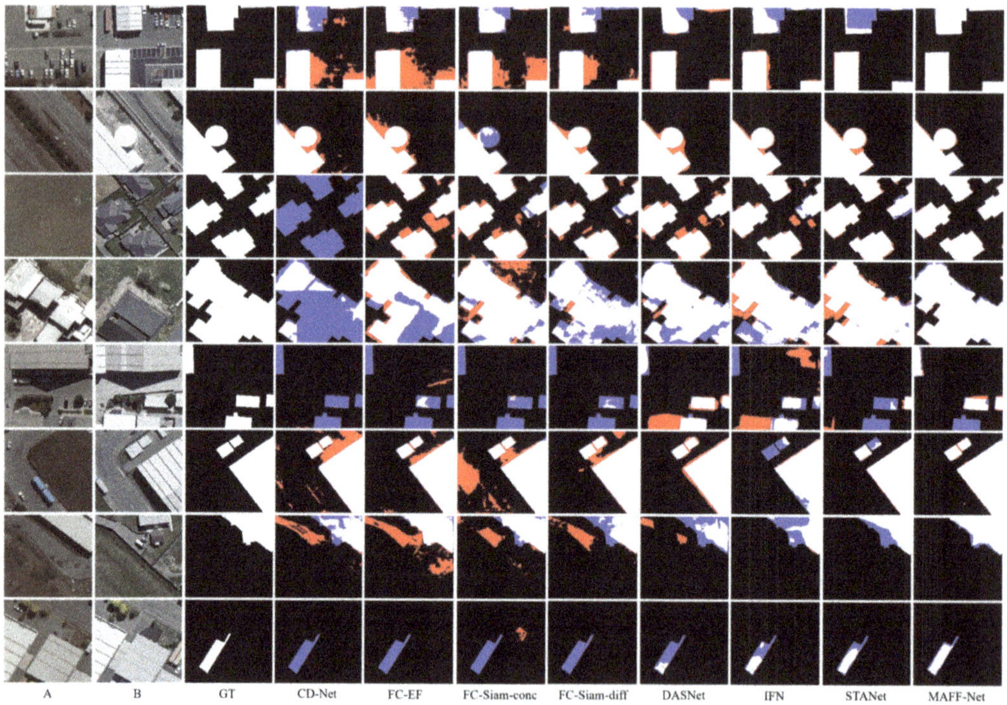

Figure 10. Illustration of a qualitative comparison on dataset WHU-CD. (**a**–**h**) indicate samples from WHU-CD and the change maps obtained with different methods. The white color indicates the changes that were correctly detected. Black indicates that no changes have been correctly detected. Red indicates false alarms. Blue indicates unpredicted changes.

Based on the attention mechanism, which can further focus on the information exchange between feature maps, DASNet works better than FC-EF. IFN pays more attention to the connection and interaction of multi-scale information. It introduces channel attention and spatial attention and uses a post-fusion strategy for deep supervision. Its *F*1 and *Kappa* scores reached 90.1% and 89.2%, respectively. STANet proposes a spatial-temporal attention module based on a feature pyramid to better adapt the network to the detection task of complex scenes, ranking second in all evaluation metrics. The proposed MAFF-Net achieves the highest level in all metrics, respectively. It is able to detect and label the change regions better than other methods because the network employs an attention-based cross-layer feature fusion strategy and also designs a refinement residual block to further improve the network detection performance.

Also, the qualitative analysis in Figure 8 allows for further analysis of the performance of each network. For visual analysis, eight challenging sets of bi-temporal images were selected and visualized. Each set of images contains different ranges of change regions or change scenes. Among the three FCN-based baselines, FC-Siam-conc and FC-Siam-diff can give better results than FC-EF. As can be seen in Figure 8, only a small number of change regions (Figure 8a) can be marked by FC-EF, but it performs poorly for smaller change regions and more complex scenes (Figure 8b–h). This is because it does not preserve the features of each original image, especially the shallow features, which makes the detected change regions significantly inaccurate. In general, the other two baselines perform better than FC-EF, as evidenced by the completeness of the information in the regions of change detected in the illustrations. However, they still suffer from many missed and false detections, such as Figure 8b–g. In particular, in Figure 8e, they do not detect the change region at all. Therefore, there is still potential for improvement. By introducing dual attention in the decoding stage, DASNet can detect most of the change regions. However, its detection performance for small change regions needs to be improved. For example, in Figure 8e, there are many missed regions in its detection results, and there are also false detection regions. This demonstrates that it is not yet quite accurate in terms of the boundaries and details of the change regions. In addition, it also does not perform well in Figure 8b,f,h with false detections and missed detections.

IFN and STANet are relatively more complete in terms of local detail because of the introduction of channels and spatial attention. However, they still have false positives and false negatives in detecting some very small target regions or edges, as shown in the red and blue regions in Figure 8c,e,g,h. The processing of some regions is too smoothed, and some edge information is ignored to some extent. The proposed MAFF-Net can better label the change regions and accurately detect the edges of the change regions. It can be seen from the exhibited samples that there are very few red and blue regions representing false and missed detections. In particular, the detection performance is well for small and complex change regions, as shown in Figure 8e–h, for example. This also demonstrates that the proposed network can detect the change regions accurately in general.

3.3.3. LEVIR-CD Dataset

As can be seen from Table 3, the difference in performance between the three baselines of FCN is not significant, where the higher score among the three indicators is FC-Siam-Diff, with *F*1 and *Kappa* scores of 83.7% and 82.8%, respectively. DASNet, by introducing dual attention, improved the *F*1-score by about 0.9% compared to the three baselines of FCN. The *F*1 score of IFN reached third place with 86.2%, while the scores of *Kappa* and *OA* also performed well. However, the scores of all metrics are lower than those of STANet, which may be because STANet pays more attention to multi-scale information while introducing attention. By introducing an attention mechanism involving multiple scales, the proposed MAFF-Net improves the *F*1 score to 89.7%, which is better than other comparative methods. Moreover, *Kappa* and *OA* reached the highest values among the compared methods with 89.1% and 98.7%, respectively.

Figure 9 also illustrates the change maps on eight selected sets of bi-temporal images. The change regions in these images cover multiple scenes, areas, shapes, and distribution ranges. For multiple regularly shaped building changes in Figure 9a,b, the overall contours of the buildings are correctly detected. However, the detection results of the CD-Net and FC-EF methods still have obvious false detection and missed detection areas. Although STANet can locate the change region, the detection of more complex and small change regions is not entirely correct. For example, as shown in Figure 9f,h, the proposed MAFF-Net is more accurate than the other methods, as seen from the fewer regions marked in red and blue. For Figure 9a,b,d, the attention-based methods DASNet and STANet and the proposed attention-based guided cross-layer feature fusion network MAFF-Net are visually closer to the GT. For the more densely distributed change regions in Figure 9c, DASNet, STANet, and MAFF-Net maintain visual correctness, while MAFF-Net has fewer errors and can accurately detect and distinguish multiple dense change regions. However, for Figure 9f–h with more complex edges and smaller change regions, IFN, DASNet, and STANet do not perform well. On the contrary, MAFF-Net shows better adaptability, and it can accurately detect changing regions with complex shapes and small objects.

3.3.4. WHU-CD Dataset

According to the data in Table 4, the performance of the methods with FCN as the baseline does not differ much. The double attention-based DASNet performs slightly better than IFN and STANet, with scores of 90.7%, 90.1%, and 99.0% for *F*1, *Kappa*, and *OA*, respectively. We attribute this to the fact that the weighted double-margin contrastive loss (WDMC) used by DASNet can solve the problem of sample imbalance. The proposed MAFF-Net achieved the best scores in all evaluation metrics compared to the other comparison methods. Compared with the method using FCN as the baseline, the proposed method obtained a 9.1%, 9.6%, and 1.0% increase in *F*1, *Kappa*, and *OA*, respectively. This also demonstrates the effectiveness of the proposed multi-attention-guided feature fusion-based method. Compared to DASNet, IFN, and STANet, the proposed method improves the gains for *F*1, *Kappa*, and *OA* by 1.7%, 2.0%, and 0.4%, respectively. Such gains are generated thanks to our fusion strategy that fully considers multi-scale features, while effectively exploiting the advantage of the attention to greatly improve the network performance.

Table 4. Comparison of WHU-CD dataset results. The best scores are highlighted in bold.

Method	*F*1 (%)	*Kappa* (%)	*OA* (%)
CDNet	80.4	79.4	98.0
FC-EF	82.3	81.4	98.2
FC-Siam-conc	82.9	82.0	98.2
FC-Siam-diff	83.3	82.5	98.4
DASNet	90.7	90.1	99.0
IFN	88.1	87.5	98.9
STANet	89.8	89.3	99.0
MAFF-Net	**92.4**	**92.1**	**99.4**

For visual comparison, Figure 10 shows some typical CD results for the test samples in the WHU-CD dataset. As shown in Figure 10a,c–e,h, there are many missed detections and false detections in the compared methods. As shown in Figure 10c,e,h, CD-Net not only has false detections but also has many missed detection regions. The performance of the FCN-based FC-EF, FC-Siam-conc, and FC-Siam-diff have been improved and the missed detection regions are significantly reduced. However, they still have the same problems as CD-Net as shown in Figure 10d,e,h. In Figure 10e,h, the attention-based DASNet, IFN, and STANet do not perform well, with significant missed detection regions and some false detection regions. In terms of consistency with the GT, the proposed MAFF-Net achieves the best visual performance. Specifically, as shown in the samples in Figure 10, MAFF-Net significantly reduces the missed detections and has a very low false detection rate compared with other methods. In addition, the change maps generated by MAFF-Net have clearer and more accurate boundaries compared with other methods.

3.4. Ablation Study

In the CD task, our proposed model achieves superior performance. To validate the effectiveness and feasibility of our proposed method, we conducted a series of ablation experiments on three datasets, CDD, LEVIR-CD, and WHU-CD, to verify that our model has advanced performance. We conducted five ablation experiments on three HR datasets, and in our experiments, the Baseline represents the ResNet18 network structure. In total, five ablation experiments were conducted in this paper: Baseline, Baseline+FEM, Baseline+FEM+FFM_S1, Baseline+FEM+FFM_S1+RRB, and the MAFF-Net (Baseline+FEM+FFM_S1+RRB+FFM_S2). As shown in Figure 11, the Baseline does not achieve good performance in detecting change regions, especially when the change region scene is more complex or the change region area is small (Figure 11d). Compared with the Baseline, the Baseline + FEM method obtains richer features after adding the FEM, which can help the network detect most of the change regions. It can be seen that the Baseline+FEM+FFM_S1 can effectively remove some irrelevant information (Figure 11f), while further capturing the change features and refining the feature representation. The FFM_S1 module adopts a cross-layer fusion strategy, which helps the model to fully fuse the features of high and low layers to achieve better feature representation. Compared with the Baseline+FEM method, the Baseline+FEM+FFM_S1 method detects more accurate and complete change regions. However, it can also be found that the method is slightly lacking when faced with small change regions or poorly characterized features (Figure 11f–1). Therefore, the Baseline+FEM+FFM_S1+RRB method aims to further refine the feature representation, which helps to detect smaller change features and improve the network performance. As can be seen by Figure 11g, the change map obtained by this method is already very close to the change region of the GT. Finally, the method proposed in this paper performs feature fusion feature maps to obtain a prediction map that is closest to the real change regions. As can be seen from Figure 11h, the change map obtained by the proposed method is very close to the GT, which also surfaces the effectiveness of the proposed method. Meanwhile, the proposed method shows good accuracy on three different datasets. By comparing the visualization results of each module, the effectiveness and accuracy of the MAFF-Net method proposed in this paper are effectively demonstrated.

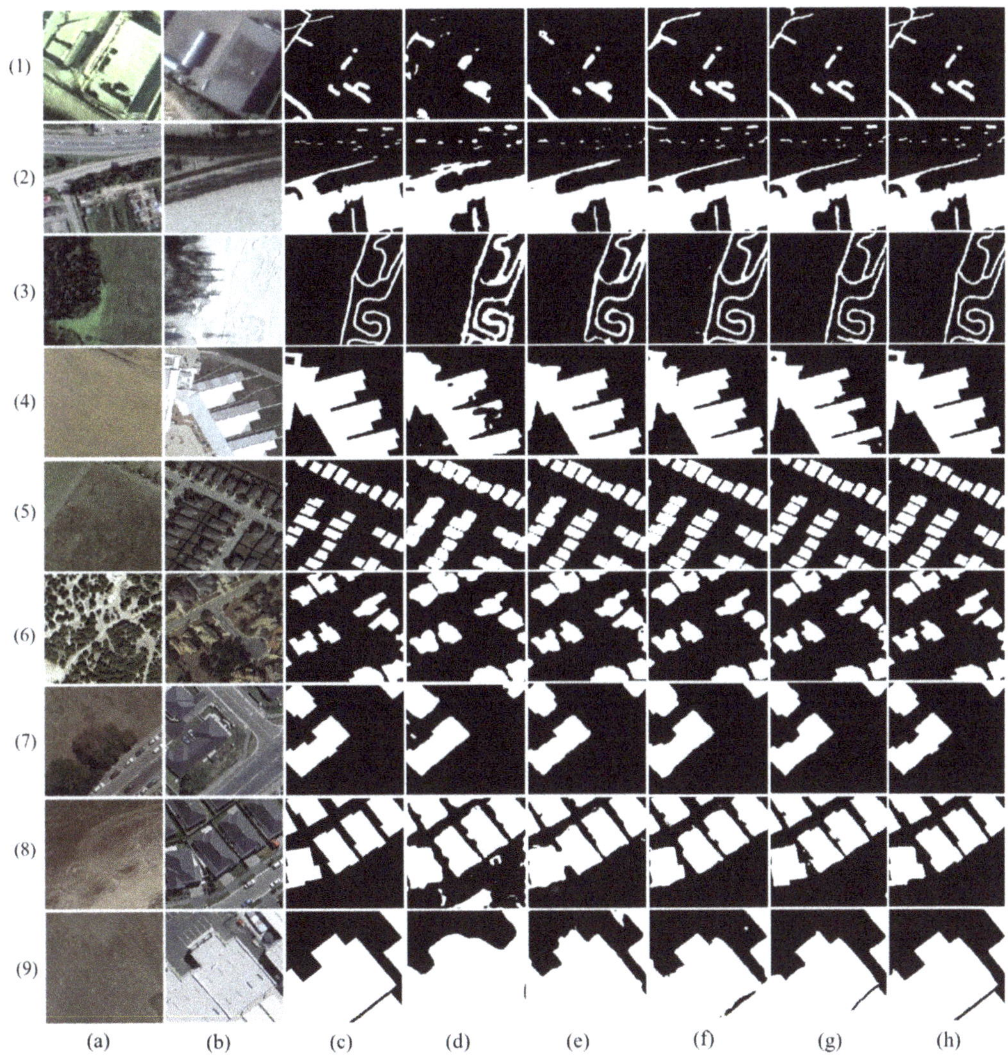

Figure 11. Visualization comparison plots of each network on different datasets in the ablation experiment. (**1–3**) indicate samples from the CDD dataset, (**4–6**) indicate samples from the LEVIR-CD dataset, and (**7–9**) indicate samples from the WHU-CD dataset. (**a**) Image T1. (**b**) Image T2. (**c**) Ground truth. (**d**) Baseline. (**e**) Baseline+FEM. (**f**) Baseline+FEM+FFM_S1. (**g**) Baseline+FEM+FFM_S1+RRB. (**h**) MAFF-Net (Baseline+FEM+FFM_S1+RRB+FFM_S2).

In addition, we also performed statistics and comparisons on the *F*1, *Kappa*, and *OA* values of different methods. As shown in Table 5, the model achieves optimal performance when all innovation modules are added, which also proves the effectiveness of our proposed innovation modules.

Table 5. Ablation study of different modules on different datasets. All the scores are described in percentage (%). The best scores are highlighted in bold.

Model					CDD			LEVIR-CD			WHU-CD		
Baseline	**FEM**	**FFM_S1**	**RRB**	**FFM_S2**	*F*1	*Kappa*	*OA*	*F*1	*Kappa*	*OA*	**F1**	*Kappa*	*OA*
√	×	×	×	×	88.0	86.3	96.9	83.3	82.4	98.2	86.0	85.3	98.8
√	√	×	×	×	93.6	92.7	98.4	87.0	86.3	98.6	89.9	89.3	99.0
√	√	√	×	×	94.6	93.8	98.7	88.2	87.6	98.7	91.4	90.9	99.1
√	√	√	√	×	95.9	95.4	99.0	88.8	88.2	**98.8**	91.9	91.5	99.2
√	√	√	√	√	**96.5**	**96.0**	**99.2**	**89.7**	**89.1**	98.7	**92.4**	**92.1**	**99.4**

In the Baseline+FEM method, as can be seen, there is a significant improvement in three indicators compared with the Baseline method. In the CDD dataset, *Kappa*, *F*1, and *OA* increased by 6.4%, 5.6%, and 1.5% compared with the Baseline, respectively. In the LEVIR-CD dataset, *Kappa*, *F*1, and *OA* were increased by 3.9%, 3.7%, and 0.4%, respectively, compared with the Baseline. In the WHU-CD dataset, *Kappa*, *F*1, and *OA* were increased by 4%, 3.9%, and 0.2%, respectively, compared with the Baseline.

In the Baseline+FEM+FFM_S1 method, it can be seen that all metrics are improved compared to the baseline+FEM method. In the CDD dataset, *Kappa*, *F*1, and *OA* improve by 1.1%, 1%, and 0.3%, respectively, compared to the Baseline. In the LEVIR-CD dataset, *Kappa*, *F*1, and *OA* improved by 1.3%, 1.2%, and 0.1%, respectively, compared to the Baseline. In the WHU-CD dataset, *Kappa*, *F*1, and *OA* improved by 1.6%, 1.5%, and 0.1%, respectively, compared to the Baseline. We can see the improvement of all metrics on all datasets, indicating the innovation and validity of our proposed FFM_S1, while the joint use of FFM_S1 and FEM achieves better performance and makes the model more accurate.

In the Baseline+FEM+FFM_S1+RRB method, it can be seen that there are improvements in all metrics compared with the Baseline+FEM+FFM_S1 method. In the CDD dataset, *Kappa*, *F*1, and *OA* improve by 1.6%, 1.3%, and 0.3%, respectively, compared to the Baseline. In the LEVIR-CD dataset, *Kappa*, *F*1, and *OA* improved by 0.6%, 0.6%, and 0.1%, respectively, compared to the Baseline. In the WHU-CD dataset, *Kappa*, *F*1, and *OA* improved by 0.6%, 0.5%, and 0.1%, respectively, compared to the Baseline+FEM+FFM_S1. We can see the improvement in all metrics on all datasets, indicating that our proposed RRB enhances the feature representation of the feature map, while the combined use of FFM_S1, FEM, and RRB leads to better performance of the model.

In the Baseline+FEM+FFM_S1+RRB+FFM_S2 method, it can be seen that all metrics are improved compared to the baseline+FEM+FFM_S1+RRB approach. In the CDD dataset, *Kappa*, *F*1, and *OA* improved by 0.6%, 0.6%, and 0.2%, respectively, compared to the Baseline. In the LEVIR-CD dataset, *Kappa* and *F*1 improved by 0.9% and 0.9%, respectively, compared to the Baseline. In the WHU-CD dataset, *Kappa*, *F*1, and *OA* improved by 0.6%, 0.5%, and 0.1%, respectively, compared to the Baseline+FEM+ FFM_S1+RRB. We can see the improvement of all the metrics on all datasets, indicating our proposal that FFM_S2 has a facilitating effect in fusing multi-scale feature information exchanges, while FFM_S1 and FFM_S2 have a mutual facilitating effect in feature extraction, and also, it is known experimentally that MAFF-Net helps the network fuse multi-scale features to achieve multi-scale information communication, which can improve the performance of the network.

3.5. Efficiency Analysis of the Proposed Network

Although the proposed network MAFF-Net achieves encouraging performance, it has some potential limitations. The computational complexity of MAFF-Net is relatively high and the number of parameters is large. This is not friendly to devices and applications with limited resources. In this section, the parameter amount (take M as the unit) and the training time of an epoch (take min/epoch as the unit) are used as quantitative indicators for evaluation. As shown in Figure 12, the number of trainable parameters of MAFF-Net is 49.08 million, which is the largest among the compared methods. However, from another perspective, the training efficiency of the proposed MAFF-Net is also relatively impressive. Compared with STANet and DASNet, the training time of the proposed method is reduced by 56.22% and 40.86%, respectively, which makes the proposed method more valuable in practical applications under the same equipment conditions.

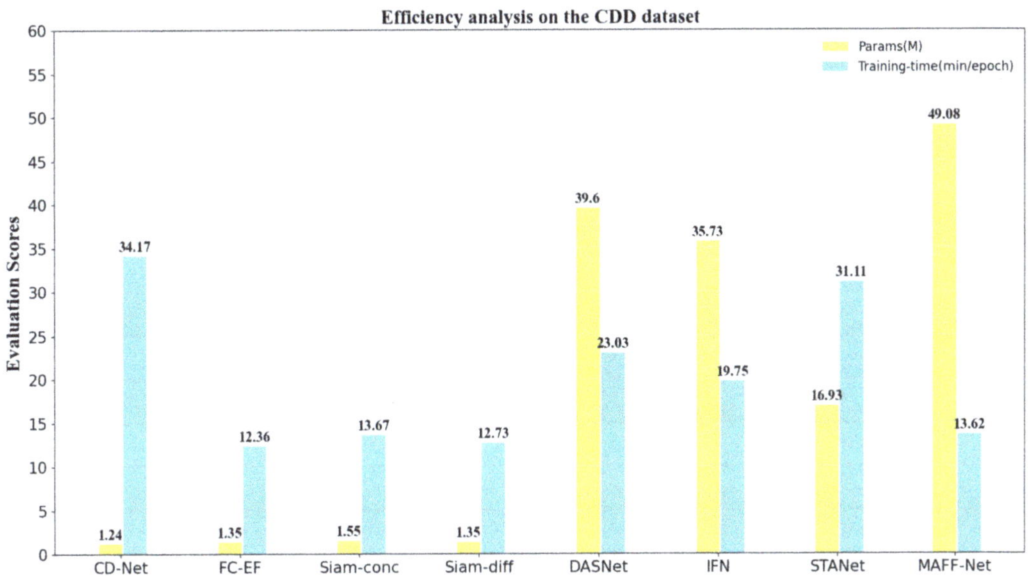

Figure 12. Illustration of an efficiency analysis of the comparison methods.

Though the number of training parameters and training time is comprehensive, the proposed method has space for improvement and enhancement in the future. For example, model compression can be performed in the proposed network, employing pruning and knowledge distillation [63,64] to reduce the size of the model.

4. Conclusions

In this paper, we propose a novel feature fusion network for remote sensing image CD tasks. To enhance the feature representation, we propose a Feature Enhancement Module (FEM), which introduces coordination attention (CA) that can capture long-range dependencies with precise location information while modeling inter-channel relationships. The FEM helps the network to further refine the features extracted by the backbone network ResNet18. The quantitative and qualitative analysis of the ablation study shows that the performance of the FEM on the Baseline is improved, which demonstrates the reasonability and effectiveness of the FEM. Considering that layer-by-layer feature fusion may lose part of the semantic information, we propose an FFM employing a cross-layer feature fusion strategy. The FFM uses semantic cues in the high-level feature map to guide feature selection in the low-level feature map. In addition, to highlight changing regions and suppress useless features, we introduce a CBAM in the FFM, which combines the advantages of channel attention and spatial attention, allowing the model to learn which region to focus on and pay more attention to critical information. Depending on the input features, we classified FFM into FFM_S1 and FFM_S2, both of which further enhance the feature fusion effect. Based on the ablation study in Section 3, we can see that the FFM significantly improves the performance of the network. To compensate for the shortcomings of using a single convolutional kernel for feature refinement, we propose a Refinement Residual Block (RRB) that employs a residual structure. The RRB changes the number of channels of the aggregated features and uses convolutional blocks to further refine the feature representation. Based on the comparison results between the proposed MAFF-Net and other methods in quantitative and qualitative analysis, the proposed method is able to efficiently detect changing regions and has a strong ability to select features through a feature fusion strategy guided by multiple attention mechanisms. On the three publicly available benchmark datasets CDD, LEVIR-CD, and WHU-CD, the $F1$ scores of MAFF-Net are improved by at least 1%, 2%, and 3%, respectively, compared to other methods. This demonstrates the better performance of our method than other SOTA methods.

However, it should be noted that, as shown in Figure 12, although the proposed model has an advantage in terms of training speed, it cannot be ignored that the number of parameters of the proposed model is relatively large, reaching 49.08 M. This has potential limitations for its practical application in the future. Therefore, in future work, we hope that the network can be made lightweight by using some model compression techniques. In addition, the proposed method solves the CD task of bi-temporal remote sensing images, and in future work, it will focus on the CD task of multi-temporal remote sensing images.

Author Contributions: Conceptualization, J.M.; methodology, J.M.; software, Y.L.; validation, J.M. and Z.Z.; formal analysis, G.S.; investigation, G.S.; resource, Y.L.; data curation, J.M.; writing-original draft preparation, J.M.; writing-review and editing, G.S., Z.Z. and Y.L. All authors have read and agreed to the published version of the manuscript.

Funding: This work was supported by the Natural Science Foundation of China No.62162059 and the National Key R & D plan project under Grant No.2018YFC0825504.

Institutional Review Board Statement: Not applicable.

Informed Consent Statement: Not applicable.

Data Availability Statement: The CDD, LEVIR-CD, WHU-CD datasets are openly available at https://drive.google.com/fifile/d/1GX656JqqOyBi_Ef0w65kDGVto-nHrNs9 (accessed on 1 December 2021), https://justchenhao.github.io/LEVIR/ (accessed on 1 December 2021), http://gpcv.whu.edu.cn/data/building_dataset.html (accessed on 1 December 2021), respectively.

Conflicts of Interest: The authors declare no conflict of interest.

Abbreviations

The following abbreviations are used in this manuscript:

CD	change detection
HR	high resolution
SOTA	state of the art
VHR	very high resolution
CNN	convolutional neural network
FCN	fully convolutional network
CA	coordinate attention
RRB	refinement residual block
FEM	feature enhancement module
FFM	feature fusion module
GT	ground truth

OA	overall accuracy
CD-Net	change detection network
FC-EF	fully convolutional early fusion
FC-Siam-conc	fully convolutional Siamese concatenation
FC-Siam-diff	fully convolutional Siamese difference
DASNet	dual attentive fully convolutional siamese networks
IFN	image fusion network
STANet	a spatial-temporal attention-based method
MAFF-Net	multi-attention guided feature fusion network

References

1. Singh, A. Review article digital change detection techniques using remotely-sensed data. *Int. J. Remote Sens.* **1989**, *10*, 989–1003. [CrossRef]
2. Radke, R.J.; Andra, S.; Al-Kofahi, O.; Roysam, B. Image change detection algorithms: A systematic survey. *IEEE Trans. Image Process.* **2005**, *14*, 294–307. [CrossRef]
3. Tison, C.; Nicolas, J.M.; Tupin, F.; Maître, H. A new statistical model for Markovian classification of urban areas in high-resolution SAR images. *IEEE Trans. Geosci. Remote Sens.* **2004**, *42*, 2046–2057. [CrossRef]
4. Papadomanolaki, M.; Vakalopoulou, M.; Karantzalos, K. A Deep Multitask Learning Framework Coupling Semantic Segmentation and Fully Convolutional LSTM Networks for Urban Change Detection. *IEEE Trans. Geosci. Remote Sens.* **2021**, *59*, 7651–7668. [CrossRef]
5. Yang, J.; Weisberg, P.J.; Bristow, N.A. Landsat remote sensing approaches for monitoring long-term tree cover dynamics in semi-arid woodlands: Comparison of vegetation indices and spectral mixture analysis. *Remote Sens. Environ.* **2012**, *119*, 62–71. [CrossRef]
6. Isaienkov, K.; Yushchuk, M.; Khramtsov, V.; Seliverstov, O. Deep Learning for Regular Change Detection in Ukrainian Forest Ecosystem With Sentinel-2. *IEEE J. Sel. Top. Appl. Earth Observ. Remote Sens.* **2021**, *14*, 364–376. [CrossRef]
7. Khan, S.H.; He, X.; Porikli, F.; Bennamoun, M. Forest Change Detection in Incomplete Satellite Images with Deep Neural Networks. *IEEE Trans. Geosci. Remote Sens.* **2017**, *55*, 5407–5423. [CrossRef]
8. Sublime, J.; Kalinicheva, E. Automatic post-disaster damage mapping using deep-learning techniques for change detection: Case study of the Tohoku tsunami. *Remote Sens.* **2019**, *11*, 1123. [CrossRef]
9. Yang, X.; Hu, L.; Zhang, Y.; Li, Y. MRA-SNet: Siamese Networks of Multiscale Residual and Attention for Change Detection in High-Resolution Remote Sensing Images. *Remote Sens.* **2021**, *13*, 4528. [CrossRef]
10. Hussain, M.; Chen, D.; Cheng, A.; Wei, H.; Stanley, D. Change detection from remotely sensed images: From pixel-based to object-based approaches. *ISPRS-J. Photogramm. Remote Sens.* **2013**, *80*, 91–106. [CrossRef]
11. Wang, L.; Li, H. Soft-change detection in optical satellite images. *IEEE Trans. Geosci. Remote Sens. Lett.* **2011**, *8*, 879–883.
12. Quarmby, N.A.; Cushnie, J.L. Monitoring urban land cover changes at the urban fringe from SPOT HRV imagery in south-east England. *Int. J. Remote Sens.* **1989**, *10*, 953–963. [CrossRef]
13. Howarth, P.J.; Wickwareg, M. Procedures for change detection using Landsat digital data. *Int. J. Remote Sens.* **1981**, *2*, 277–291. [CrossRef]
14. Ludeke, A.K.; Maggio, R.C.; Reid, L.M. An analysis of anthropogcnic deforcstation usinglogistic regression and GIS. *J. Environ. Manag.* **1990**, *31*, 247–259. [CrossRef]
15. Zhang, J.; Wang, R. Multi-temporal remote sensing change detection based on independent component analysis. *Int. J. Remote Sens.* **2006**, *27*, 2055–2061. [CrossRef]
16. Nielsen, A.A.; Conradsen, K.; Simpson, J.J. Multivariate alteration detection (MAD) and MAF postprocessing in multispectral, bitemporal image data: New approaches to change detection studies. *Remote Sens. Environ.* **1998**, *64*, 1–19. [CrossRef]
17. Nielsen, A.A. The Regularized Iteratively Reweighted MAD Method for Change Detection in Multi- and Hyperspectral Data. *IEEE Trans. Image Process.* **2007**, *16*, 463–478. [CrossRef]
18. Bovolo, F.; Bruzzone, L. A theoretical framework for unsupervised change detection based on change vector analysis in the polar domain. *IEEE Trans. Geosci. Remote Sens.* **2007**, *45*, 218–236. [CrossRef]
19. Bovolo, F.; Marchesi, S.; Member, S. A framework for automatic and unsupervised detection of multiple changes in multitemporal images. *IEEE Trans. Geosci. Remote Sens.* **2012**, *50*, 2196–2212. [CrossRef]
20. Liu, S.; Bruzzone, L.; Bovolo, F.; Du, P. Hierarchical unsupervised change detection in multitemporal hyperspectral images. *IEEE Trans. Geosci. Remote Sens.* **2015**, *53*, 244–260.
21. Liu, S.; Bruzzone, L.; Bovolo, F.; Zanetti, M.; Du, P. Sequential spectral change vector analysis for iteratively discovering and detecting multiple changes in hyperspectral images. *IEEE Trans. Geosci. Remote Sens.* **2015**, *53*, 4363–4378. [CrossRef]
22. Thonfeld, F.; Feilhauer, H.; Braun, M.; Menz, G. Robust change vector analysis (RCVA) for multi-sensor very high resolution optical satellite data. *Int. J. Appl. Earth Obs. Geoinf.* **2016**, *50*, 131–140. [CrossRef]
23. Blaschke, T.; Hay, G.J.; Kelly, M.; Lang, S.; Hofmann, P.; Addink, E.; Feitosa, R.; Meer, F.; Werff, H.; Coillie, F.; et al. Geographic object-based image analysis–Towards a new paradigm. *ISPRS-J. Photogramm. Remote Sens.* **2014**, *87*, 180–191. [CrossRef]

24. Ma, L.; Li, M.; Blaschke, T.; Ma, X.; Tiede, D.; Cheng, L.; Chen, D. Object-based change detection in urban areas: The effects of segmentation strategy, scale, and feature space on unsupervised methods. *Remote Sens.* **2016**, *8*, 761. [CrossRef]
25. Zhang, Y.; Peng, D.; Huang, X. Object-based change detection for VHR images based on multiscale uncertainty analysis. *IEEE Geosci. Remote Sens. Lett.* **2017**, *15*, 13–17. [CrossRef]
26. Zhang, C.; Li, G.; Cui, W. High-resolution remote sensing image change detection by statistical-object-based method. *IEEE J. Sel. Top. Appl. Earth Observ. Remote Sens.* **2018**, *11*, 2440–2447. [CrossRef]
27. Gil-Yepes, J.L.; Ruiz, L.A.; Recio, J.A.; Balaguer-Beser, Á.; Hermosilla, T. Description and validation of a new set of object-based temporal geostatistical features for land-use/land-cover change detection. *ISPRS J. Photogramm. Remote Sens.* **2016**, *121*, 77–91. [CrossRef]
28. Qin, Y.; Niu, Z.; Chen, F.; Li, B.; Ban, Y. Object-based land cover change detection for cross-sensor images. *Int. J. Remote Sens.* **2013**, *34*, 6723–6737. [CrossRef]
29. Tang, D.; Wei, F.; Yang, N.; Zhou, M.; Liu, T.; Qin, B. Learning sentiment-specific word embedding for twitter sentiment classification. In Proceedings of the 52nd Annual Meeting of the Association for Computational Linguistics, Baltimore, MD, USA, 23–25 June 2014; pp. 1555–1565.
30. Kim, Y.; Jernite, Y.; Sontag, D.A.; Rush, A.M. Character-aware neural language models. In Proceedings of the Thirtieth AAAI Conference on Artifcial Intelligence, Phoenix, AZ, USA, 12–17 February 2016; pp. 2741–2749.
31. Lei, T.; Zhang, Q.; Xue, D.; Chen, T.; Meng, H.; Nandi, A.K. End-to-end Change Detection Using a Symmetric Fully Convolutional Network for Landslide Mapping. In Proceedings of the IEEE International Geoscience and Remote Sensing Symposium (IGARSS), Brighton, UK, 12–17 May 2019; pp. 3027–3031.
32. Li, X.; Yuan, Z.; Wang, Q. Unsupervised Deep Noise Modeling for Hyperspectral Image Change Detection. *Remote Sens.* **2019**, *11*, 258. [CrossRef]
33. Xu, Q.; Chen, K.; Zhou, G.; Sun, X. Change Capsule Network for Optical Remote Sensing Image Change Detection. *Remote Sens.* **2021**, *13*, 2646. [CrossRef]
34. LeCun, Y.; Bottou, L.; Bengio, Y.; Haffner, P. Gradient-based learning applied to document recognition. *Proc. IEEE* **1998**, *86*, 2278–2324. [CrossRef]
35. He, K.; Zhang, X.; Ren, S.; Sun, J. Deep residual learning for image recognition. In Proceedings of the IEEE Conference on Computer Vision and Pattern Recognition (CVPR), Las Vegas, NV, USA, 27–30 June 2016; pp. 770–778.
36. Shelhamer, E.; Long, J.; Darrell, T. Fully Convolutional Networks for Semantic Segmentation. *IEEE Trans. Pattern Anal. Mach. Intell.* **2017**, *39*, 640–651. [CrossRef] [PubMed]
37. Ronneberger, O.; Fischer, P.; Brox, T. U-net: Convolutional networks for biomedical image segmentation. In Proceedings of the International Conference on Medical Image Computing and Computer-Assisted Intervention (MICCAI), Munich, Germany, 5–9 October 2015; pp. 234–241.
38. Caye Daudt, R.; Le Saux, B.; Boulch, A. Fully Convolutional Siamese Networks for Change Detection. In Proceedings of the 25th IEEE International Conference on Image Processing (ICIP), Athens, Greece, 7–10 October 2018; pp. 4063–4067.
39. Daudt, R.C.; Le Saux, B.; Boulch, A.; Gousseau, Y. High Resolution Semantic Change Detection. *arXiv* **2018**, arXiv:1810.08452v1.
40. Lei, T.; Zhang, Y.; Lv, Z.; Li, S.; Liu, S.; Nandi, A.K. Landslide Inventory Mapping from Bi-temporal Images Using Deep Convolutional Neural Networks. *IEEE Geosci. Remote Sens. Lett.* **2019**, *16*, 982–986. [CrossRef]
41. Zhang, Y.; Zhang, S.; Li, Y.; Zhang, Y. Coarse-to-Fine Satellite Images Change Detection Framework via Boundary-Aware Attentive Network. *Sensors* **2020**, *20*, 6735. [CrossRef]
42. Chen, H.; Shi, Z. A Spatial-Temporal Attention-Based Method and a New Dataset for Remote Sensing Image Change Detection. *Remote Sens.* **2020**, *12*, 1662. [CrossRef]
43. Chen, H.; Qi, Z.; Shi, Z. Remote Sensing Image Change Detection With Transformers. *IEEE Trans. Geosci. Remote Sens.* **2021**, *60*, 1–14. [CrossRef]
44. Chen, J.; Yuan, Z.; Peng, J.; Chen, L.; Huang, H.; Zhu, J.; Liu, Y.; Li, H. DASNet: Dual attentive fully convolutional siamese networks for change detection of high resolution satellite images. *IEEE J. Sel. Top. Appl. Earth Obs. Remote Sens.* **2020**, *14*, 1194–1206. [CrossRef]
45. Zhang, C.; Yue, P.; Tapete, D.; Jiang, L.; Shangguan, B.; Huang, L.; Liu, G. A deeply supervised image fusion network for change detection in high resolution bi-temporal remote sensing images. *ISPRS J. Photogramm. Remote Sens.* **2020**, *166*, 183–200. [CrossRef]
46. Hou, Q.; Zhou, D.; Feng, J. Coordinate Attention for Efficient Mobile Network Design. In Proceedings of the IEEE Conference on Computer Vision and Pattern Recognition (CVPR), Virtually, Nashville, TN, USA, 19–25 June 2021.
47. Zhang, Y.; Fu, L.; Li, Y.; Zhang, Y. HDFNet: Hierarchical Dynamic Fusion Network for Change Detection in Optical Aerial Images. *Remote Sens.* **2021**, *13*, 1440. [CrossRef]
48. Lin, M.; Chen, Q.; Yan, S. Network in network. In Proceedings of the International Conference on Learning Representations (ICLR), Banff, AB, Canada, 14–16 April 2014; pp. 1–10.
49. Szegedy, C.; Liu, W.; Jia, Y.; Sermanet, P.; Rabinovich, A. Going deeper with convolutions. In Proceedings of the IEEE Conference on Computer Vision and Pattern Recognition (CVPR), Boston, MA, USA, 8–10 June 2015; pp. 1–9.
50. Yang, L.; Chen, Y.; Song, S.; Li, F.; Huang, G. Deep Siamese Networks Based Change Detection with Remote Sensing Images. *Remote Sens.* **2021**, *13*, 3394. [CrossRef]

51. Wang, D.; Chen, X.; Jiang, M.; Du, S.; Xu, B.; Wang, J. ADS-Net:An Attention-Based deeply supervised network for remote sensing image change detection. *Int. J. Appl. Earth Obs. Geoinf.* **2021**, *101*, 102348. [CrossRef]
52. Zeiler, M.D.; Krishnan, D.; Taylor, G.W.; Fergus, R. Deconvolutional networks. In Proceedings of the IEEE Conference on Computer Vision and Pattern Recognition (CVPR), San Francisco, CA, USA, 13–18 June 2010; pp. 2528–2535.
53. Dumoulin, V.; Visin, F. A guide to convolution arithmetic for deep learning. *arXiv* **2016**, arXiv:1603.07285.
54. Augustus, O.; Vincent, D.; Chris, O. Deconvolution and Checkerboard Artifacts. *Distill* **2016**, *1*, e3.
55. Woo, S.; Park, J.; Lee, J.Y.; So Kweon, I. Cbam: Convolutional block attention module. In Proceedings of the European Conference on Computer Vision (ECCV), Munich, Germany, 8–14 September 2018; pp. 3–19.
56. Ioffe, S.; Szegedy, C. Batch Normalization: Accelerating Deep Network Training by Reducing Internal Covariate Shift. In Proceedings of the International Conference on Machine Learning (ICML), Lille, France, 6–11 July 2015; pp. 448–456.
57. Glorot, X.; Bordes, A.; Bengio, Y. Deep sparse rectifier neural networks. In Proceedings of the International Conference on Artificial Intelligence and Statistics (AISTATS), Fort Lauderdale, FL, USA, 11–13 April 2011; pp. 315–323.
58. Gulcehre, C.; Moczulski, M.; Denil, M.; Bengio, Y. Noisy activation functions. In Proceedings of the International Conference on Machine Learning (ICML), New York, NY, USA, 19–24 June 2016; pp. 3059–3068.
59. Yu, C.; Wang, J.; Peng, C.; Gao, C.; Yu, G.; Sang, N. Learning a discriminative feature network for semantic segmentation. In Proceedings of the IEEE Conference on Computer Vision and Pattern Recognition (CVPR), Salt Lake City, UT, USA, 18–22 June 2018; pp. 1857–1866.
60. Lebedev, M.; Vizilter, Y.V.; Vygolov, O.; Knyaz, V.; Rubis, A.Y. Change Detection in Remote Sensing Images Using Conditional Adversarial Networks. *Int. Arch. Photogram. Remote Sens. Spat. Inf. Sci.* **2018**, *42*, 565–571. [CrossRef]
61. Ji, S.; Wei, S.; Lu, M. Fully convolutional networks for multisource building extraction from an open aerial and satellite imagery data set. *IEEE Trans. Geosci. Remote Sens.* **2018**, *57*, 574–586. [CrossRef]
62. Alcantarilla, P.F.; Simon, S.; Germán, R.; Roberto, A.; Riccardo, G. Street-view change detection with deconvolutional networks. *Auton. Robot.* **2018**, *42*, 1301–1322. [CrossRef]
63. Li, H.; Kadav, A.; Durdanovic, I.; Samet, H.; Graf, H.P. Pruning filters for efficient convnets. *arXiv* **2016**, arXiv:1608.08710.
64. Vadera, M.P.; Marlin, B.M. Challenges and Opportunities in Approximate Bayesian Deep Learning for Intelligent IoT Systems. *arXiv* **2021**, arXiv:2112.01675.

sensors

MDPI

Article

Mapping and Discriminating Rural Settlements Using Gaofen-2 Images and a Fully Convolutional Network

Ziran Ye [1], Bo Si [1], Yue Lin [1], Qiming Zheng [1,2], Ran Zhou [1], Lu Huang [3,*] and Ke Wang [1,3]

1 Institute of Applied Remote Sensing and Information Technology, College of Environmental and Resource Sciences, Zhejiang University, Hangzhou 310058, China; smd_ye@zju.edu.cn (Z.Y.); bo_si@zju.edu.cn (B.S.); joyelin_2018@zju.edu.cn (Y.L.); qmzheng@zju.edu.cn (Q.Z.); 3160100350@zju.edu.cn (R.Z.); kwang@zju.edu.cn (K.W.)

2 Department of Biological Sciences, National University of Singapore, 14 Science Drive 4, Singapore 117543, Singapore

3 The Rural Development Academy, Zhejiang University, Hangzhou 310058, China

* Correspondence: haoyubailu@zju.edu.cn

Received: 21 September 2020; Accepted: 23 October 2020; Published: 25 October 2020

Abstract: New ongoing rural construction has resulted in an extensive mixture of new settlements with old ones in the rural areas of China. Understanding the spatial characteristic of these rural settlements is of crucial importance as it provides essential information for land management and decision-making. Despite a great advance in High Spatial Resolution (HSR) satellite images and deep learning techniques, it remains a challenging task for mapping rural settlements accurately because of their irregular morphology and distribution pattern. In this study, we proposed a novel framework to map rural settlements by leveraging the merits of Gaofen-2 HSR images and representation learning of deep learning. We combined a dilated residual convolutional network (Dilated-ResNet) and a multi-scale context subnetwork into an end-to-end architecture in order to learn high resolution feature representations from HSR images and to aggregate and refine the multi-scale features extracted by the aforementioned network. Our experiment in Tongxiang city showed that the proposed framework effectively mapped and discriminated rural settlements with an overall accuracy of 98% and Kappa coefficient of 85%, achieving comparable and improved performance compared to other existing methods. Our results bring tangible benefits to support other convolutional neural network (CNN)-based methods in accurate and timely rural settlement mapping, particularly when up-to-date ground truth is absent. The proposed method does not only offer an effective way to extract rural settlement from HSR images but open a new opportunity to obtain spatial-explicit understanding of rural settlements.

Keywords: rural settlements; fully convolutional network; multi-scale context; high spatial resolution images

1. Introduction

Since the reform and opening-up, drastic urbanization has been taking place in China. In a stark contrast, the development of rural areas, however, is not in concert with that of urban areas, but is greatly lagging behind and restricted. Mass population migration, from rural to urban areas, has given rise to a succession of impacts on rural areas, including population decline, industry recession and land abandonment [1,2]. In 2018, China stepped up its efforts to revitalize rural regions. Building the new style of rural community with better infrastructure is one of the important measures to improve the wellbeing of rural people. Thus, a spatial-explicit understanding of rural settlements regarding their distributions is of critical essence to effective land management and policy making.

Satellite-based earth observation is a key enabler for capturing spatial information of buildings in rural areas. High spatial resolution (HSR) images open new opportunities for slums and informal settlement detection and rural land cover mapping [3,4]. Compared with medium resolution image which mainly offers spectral information (in terms of a single image) [5], using HSR images can leverage both spectral and spatial information. HSR image analysis basically relies on image classification (e.g., pixel-based) and segmentation (e.g., Object-Based Image Analysis (OBIA)) techniques [6,7], with the help of handcrafted features extracted from spectral (e.g., reflectance and spectral indices, like Normalized Difference Vegetation Index (NDVI)) and spatial (texture statistics, morphological profiles, and oriented gradients) [8,9]. With an ever-increasing focus on rural areas, satellite images have been extensively used for rural settlement mapping [10,11]. Nevertheless, applying HSR images to rural settlement detection remains a challenging task due to the following issues. First, the size and spatial distribution of rural settlements varies significantly, e.g., clustered or scattered, because rural planning is changing over time. Second, the intra-class variation makes it difficult to distinguish rural settlements from construction materials when using spectral information alone. Third, when considering large spatial areas, the spectral and spatial responses from ground objects present an extremely complex pattern [8]. In order to discriminate rural settlements, more context information is required in the classification. In previous studies, such as [12,13], landscape metrics were used as the spatial contextual information to identify rural settlements from HSR satellite imagery. These methods exploit tailored segment-based features and have achieved acceptable performance. However, parameters optimization and handcrafted features selection are laborious tasks and are highly hinged upon expert experience, and trial-and-error tests.

Deep learning methods, such as convolutional neural networks (CNNs), have shown great potential for automatically features learning without human intervention. CNNs are able to generate robust feature representations hierarchically and have become increasingly popular in image classification and semantic segmentation [14]. Semantic segmentation for remote sensing data usually refers to extracting terrestrial objects from earth observation images using CNNs model, that is, each pixel is assigned a semantic label in pixel-based classification [15]. The fully convolutional network (FCN) [16] extends CNNs to segmentation, emerging as the preferred scheme for semantic labeling tasks. FCN inputs images of arbitrary sizes into a standard CNN, extract feature maps using layer-wise activation and abstraction, and then output high resolution predictions in an end-to-end fashion. The essential advantages of FCNs are the intrinsic ability to enhance feature representation and the flexibility to accept input images of any size. Previous studies have applied FCN and its variants to detect buildings and settlements [17–19]. It is further found that incorporating contextual relations in CNNs can improve classification accuracy [20,21]. Nevertheless, most of the above-mentioned approaches are designed to extract target objects in urban areas from the standard datasets [22]. In rural areas, built-up areas tend to be sparse and can be easily omitted [23]. Due to the significant differences in the appearance of urban and rural buildings, directly employing existing deep approaches to map rural settlements does not guarantee good performance. In addition, the difficulty in image interpretation increases sharply as the spatial resolution increases. Therefore, we wish to make use of the advantages of deep learning technique to contribute to the area of rural settlements identification in HSR images. By far, only a few studies applied FCNs to extract rural residential areas [24,25], and most of them were limited by the spatial resolution of images or the extent of application. The effectiveness of FCNs in rural settlement mapping using HSR images requires further in-depth examination. In short, it is imperative to develop an effective method to buttress automatic extraction of rural settlements using HSR images.

The overall objective of this paper is to develop a framework for automatically identifying rural settlements in HSR satellite images based on deep learning technique. Our main contributions are: (1) This application introduces a deep FCN method to recognize rural settlements. Specifically, dilated convolutions are used to extract deep features at high spatial resolution. (2) A multiple scale context subnetwork, which adopts a popular squeeze and excitation (SE) module [26] to aggregate multi-scale context, is exploited to generate discriminative representations. The proposed deep learning-based rural settlement extraction scheme can flexibly take multi-spectral HSR images as input to distinguish different types of rural settlements.

2. Study Area and Data

In this research, eleven towns of Tongxiang County were selected as study area, a typical rural region undergoing rapid rural development and transformation in the Yangtze River delta of China (120°30′13″E, 30°41′10″W). Tongxiang, located in the Hangjiahu plain, has a temperate climate with distinct seasonality. Since 2000, several land consolidation projects have been carried out to promote the construction of new countryside. Currently, the construction and renovation of countryside are still ongoing in Tongxiang, so the old scattered low-rise houses are mixed with uniformly planned residential buildings. Therefore, this area is an ideal study area to examine our proposed method. We preliminarily divide these settlements into two categories. Figure 1 shows examples of two types of rural settlements in the study area—low-density settlements and high-density settlements.

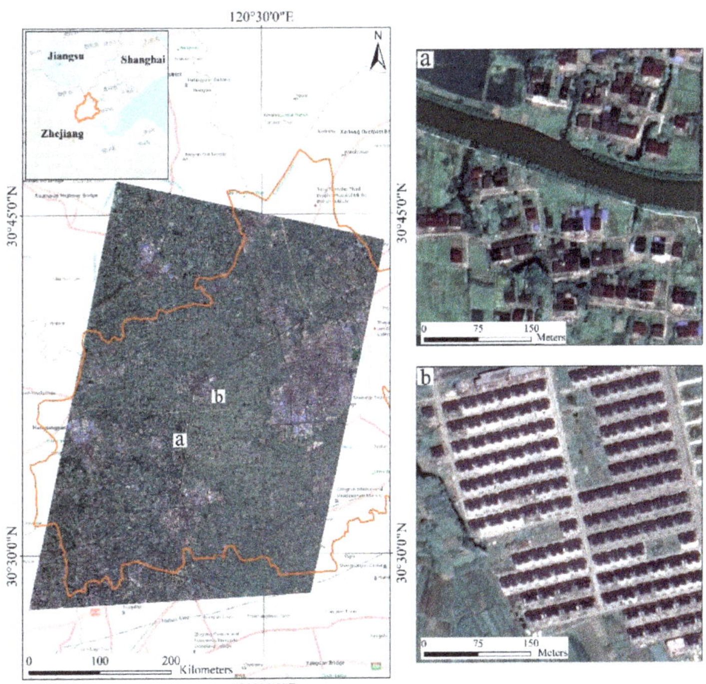

Figure 1. GaoFen-2 image of Tongxiang study area on July 2016. Example of (**a**) low-density rural settlement and (**b**) high-density rural settlement.

1. Low-density settlements (LDS): most of LDS are old-style rural settlements which are scattered and disorderly distributed and have different orientations. These low-density rural settlements are mainly located close to rivers and streams in support of farming and transportation of smallholders. The boundaries of low-density settlements are obscured by the surrounding vegetation.
2. High-density settlements (HDS): newly built residential areas where multi-story buildings accommodate several families. Such settlements have a higher building density than low-density settlements, and buildings inside these settlements have an identical spacing and the same surface. High-density settlements mainly distribute adjacent to the newly built transportation roads, providing easy access to nearby towns.

China's GaoFen-2 (GF-2) HSR images were used, comprising four multispectral bands (MSS) with a spatial resolution of 4 m and a panchromatic band (PAN) with a spatial resolution of 1 m. The acquisition time of two images was on July 2016. And we collected the land use data of the study area in 2015 (provided by the Bureau of Land and Resources, Tongxiang, China) to generate ground truth data.

3. Methods

Figure 2 demonstrates the flowchart of our proposed method. First, the GF-2 image was pre-processed and split into training set and test set. Second, the trained model was used to classify the rural settlements. Finally, accuracy assessment was conduct on the test set. The details of the proposed method are described in the following subsections.

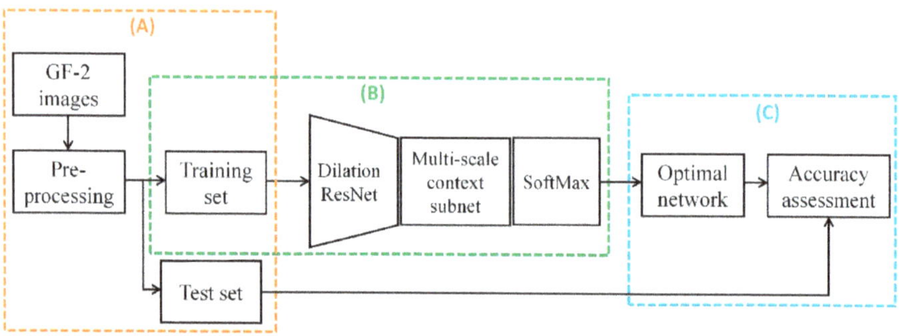

Figure 2. Flowchart of the proposed research framework: (**A**) generate data sets, (**B**) model training, and (**C**) accuracy assessment.

3.1. Data Preprocessing

For the cloud-free and haze-free atmospheric condition in the acquired image, there was no need for atmospheric correction in the preprocessing. After orthorectification, the MSS image and PAN images were then fused using the Gram–Schmidt pan-sharpening method [27]. The fusion image (1 m) had a dimension of 29,970 × 34,897 pixels, equivalent to about 700 km^2. The reference map was generated based on (1) the land-use change surveying map and (2) visual interpretation by local experts. Note that the ground truth data in our study was spatially sparse, which thereby was more in line with real-world scenarios, where densely annotated data is rarely available.

3.2. Rural Settlement Detection Using FCN

3.2.1. Dilated Residual Convolutional Network

The task of automatically extracting settlement information in a large rural region can be formulated as a semantic labeling problem to distinguish pixels of categories. In this section, we wish to put forward an end-to-end method based on semantic segmentation scheme to identify rural settlements. Our approach used ResNet50 architecture as the feature extractor of FCN-based method. The ResNet consists of five stages of convolutional layers. In the first stage, a convolution layer performs 7 × 7 convolution and is followed by a maxpooling operation, which outputs the features that are a quarter of the size of the original image. In the remaining four stages, each stage contains several blocks, which is a stack of two 3 × 3 convolutional layers. Moreover, two types of shortcut connection are introduced in the blocks to fuse input feature maps with output feature maps according to the size of input and output features. More details about ResNet50 can be found in [28].

As the network goes deeper, the resolution of feature maps becomes smaller while the channels increase. For example, the output feature maps of the last stage are reduced to 1/32 of the size of the original input. Compared with the complex background in HSR images, the objects of our interest (i.e., rural settlements) are smaller and sparser. Besides, the loss of spatial information due to the progressive down-sampling in the network is harmful for identifying small objects. To retain the large receptive field and increase spatial resolution in higher layers of network simultaneously, we adopt convolutions with dilated kernels into the ResNet. In the last two stages of original ResNet50, the strided convolution layer, which is used to reduce the output resolution at the beginning of each stage, is substituted by a convolution layer with the stride of 1 (meaning no downsampling). Recent studies [29,30] indicate this conversion does not affect the receptive field of the first layer of the stage, but it reduces the receptive field of subsequent layers by half. In order to increase the receptive field of those layers, convolutions with different dilation factors were adopted. Specifically, the dilation ratio of convolutional kernel was set as 2 and 5 in the fourth and the fifth stage, respectively. Dilated convolutions were thus expected to enlarge the receptive field of layers and to generate features with high spatial resolution. As a result, the output size would increase from 1/32 to 1/8 of the input image.

3.2.2. Multi-Scale Context Subnetwork

Some upgraded low-density houses and high-density buildings may have used similar roofing materials. In order to distinguish between the two categories of rural settlements, their spatial distribution and context need to be fully considered. Due to a great variety of the size of rural settlements, it is necessary to capture multiple scales information to identify objects in rural residential areas. Instead of using multiple rescaled versions of an image as input to obtain multi-scale context information, we introduced a multi-scale spatial context structure to handle the scale variation of rural residential objects. Commonly, the deep layers in CNNs respond to global context information and the shallow layers are more likely to be activated by local texture and patterns. Benefit from the dilation convolution maintaining spatial resolution, the three scale-levels features extracted by the backbone ResNet50 can be utilized at the same time. Our structure further enhanced the information propagation across layers. As shown in Figure 3, the output features of last three stages were filtered by 1×1 convolution layers to shrink the channel to 256 and then concatenated together. It is notable that we appended 3×3 convolution on the merged map to generate the subsequent feature map, which was to reduce the misalignment when fusing features of different levels. Secondly, a residual correction was introduced to alleviate the lack of information during feature fusion. Finally, feature selection was conducted by employing an advanced channel encoding module named "squeeze and excitation" block (SE block) [26], which adaptively recalibrated channel-wise feature responses. Once features were input into the module, global average pooling was used to generate a vector as channel-wise descriptors of the input features. Subsequently, two fully connected layers were applied to the vector to learn nonlinear interaction among channels. The sigmoid activation function would then generate a weight vector as a scale factor for the class-dependent features. The features refined by the above reweighting process had discriminative representations, which were helpful for object identification. Based on abundant positioning and identification information, the successive 3×3 convolution layer was expected to produce more accurate features. Finally, the refined deep features were then concatenated with the corresponding low-level features (e.g., Res-layer1 in ResNet50) in order to restore spatial details. After the fusion, we applied another 3×3 convolution layer and a simple bilinear upsampling to get the final segmentation. Table S1 shows the specific design of our segmentation network.

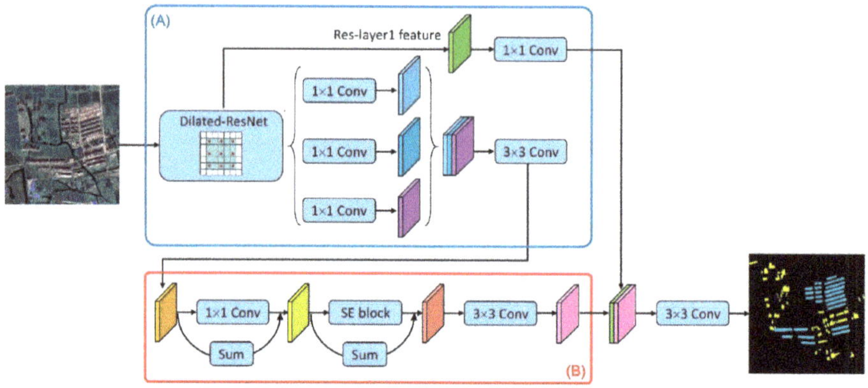

Figure 3. Overview of the proposed detection architecture. (**A**) the Dilated-ResNet extracted multi-level features with high spatial resolution; (**B**) the context subnetwork exploited the multi-scale context and mapped features to desired outputs.

3.2.3. Multi-Spectral Images-Based Transfer Learning

CNNs are generally data-driven approach and are usually trained on large datasets. In practice, a sufficiently large data set is rare. Instead, it is more practical to use a deep network previously trained on a big dataset (e.g., ImageNet) as an initial model or a feature extractor for the target task. This scheme is known as transfer learning [31]. In brief, the idea of transfer learning is to leverage knowledge from the source domain to boost learning in the target domain, as features of CNNs are more generic in early layers. Compared with training from scratch, the cost of fine-tuning the pre-trained network is much lower. Several attempts have been made to improve the learning task in remote sensing datasets by using transfer learning [32–34].

The ResNet50 is initially designed for RGB images [28]. To better adapt to multispectral remote sensing data which have the red (R), green (G), blue (B) and near-infrared (NIR) bands, the network was expanded to take advantage of more input bands than RGB. Different from the idea of using an additional convolution layer at the beginning of network [35] or adding a branch to accept multiband inputs [34], we directly modified the original 7 × 7 convolution layer in the first stage of ResNet to make it flexible to receive multispectral images and output 64 features.

3.3. *Method Implementation and Accuracy Assessment*

A total of 7605 tiles of a size of 256 × 256 pixels were cropped from the training area of the preprocessed GF-2 imagery, and we randomly selected about 20% of image patches as the validation set. Data augmentations consisting of flipping and rotation of 90 degrees were applied to enlarge the training set. The proposed network was trained on a 24 GB Nvidia P6000 GPU. The weights of network were initialized using the pre-trained ResNet50 model. We copied the weights of the first channel to initialize the newly added channel in the first convolution layer. An adaptive algorithm Adam [36] was employed as the optimizer, and the learning rate was set to 0.001. A batch size of 8 was used, running the optimizer for 30 epochs with an early stopping strategy which stopped training process when the monitored quantity (i.e., validation loss) had stopped improving for 5 epochs. The proposed method was implemented on the Pytorch framework.

Figure 4 shows the training area and the test area in the experiment. The point test samples were all over the entire study area except the training area. In order to further evaluate the area accuracy, we selected a small area in the test area as the polygon test subset, and rural settlements in the polygon test subset were densely labeled. The random point generating algorithm in ArcGIS [37] was applied to generate a total of 11,628 sample points. After that, we manually annotated these sample points based on higher resolution images of Google Earth and visual interpretation. In addition to the two types of settlements about which we were concerned, all other objects in the image were included in the background category. Table 1 lists the number of test set samples.

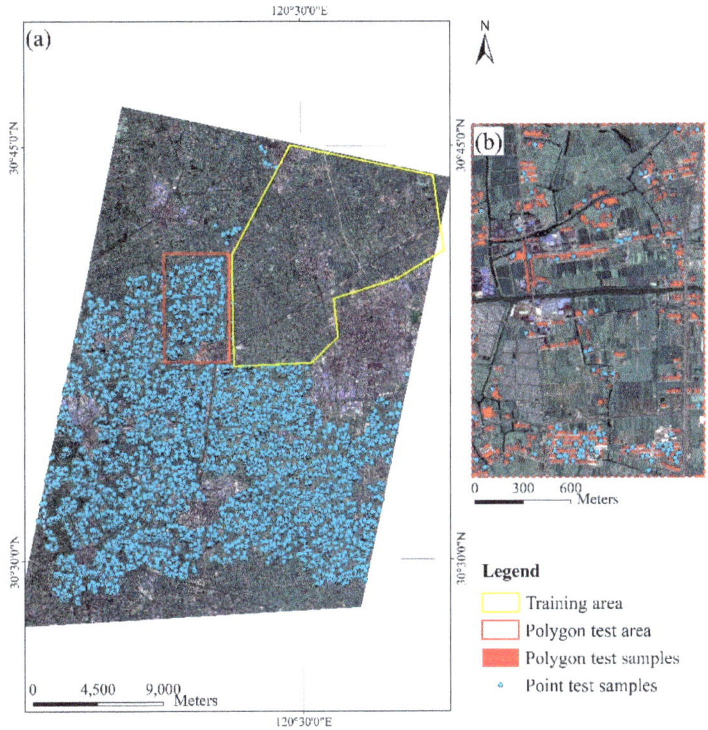

Figure 4. The (**a**) Tongxiang data set used in the experiments. (**b**) Example of test samples.

Table 1. The number of testing samples.

	LDS	HDS	Backgrounds	Sum
Point-based testing samples	6125	2616	2887	11,628
Polygon-based testing samples	1831	438	/	2269

Following the previous studies, the overall accuracy (OA), producer's accuracy (PA), user's accuracy (UA) and Kappa coefficient (Kappa) [38] were used to assess the performances of methods. The producer's accuracy represents the probability that pixels of a category are correctly classified, whereas the user's accuracy indicates the probability that the classified pixels belong to this category. Overall accuracy is the percentage of correctly classified pixels. The Kappa analysis is a discrete multivariate technique used in accuracy assessment to test whether one error matrix is significantly different from another [39], and Kappa coefficient calculated based on the individual error matrices can be regarded as another measure of accuracy.

4. Results and Discussions

4.1. Rural Settlements Identification

Figure 5 shows the resulting rural settlements of our study area. Tables 2 and 3 present confusion matrices on test sets. The proposed method achieved the OA of 98.31% with a Kappa coefficient of 0.9724 on the point test set, and the UA and PA of two settlements classes reached about 98%. The classification accuracy on polygon-based testing samples was different, the accuracies of low-density class (UA of 88.00% and PA of 84.97%) were higher than those of high-density class (UA of 85.22% and PA of 84.68%). In terms of overall classification, the Kappa coefficient of 0.8591 in the polygon-based testing method were lower than that of the point-based test set. This was because the polygon-based test method had strict requirements on the object boundary. Visual interpretation indicates that the proposed method can effectively distinguish rural residential areas from other man-made structures (white circle in Figure 5). It was observed that the footprints of HDS were more smoothed than the LDS, where the latter ones were inclined to be obscured by the surroundings, e.g., tress and shadows. The introduction of multi-scale context made it easier for HDS with relatively uniform scales to be detected, which was reflected by the PA. In addition, a few LDS houses on the edge of HDS were misclassified into isolated houses within HDS. This was caused by a similar roofs and ambient vegetation (red circle in Figure 5). It further suggests that the polygon-based testing method is necessary. Previous studies considered recognition accuracy, but sometimes did not include area accuracy of rural settlements.

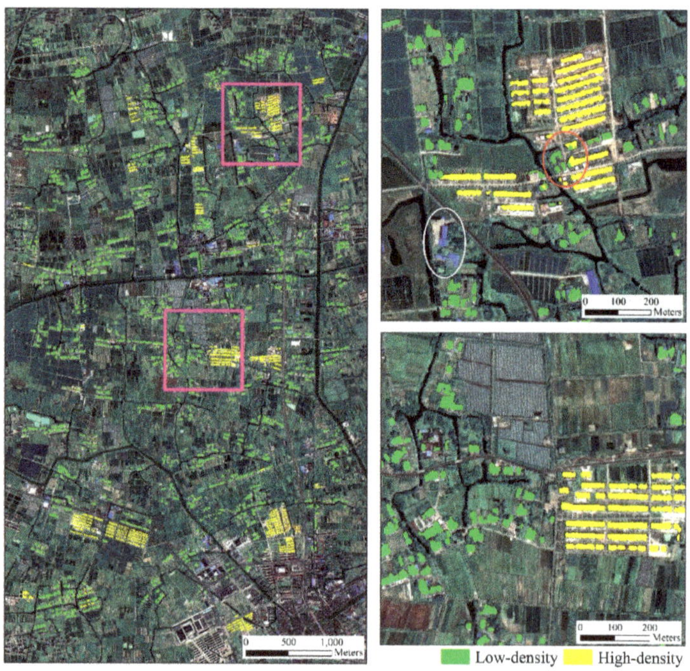

Figure 5. Classification result of the polygon test area.

Table 2. Confusion matrix of point test set.

		Predicted Class			
		LDS	**HDS**	**Backgrounds**	**Sum**
Ground truth	LDS	5997	3	125	6125
	HDS	4	2551	61	2616
	Backgrounds	4	0	2883	2887
	Sum	6005	2554	3069	11,628
	UA	99.87%	99.88%	93.94%	
	PA	97.91%	97.52%	99.86%	
	OA	98.31%			
	Kappa	0.9724			

Table 3. Confusion matrix of polygon test set (m^2).

		Predicted Class			
		LDS	**HDS**	**Backgrounds**	**Sum**
Ground truth	LDS	720,551	9228	118,198	847,977
	HDS	2673	349,060	60,476	412,209
	Backgrounds	95,539	51,323	24,231,862	24,378,724
	Sum	818,763	409,611	24,410,536	25,638,910
	UA	88.00%	85.22%	99.27%	
	PA	84.97%	84.68%	99.40%	
	OA	98.68%			
	Kappa	0.8591			

4.2. Ablation Experiments of Model

This study proposed a deep learning-based approach to extract rural settlements using HSR images. Experiments were carried out to explore the contribution of each part of the proposed deep method. Table 3 compares the performance of models with different settings based on the polygon test set. As showed in Table 4, when applying the original ResNet50 for segmentation, the accuracies of low-density class (UA of 82.50% and PA of 83.30%) were higher than those of high-density class (UA of 80.45% and PA of 67.75%). The low classification of PA indicates extracting HDS is rather challenging than LDS. When the last two stages of the baseline network were replaced by dilated convolutions, the PA index of high-density class was increased significantly by about 9%, while the UA of high-density class and the PA of low-density class had a moderate decrease. These indicated that the sub-module (+Dilation) was still insufficient. The possible reasons for the inconsistent changes in accuracies are the contradiction between the improvements brought by dilated convolutions and the defects of using single-scale feature. When comparing with the sub-module (+Dilation), another sub-module (+Dilation+Multiscale) yielded better accuracy on high-density class (UA of 84.88% and PA of 83.19%), with a slight increase in PA of low-density class, indicating that multi-scale context information enhanced the recognition power of the model. From Table 4, it can be seen that the proposed model achieved the largest OA of 98.68% with a Kappa coefficient of 0.8591. At the top of the aggregation layer, SE block captured feature dependencies in the channel dimension, and such feature selection process further improved the model performance. Figure 6 shows the visualization results of test set samples before and after recalibration with the SE block, implemented by t-SNE [40] technique. After the SE block, some samples of rural settlements classes gathered and were away from the background group, implying that the output of the channel relation module is more helpful for this classification task.

Table 4. Model comparisons with baseline, where values in bold are the best.

	OA	UA		PA		Kappa
		LDS	**HDS**	**LDS**	**HDS**	
Res50Seg (Baseline)	98.36%	82.50%	80.45%	83.30%	67.75%	0.8329
+Dilation	98.39%	84.25%	78.76%	80.53%	76.90%	0.8363
+Dilation+Multiscale	98.53%	87.24%	84.88%	81.90%	83.19%	0.8513
+Dilation+Multiscale+SE (Ours)	**98.68%**	**88.00%**	**85.22%**	**84.97%**	**84.68%**	**0.8591**

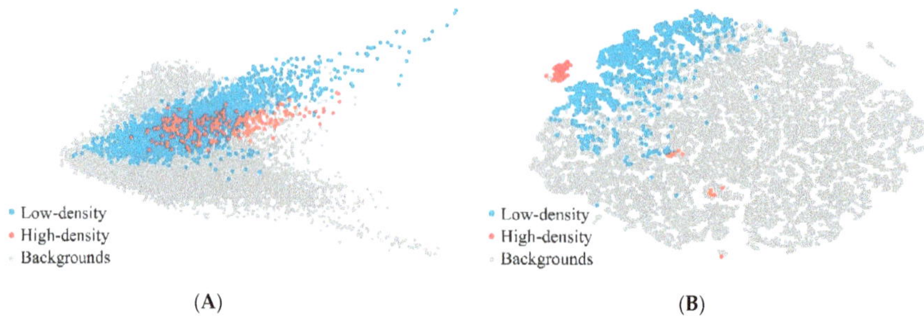

Figure 6. Visualization of test set samples before (**A**) and after recalibration (**B**) with SE block. Different colors represent different categories.

4.3. Data Input Strategies

Further experiments on two data input strategies, i.e., four channels and three channels, were conducted on the polygon test set. It was found that the classification accuracy of NIR-R-G-B composite images was slightly better than that of the R-G-B, but no significant difference was observed (Figure 7). It indicates that additional information of NIR band has positive effects on rural settlement extraction, while the powerful ability of CNNs to extract texture information from R-G-B images offset the gap between the two input strategies. Although the NIR band did not provide as great an improvement in accuracy as the DSM information [34], the strategy of using pre-trained weights of RGB data to initialize multispectral remote sensing images could be extended in the future.

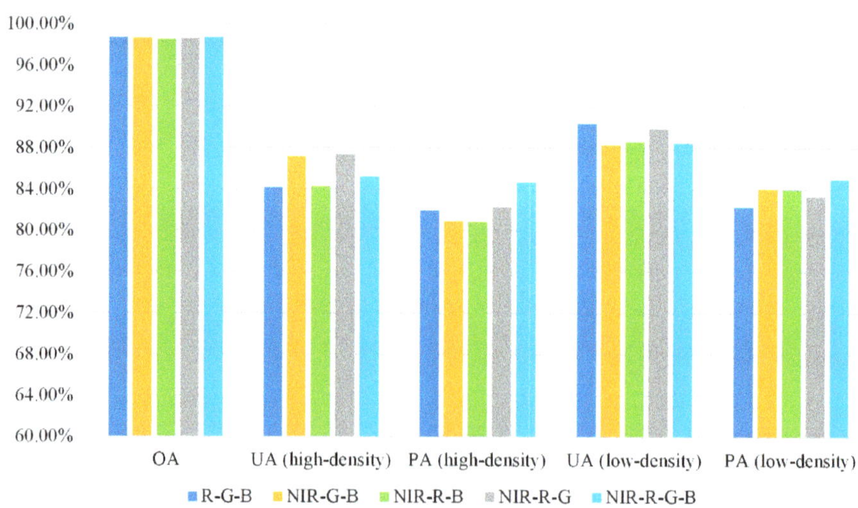

Figure 7. Accuracy assessment of different data input strategies.

4.4. Comparative Studies with Different Methods

Five state-of-the-art methods were compared, including an object-based image analysis (OBIA) method and four FCN based deep models. These methods have been proven effective in delineation of settlements and/or object detection for satellite images. The detailed information of each method can be found in the publication and we just briefly summarized their key technologies.

1. OBIA [12]: a novel object-based image classification method which integrates hierarchical multi-scale segmentation and landscape analysis. This method makes use of spatial contextual information and subdivides different types of rural settlements with high accuracy.

2. FCN [25]: a proposed fully convolutional network which comprises an encoder based on the VGG-16 network and a decoder consists of three stacked deconvolution layers. As far as we know, this is the first time that a deep learning FCN model has been used for rural residential areas extraction.
3. UNet [41]: a robust CNN architecture which consists of two symmetric contracting and expansive paths, which are made up of successive convolution layers. UNet is one of the deep learning methods often applied in the remote sensing field due to its efficiency and simplicity [42].
4. SegNet [43]: an encoder-decoder architecture uses the pooling indices to perform upsampling. It is a classic and efficient model that is often used as a baseline for semantic segmentation. Persello et al. [44] successfully delineated agricultural fields in smallholder farms from satellite images using SegNet.
5. DeeplabV3+ [20]: a state-of-the-art semantic segmentation model combining spatial pyramid pooling module and encode-decoder structure. It has achieve a performance of 89% on the PASCAL VOC 2012 semantic segmentation dataset.

Figure 8 demonstrates samples selected from classification results of all six methods based on the polygon test set. Quantitative results are presented in Table 5. In terms of overall performance, all six methods exhibited a high accuracy (OA > 0.97), and the results of the Kappa coefficient were consistent with OA. However, there were obvious differences about class-specific measures among the methods. With regards to UA, PA, the proposed method achieved the best accuracies, slightly better than the accuracies of DeeplabV3+. The UA and PA of SegNet and UNet were relatively close, but not as good as the proposed method. Unfortunately, the PA of FCN was lowered than other methods, indicating FCN is not the best choice to distinguish settlements pixels. Finally, the results of OBIA indicate that, for high-density class, the object-based method performs better than SegNet and UNet in PA significantly and slightly worse in UA, but lags far behind in Kappa values.

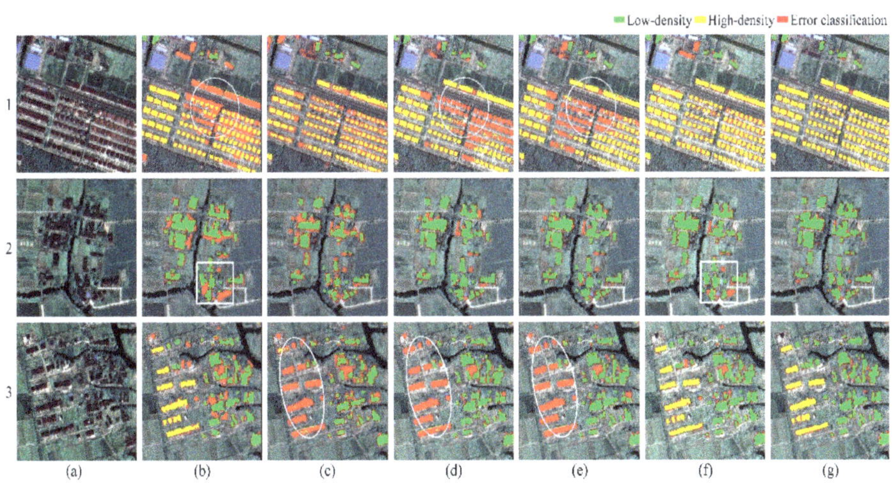

Figure 8. Example of results on Tongxiang polygon test set. (**a**) Original images, (**b**) OBIA, (**c**) FCN, (**d**) UNet, (**e**) SegNet, (**f**) DeeplabV3+, (**g**) The proposed method.

Table 5. Accuracy assessment of different methods, where values in bold are the best.

Method	OA	UA		PA		Kappa
		LDS	HDS	LDS	HDS	
OBIA	97.54%	75.24%	71.44%	72.24%	79.95%	0.7397
FCN	97.46%	73.11%	75.44%	70.28%	55.46%	0.7205
UNet	98.39%	84.58%	77.08%	80.32%	66.45%	0.8245
SegNet	98.37%	84.06%	78.51%	80.20%	68.79%	0.8232
DeeplabV3+	98.69%	87.92%	83.43%	**85.51%**	82.93%	0.8520
Ours	**98.68%**	**88.00%**	**85.22%**	84.97%	**84.68%**	**0.8591**

For the low-density class, all deep techniques, except FCN, achieved satisfying performance because the number of low-density pixels was relatively large in the training data, which was an advantage for data-driven deep learning methods. The FCN model only used deep features for classification, and the loss of spatial information led to blurred building boundaries. In contrast, the object-based method performed better for HDS identification. Unlike the end-to-end deep methods, the performance of object-based method was heavily depended on the scale parameter of segmentation. The new-style HDS' scale was relatively uniform and could be effectively extracted using OBIA method, even with a small sample size. Comparatively, LDS had a large size variation and were more sensitive to the choice of segmentation scale. Although the multi-context OBIA method exploited multiple segmentation scales to obtain the objects to be classified, it was still insufficient to separate the LDS of different sizes from the surrounding vegetation. Figure 8b shows that the OBIA method tends to intermingle the adjacent houses with vegetation or ground due to an improper segmentation scale selection. Moreover, manually designed features reduced the generalizability of methods in a large region. SegNet and UNet struggled in scenes where LDS and HDS are co-existed and mixed (Figure 8d,e). Compared with SegNet and UNet, using multi-scale context information helped the proposed method and DeeplabV3+ to reduce the misclassification of HDS. However, it inevitably induces some ambiguities on the boundaries of polygons (Figure 8f,g).

Table 6 lists the computing time of the proposed method and other methods. For the OBIA method, the segmentation and classification were conducted separately, and thereby showed the least time consumption. Instead, deep learning methods were end-to-end approaches. Among deep learning methods, FCN consumed fewer computing resources and had the shortest inference time because FCN had abandoned the full connection layers with lots of parameters. Therefore, the lack of feature representation capability limited the performance of FCN in this task. The proposed model showed similar model size and inference time with SegNet, but it took less training time to reach convergence. UNet and DeeplabV3+ have more parameters and they take longer to converge. Overall, the proposed method is more efficient.

Table 6. The efficiency of different methods.

Method	Parameters	Training Time	Inference Time
OBIA		~0.5 h	~10 m
FCN	12.38 million	~3.1 h	0 m 17 s
UNet	33.40 million	~11.8 h	0 m 39 s
SegNet	29.44 million	~ 8.2 h	0 m 31 s
DeeplabV3+	39.76 million	~12.9 h	0 m 32 s
Ours	28.04 million	~5.8 h	0 m 25 s

4.5. Analysis and Potential Improvements

In our analysis, we found that all selected deep methods, except the proposed method and DeeplabV3+, were not as effective in the high-density category as in the low-density category. One possible reason was that the downsampling operation of the comparative methods was aggressive.

Instead, using dilated residual convolutional network retained the spatial resolution of features. Given the input image patch (256 × 256), the deepest feature map of the proposed network maintains an appropriate size (32 × 32), which helps to restore the geometry of settlements. In this way, the accuracy of HDS increased greatly. However, the problem of scale selection remained. Unsynchronized scales of different types of settlements made it difficult to determine the optimal scale. The proposed multi-scale context subnetwork involved multiple scales, thereby reducing the dependence on a single optimal scale to a certain extent. However, the minimum scale (32 × 32) of representations applicable in the Tongxiang dataset may not match other HSR data. Thus, if the proposed method is applied to other data, determining an appropriate scale range would depend on the size of settlements objects and input images.

In some areas, HDS and LDS could not be easily recognized as they were in similar shapes, structures. Deep features at multiple scale could handle such complex patterns of settlements objects of different sizes, and the SE block modeled the global contextual relation of fused features, enabling feature selection in the channel dimension. The multi-scale context subnetwork gave more confident predictions at pixel level. The way that DeeplabV3+ uses the spatial pyramid module to encode multi-scale context information has achieved similar effects as our context subnetwork. The experimental results demonstrated that the proposed multi-scale network distinguish two types of settlements objects effectively. Nevertheless, contours of rural settlements needed to be further refined. Blurred object boundaries were an inherent and common defect of CNN-based semantic segmentation models. The downsampling process in the CNN model inevitably lost spatial details, which was detrimental to the preservation of edge information. However, this was a trade-off between spatial resolution and semantic feature representation of segmentation models. Our results showed that in this application, the use of dilated convolution instead of downsampling alleviated the loss of boundary details.

Segmentation and classification are conducted separately in OBIA method, which makes the classification result greatly affected by the performance of segmentation algorithm. Besides, handcrafted features used in OBIA are difficult to achieve an optimal balance between discriminability and robustness, since these features cannot easily consider the details of real data, especially in the case of HSR images that images can change a lot in large extent [45]. Instead, deep learning methods conduct segmentation and classification at the same time, and the classification results in Table 5 prove the superiority of the proposed method. Though deep learning methods take longer to train, it takes only a few seconds for a trained network to classify images. From the perspective of application, this is more applicable to the situation of big data of HSR images. Moreover, observation from the OBIA results, image segments could preserve the precise edges if under the appropriate segmentation scale. According to this observation, it is promising to combine the segmentation of OBIA and the feature representation of deep learning to classify rural settlements. Furthermore, this leaves open the question of whether a non-differentiable segmentation algorithm can be integrated into CNNs. In future, we hope to find a way to integrate the advantage of OBIA segmentation into the proposed framework of a deep network for rural settlement mapping.

5. Conclusions

Rural settlements classification using HSR remotely sensed image remains a challenging task, due to the intra-class spectral variation and spatial scale variation. This paper presents an effective rural settlements extraction method based on a deep fully convolutional network (FCN) from HSR satellite images. In the proposed multi-scale FCN model, dilated convolution was utilized to extract feature representations with high spatial resolution. A subnetwork improved the discrimination power of the network by aggregating and re-weighting multi-scale context information across layers. High spatial resolution representations and multi-scale context information helped to locate and further subdivide rural settlements. Experimental results on GF-2 images acquired over a typical rural area located in Tongxiang, China, showed the proposed method produced the most accurate classification results of rural settlements, comparing with other state-of-the-art methods and the sub-modules. In summary, our proposed method was promising in terms of its potential for rural settlements extraction from HSR images. From a rural management perspective, this work describes a scheme for rapid identification of rural settlements in a large region by using HSR images. The classification method presented here could be extended to the identification of rural settlements in a larger area, and the results can be used as a guide for on-site verification or enforcement in cadastral inventory.

In future works, further improvements could be made by integrating multi-temporal HSR images and multi-modal data, so that the dynamics of rural settlements can be characterized.

Supplementary Materials: The following are available online at http://www.mdpi.com/1424-8220/20/21/6062/s1, Table S1: Specification of our network architecture.

Author Contributions: Conceptualization, Z.Y.; methodology, Z.Y.; software, Z.Y.; validation, R.Z.; investigation, B.S.; resources, B.S.; data curation, R.Z.; writing—original draft, Z.Y.; writing—review and editing, Y.L. and Q.Z.; visualization, Y.L.; supervision, L.H. and K.W.; funding acquisition, K.W. All authors have read and agreed to the published version of the manuscript.

Funding: This work was supported by the National Key Research and Development Program of China, grant no. 2016YFC0502704, the National Natural Science Foundation of China, grant no. 41701638, the National Natural Science Foundation of China, grant no. 41971236, the Basic Public Welfare Research Program of Zhejiang Province, grant no. LGJ19D010001, and Zhejiang University Student Research Training Program (2019).

Acknowledgments: The authors appreciate the reviewers and the editor for their constructive comments and suggestions. We would especially like to thank Dr. Xinyu Zheng for his help.

Conflicts of Interest: The authors declare no conflict of interest.

References

1. Long, H.; Liu, Y.; Wu, X.; Dong, G. Spatio-temporal dynamic patterns of farmland and rural settlements in Su–Xi–Chang region: Implications for building a new countryside in coastal China. *Land Use Policy* **2009**, *26*, 322–333. [CrossRef]
2. Shan, Z.; Feng, C. The Redundancy of Residential Land in Rural China: The evolution process, current status and policy implications. *Land Use Policy* **2018**, *74*, 179–186. [CrossRef]
3. Kit, O.; Ludeke, M.K.B.; Reckien, D. Texture-based identification of urban slums in Hyderabad, India using remote sensing data. *Appl. Geogr.* **2012**, *32*, 660–667. [CrossRef]
4. Conrad, C.; Rudloff, M.; Abdullaev, I.; Thiel, M.; Löw, F.; Lamers, J. Measuring rural settlement expansion in Uzbekistan using remote sensing to support spatial planning. *Appl. Geogr.* **2015**, *62*, 29–43. [CrossRef]
5. Yang, C.; Wang, X.; Huang, H. Comparison of Extracting Rural Residential Area from Satellite Images with Multiresolution. In Proceedings of the 2008 IEEE International Geoscience and Remote Sensing Symposium (IGARSS 2008), Boston, MA, USA, 8–11 July 2008.
6. Weiss, M.; Jacob, F.; Duveiller, G. Remote sensing for agricultural applications: A meta-review. *Remote. Sens. Environ.* **2020**, *236*, 111402. [CrossRef]
7. Blaschke, T.; Hay, G.J.; Kelly, M.; Lang, S.; Hofmann, P.; Addink, E.; Feitosa, R.Q.; Van Der Meer, F.; Van Der Werff, H.; Van Coillie, F.; et al. Geographic Object-Based Image Analysis—Towards a new paradigm. *ISPRS J. Photogramm. Remote. Sens.* **2014**, *87*, 180–191. [CrossRef] [PubMed]
8. Zhao, W.; Du, S.; Emery, W.J. Object-Based Convolutional Neural Network for High-Resolution Imagery Classification. *IEEE J. Sel. Top. Appl. Earth Obs. Remote. Sens.* **2017**, *10*, 3386–3396. [CrossRef]
9. Persello, C.; Stein, A. Deep Fully Convolutional Networks for the Detection of Informal Settlements in VHR Images. *IEEE Geosci. Remote. Sens. Lett.* **2017**, *14*, 2325–2329. [CrossRef]
10. Hoffman-Hall, A.; Loboda, T.V.; Hall, J.V.; Carroll, M.L.; Chen, D. Mapping remote rural settlements at 30 m spatial resolution using geospatial data-fusion. *Remote. Sens. Environ.* **2019**, *233*, 111386. [CrossRef]
11. Zhang, B.; Liu, Y.; Zhang, Z.; Shen, Y. Land use and land cover classification for rural residential areas in China using soft-probability cascading of multifeatures. *J. Appl. Remote. Sens.* **2017**, *11*, 1. [CrossRef]
12. Zheng, X.; Wu, B.; Weston, M.; Zhang, J.; Gan, M.; Zhu, J.; Deng, J.; Wang, K.; Teng, L. Rural Settlement Subdivision by Using Landscape Metrics as Spatial Contextual Information. *Remote. Sens.* **2017**, *9*, 486. [CrossRef]

13. Zheng, X.; Wang, Y.; Gan, M.; Zhang, J.; Teng, L.; Wang, K.; Shen, Z.; Zhang, L. Discrimination of Settlement and Industrial Area Using Landscape Metrics in Rural Region. *Remote. Sens.* **2016**, *8*, 845. [CrossRef]
14. Ma, L.; Liu, Y.; Zhang, X.; Ye, Y.; Yin, G.; Johnson, B.A. Deep learning in remote sensing applications: A meta-analysis and review. *ISPRS J. Photogramm. Remote. Sens.* **2019**, *152*, 166–177. [CrossRef]
15. Volpi, M.; Tuia, D. Dense Semantic Labeling of Subdecimeter Resolution Images with Convolutional Neural Networks. *IEEE Trans. Geosci. Remote. Sens.* **2016**, *55*, 881–893. [CrossRef]
16. Long, J.; Shelhamer, E.; Darrell, T. Fully convolutional networks for semantic segmentation. In Proceedings of the IEEE Conference on Computer Vision and Pattern Recognition, Boston, MA, USA, 7–12 June 2015; pp. 3431–3440.
17. Liu, Y.; Fan, B.; Wang, L.; Bai, J.; Xiang, S.; Pan, C. Semantic labeling in very high resolution images via a self-cascaded convolutional neural network. *ISPRS J. Photogramm. Remote. Sens.* **2018**, *145*, 78–95. [CrossRef]
18. Wang, H.; Wang, Y.; Zhang, Q.; Xiang, S.; Pan, C. Gated Convolutional Neural Network for Semantic Segmentation in High-Resolution Images. *Remote. Sens.* **2017**, *9*, 446. [CrossRef]
19. Wu, G.; Shao, X.; Guo, Z.; Chen, Q.; Yuan, W.; Shi, X.; Xu, Y.; Shibasaki, R. Automatic Building Segmentation of Aerial Imagery Using Multi-Constraint Fully Convolutional Networks. *Remote. Sens.* **2018**, *10*, 407. [CrossRef]
20. Encoder-Decoder with Atrous Separable Convolution for Semantic Image Segmentation. Available online: https://arxiv.org/abs/1802.02611 (accessed on 25 October 2020).
21. Ye, Z.; Fu, Y.; Gan, M.; Deng, J.; Comber, A.; Wang, K. Building Extraction from Very High Resolution Aerial Imagery Using Joint Attention Deep Neural Network. *Remote. Sens.* **2019**, *11*, 2970. [CrossRef]
22. Zhu, X.X.; Tuia, D.; Mou, L.; Xia, G.-S.; Zhang, L.; Xu, F.; Fraundorfer, F. Deep Learning in Remote Sensing: A Comprehensive Review and List of Resources. *IEEE Geosci. Remote. Sens. Mag.* **2017**, *5*, 8–36. [CrossRef]
23. Qiu, C.; Schmitt, M.; Geiß, C.; Chen, T.-H.K.; Zhu, X.X. A framework for large-scale mapping of human settlement extent from Sentinel-2 images via fully convolutional neural networks. *ISPRS J. Photogramm. Remote. Sens.* **2020**, *163*, 152–170. [CrossRef]
24. Gevaert, C.M.; Persello, C.; Sliuzas, R.; Vosselman, G. Monitoring household upgrading in unplanned settlements with unmanned aerial vehicles. *Int. J. Appl. Earth Obs. Geoinform.* **2020**, *90*, 102117. [CrossRef]
25. Lu, C.; Yang, X.; Wang, Z.; Liu, Y. Extracting Rural Residential Areas from High-Resolution Remote Sensing Images in the Coastal Area of Shandong, China Based on Fast Acquisition of Training Samples and Fully Convoluted Network. In Proceedings of the 2018 10th IAPR Workshop on Pattern Recognition in Remote Sensing (PRRS), Beijing, China, 19–20 August 2018; pp. 1–4.
26. Hu, J.; Shen, L.; Albanie, S.; Sun, G.; Wu, E. Squeeze-and-Excitation Networks. *IEEE Trans. Pattern Anal. Mach. Intell.* **2019**, *42*, 2011–2023. [CrossRef] [PubMed]
27. Process for Enhancing the Spatial Resolution of Multispectral Imagery Using Pan-Sharpening. Available online: https://www.lens.org/lens/patent/US_6011875_A?locale=es (accessed on 25 October 2020).
28. Deep Residual Learning for Image Recognition. Available online: https://arxiv.org/abs/1512.03385 (accessed on 25 October 2020).
29. Yu, F.; Koltun, V.; Funkhouser, T. Dilated Residual Networks. Available online: https://arxiv.org/abs/1705.09914 (accessed on 25 October 2020).
30. Multi-Scale Context Aggregation by Dilated Convolutions. Available online: https://arxiv.org/abs/1511.07122 (accessed on 25 October 2020).
31. Razavian, A.S.; Azizpour, H.; Sullivan, J.; Carlsson, S. CNN Features Off-the-Shelf: An Astounding Baseline for Recognition. In Proceedings of the 2014 IEEE Conference on Computer Vision and Pattern Recognition Workshops, Columbus, OH, USA, 23–28 June 2014; pp. 512–519.
32. Marmanis, D.; Datcu, M.; Esch, T.; Stilla, U. Deep Learning Earth Observation Classification Using ImageNet Pretrained Networks. *IEEE Geosci. Remote. Sens. Lett.* **2015**, *13*, 105–109. [CrossRef]
33. Zhao, B.; Huang, B.; Zhong, Y. Transfer Learning With Fully Pretrained Deep Convolution Networks for Land-Use Classification. *IEEE Geosci. Remote. Sens. Lett.* **2017**, *14*, 1436–1440. [CrossRef]
34. Marmanis, D.; Schindler, K.; Wegner, J.D.; Galliani, S.; Datcu, M.; Stilla, U. Classification with an edge: Improving semantic image segmentation with boundary detection. *ISPRS J. Photogramm. Remote. Sens.* **2018**, *135*, 158–172. [CrossRef]

35. Huang, J.; Zhang, X.; Xin, Q.; Sun, Y.; Zhang, P. Automatic building extraction from high-resolution aerial images and LiDAR data using gated residual refinement network. *ISPRS J. Photogramm. Remote. Sens.* **2019**, *151*, 91–105. [CrossRef]
36. Adam: A Method for Stochastic Optimization. Available online: https://arxiv.org/abs/1412.6980 (accessed on 25 October 2020).
37. Create Random Points. Available online: https://desktop.arcgis.com/en/arcmap/10.3/tools/data-management-toolbox/create-random-points.htm (accessed on 4 May 2020).
38. Congalton, R.G. A review of assessing the accuracy of classifications of remotely sensed data. *Remote. Sens. Environ.* **1991**, *37*, 35–46. [CrossRef]
39. Bishop, Y.M.M.; Fienberg, S.E.; Holland, P.W. *Discrete Multivariate Analysis*; Springer: New York, NY, USA, 2007.
40. Maaten, L.V.D.; Hinton, G. Visualizing Data using t-SNE. *J. Mach. Learn. Res.* **2008**, *9*, 2579–2605.
41. Ronneberger, O.; Fischer, P.; Brox, T. U-Net: Convolutional Networks for Biomedical Image Segmentation. In Proceedings of the Medical Image Computing and Computer-Assisted Intervention(MICCAI 2015), Munich, Germany, 5–9 October 2015; pp. 234–241.
42. Flood, N.; Watson, F.; Collett, L. Using a U-net convolutional neural network to map woody vegetation extent from high resolution satellite imagery across Queensland, Australia. *Int. J. Appl. Earth Obs. Geoinform.* **2019**, *82*, 101897. [CrossRef]
43. Badrinarayanan, V.; Kendall, A.; Cipolla, R. SegNet: A Deep Convolutional Encoder-Decoder Architecture for Image Segmentation. *IEEE Trans. Pattern Anal. Mach. Intell.* **2017**, *39*, 2481–2495. [CrossRef]
44. Persello, C.; Tolpekin, V.; Bergado, J.; De By, R. Delineation of agricultural fields in smallholder farms from satellite images using fully convolutional networks and combinatorial grouping. *Remote. Sens. Environ.* **2019**, *231*, 111253. [CrossRef] [PubMed]
45. Zhang, L.; Zhang, L.; Dua, B. Deep Learning for Remote Sensing Data: A Technical Tutorial on the State of the Art. *IEEE Geosci. Remote. Sens. Mag.* **2016**, *4*, 22–40. [CrossRef]

sensors

Article

Automatic Changes Detection between Outdated Building Maps and New VHR Images Based on Pre-Trained Fully Convolutional Feature Maps

Yunsheng Zhang [1], Yaochen Zhu [2], Haifeng Li [1], Siyang Chen [1], Jian Peng [1] and Ling Zhao [1,*]

1 School of Geoscience and Info-Physics, Central South University, Changsha 410083, China; zhangys@csu.edu.cn (Y.Z.); lehaifeng@csu.edu.cn (H.L.); siyangchen@csu.edu.cn (S.C.); PengJ2017@csu.edu.cn (J.P.)

2 School of Remote Sensing and Information Engineering, Wuhan University, Wuhan 430072, China; 0107150110@csu.edu.cn

* Correspondence: zhaoling@csu.edn.cn

Received: 27 August 2020; Accepted: 23 September 2020; Published: 27 September 2020

Abstract: Detecting changes between the existing building basemaps and newly acquired high spatial resolution remotely sensed (HRS) images is a time-consuming task. This is mainly because of the data labeling and poor performance of hand-crafted features. In this paper, for efficient feature extraction, we propose a fully convolutional feature extractor that is reconstructed from the deep convolutional neural network (DCNN) and pre-trained on the Pascal VOC dataset. Our proposed method extract pixel-wise features, and choose salient features based on a random forest (RF) algorithm using the existing basemaps. A data cleaning method through cross-validation and label-uncertainty estimation is also proposed to select potential correct labels and use them for training an RF classifier to extract the building from new HRS images. The pixel-wise initial classification results are refined based on a superpixel-based graph cuts algorithm and compared to the existing building basemaps to obtain the change map. Experiments with two simulated and three real datasets confirm the effectiveness of our proposed method and indicate high accuracy and low false alarm rate.

Keywords: changes detection; fully convolutional feature maps; outdated building map; VHR images

1. Introduction

Developing countries have witnessed a rapid expansion of urban areas during the last decades. With the fast urbanization, updating buildings geo-database plays an important role in urban planning, as it provides valuable information regarding, e.g., land use/cover monitoring [1], evaluation of agricultural lands decline [2], disaster assessment [3], civil BIM updating [4]. Such information also enables the government to adopt suitable and sustainable development strategies. Automatic building geo-database updating relies on identifying the areas, where changes occurred. Currently, change identification is mainly a labor-intensive work, especially in urban environments, due to its complexity. Therefore, automatic geo-database updating based on remote sensing images remains an open and unsolved issue.

During the past decades, several methods have been proposed to increase the level of automation in change detection. According to their comparison basis, the change detection methods can be categorized into two classes: (1) Image-image comparison; and (2) image-map comparison [5]. The former approach aims at direct recognition of differences between multi-temporal remotely sensed images [6,7]. The image-map comparison-based method, however, detects changes between existing data and newly acquired images, where the semantic classification of the newly acquired images is also required. For image-map comparison, supervised machine learning methods are employed, see, e.g., Reference [8]. However, for an accurate classifier to be trained, a large enough set of labeled samples is required. Labeling samples, however, need expensive manual work and a high level of expertise and knowledge on image interpretation.

To address this issue, existing GIS data or online maps, such as Open Street Map (OSM) data, and Google maps, are employed to provide prior information. For example, Bouziani et al. obtain prior class knowledge from the existing geo-database to identify the change of buildings based on transitional probability between classes, and to change map segmentation [5]. Kaiser et al. exploit the online map to guide aerial image segmentation, although they simply ignore the temporal inconsistencies between the used map and aerial images, and simply count on human interaction to remove the mis-registrations between the map and the roof images of buildings [9]. Wan et al. employ OSM data to obtain initial samples for training SVM to classify HRS images [10]. To alleviate the effect of intrinsic errors caused by incorrect labeling by volunteers, they further use a cluster analysis to filter out the possible errors. Gevaert et al. provide a model for outdated base-maps as noisy labels of newly acquired UAV images, and then utilize data cleansing methods to filter out the potentially mislabeled samples, and further re-predict their labels by supervised classification [11]. Chen et al. treat historical digital line graph (DLG) data as the source of initial noisy labels, and then the pure part is selected by an iterative training method [12]. For highly accurate classification, they also use several hand-crafted image-based and point-cloud based features for the supervised classification task. The elevation feature is also very useful to distinguish buildings; however, it is not always available.

In addition to the availability of a large enough set of labeled samples, selecting proper discriminable features is another key point for classification. Some carefully hand-crafted features are heuristically proposed and combined to classify VHR images. Most of the existing methods employ spectral and textural features, or DEM data, as feature descriptors, see References [11,13,14]. Although the hand-crafted features are designed to describe a specific image pattern, their performance depends on the available training data. Different from hand-crafted features, the recently developed deep learning techniques directly learn features from the original data. Deep learning is widely used in various research areas, e.g., natural image classification [15], object detection [16], and semantic segmentation [17]. Deep learning methods are also used to learn features from remote sensing (RS) images for classification [18]. For instance, autoencoder-based techniques are used in RS for extracting features from images [19–21]. Such methods learn to extract feature encodings in an unsupervised setting, which can then be reconstructed back to the input with minimum error [21]. Different variations of autoencoders are applied to various tasks in the RS field. By increasing the spatial resolution of the RS images, the training of such autoencoders becomes time-consuming and further requires large memory.

In practice, a large set of accurately labeled data is often unavailable. In recent works, this issue is addressed in the RS domain by training deep convolutional neural networks (DCNNs) from scratch. Feature extraction DCNNs is also widely used in computer vision research, where the training is based on large open-source datasets, see References [22,23]. The intuition behind DCNNs is that with strong learning abilities, DCNNs can learn to respond to various kinds of feature patterns in different abstract-levels from large and complex datasets. The learned features can then be generalized to be used for smaller datasets, even if those datasets are remarkably different from the training datasets [24]. Much research has been done to generate a single feature descriptor for the whole image with high-level activations of pre-trained DCNNs [25]. In these methods, the size of the input is strictly fixed, so interpolations are needed to resize the images to a specified scale. To extract dense feature maps in a pixel-wise fashion, such methods need to crop window, resize, and do forward propagation at the center of each pixel [20,26]. Since most of the computation in the neighboring windows are shared through the convolution, they are computationally redundant and limited to small/moderate-size images. Many existing methods focus on extracting features from the back part of DCNNs (i.e., the last convolutional layer and fc layers) and generate one single feature description for the whole image.

To improve classification performance, the spatial context of the images has to be fully used [23,27]. Single-pixel based methods are unable to take a large enough image field to distinguish the building objects from the background information and ensure a consistent classification result in the global context. Several pixel-based methods are proved to be successful for change detection of low- and moderate-resolution remotely sensed images [7]. Nevertheless, with the emergence of high-resolution remote sensing (HRS) data, such methods are not effective, since the results can easily keep salt-and-pepper noise, due to increasing (decreasing) intra-(inter-)class variance [28]. To address this issue, object-based methods are adopted in References [29–32]. Such object-based change detection methods significantly reduce the required amount of data to be processed, and further generate change recognition result with shape and boundary information that can be directly used to update geo-databases, see Reference [33]. This however may lead to new problems as object segmentation is intrinsically challenging for remote sensing images [34].

In this paper, we propose to cast the image-map change detection problem into the identification and correction of noisy labels. For extracting discriminable features, a fully convolutional network (FCN) pre-trained on the PASCAL VOC dataset [17] is treated as a fully convolutional feature extractor (FCFE). Since the long-range relationship comparatively is trivial in the HRS images, and spatial information is severely lost by down-sampling in the last convolutional layers, only first two groups of convolutional layers (4 layers) are preserved. The tensors from all convolutional layers are then up-sampled to the same size of the input and fused together by concatenation as pixel-wise features. Through FCFE, the feature computation of all pixels is achieved by a single forward propagation. Therefore, it is more efficient than that of the most window-based feature extractors. However, directly concatenated and up-sampled pixel-wise features are redundant and have a high dimension for subsequent processing. Therefore, a noise label guided feature selection is proposed to select the most informative features for building extraction. As pixel-wise re-predicted labels of newly acquired HRS images are usually fragmented, especially in areas with a similar spectral, textural characteristic, such as buildings, roads, and bare soil. To alleviate this problem, new HRS images are segmented into superpixels, and then superpixel-based graph cuts are used to refine the initial classification result. For further performance improvement, we also propose a new label uncertainty calculation technique for each superpixel.

The contribution of our work are the following: (1) We present a novel framework with the combination of pixel-wise and object-based analysis for image-map change detection based on data cleaning method; (2) FCN pre-trained on the PASCAL VOC dataset for semantic segmentation is then used to reconstruct the proposed fully convolutional feature extractors to extract dense features of HRS images; and (3) outdated noise label is then used to guide the feature selection for eliminating the redundancy of the features.

The remainder of this paper is organized as the following. Section 2 provides the details of the proposed image-map change detection framework. Section 3 analyses the performance of experiments conducted on two simulated, and three real datasets. Finally, conclusions are presented in Section 4.

2. Methods

2.1. Overview of the Method

The workflow of the proposed approach is illustrated in Figure 1, where the three main components are:

(1) Feature calculation, which is a fully convolutional feature extractor reconstructed from FCN-8s [17] and pre-trained on the PASCAL VOC dataset. Feature calculation extracts multi-scale pixel-wise features from newly acquired HRS images. An RF classifier is then trained to rank the importance of the extracted features based on the outdated basemap. After that, representative features are selected as feature descriptors for each pixel.

(2) Initial classification, where the label uncertainty for each pixel is estimated through cross-validation based on selected features. The reliable (unchanged) pixels are then separated as training samples to train the new RF classifier, and potentially changed pixels are re-predicted.

(3) Post optimization and change map computing, where the SLIC (Simple Linear Iterative Cluster) algorithm [35] is used to segment HRS images into superpixels, and the probability of superpixels for each label is estimated. The negative logarithm of probability is then used to construct the data term. A Gaussian kernel of normalized RGB feature is then used to construct a smooth term of the energy function. After that, the graph cuts algorithm is used to minimize the energy function and obtain the optimized, updated label. The updated labels are finally compared with the outdated basemap to compute the change map.

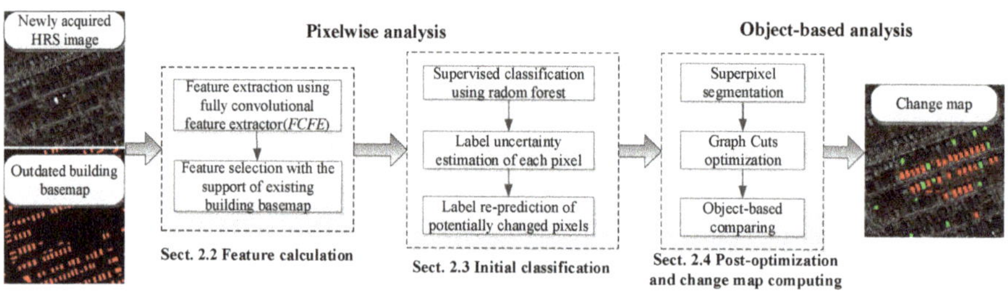

Figure 1. Flowchart of the proposed change detection framework. HRS, resolution remotely sensed.

2.2. Feature Extraction through Fully Convolutional Feature Extractor

Although the last layers of CNNs are more effective in capturing semantics, they are ineffective in capturing fine-grained spatial details, which are needed for spatial feature extraction [36]. Two obstacles that hinder the direct transformation of DCNNs into dense feature extractors are: (1) Pooling layers shrink features maps exponentially, and this depresses valuable spatial information; (2) fc layers map fix-size feature tensors into activation vectors, this constrains the input size. In computer vision, images are relatively small and contain only a few salient objects and/or one main scene. This makes cascaded down-sampling important to extract relationships within the main objects. However, HRS images contain objects this belong to different categories, and there exists no single subject being able to globally determine the theme of HRS images. Therefore, long-range relationships captured by stacked pooling layers seem trivial, but the local response captured by the early convolutional layers (convlayer) is much more important.

Convolutional kernels in DCNNs pre-trained on a very-large dataset are considerably rich filter banks capturing various kinds of features. Zeiler and Fergus demonstrate that the early convlayer encodes low-level features, such as edges, corners, shapes, or textures, while the deeper layers extract high-level information, such as objects, or categories [37]. Kemker et al. assert that the features extracted by the convlayer of the pre-trained DCNNs can produce Gabor-like results [38]. Generally, feature maps extracted by the deeper convlayer are coarse and abstract, suffer from a severe size reduction, and contain more information of the source datasets, which is irrelevant when transferring to a new target dataset. Nevertheless, feature maps extracted from the earlier layers are fine-grained and adhere better to the boundaries. Therefore, one can assume that the features from early convlayers of pre-trained DCNNs have stronger generalization abilities [39]. Since convlayers also accepts arbitrary input size and intrinsically preserves spatial information, fully convolutional networks (FCN) reconstructed by the early part of pre-trained DCNNs are more efficient to extract dense features.

FCN-8s [17] is an FCN pre-trained on the PASCAL VOC dataset for 20-class semantic segmentation, is used to reconstruct the proposed fully convolutional feature extractors (FCFE). The used FCN-8s is trained on the PASCAL VOC 2011 segmentation challenge training set, which includes 11,530 images and 5034 segmentations. It is reconstructed and fine-tuned from VGGNet [40] that is pre-trained on ImageNet. FCN-8s consists of five groups of convlayers with pooling layers that encode the input image into high-dimensional dense feature maps. It also has three deconvolutional layers that up-sample and fuse activations from the last three pooling layers to the size of the input as the predictions. The structure of the original FCN-8s is illustrated in Figure 2.

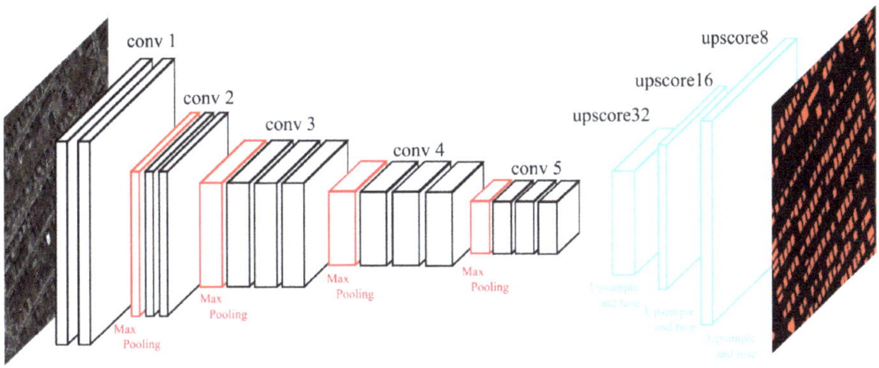

Figure 2. Structure of the original fully convolutional network (FCN)-8s [17].

2.2.1. Structure of the Proposed Fully Convolutional Feature Extractor

The structure of the proposed fully convolutional feature extractor is illustrated in Figure 3. To reconstruct pre-trained FCN-8s for dense feature extraction tasks, we make the following three modifications: (1) The feature maps extracted by convlayers after the pool2 layer are coarse (i.e., one-sixteenth the size of original image), and assumed to contain more information about source dataset. Therefore, only the first two groups of convlayers with the first pooling layers are preserved. This modification is aimed to exploit multi-level well-generalized features, while preserving valuable spatial information. (2) In the original FCN-8s, the first convlayer zero-pads the input image with 100 pixels to prevent severe size-reduction imposed by cascaded pooling layers. Other convlayers also pad the input feature map with 1 pixel. Note that all convolution kernels in FCN-8s are 3×3 in size, and their output has exactly the same spatial dimension as the input. In our fully-convolutional feature extractor (FCFE), all convlayers are set to pad input the feature map with 1 pixel. Therefore, feature maps from the first group of convlayers have the same size as the input image, while feature maps from the last convlayers are two-times downsampled. (3) The feature map extracted from the last group of convlayers is upsampled to the input size using bilinear interpolation. All feature maps are then concatenated to multi-scale deep features.

In Figure 3, the multi-scale features extracted by FCFE are up-sampled and fused feature maps from conv1_1, conv1_2, conv2_1, and conv2_2 layers of PASCAL VOC dataset-pretrained FCN-8s model, with 64, 64, 128, and 128 channels, respectively. Layer deconv2 uses bilinear interpolation to upsample feature maps from conv2_1 and conv2_2 to the size of the input image and fuse them together. The fusing1 layer concatenates the feature maps from conv1_1, conv1_2, and deconv2 to obtain the final 384-dimensional multi-scale features.

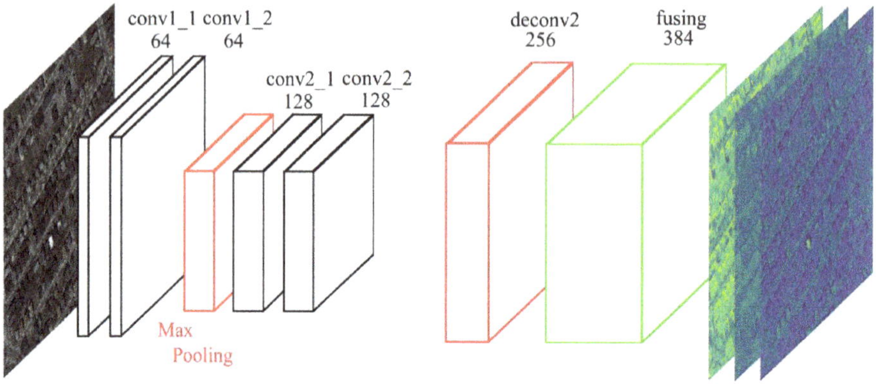

Figure 3. Structure of the proposed FCFE.

2.2.2. Feature Selection Guided by the Existing Basemaps Using Random Forest

Only part of the features directly extracted by the FCFE is highly discriminative for buildings, and the rest are redundant and high-dimensional. Therefore, direct feeding of the features into the subsequent data cleaning pipeline demands excessive computation, and also harms the data cleaning effects. According to the study in Reference [41], each feature layer generated by DCNN responds to a major class. Thus, the feature selection processing is performed to select the most informative features and ensure the classification result. Feature selection is the process of removing redundant and irrelevant features, often accomplished by determining the usefulness of all feature variables [42]. Feature selection methods can be generally classified into three categories, including supervised, semi-supervised, and unsupervised methods. The existing building basemaps may contain erroneously labeled areas, due to time-lapse with the newly acquired HRS image, however, the majority of the labels remain correct and can be used in the feature selection schemes.

Here we employ RF classifiers to select features in our proposed method. RF classifier trains multiple decision trees with a random subset of samples based on a random subset of features [43,44]. RF algorithm can be trained efficiently to process the multiple label classification problems, and it is widely used in RS image classification tasks [43]. RF also provides the importance of the used features. Therefore, the feature importance estimated by RF is the average importance of each decision tree.

In order to select the salient feature that discriminates well from the building to background pixels, 384-dimensional FCFE extracted features and existing building basemaps, as pixel-wise labels, are considered as the training set to fit an RF classifier. The features' importance is then evaluated, and nch (experimentally set to be 20) most important features are selected chosen to form the feature descriptor of the newly acquired HRS image.

To visually analyze the features extracted by the proposed method, an image, as shown in Figure 4, is used to perform the FCFE and feature selection processing. To display and compare features inner-layer- and cross-layer-wise, eight features are randomly chosen from each layer, and a total number of 32 feature maps are illustrated in Figure 5.

By carefully examining Figure 5, three characteristics of the feature extracted by FCFE can be concluded: (1) A small part of the features is highly discriminative between buildings and background, with the corresponding feature maps showing salient contrast between the two classes; (2) a large number of features are less useful; with feature maps being ambiguous and showing inconspicuous differences; (3) features from early convlayers are fine-grained and adhere better to the boundaries, whereas features from latter convlayers are comparatively coarse and more abstract.

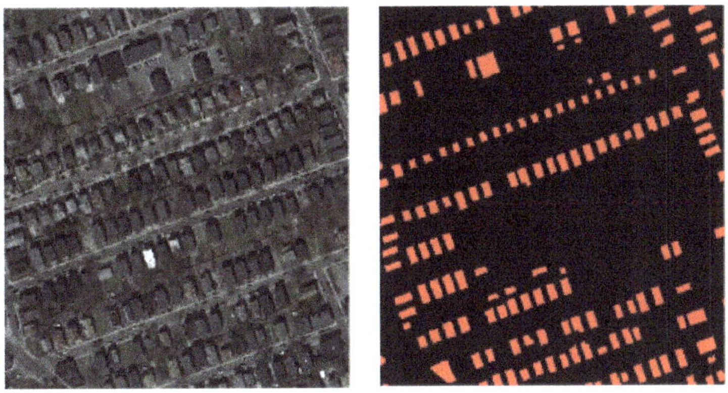

(a) (b)

Figure 4. Example data for illustration of the proposed feature extraction and selection techniques. (**a**) Example image, and (**b**) outdated map.

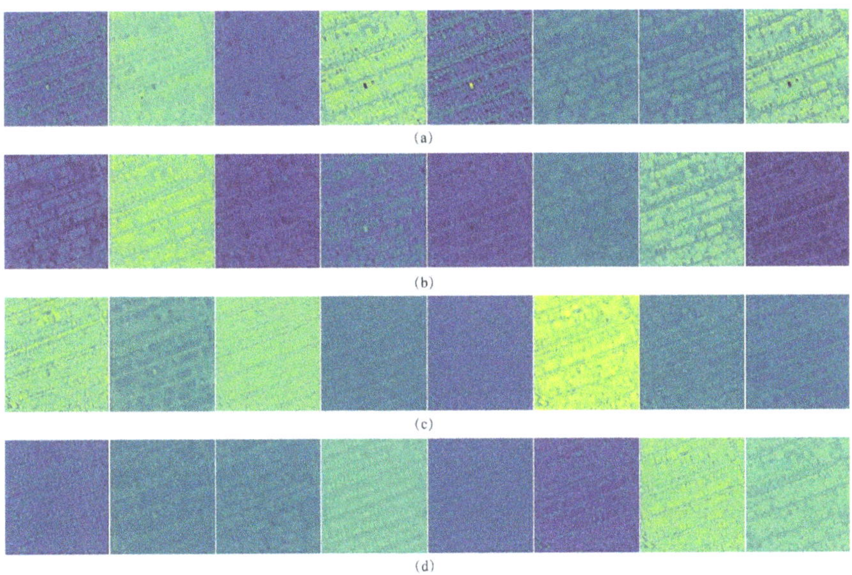

Figure 5. Eight randomly selected feature maps from each layer of the FCFE; (**a**) conv1_1; (**b**) conv1_2; (**c**) conv2_1; (**d**) conv2_2.

Sixteen most important features chosen after feature selection are shown in Figure 6. Three properties of selected features can be seen in Figure 6: (1) By filtering the ineffective features out, the remaining features are more representative and visually separable; (2) selected feature maps are functionally versatile. It is also seen that (a,d,e,h,o) positively respond to the buildings, whereas (b,c,f,j,k) negatively respond to the buildings; and (I,m,p) strongly respond to shadows and are actually shadow detectors. Since the buildings are supposed to be near, where the shadows appear, the detection of shadows can positively support the recognition of buildings. (3) Features from four convlayers are all selected to form the multi-scale features. As stated before, features from early layers contain low-level knowledge, such as positions and boundaries, while features from latter layers encode high-level intuitions, such as neighboring and contextual information. Based on that, the selected features are complementary and representative, and they are combined into a feature descriptor for HRS images.

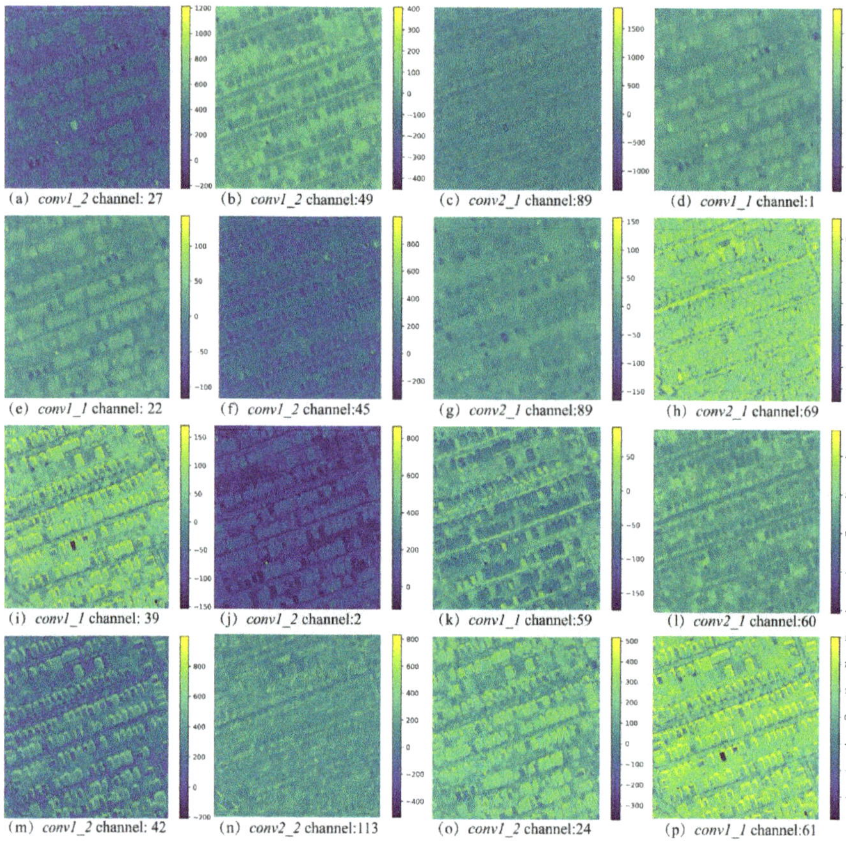

Figure 6. Sixteen most important features were selected by RF with the support of existing building basemaps.

2.3. Initial Classification by Automatic Sample Selection Using RF

As noise label is used to guide the feature selection. This however may harm the classification result compared to the pure label. Therefore, the existing basemaps are viewed as noisy labels of newly acquired HRS image; then, the selected deep features are utilized to purify the initial labels through a data cleaning procedure.

In the field of machine learning, data cleaning is often introduced in the classification task with noisy labels, and intends to identify and correct mislabeled samples [45]. The core of the data cleaning idea lies in estimating the label uncertainty of each sample. Note that in the label uncertainty estimation step, the training data is also noisy. Therefore, classifiers that are robust to label noise are preferable. Most classifiers are highly sensitive to the label noise, such as SVM and AdaBoost. However, some algorithms can avoid the effect of label noise to an extent. As mentioned before, the random forest is an ensemble decision tree classifier that introduces randomness in both samples and features selection, which makes it more robust, thus suitable for data cleaning tasks.

Inspired by the work in Reference [46], we use a cross-validation algorithm to estimate the uncertainty of the samples' labels. The pseudocode for estimating the uncertainty of the initial labels is given in Algorithm 1.

Algorithm 1. Label uncertainty estimation

Input: S (sample set, i.e., pixel index from HRS image) with F (features from Section 2.2), L (noisy label acquired from the existing basemaps); k_{max} (pre-defined times of dataset partition); N_{est} (number of RF meta-estimators); D_{max} (max depth of the decision trees in RF)
Procedure:

(1) Divide S into S_{pos}, and S_{neg} according to L.
(2) Initialize M_u as N-dimensional zero vectors as the label uncertainty estimator, N is sample capacity.

For k in range(k_{max}):

(3) Randomly divide S_{pos} into equally-sized $S^k{}_{pos1}$ and $S^k{}_{pos2}$. Almost equally-sized $S^k{}_{neg}$ are randomly chosen from S_{neg}.
(4) Train RF classifier, $RF^k{}_{pos1neg}$, with $S^k{}_{pos1}$ and $S^k{}_{neg}$. Predict the label of $S^k{}_{pos2}$, $C^k{}_{pos2}$. Update M_u for negative $C^k{}_{pos}$
(5) Estimate the label uncertainty of $S^k{}_{pos1}$ that is similar to step (4).
(6) Estimate the label uncertainty of S_{neg} as (4), (5).

End for
Output: Accumulator M_u indicating the label uncertainty of S.

For supervised machine learning, equally-sized training samples for each class are preferable. However, in satellite images, the background usually occupies more space than that of the buildings. In order to adjust the bias introduced by unbalancing distribution of samples, a larger penalty is imposed on inconsistent label prediction results of the background samples, i.e.,

$$M_u[L(S) \neq L_p(S)] = \begin{cases} 1 & \text{if } L(S) = \text{pos} \\ \sqrt{N_{neg}/N_{pos}} & \text{otherwise} \end{cases}, \tag{1}$$

where M_u is an accumulative matrix describing label uncertainty of each sample, $L(S)$ is the noisy label of S, $L_p(S)$ is the label predicted by the classifier, N_{neg}, and N_{pos} are the number of background, and building pixels, respectively.

After obtaining Mu, $r = M_u/k$ is calculated for each pixel, then a pixel with $r > 0.5$ is a possible mislabeled sample. Otherwise, it is considered as a clean sample. Finally, these cleaned samples are used to train an RF classifier, rFfinal, to predict the label of potentially changed samples to building or other class. The label probability of each sample is also obtained by Rffinal, which is then used for subsequent post-processing.

2.4. Post-Optimization Using Graph Cuts and Change Map Computing

Since the data cleaning processing is conducted pixel-wise, and little contextual information is taken into account, the initial classification result is fragmented. To ensure neighborhood consistency, post-optimization processing is formulated as an energy minimization problem, and graph cuts [47] algorithm that are performed on superpixels instead of entire pixels are used to find the solution and ensure the efficiency.

Here we use the SLIC algorithm to segment the HRS image into superpixels. It is shown that SLIC generates compact superpixels adhering tightly to the boundary [35]. The probability of the superpixel belonging to each class (building or other) is then calculated using Equation (2). It includes two aspects: (1) The averaged label probability of pixels in the superpixel; and (2) the proportion of pixels belongs to the current class.

$$\mathrm{p}(L(Spix) = c) = 0.5 \times (\sum\nolimits_{pix \in Spix} \mathrm{p}(L(pix) = c) + \frac{|pix \in Spix, L(pix) = c|}{|pix \in Spix|}) \tag{2}$$

where *Spix* is the superpixel, *pix* are the pixels belonging to *Spix*, *c* is the label of two defined classes, $L(x)$ returns the label of x, and |s| is the number of elements in set s.

The basic idea of graph cuts is to incorporate prior knowledge of label assignment, and the penalty imposed on adjacent superpixels with different labels, into a weighted graph. We then construct an energy function on the graph, and the optimal label assignment is obtained by optimizing the energy function defined as:

$$E = \sum_{i} D(c_i) + \lambda \sum_{i<j} S(c_i, c_j). \tag{3}$$

The first term, $D(c_i)$, is the data term which is determined by the negative logarithm of the probability obtained from Equation (3) and defined as

$$D(c_i) = -\log(p(L(Spix_i) = c_i)) \tag{4}$$

The second term in Equation (3), $S(c_i, c_j)$, is the smooth term, imposing a penalty on adjacent superpixels with different labels according to their similarity. Metric of spectral difference, i.e., Gaussian kernel of the averaged RGB feature, is utilized as the similarity measurement. Since the longer boundary is shared between the two superpixels, the higher their influence will be on each other, the penalty is weighted on the mutual border length. The smooth term employed in this paper is defined as:

$$S(c_i, c_j) = w(i, j) \times \exp \frac{(\|f_i - f_j\|)}{\sigma^2} \times \delta(i, j), \tag{5}$$

where

$$w(i, j) = \frac{bon(i, j) \times |N(i)|}{\sum\limits_{j \in N(i)} bon(i, j)}, \tag{6}$$

$$\delta(i, j) = \begin{cases} 1 & \text{if } c_i \neq c_j \\ 0 & \text{otherwise} \end{cases}, \tag{7}$$

σ is the standard deviation of Gaussian Kernel; f_i, f_j are the averaged RGB feature of ith and jth superpixels, respectively; $bon(i, j)$ is the shared border length of the ith and jth superpixels; $|N(i)|$ is the number of neighbors of superpixel i; and c_i is the label of superpixel i.

The parameter, λ, in Equation (3) controls the proportion of smooth term in the energy function. The larger the value of λ, the heavier will be the penalty imposed on the adjacent superpixels with different labels. This leads to more smoothing effects. The value of λ is related to the size of buildings in HRS image. If most buildings are small, consisting of only a few superpixels, λ needs to be reduced to avoid over-smoothing of the building superpixels by the surrounding background superpixels. Otherwise, λ, is set to a larger value to introduce a better smoothing effect.

After building the energy function, the maximum flow of the graph [48] is obtained to get the minimum cuts and obtain the optimal label for each superpixel. After obtaining the final classification result of the new HRS images, the labels of the images are compared to the existing map to obtain the change map.

3. Experimental Results and Discussion

The proposed framework is implemented using python language. Pre-trained model weights of FCN-8s are obtained from (https://github.com/shelhamer/fcn.berkeleyvision.org) under caffe [49] framework and then transformed into tensorflow (https://www.tensorflow.org/) readable form, and reconstructed into fully convolutional feature extractor (FCFE). Graph cuts are implemented using PyMaxflow (https://github.com/pmneila/PyMaxflow).

3.1. Experiment Setup

3.1.1. Datasets Description

To evaluate the proposed method, we use five datasets as shown in Figure 7, they include two sets, including ISPRS simulated dataset, and Boston real dataset—for details, see Table 1:

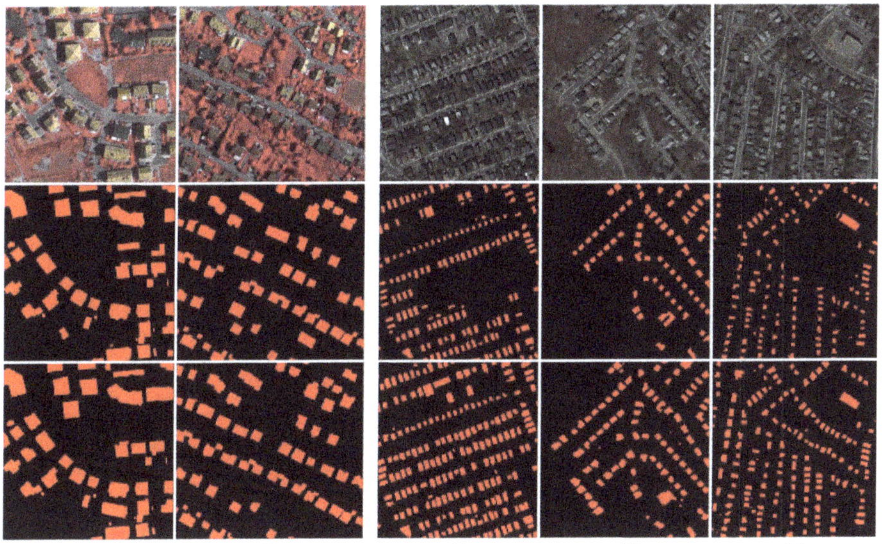

Figure 7. Experimental data sets: (**a**,**b**) ISPRS simulated dataset, (**c**–**e**) Boston real dataset (the first row is the newly acquired HRS image, the middle row is the outdated building map, and the third row is the ground truth building map for new HRS images).

Table 1. Details of newly acquired HRS images in five datasets.

Dataset		Source	Size (pixels)	Spatial Resolution (m)
ISPRS simulated dataset	a	Aerial	1996 × 1995	0.09
	b	Aerial	2818 × 2558	0.09
Boston real dataset	c	Google Earth	1031 × 1097	1
	d	Google Earth	1132 × 1139	1
	e	Google Earth	1159 × 1179	1

ISPRS simulated dataset: Two airborne images from ISPRS 2D semantic segmentation benchmarks (downloaded from http://www2.isprs.org/commissions/comm3/wg4/2d-sem-label-vaihingen.html) are employed to simulate two synthetic datasets as newly acquired HRS images. Approximately 10% of new building labels are randomly added. To simulate the outdated basemaps, 15% of the existing labels are deleted from the ground truth.

Boston real dataset: Three real datasets are selected from the urban areas of Boston, USA. The outdated basemaps are obtained from an existing classification dataset [50] (downloaded from http://www.cs.utoronto.ca/~vmnih/data/), and regions that contain obvious changes are cropped. Then the corresponding newly acquired HRS images are downloaded from Google Earth. The main challenges with this dataset are: (1) Backgrounds are heterogeneous and share spectral similarity with the buildings; therefore, pure pixel-based change detection may result in a high false-positive rate. (2) Buildings are relatively small; therefore, object-based strategies may suffer from instability of random classifiers. This may lead to false-negative outcomes. (3) Labels of the existing buildings suffer from severe mis-registration error, which makes information about building samples inaccurate. In order to evaluate the effectiveness of the proposed framework, an expert person is also invited to delineate the buildings' boundaries from the HRS images. The results are then reviewed by another expert, both independent of the experiment.

3.1.2. Assessment Criteria

In image-image change detection, the recognition result is a change map indicating the location of pixels that are notably different between multiple images. The result of image-map comparison is the updated label map. Similar criteria can be used to assess the accuracy assessment in both change detection techniques. In this paper, three evaluating indexes are obtained in pixel-wise fashion to evaluate the accuracy of the change detection result, including, completeness (Comp), false detection rate (FDR), and overall accuracy (OA):

$$Completeness = \frac{C_d}{C_t}, \tag{8}$$

$$FDR = 1 - \frac{C_d}{C_a}, \tag{9}$$

$$OA = \frac{C_d + C_n}{C}, \tag{10}$$

where C_d is the number of changed pixels (both background to building and building to background) that are correctly detected, C_t is the number of really changed pixels between newly acquired HRS image and the outdated basemap, C_a is the number of all the pixels that are labeled differently in the new labeled map, and the outdated basemap, C_n is the number of unchanged pixels that are correctly detected, and C is the number of pixels in the HRS image. Completeness measures the percentage of successfully corrected changed pixels among all changed pixels, whereas *FDR* reflects the proportion of false change pixels that are labeled as changed by the proposed algorithm. The *OA* also determines the comprehensive detection capability by taking both changed and unchanged pixels into account.

3.1.3. Parameters Setting

There are three parameters having a high impact on the results. All these parameters are set based on trial and error. Unless otherwise stated, these parameters are used in our experiments.

The first one is a max depth of the RF classifier, D_{max}, which determines the degree to which RF fits the training set. For a small D_{max}, RF is under-fit to the training set resulting in a high variance. If D_{max} is set to a large value, RF tends to over-fit to the mislabeled data in the training sets, resulting in a high bias. To balance the completeness and FDR, we set $D_{max} = 11$.

Compared to D_{max}, a number of decision tree estimators, N_{est}, in RF has trivial effects on the data cleansing accuracy. For $N_{est} < 5$, OA and FDR slightly fluctuate, due to the intrinsic randomness of the meta-classifiers, whereas for $N_{est} > 5$, OA and FDR converge to a fixed level. Since the computational demands are linearly proportional to N_{est}, we set its value to the minimum stable value of 5.

The main parameters of the post-optimization are the proportion of smooth term, λ, and the standard deviation of Gaussian kernel, σ.

Parameter λ controls the smoothness of the classification result. For a small λ, graph cuts tend to undersmooth the label results, and thus, holes and gaps of building labels and spurious fragmentations are under smoothened, causing a low completeness and OA, and a high FDR. For a very large λ, the label results are over smoothened and lots of existing buildings are obliterated, causing the bounce of FDR and re-sink of completeness and OA. Here, we set λ equal to 1.0 for ISPRS datasets, and 0.3 for Boston datasets. The value of σ is also set to 10.

3.2. *Results of ISPRS Simulated Data*

3.2.1. Change Detection Results

The detection results of the ISPRS datasets are presented in Figure. 8. The middle row of Figure 8 presents the initial classification results. The bottom row of Figure 8 shows the results after optimization by using a graph cuts algorithm. Initial results show that most of the new buildings are detected. However, these building labels have holes and gaps that undermine OA. Moreover, in areas that share similar spectral textual characteristics with the buildings, such as bare soil and roads, spurious and fragmented building labels occur. This results in a high FDR. After optimization, more pure building extraction results are obtained.

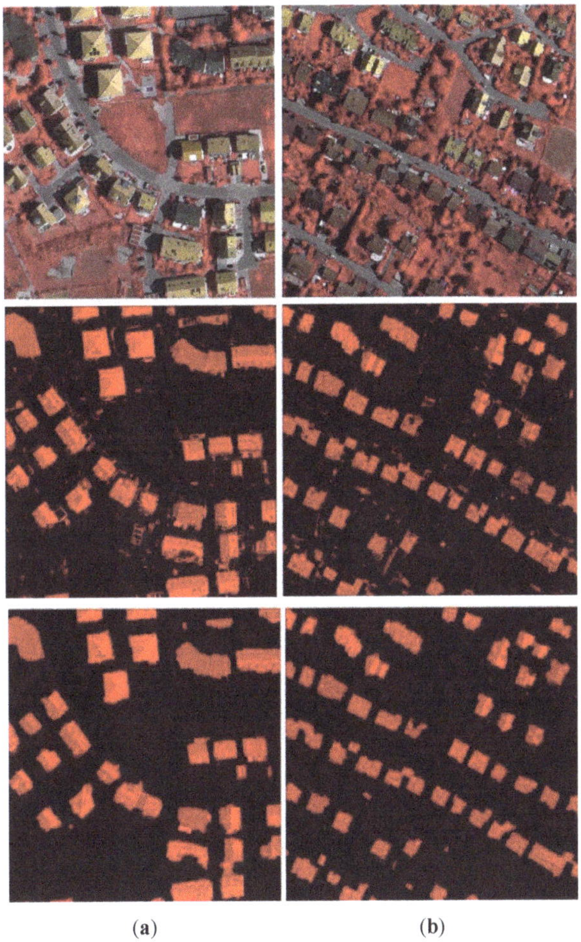

Figure 8. Experiment results: (**a**) results of the ISPRS simulated dataset a, (**b**) results of the ISPRS simulated dataset b (the first row is the HRS images, the second row is the initial classification results, and the third row is the final classification results).

3.2.2. Results with Different Label Noise Levels

Here we analyze the performance of the proposed method on data sets with different levels of label noise and the overall accuracy w.r.t. different settings are explored. The HRS images, as shown in

Figure 7a,b, are segmented into superpixels with the approximate size of the buildings. The labels of specified proportions of superpixels (ranging from 5% to 50%) are then selected randomly and flipped to introduce different levels of noise. The whole procedure of the proposed method is then performed on these modified data sets, and the results are presented in Figure 9.

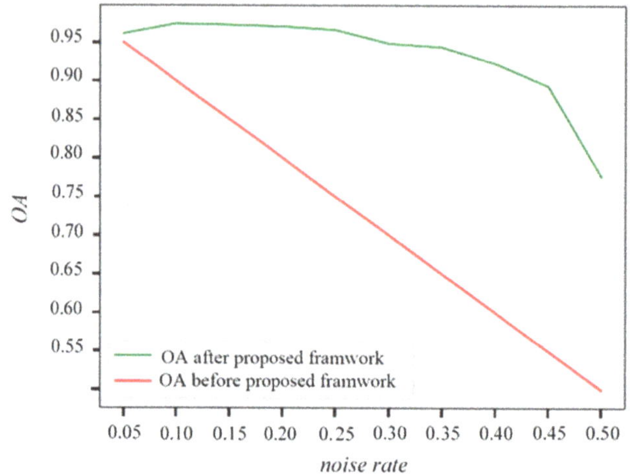

(a)

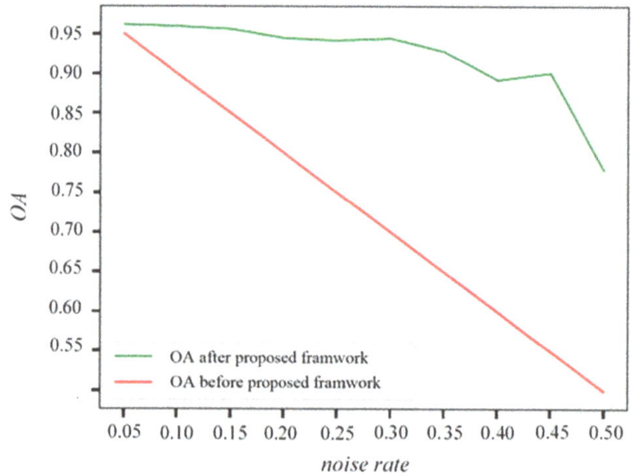

(b)

Figure 9. Overall accuracy w.r.t different simulated noise levels: (**a**) results of ISPRS dataset a, (**b**) results of ISPRS dataset b.

The results indicate that for noise rates up to 40%, the overall accuracy of the proposed method is above 90%. Even in cases where the original noise rate reaches as high as 50% (which means the information provided by outdated basemaps are mixed), the proposed framework is able to obtain an accuracy of 75%. This indicates the effectiveness of the proposed method.

3.3. *Results of Boston Real Dataset*

3.3.1. Detection Results

Figure 10 shows the outcomes of the initial classification results of Boston real datasets. Comparing the results obtained by the proposed method (the middle row of Figure 10) and the ground truth map (the bottom row of Figure 10), it is seen that most of the new buildings are correctly detected, and mis-registration errors are corrected. However, these building labels have holes and gaps that undermine OA.

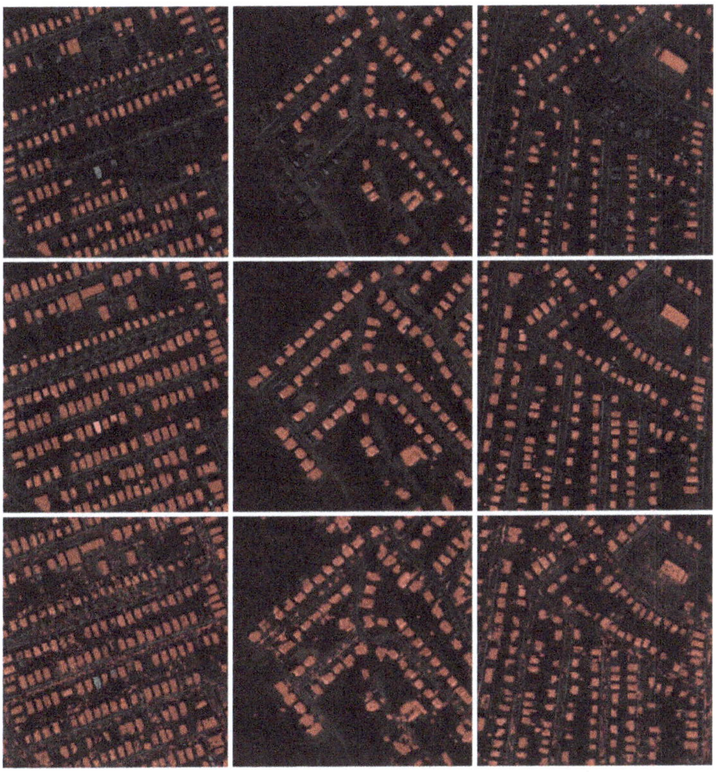

(a) (b) (c)

Figure 10. Initial classification result: (**a**) results of Boston real dataset c, (**b**) results of Boston real dataset d, (**c**) results of Boston real dataset e (first row—outdated basemap; middle row—groundtruth; third row—data cleansing result).

After optimization using graph cuts, the results are presented in the third row of Figure 11. Compared with the first row in Figure 11, it is seen that the phenomenon of small segments is removed, and the building extraction results are more accurate. Based on the optimized classification results, we obtain the change maps and compare them with the ground truth of the change map. The results are shown in the fourth row of Figure 11, where the red color means the changes are correctly detected, and the green means the changes are not detected.

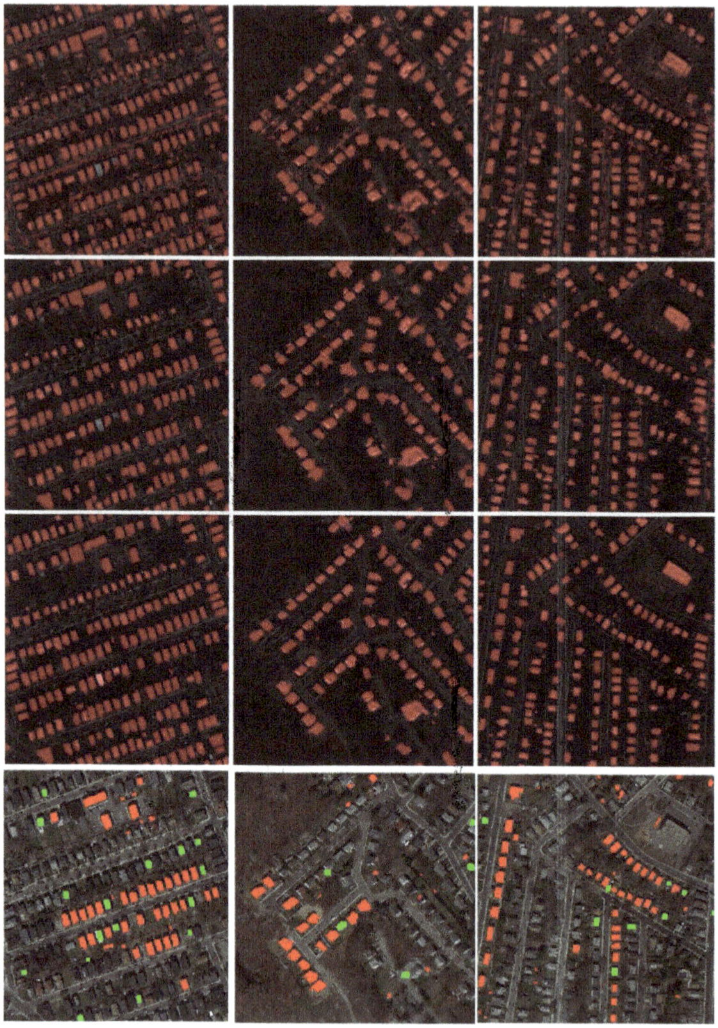

(a) (b) (c)

Figure 11. Results after post-optimization: (**a**) results of Boston real dataset c, (**b**) results of Boston real dataset d, (**c**) results of Boston real dataset e (first row—building label maps before optimization by object-based analysis and graph cuts; middle row—building label maps optimized by object-based analysis and graph cuts; third row—building map ground truth; fourth row—change map, where red means the changes are correctly detected, green means they are not).

3.3.2. Performance Comparison

In order to demonstrate the effectiveness of the proposed method, comparisons are made to three benchmarking methods, namely, A, B, and C. Method A employs the same framework as the proposed method, but uses conventional spatial-spectral features by combing GLCM textural features and normalized RGB, to replace the feature detector in our method. Method B employs a deep feature extractor as in Reference [24], and then follows the following steps: (1) Segmentation of the HRS images into superpixels; (2) cropping the bounding box of each superpixel, feeding it into ImageNet pre-trained VGGNet, extracting 4096-dimensional features from fc7, and reducing them to 100-dimensional using principal component analysis; (3) cleansing the data using graph cuts optimization. Method C is a fully pixel-based method that directly uses pixel-wise re-predicted label map for graph cuts optimization.

For the four methods to be comparable, the receptive field of features is set to 15, which is the same as the proposed method. Meanwhile, all the hyperparameters are determined through a grid search to obtain the highest performance. The accuracy results are shown in Table 2. The results confirm that the proposed method overperforms methods A, B, and C.

Table 2. Comparison Results.

Method	Dataset (c)			Dataset (d)			Dataset (e)		
	Comp	FDR	OA	Comp	FDR	OA	Comp	FDR	OA
Proposed	**0.861**	**0.269**	**0.942**	**0.878**	**0.268**	**0.966**	**0.890**	**0.223**	**0.963**
A	0.736	0.645	0.798	0.784	0.732	0.822	0.762	0.733	0.761
B	0.419	0.495	0.874	0.246	0.594	0.919	0.304	0.600	0.887
C	0.746	0.431	0.896	0.759	0.372	0.948	0.755	0.468	0.907

Compared with the proposed method, Method A shows a lower AR and a higher FDR. This shows that the deep features perform better than the hand-crafted features. Method B employs an earlier deep feature extraction strategy, however its performance on the experiment data is very low. The reason is that the buildings in the used datasets are generally small; this leads to two problems in direct segmentation of the HRS images into objects and in data cleansing: (1) The number of building samples is severely decreased, therefore, enough information is unavailable to distinguish background from the building; (2) a single building only consists of few superpixels, this makes the building objects vulnerable to the instability of random classifiers and/or over-smoothing by surrounding background objects. Nevertheless, with additional pixel-wise graph cuts post-processing in Method C, the accuracy remains low compared to the initial classification result. This is because the graph cuts algorithm punishes adjacent pixels with different labels and the correction of spurious clique needs lots of energies. Therefore, they cannot be corrected through max-flow optimization of the energy function. On the contrary, holes in building labels and fragmentations in non-building areas may dilate, leading to decreasing AR and OA.

All the experiments were performed on a laptop computer with Intel Core i7-7700HQ at a 2.8 GHz CPU with 32 GB memory, and an NVIDIA GTX1060MAXQ GPU (with 6.0 GB memory). The processing time is about five minutes for the three real data sets.

4. Conclusions and Future Works

In this paper, we proposed a novel framework for image-map building change detection. First, we demonstrated the representative ability of the features extracted from the early convlayer of pre-trained DCNNs and proved the feasibility of selecting important features using outdated building basemaps. Then, a random forest-based data cleansing method was implemented to preliminarily detect and correct changed pixels. The pixel-level re-predicted label maps were, however, fragmented, therefore, we adopted object-based analysis to introduce contextual information and ameliorate spurious predictions. We then used a graph cuts algorithm to optimize the label assignment results.

There are some limitations in the proposed method; for instance, a sparse distribution of the buildings may result in omission errors. Since FCFE demonstrates high efficiency in dense feature descriptors, it can be used in other tasks, such as classification and image registration [51].

Author Contributions: Conceptualization, Y.Z. (Yunsheng Zhang), J.P.; Methodology, Y.Z. (Yunsheng Zhang), Y.Z. (Yaochen Zhu), J.P.; Software, Y.Z. (Yaochen Zhu), S.C.; Validation, Y.Z. (Yaochen Zhu), S.C.; Resources, H.L., L.Z.; Writing—Original Draft Preparation, Y.Z. (Yunsheng Zhang), Y.Z. (Yaochen Zhu), J.P.; Writing—Review & Editing, Y.Z. (Yunsheng Zhang), H.L., L.Z. All authors have read and agreed to the published version of the manuscript.

Funding: This research was funded by Hunan Provincial Natural Science Foundation of China (No. 2018JJ3637), Natural Science Foundation of China (No. 51978283), Open Fund of Key Laboratory of Urban Land Resource Monitoring and Simulation, Ministry of Land and Resource (No. KF-2018-03-047).

Conflicts of Interest: The authors declare no conflict of interest.

References

1. Chen, J.; Lu, M.; Chen, X.; Chen, J.; Chen, L. A spectral gradient difference based approach for land cover change detection. *ISPRS J. Photogram. Remote Sens.* **2013**, *85*, 1–12. [CrossRef]
2. Kalnay, E.; Cai, M. Impact of urbanization and land-use change on climate. *Nature* **2003**, *423*, 528–531. [CrossRef] [PubMed]
3. Dong, L.; Shan, J. A comprehensive review of earthquake-induced building damage detection with remote sensing techniques. *ISPRS J. Photogram. Remote Sens.* **2013**, *84*, 85–99. [CrossRef]
4. Han, D. Construction monitoring of civil structures using high resolution remote sensing images. In Proceedings of the 13th SGEM GeoConference on Informatics, Geoinformatics and Remote Sensing, Albena, Bulgaria, 16–22 June 2013; pp. 595–600.
5. Bouziani, M.; Goïta, K.; He, D.C. Automatic change detection of buildings in urban environment from very high spatial resolution images using existing geodatabase and prior knowledge. *ISPRS J. Photogram. Remote Sens.* **2010**, *65*, 143–153. [CrossRef]
6. Dianat, R.; Kasaei, S. Change detection in optical remote sensing images using difference-based methods and spatial information. *IEEE Geosci. Remote Sens. Lett.* **2009**, *7*, 215–219. [CrossRef]
7. Tewkesbury, A.P.; Comber, A.J.; Tate, N.J.; Lamb, A.; Fisher, P.F. A critical synthesis of remotely sensed optical image change detection techniques. *Remote Sens. Environ.* **2015**, *160*, 1–14. [CrossRef]
8. Guo, Z.; Du, S. Mining parameter information for building extraction and change detection with very high-resolution imagery and GIS data. *GISci. Remote Sens.* **2017**, *54*, 3–63. [CrossRef]
9. Kaiser, P.; Wegner, J.D.; Lucchi, A.; Jaggi, M.; Hofmann, T.; Schindler, K. Learning aerial image segmentation from online maps. *IEEE Trans. Geosci. Remote Sens.* **2017**, *55*, 6054–6068. [CrossRef]
10. Taili, W.; Hongyang, L.; Qikai, L.; Nianxue, L. Classification of high-resolution remote-sensing image using openstreetmap information. *IEEE Geosci. Remote Sens. Lett.* **2017**, *14*, 2305–2309.
11. Gevaert, C.M.; Persello, C.; Elberink, S.O.; Vosselman, G.; Sliuzas, R. Context-based filtering of noisy labels for automatic basemap updating from UAV data. *IEEE J. Sel. Top. Appl. Earth Obs. Remote Sens.* **2017**, *11*, 2731–2741. [CrossRef]
12. Chen, S.; Zhang, Y.; Nie, K.; Li, X.; Wang, W. Extracting building areas from photogrammetric DSM and DOM by automatically selecting training samples from historical DLG data. *ISPRS Int. J. Geo-Inf.* **2020**, *9*, 18. [CrossRef]
13. Ojala, T.; Pietikainen, M.; Maenpaa, T. Multiresolution gray-scale and rotation invariant texture classification with local binary patterns. *IEEE Trans. Pattern Anal. Mach. Intell.* **2002**, *24*, 971–987. [CrossRef]
14. Mirzapour, F.; Ghassemian, H. Using GLCM and Gabor filters for classification of PAN images. In Proceedings of the 2013 21st Iranian Conference on Electrical Engineering (ICEE), Mashhad, Iran, 14–16 May 2013; pp. 1–6.
15. Krizhevsky, A.; Sutskever, I.; Hinton, G. ImageNet classification with deep convolutional neural networks. In Proceedings of the Advances in Neural Information Processing Systems (NIPS), Lake Tahoe, NV, USA, 3–6 December 2012; pp. 1097–1105.
16. Ren, S.; He, K.; Girshick, R.; Sun, J. Faster R-CNN: Towards real-time object detection with region proposal networks. *IEEE Trans. Pattern Anal. Mach. Intell.* **2017**, *39*, 1137–1149. [CrossRef] [PubMed]
17. Long, J.; Shelhamer, E.; Darrell, T. Fully convolutional networks for semantic segmentation. In Proceedings of the IEEE Conference on Computer Vision and Pattern Recognition (CVPR), Boston, MA, USA, 8–10 June 2015; pp. 3431–3440.
18. Yao, C.; Luo, X.; Zhao, Y.; Zeng, W.; Chen, X. A review on image classification of remote sensing using deep learning. In Proceedings of the 2017 3rd IEEE International Conference on Computer and Communications (ICCC), Chengdu, China, 13–16 December 2017; pp. 1947–1955.
19. Cheriyadat, A.M. Unsupervised feature learning for aerial scene classification. *IEEE Trans. Geosci. Remote Sens.* **2014**, *52*, 439–451. [CrossRef]

20. Zhang, P.; Gong, M.; Su, L.; Liu, J.; Li, Z. Change detection based on deep feature representation and mapping transformation for multi-spatial-resolution remote sensing images. *ISPRS J. Photogram. Remote Sens.* **2016**, *116*, 24–41. [CrossRef]
21. Vincent, P.; Larochelle, H.; Lajoie, I.; Bengio, Y.; Manzagol, P.A. Stacked denoising autoencoders: Learning useful representations in a deep network with a local denoising criterion. *J. Mach. Learn. Res.* **2010**, *11*, 3371–3408.
22. Marmanis, D.; Datcu, M.; Esch, T.; Stilla, U. Deep Learning Earth Observation Classification Using ImageNet Pretrained Networks. *IEEE Geosci. Remote Sens. Lett.* **2015**, *13*, 105–109. [CrossRef]
23. Bei, Z.; Bo, H.; Zhong, Y. Transfer learning with fully pretrained deep convolution networks for land-use classification. *IEEE Geosci. Remote Sens. Lett.* **2017**, *14*, 1436–1440.
24. Penatti, O.A.B.; Nogueira, K.; Dos Santos, J.A. Do deep features generalize from everyday objects to remote sensing and aerial scenes domains? In Proceedings of the 2015 IEEE Conference on Computer Vision and Pattern Recognition Workshops (CVPRW), Boston, MA, USA, 7–12 June 2015; pp. 44–51.
25. Fan, H.; Xia, G.S.; Hu, J.; Zhang, L. Transferring deep convolutional neural networks for the scene classification of high-resolution remote sensing imagery. *Remote Sens.* **2015**, *7*, 14680–14707.
26. Paoletti, M.E.; Haut, J.M.; Plaza, J.; Plaza, A. A new deep convolutional neural network for fast hyperspectral image classification. *ISPRS J. Photogram. Remote Sens.* **2018**, *145*, 120–147. [CrossRef]
27. Gong, J.; Hu, X.; Pang, S.; Li, K. Patch matching and dense CRF-based co-refinement for building change detection from Bi-temporal aerial images. *Sensors* **2019**, *19*, 1557. [CrossRef] [PubMed]
28. Huang, X.; Zhang, L. An SVM ensemble approach combining spectral, structural, and semantic features for the classification of high-resolution remotely sensed imagery. *IEEE Trans. Geosci. Remote Sens.* **2013**, *51*, 257–272. [CrossRef]
29. Chen, G.; Hay, G.J.; Carvalho, L.M.T.; Wulder, M.A. Object-based change detection. *Int. J. Remote Sens.* **2012**, *33*, 4434–4457. [CrossRef]
30. Griffiths, P.; Hostert, P.; Gruebner, O.; Linden, S.V.D. Mapping megacity growth with multi-sensor data. *Remote Sens. Environ.* **2010**, *114*, 426–439. [CrossRef]
31. Du, P.; Liu, S.; Gamba, P.; Tan, K.; Xia, J. Fusion of difference images for change detection over urban areas. *IEEE J. Sel. Top. Appl. Earth Obs. Remote Sens.* **2012**, *5*, 1076–1086. [CrossRef]
32. Hussain, M.; Chen, D.; Cheng, A.; Wei, H.; Stanley, D. Change detection from remotely sensed images: From pixel-based to object-based approaches. *ISPRS J. Photogram. Remote Sens.* **2013**, *80*, 91–106. [CrossRef]
33. Ma, L.; Li, M.; Thomas, B.; Ma, X.; Dirk, T.; Liang, C.; Chen, Z.; Chen, D. Object-Based Change Detection in Urban Areas: The effects of segmentation strategy, scale, and feature space on unsupervised methods. *Remote Sens.* **2016**, *8*, 761. [CrossRef]
34. Blaschke, T. Object based image analysis for remote sensing. *ISPRS J. Photogram. Remote Sens.* **2010**, *65*, 2–16. [CrossRef]
35. Achanta, R.; Shaji, A.; Smith, K.; Lucchi, A.; Fua, P.; Süsstrunk, S. SLIC Superpixels compared to state-of-the-art superpixel methods. *IEEE Trans. Pattern Anal. Mach. Intell.* **2012**, *34*, 2274–2282. [CrossRef]
36. Cheng, G.; Li, Z.; Han, J.; Yao, X.; Guo, L. Exploring hierarchical convolutional features for hyperspectral image classification. *IEEE Trans. Geosci. Remote Sens.* **2018**, *56*, 1–11. [CrossRef]
37. Matthew, D.; Fergus, R. Visualizing and understanding convolutional neural networks. In Proceedings of the 13th European Conference Computer Vision and Pattern Recognition (ECCV), Zurich, Switzerland, 5–12 September 2014; pp. 6–12.
38. Kemker, R.; Salvaggio, C.; Kanan, C. Algorithms for semantic segmentation of multispectral remote sensing imagery using deep learning. *ISPRS J. Photogram. Remote Sens.* **2018**, *145*, 60–77. [CrossRef]
39. Lin, G.; Milan, A.; Shen, C.; Reid, I. RefineNet: Multi-path Refinement Networks for High-Resolution Semantic Segmentation. In Proceedings of the 2017 IEEE Conference on Computer Vision and Pattern Recognition (CVPR), Honolulu, HI, USA, 21–26 July 2017; pp. 5168–5177.
40. Simonyan, K.; Zisserman, A. Very Deep Convolutional Networks for Large-Scale Image Recognition. In Proceedings of the International Conference on Learning Representations (ICLR), San Diego, CA, USA, 7–9 May 2015.
41. Chen, J.J.; Yuan, C.; Deng, M.; Tao, C.; Peng, J.; Li, H. On the Selective and Invariant Representation of DCNN for High-Resolution Remote Sensing Image Recognition. *arXiv* **2017**, arXiv:1708.01420.

42. Guyon, I.; Elisseeff, A. An introduction to variable and feature selection. *J. Mach. Learn. Res.* **2003**, *3*, 1157–1182.
43. Belgiu, M.; Dragut, L. Random forest in remote sensing: A review of applications and future directions. *ISPRS J. Photogram. Remote Sens.* **2016**, *114*, 24–31. [CrossRef]
44. Breiman, L. Random Forest. In *Machine Learning*; Springer: Berlin, Germany, 2001; Volume 45, pp. 5–32.
45. Zhu, X.; Wu, X. Class Noise vs. Attribute Noise: A Quantitative Study. *Artif. Intell. Rev.* **2004**, *22*, 177–210. [CrossRef]
46. Zhuqiang, L.; Liqiang, Z.; Ruofei, Z.; Tian, F.; Liang, Z.; Zhenxin, Z. Classification of urban point clouds: A robust supervised approach with automatically generating training data. *IEEE J. Sel. Top. Appl. Earth Obs. Remote Sens.* **2017**, *10*, 1–14.
47. Boykov, Y.Y.; Jolly, M. Interactive graph cuts for optimal boundary & region segmentation of objects in N-D images. In Proceedings of the Eighth IEEE International Conference on Computer Vision, ICCV 2001, Vancouver, BC, Canada, 7–14 July 2001; pp. 105–112.
48. Boykov, Y.; Kolmogorov, V. An experimental comparison of min-cut/max- flow algorithms for energy minimization in vision. *IEEE Trans. Pattern Anal. Mach. Intell.* **2004**, *26*, 1124–1137. [CrossRef]
49. Jia, Y.; Shelhamer, E.; Donahue, J.; Karayev, S.; Long, J.; Girshick, R.; Guadarrama, S.; Darrell, T. Caffe: Convolutional Architecture for Fast Feature Embedding. In Proceedings of the 22nd ACM International Conference on Multimedia, Orlando, FL, USA, 3–7 November 2014; pp. 675–678.
50. Mnih, V. Machine Learning for Aerial Image Labeling. Ph.D. Thesis, University of Toronto, Toronto, ON, Canada, 2013.
51. Zhang, L.; Zhang, L.; Bo, D. Deep Learning for Remote Sensing Data: A Technical Tutorial on the State of the Art. *IEEE Geosci. Remote Sens. Mag.* **2016**, *4*, 22–40. [CrossRef]

sensors

MDPI

Letter

Chimney Detection Based on Faster R-CNN and Spatial Analysis Methods in High Resolution Remote Sensing Images

Chunming Han [1,2], Guangfu Li [1,2], Yixing Ding [1,*], Fuli Yan [1] and Linyan Bai [1]

1 Key Laboratory of Digital Earth Science, Aerospace Information Research Institute, Chinese Academy of Sciences, Beijing 100094, China; hancm@radi.ac.cn (C.H.); ligf@radi.ac.cn (G.L.); yanfl@radi.ac.cn (F.Y.); baily@radi.ac.cn (L.B.)

2 School of Electronic Electrical and Communication Engineering, University of Chinese Academy of Sciences, Beijing 100049, China

* Correspondence: dingyx@radi.ac.cn

Received: 2 July 2020; Accepted: 1 August 2020; Published: 5 August 2020

Abstract: Spatially location and working status of pollution sources are very important pieces of information for environment protection. Waste gas produced by fossil fuel consumption in the industry is mainly discharged to the atmosphere through a chimney. Therefore, detecting the distribution of chimneys and their working status is of great significance to urban environment monitoring and environmental governance. In this paper, we use an open access dataset BUAA-FFPP60 and the faster regions with convolutional neural network (Faster R-CNN) algorithm to train the preliminarily detection model. Then, the trained model is used to detect the chimneys in three high-resolution remote sensing images of Google Maps, which is located in Tangshan city. The results show that a large number of false positive targets are detected. For working chimney detection, the recall rate is 77.27%, but the precision is only 40.47%. Therefore, two spatial analysis methods, the digital terrain model (DTM) filtering, and main direction test are introduced to remove the false chimneys. The DTM is generated by ZiYuan-3 satellite images and then registered to the high-resolution image. We set an elevation threshold to filter the false positive targets. After DTM filtering, we use principle component analysis (PCA) to calculate the main direction of each target image slice, and then use the main direction to remove false positive targets further. The results show that by using the combination of DTM filtering and main direction test, more than 95% false chimneys can be removed and, therefore, the detection precision is significantly increased.

Keywords: target detection; high resolution remote sensing image; chimney; faster R-CNN; spatial analysis

---

1. Introduction

In recent decades, rapid economic development has led to a significant increase in energy consumption. In China's primary energy share in 2019, the proportion of fossil energy consumption was still more than 85%, according to the BP Statistical Review of World Energy. The burning of fossil fuels will release a large amount of pollutants into the atmosphere, which will cause serious environmental problems and endanger the health of nearby residents. Among different pollutant discharge sources, the industry discharge contributes the most. The waste gas produced by fossil fuel consumption in industry is mainly discharged to the atmosphere through the chimney. Therefore, the distribution of working chimneys serve as a very important indicator of local air pollution situation. Detecting the number of chimneys and their working status is of great significance to urban environment monitoring and environmental governance.

Target detection on high-resolution remote sensing image provides an efficient and accurate way to detect the position and status of the chimney. There are two types of target detection algorithms: traditional algorithms and algorithms based on deep learning. The traditional algorithms, such as the Local Binary Pattern (LBP) [1] algorithm, scale-invariant feature transform (SIFT) [2] algorithm, and the Support Vector Machine (SVM) [3] algorithm, do not perform well in accuracy and robustness when used for dealing with complex recognition problems [4]. To increase the detection accuracy, a deep learning algorithm, convolutional neural network (CNN) [5], has been proposed to imitate the human brain neuron connection and transfer message mechanism. This kind of deep learning algorithm can be divided into two categories, the one-step algorithm and the two-step algorithm. The one-step algorithm, such as Single Shot MultiBox Detector (SSD) [6], and You Only Look Once (YOLO) [7], has less accuracy as well as lower computational cost. The two-step algorithm, such as region-based convolutional neural networks (R-CNNs) [8], Fast R-CNN [9], and Faster R-CNN [10], is characterized by its high accuracy and high time cost.

At present, deep learning has been successfully applied in remote sensing images in aircraft detection [11–13], ship detection [14–16], oil tank [17–19] detection with good performance. Several experiments on chimney detection have also been reported. Yao et al. [20] used the Faster R-CNN to detect the chimney and condensing tower. Zhang et al. [21] established the BUAA-FFPP60 dataset, which can be used not only to detect the targets, but also to confirm their working status. Comparison among different deep learning algorithms [6,10,22–27] is also made based on performance indicators, such as accuracy, model memory size, and running time, and results show that no single algorithm performs well in all aspects. Deng et al. [28] increased the number and scale of feature pyramids, based on the original Feature Pyramid Network (FPN), to improve the detection accuracy.

In practical application, the image always contains various artificial targets. Some targets are very close to the chimney in textures and geometric features, such as roads, building edges, and oil tanks. The Faster R-CNN for chimney detection in the aforementioned references is based on specific datasets that only contain manually selected chimneys. When the Faster R-CNN is used in a large-scale scene, there will be a large number of chimney-like targets that are misclassified into chimneys, leading to a significant decrease in precision. In order to improve the precision, we use two spatial analysis methods. The digital terrain model (DTM) is first introduced. DTM reflects the height fluctuation of ground objects. The chimney is a vertical object and appears elongated in the image. Therefore, where there is a chimney, the DTM will change dramatically. It can be used as a condition to determine whether there is a chimney by detecting the severity of the changes. In addition, in a high-resolution image, the field of view is relatively small, so the changes in observing angle in one image is small. Consequently, the chimneys in one image show the same pointing direction. In this paper, we call this direction the main direction of this image. Therefore, the detected objects that are not in accordance with main direction can be considered as false detections.

In this paper, we use BUAA-FFPP60 dataset [21] and Faster R-CNN algorithm to train the preliminarily detection model. Then, two spatial analysis methods, the DTM filtering and main direction test, are introduced to remove the false chimneys. The detailed description of the method is in Section 2, and the result discussion in Section 3. The results show that the elevation filtering and main direction test are both very effective in reducing false detection rate. Furthermore, the combination of these two methods show extremely good performance in increasing detecting precision.

2. Methodology

The method proposed in this paper consists of three parts: (1) the preliminary detection on enhanced images by Faster R-CNN, (2) the elevation filtering using local DTM, (3) the main direction test. The overall process diagram is given in Figure 1. Considering that the condensing tower is detected in former studies, its experimental results are preserved as comparative references. Furthermore, although the thermal infrared data are helpful for detecting the working chimneys, the resolutions of commonly accessible data are too low. Therefore, they are not used in this paper.

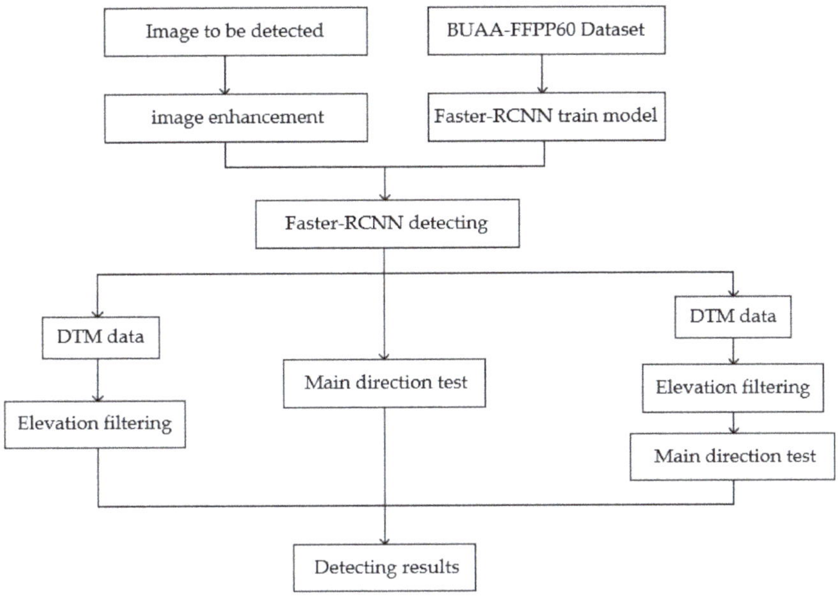

Figure 1. Process diagram.

2.1. Faster R-CNN for Target Detection

The Faster R-CNN is chosen for preliminary detection for its high accuracy in chimney detection compared with other methods [21]. As mentioned before, the Faster R-CNN contains two steps [10]. The first step is Region Proposal Network (RPN). RPN takes an image as input and outputs a set of rectangular object proposal regions, each with an objectness score. The second step is Fast R-CNN detection in the proposed regions. Both RPN and Fast R-CNN share the same convolutional layers, rather than learning two separate networks. Figure 2 shows the process structure of Faster R-CNN. It first performs the deep fully convolution on the input image to obtain feature maps. Then, the feature maps are used by RPN to generate proposal regions. Fast R-CNN uses feature map and proposal regions to generate region of interest (ROI) pooling. After that, the fully connected layer is used for classification and regression operations.

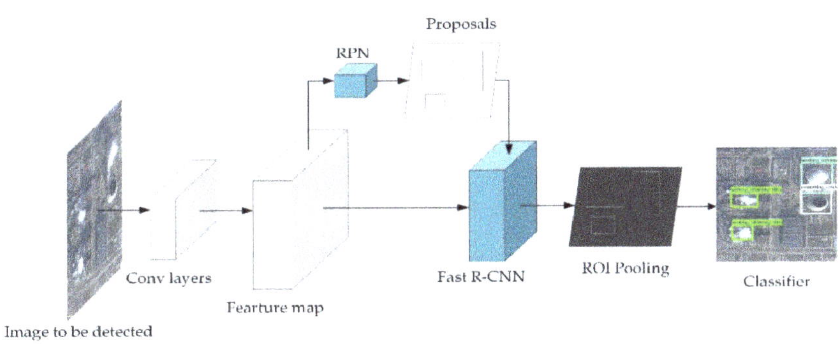

Figure 2. Faster region-based convolutional neural networks (R-CNN) structure diagram.

Different types of targets correspond to different anchors, which are a serious reference boxes in each sliding-window when region proposals are generated. Anchor size can be obtained from previous experience. In order to fit chimney and condensing tower detection, we set four types anchors of scales: 32^2, 64^2, 128^2, and 256^2, and five aspect types of ratios: 1:1, 1:2, 1:3, 2:1, and 3:1. The resnet101 [29] trained on coco [30] is selected as the pre-training model. This model is one of widely used model in the field of target detection because of high accuracy and speed.

2.2. The Elevation Filtering Using Local DTM

DTM is a digital description of the shape, size, and elevation of terrain. The chimney and condensing tower are usually higher than the surrounding features. In the place where there is a chimney or a condensing tower, the value of DTM shows obvious fluctuations, and the height difference can achieve as large as 20 m. In place where false detection appears, the value of DTM changes more gradually.

To get the DTM slice images, which are pieces of DTM image cut from whole DTM image correspondent to the target bonding box, the detection results of Faster R-CNN are registered to DTM first. Then, the bounding boxes are used to cut several slices from the DTM. Then statistical operations are performed in slices. The max and mean height of each DTM slice are calculated as follows:

$$V_{mean} = \frac{1}{m \times n} \sum_{i=1}^{m} \sum_{j=1}^{n} f(x_i, y_j) \tag{1}$$

$$V_{max} = Max(f(x_1, y_1), f(x_2, y_2) \ldots \ldots f(x_m, y_n)) \tag{2}$$

Where V_{mean} is average value of slice, V_{max} is the maximum value of slice, $f(x_i, y_j)$ is the pixel value of the slice, m and n are the number of rows and columns of the slice, respectively. The filter condition is given by:

$$\begin{cases} V_{max} - V_{mean} > T \\ V_{max} - V_{mean} < T \end{cases} \tag{3}$$

T is threshold value. The difference between the max height and mean height in the slice should be larger than the threshold, or else the detected object will be considered as false positive and removed from the set of detected chimneys. The value of threshold is set to be 20 m according to the National Standard of China, the Emission Standard of air pollutants for boiler [31], in which states that the coal combustion chimney should not be less than 20 m. Moreover, we also experimentally test 5 threshold values. The experiment results are shown in Table 1. When the threshold is 16 m or 18 m, the number of false positive targets is still too large. When the threshold increases to 20 m, although 3 chimneys are mis-removed, the number of false positive targets is greatly reduced. When the threshold is 22 m or 24 m, there will be too many mis-removed chimneys. Thus, a 20 m-threshold seems reach a good compromise between low mis-removal and effective deletion of false positive targets.

Table 1. Threshold experiments.

Threshold	Chimneys	Condensing Tower	False Positive Targets
0	79	9	178
16	79	9	81
18	77	9	63
20	76	9	25
22	70	9	21
24	62	8	18

2.3. Main Direction Test

The chimney is a long and vertical object. In the bounding box, the image slice, which contains a chimney, will show obvious directional texture features. Moreover, the chimney and the condensing tower in one high-resolution remote sensing image are all approximately pointed to the same direction, which is called main direction in this paper. We found that a lot of the mis-detected targets do not have the same feature. Therefore, the false chimneys can be further removed by testing its consistency with the main direction. The principle component analysis (PCA) is used to calculate the main direction of each image slice. The processing flow is:

(1) Gaussian filtering the image slice to remove noise interference;
(2) converting the image slice into a grayscale image;
(3) binarizing and extracting the position coordinates of non-zero pixels to construct a position matrix, and then calculating its covariance matrix;
(4) calculating the eigenvector corresponding to the max eigenvalue of covariance matrix;
(5) calculating the main direction angle of each slice according to the eigenvector.

Figure 3 shows two examples of using this method to find the main direction of each detected target. After calculating the main directions of all slices, the distribution histogram will be mapped at intervals of 5 degrees. The maximum value in the histogram is considered as the main direction d of the entire image. Then, the detected target whose main direction is close to the main direction of the image will be considered as true detection. The decision criteria is set to be $d \pm 5°$ for chimney, and $d \pm 8°$ for condensing tower since the condensing tower is much wider than the chimney in the image.

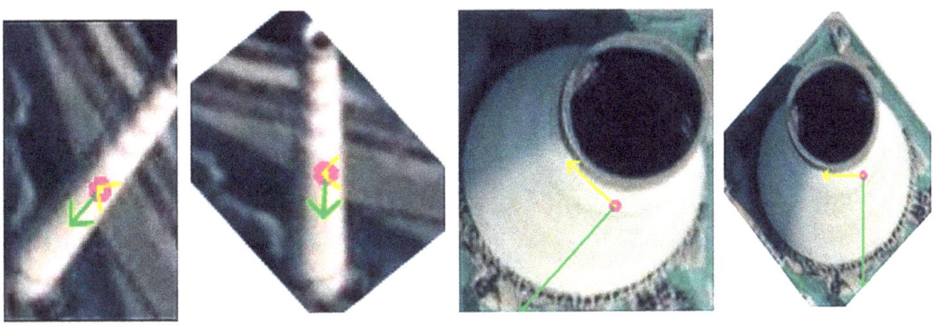

Figure 3. Main direction rotation image. The green arrow represents the main direction, while the yellow arrow represents the direction perpendicular to the main direction.

3. Results

3.1. Dataset, Experimental Area, and Data

The dataset used in this experiment is BUAA-FFPP60, which is collected and produced by Beihang University. The dataset is composed of chimneys and condensation towers distributed in the 123-km^2 power plant in the Beijing–Tianjin–Hebei area. There are 318 original pictures, of which 31 are test pictures. The remaining 287 pictures are mirrored or rotated by 90° to generate 861 training pictures. The pictures come from Google map with a resolution of 1 m, ranging in size from 500 × 500 to 700 × 1250 pixels. The working state of the chimney and condensation tower is determined by whether there is smoke. The four labels in the dataset are working chimney, non-working chimney, working condensation tower, and non-working condensation tower. Figure 4 shows some examples of dataset.

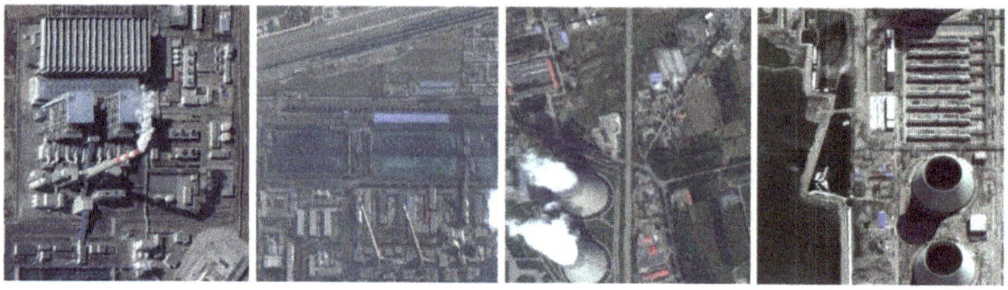

Figure 4. BUAA-FFPP60 dataset samples. Four subfigures indicate the working chimneys, non-working chimneys, working condensing towers, non-working condensing towers, respectively.

The area selected for this experiment is Tangshan City, Hebei Province, located 180 km southeast of Beijing. It is a regional core city of Beijing-Tianjin-Tangshan city group, and burdens the task of releasing the industrial pressure of Beijing, the capital of China. Tangshan City is a typical industrial city in North China, and the total crude steel production in 2018 is 133 million tons, about 7.35% of world's total production. Meanwhile, it is also one of the cities with the worst air quality in the country. According to the "Tangshan City Environmental Status Report", in 2011, the emissions of sulfur dioxide and nitrogen oxides in Tangshan City were 336.54 thousand tons and 40.59 thousand tons, respectively [32]. Numerous steel factories and power plants with a large number of chimneys and condensation towers in Tangshan have contributed the most to the hazardous air pollutants. Therefore, investigating the position and working status of chimneys and condensation towers is very important to region environmental governance.

Three Google Maps images with 1-m resolution covering about 600 km^2 are used for final detection. Sizes of images are 16,000 × 25,000 pixels, 10,000 × 10,000 pixels and 10,000 × 10,000 pixels, respectively. The images cover Lubei District, Guye District, Kaiping District, and Fengrui District. The images from ZiYuan-3 satellite with size of 24,500 × 20,000 is used to generate DTM.

3.2. Experimental Results and Analysis

3.2.1. Accuracy of Faster R-CNN Trained Model

We performed the experiments on a computer with a 2.5 GHz Central Processing Unit (CPU) and a NVIDIA GeForce GTX 2080Ti Graphics Processing Unit (GPU). The memory sizes of CPU and GPU are 8 GB and 11 GB, respectively. The TensorFlow [33] deep learning framework was selected to train 861 Google map images of the BUAA-FFPP60 dataset. The pre-training model is the resnet101 [29] model trained on coco [30]. The number of training iterations is 170,000 and the learning rate is 0.001.

To evaluate the detection accuracy of the Faster R-CNN models, we test the trained model on test image of BUAA-FFPP60 dataset. When the detect target is true, the test result is a true positive (TP), and when the detect targets is false, the test result is false positive (FP). The false negative (FN) indicates the number of undetected true target in the image. Then, we can combine these into three metrics, precision (P), recall (R), and quality (Q):

$$\mathrm{P} = \frac{TP}{TP + FP} \tag{4}$$

$$\mathrm{R} = \frac{TP}{TP + FN} \tag{5}$$

$$\mathrm{Q} = \frac{TP}{TP + FP + FN} \tag{6}$$

For test samples, the precisions of working chimney, non-working chimneys, condensing tower, and non-condensing tower are 0.7210, 0.7326, 0.9482, and 0.9551, respectively. The recall rates are 0.8674, 0.8642, 0.9707and 0.9659 respectively. The qualities are 0.6451, 0.6629, 0.9423, and 0.9473, respectively.

3.2.2. The Results from Faster R-CNN

After, we get the trained model. The Google images were input to the trained Faster R-CNN network. Due to the large area, the entire image is detected by window. The window size is 700×700 pixels and the step length is 500 pixels. The overlapped area in each step is as wide as 200 pixels, which is wide enough to prevent missing detection of chimneys at the edge of image. In order to detect more targets, we add an image enhancement method by adjusting the brightness and contrast ratio before Faster R-CNN detection. We also set a low network detection probability threshold, which is 0.3, to reduce the false negative and increase the recall rate.

In order to analyze the detection accuracy, we divide the detection results into nine types: working chimneys, non-working chimneys, working condensing towers, non-working condensing towers, road, architecture, tank, lake, topography. Figure 5 shows some examples of false detection.

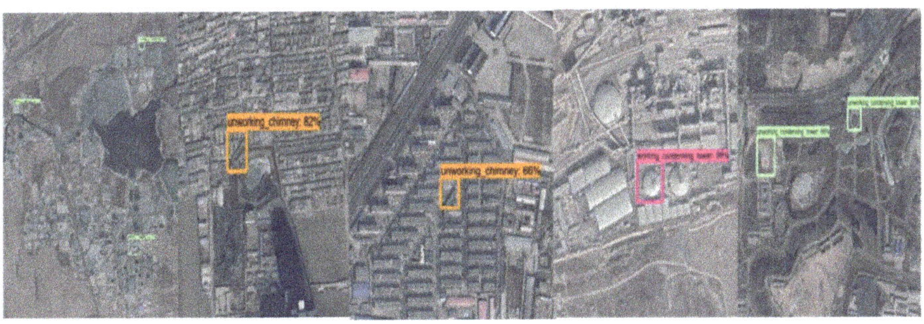

Figure 5. The false detection objects are divided into five categories: lake, road, architecture, tank, and other objects. The pink boxes represent working condensing tower, the green boxes represent non-working condensing tower, and the yellow boxes represent non-working chimney.

It can be found from Table 2 that the road and architecture are most likely to be mis-detected as chimneys, the number of which are 45 and 59 respectively. Condensing towers are most likely to be mixed up by tanks and lakes. The false detection rate of working chimneys, non-working chimneys, working condensing towers and non-working condensing towers are 0.5952, 0.5810, 0.8214, and 0.9166, respectively.

Table 2. Faster R-CNN detection result.

	Working Chimneys	Non-Working Chimneys	Working Condensing Towers	Non-Working Condensing Towers
working chimneys	17	1	0	0
non-working chimneys	0	62	0	0
working condensing towers	0	0	5	0
non-working condensing towers	0	0	0	4
roads	4	41	0	0
architectures	19	40	5	0
tanks	0	0	8	7
lakes	0	0	3	31
other objects	2	5	7	6
false detection rate	0.5952	0.5810	0.8214	0.9166

3.2.3. The Results from Faster R-CNN, Elevation Filtering, and Main Direction Test

It can be found from Table 3 that by using the detection and test method—most of the false chimneys are removed. The false detection rate of working chimneys, non-working chimneys, working condensing towers and non-working condensing towers are significantly reduced to 0.0555, 0.0634, 0.1667, and 0.2, respectively. Meanwhile, only three non-working chimneys are mis-removed. That means after processing the true chimneys have been well retained.

Table 3. Faster R-CNN+ elevation filtering + main direction detection result.

	Working Chimneys	Non-Working Chimneys	Working Condensing Towers	Non-Working Condensing Towers
Working chimneys	17	1	0	0
non-working chimneys	0	59	0	0
working condensing towers	0	0	5	0
non-working condensing towers	0	0	0	4
road	0	0	0	0
architecture	1	3	0	0
tank	0	0	1	1
lake	0	0	0	0
other objects	0	1	0	0
false detection rate	0.0555	0.0634	0.1667	0.2

3.2.4. Discussion

Five false targets are shown in Table 4. The first line shows that cable tower is detected as non-working chimney. The cable tower is highly similar to chimney in both texture feature and three-dimensional structure. The main direction of the image slice is 30.19°, while the main direction of the whole image is 42.23°. This difference may be caused by some decorative or structural curves on the cable tower, which makes it not so straight in the image. However, similar loaded or decorative component is seldom attached on a chimney, so the true chimney is unlikely to be mis-removed. In the second line, a big tank is mistakenly detected as a working condensing tower. They are similar in height, so cannot be distinguished by only introducing the DTM. However, its aspect ratio, which is much smaller than true condensing tower, make the calculation of main direction after binarization unstable, leading to a large different with the image main direction. For the chimney like objects (including condensing tower), which has large aspect ratio, the main direction is determined by the pixel value distribution of wall. For those with low aspect ratio, such as the oil tank, the main direction is highly affected by the pixel distribution of its top cover. Therefore, the main direction test is also useful to distinguish some objects with different aspect ratio. In line 3, a complex scene with working and non-working chimneys, oil tanks, and steam vents is shown. There are only two chimneys in this image, one undetected working chimney in the red circle. The reason why the working chimney in the red circle remains undetected is that the two spatial analysis methods introduced in this paper are ineffective to reduce the false negatives. We think that the improvement in detection ability of neural network and completeness of the training dataset might be helpful. The detected non-working chimney is in the upper left corner. The rest of the detected objects are all false. The objects with lower height, including a steam vent, can be removed by DTM filtering. The main direction test can remove all false target in line 3 because the main directions of most interfering targets are randomly distributed except some high vertical objects. However, it is possible that the main direction of interfering target is coincidently consistent with the main direction of the image. Two examples show in line 4 and 5. The false targets cannot be removed by main direction test are mainly ground texture, shadows or structure that caused by overlapping.

Table 4. Examples of four-class detection method results. The pink boxes represent working condensing tower, the green boxes represent non-working condensing tower, the blue boxes represent working chimney, and the yellow boxes represent non-working chimney.

NO.	Faster R-CNN Detection	Combination of Faster R-CNN and Elevation Filtering Detection	Combination of Faster R-CNN and Main Direction Detection	Combination of Faster R-CNN and Elevation Filtering and Main Direction Detection
1				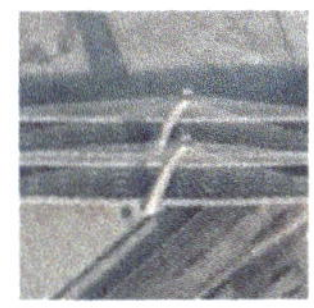
2	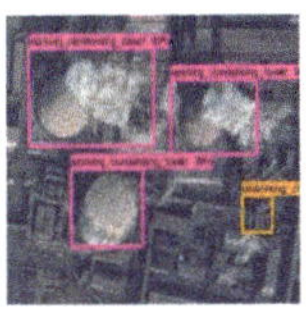	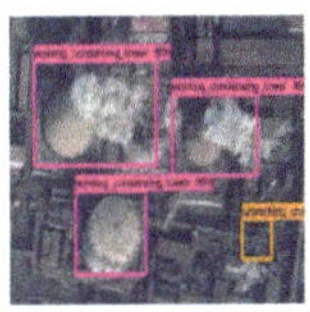	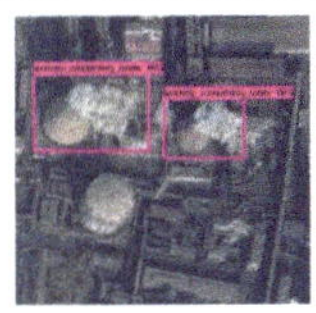	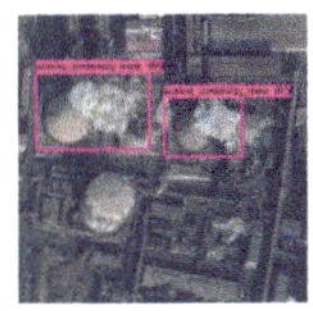
3	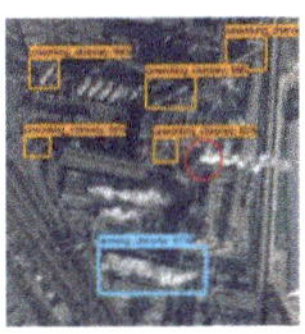	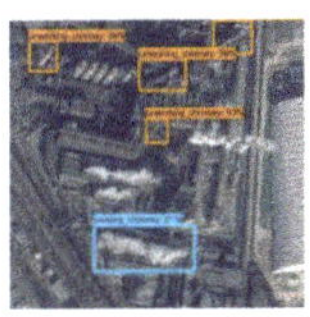		
4	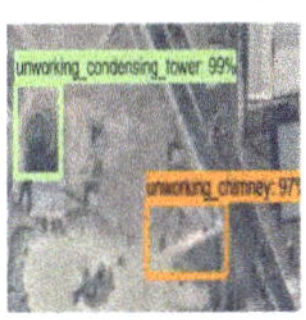			
5	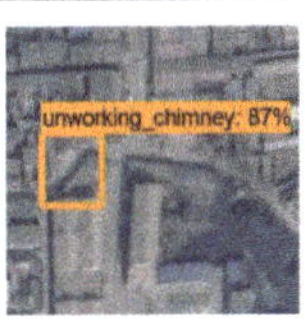		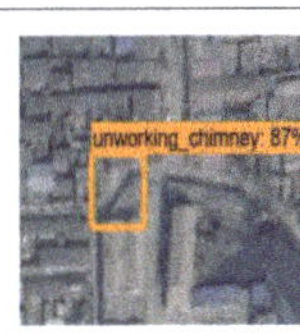	

The final evaluation indexes are shown in Table 5. The total target number (N) indicates the total chimneys in 3 images. The recall rates of four kinds of targets are 0.7727, 0.7662, 1, and 1, respectively. These values are much closed to the testing accuracies on BUAA-FFPP60 dataset. However, in practice, there is a large number of FPs, causing a very low precision. The original precisions are only 0.047, 0.4048, 0.2173, and 0.0833 for four kinds of target, respectively. After using two spatial analysis method, the FPs are largely removed. The precisions are increased to 0.9444, 0.9365, 0.833, and 0.8, respectively. The final qualities are 0.7391, 0.7108, 0.8333, and 0.8, respectively. The final qualities of working and nonworking chimneys are both significantly higher than the qualities calculated on testing samples. It can be concluded that the spatial analysis methods are very effective to increase the final precision and final quality.

Table 5. The accuracy of the experiment.

Target Type	Working Chimneys	Non-Working Chimneys	Working Condensing Towers	Non-Working Condensing Towers
N	22	77	5	4
TP	17	62/59 *	5	4
FP	25	86	23	44
FN	5	18	0	0
Recall	0.7727	0.7662	1	1
Precision	0.4047	0.4048	0.2173	0.0833
Final Precision	0.9444	0.9365	0.833	0.8
Final Quality	0.7391	0.7108	0.8333	0.8

* Three non-working chimneys are mis-removed.

In terms of category, chimneys have relatively low recall rate but high final precision. That is because the chimney is narrow in the image, and easily be interfered by noise, such as shadow, road, and build. Meanwhile, its unique contour makes it easy to distinguish with false chimney by spatial analysis method. In contrary, the condensing tower is easy to be detected by image-processing-based method, the Faster R-CNN, for its integrality appearance in image. Its relatively low final precision may partly result from the small number of samples.

4. Conclusions

In this paper, we use the Faster R-CNN to train the detection model on an open access dataset, BUAA-FFPP60. After the model is trained and tested, we used the model to detect the chimneys in three high-resolution remote sensing images of Google Maps, which is located in Tangshan city. The recall rates for working chimneys, non-working chimneys, working condensing towers, and non-working condensing towers are 77.27%, 76.62%, 100%, and 100%, respectively. However, the precisions for these targets are only 40.47%, 40.48%, 21.73%, and 8.3%, respectively. To increase the precision of detection, two spatial analysis methods, the DTM filtering and main direction test, are introduced to remove the false positive targets. The results show that more than 95% false chimneys can be removed, and the final precision of detection are 94.44%, 93.65%, 83.3%, and 80% respectively. There also exists a possibility that truly detected chimneys might be removed by these spatial analysis methods. However, in our experiment, only three non-working chimneys have been mistakenly removed. Therefore, DTM filtering and main direction tests are very effective methods to remove the false chimneys in detection results from Faster R-CNN. Although the two spatial analysis methods are very effective and robust to remove false positives, they are not useful to reduce the false negative. To reduce the false negative or increase the recall rate, we use an image enhancement method and a low Faster R-CNN threshold. We also suggest that further studies focus on more methods to reduce the false negatives, such as introducing more pre-processing, constructing new architecture of neural networks, and improving the completeness of the training dataset.

Author Contributions: Conceptualization, C.H. and Y.D.; methodology, F.Y. and Y.D.; software, G.L.; validation, G.L., Y.D., and F.Y.; formal analysis, C.H.; investigation, L.B.; resources, G.L. and F.Y.; data curation, G.L. and C.H.; writing—original draft preparation, G.L. and C.H.; writing—review and editing, Y.D.; visualization, G.L.; supervision, C.H.; project administration, C.H.; funding acquisition, C.H. All authors have read and agreed to the published version of the manuscript.

Funding: This research was funded by the National Key Research and Development Program of China, grant number 2018YFC0213600, and the Strategic Priority Research Program of the Chinese Academy of Sciences, grant number GranXDA19030101 and the National Natural Science Foundation of China, grant number 41590853.

Conflicts of Interest: The authors declare no conflict of interest.

References

1. Ojala, T.; Pietikainen, M.; Maenpaa, T. Multiresolution gray-scale and rotation invariant texture classification with local binary patterns. *IEEE Trans. Pattern Anal. Mach. Intell.* **2002**, *24*, 971–987. [CrossRef]

2. Moranduzzo, T.; Melgani, F. A SIFT-SVM method for detecting cars in UAV images. In Proceedings of the 2012 IEEE International Geoscience and Remote Sensing Symposium, Munich, Germany, 22–27 July 2012; pp. 6868–6871.
3. Melgani, F.; Bruzzone, L. Classification of hyperspectral remote sensing images with support vector machines. *IEEE Trans. Geosci. Remote Sens.* **2004**, *42*, 1778–1790. [CrossRef]
4. Cheng, G.; Han, J. A survey on object detection in optical remote sensing images. *ISPRS J. Photogramm. Remote Sens.* **2016**, *117*, 11–28. [CrossRef]
5. Sermanet, P.; Chintala, S.; LeCun, Y. Convolutional neural networks applied to house numbers digit classification. In Proceedings of the 21st International Conference on Pattern Recognition (ICPR2012), Tsukuba, Japan, 11–15 November 2012; pp. 3288–3291.
6. Liu, W.; Anguelov, D.; Erhan, D.; Szegedy, C.; Reed, S.; Fu, C.-Y.; Berg, A.C. Ssd: Single shot multibox detector. In *Proceedings of European Conference on Computer Vision*; Springer: Cham, Switzerland, 2016; pp. 21–37.
7. Redmon, J.; Divvala, S.; Girshick, R.; Farhadi, A. You only look once: Unified, real-time object detection. In Proceedings of the IEEE Conference on Computer Vision and Pattern Recognition, Las Vegas, NV, USA, 27–30 June 2016; pp. 779–788.
8. Girshick, R.; Donahue, J.; Darrell, T.; Malik, J. Rich feature hierarchies for accurate object detection and semantic segmentation. In Proceedings of the IEEE Conference on Computer Vision and Pattern Recognition, Columbus, OH, USA, 23–28 June 2014; pp. 580–587.
9. Girshick, R. Fast r-cnn. In Proceedings of the IEEE International Conference on Computer Vision, Santiago, Chile, 7–13 December 2015; pp. 1440–1448.
10. Ren, S.; He, K.; Girshick, R.; Sun, J. Faster r-cnn: Towards real-time object detection with region proposal networks. In Proceedings of the Advances in Neural Information Processing Systems, Montreal, Quebec, Canada, 7–12 December 2015; pp. 91–99.
11. Wu, H.; Zhang, H.; Zhang, J.; Xu, F. Fast aircraft detection in satellite images based on convolutional neural networks. In Proceedings of the 2015 IEEE International Conference on Image Processing (ICIP), Quebec City, QC, Canada, 27–30 September 2015; pp. 4210–4214.
12. Zhang, F.; Du, B.; Zhang, L.; Xu, M. Weakly Supervised Learning Based on Coupled Convolutional Neural Networks for Aircraft Detection. *IEEE Trans. Geosci. Remote. Sens.* **2016**, *54*, 5553–5563. [CrossRef]
13. Wang, W.; Nie, T.; Fu, T.; Ren, J.; Jin, L. A novel method of aircraft detection based on high-resolution panchromatic optical remote sensing images. *Sensors* **2017**, *17*, 1047. [CrossRef] [PubMed]
14. Zou, Z.; Shi, Z. Ship Detection in Spaceborne Optical Image With SVD Networks. *IEEE Trans. Geosci. Remote. Sens.* **2016**, *54*, 5832–5845. [CrossRef]
15. Yao, Y.; Jiang, Z.; Zhang, H.; Zhao, D.; Cai, B. Ship detection in optical remote sensing images based on deep convolutional neural networks. *J. Appl. Remote. Sens.* **2017**, *11*, 042611. [CrossRef]
16. Yang, X.; Sun, H.; Fu, K.; Yang, J.; Sun, X.; Yan, M.; Guo, Z. Automatic ship detection in remote sensing images from google earth of complex scenes based on multiscale rotation dense feature pyramid networks. *Remote Sens.* **2018**, *10*, 132. [CrossRef]
17. Zhu, C.; Liu, B.; Zhou, Y.; Yu, Q.; Liu, X.; Yu, W. Framework design and implementation for oil tank detection in optical satellite imagery. In Proceedings of the 2012 IEEE International Geoscience and Remote Sensing Symposium, Munich, Germany, 22–27 July 2012; pp. 6016–6019.
18. Zhang, L.; Shi, Z.; Wu, J. A Hierarchical Oil Tank Detector With Deep Surrounding Features for High-Resolution Optical Satellite Imagery. *IEEE J. Sel. Top. Appl. Earth Obs. Remote. Sens.* **2015**, *8*, 4895–4909. [CrossRef]
19. Liu, Z.; Zhao, D.; Shi, Z.; Jiang, Z. Unsupervised Saliency Model with Color Markov Chain for Oil Tank Detection. *Remote Sens.* **2019**, *11*, 1089. [CrossRef]
20. Yao, Y.; Jiang, Z.; Zhang, H.; Cai, B.; Meng, G.; Zuo, D. Chimney and condensing tower detection based on faster R-CNN in high resolution remote sensing images. In Proceedings of the 2017 IEEE International Geoscience and Remote Sensing Symposium (IGARSS), Fort Worth, TX, USA, 23–28 July 2017; pp. 3329–3332.
21. Zhang, H.; Deng, Q. Deep Learning Based Fossil-Fuel Power Plant Monitoring in High Resolution Remote Sensing Images: A Comparative Study. *Remote Sens.* **2019**, *11*, 1117. [CrossRef]
22. Dai, J.; Li, Y.; He, K.; Sun, J. R-fcn: Object detection via region-based fully convolutional networks. In Proceedings of the Advances in Neural Information Processing Systems, Barcelona, Spain, 5–10 December 2016; pp. 379–387.

23. Dai, J.; Qi, H.; Xiong, Y.; Li, Y.; Zhang, G.; Hu, H.; Wei, Y. Deformable convolutional networks. In Proceedings of the IEEE International Conference on Computer Vision, Venice, Italy, 22–29 October 2017; pp. 764–773.
24. Fu, C.-Y.; Liu, W.; Ranga, A.; Tyagi, A.; Berg, A.C. Dssd: Deconvolutional single shot detector. *arXiv* **2017**, arXiv:1701.06659.
25. Lin, T.-Y.; Dollár, P.; Girshick, R.; He, K.; Hariharan, B.; Belongie, S. Feature pyramid networks for object detection. In Proceedings of the IEEE Conference on Computer Vision and Pattern Recognition, Honolulu, HI, USA, 21–26 July 2017; pp. 2117–2125.
26. Lin, T.-Y.; Goyal, P.; Girshick, R.; He, K.; Dollár, P. Focal loss for dense object detection. In Proceedings of the IEEE International Conference on Computer Vision, Venice, Italy, 22–29 October 2017; pp. 2980–2988.
27. Redmon, J.; Farhadi, A. Yolov3: An incremental improvement. *arXiv* **2018**, arXiv:1804.02767.
28. Deng, Q.; Zhang, H.; Bruzzone, L.; Bovolo, F.; Benediktsson, J.A. Chimney and condensing tower detection based on FPN in high-resolution remote sensing images. In Proceedings of the Image and Signal Processing for Remote Sensing XXV, Strasbourg, France, 9–11 September 2019.
29. He, K.; Zhang, X.; Ren, S.; Sun, J. Deep residual learning for image recognition. In Proceedings of the IEEE Conference on Computer Vision and Pattern Recognition, Las Vegas, NV, USA, 27–30 June 2016; pp. 770–778.
30. Lin, T.-Y.; Maire, M.; Belongie, S.; Hays, J.; Perona, P.; Ramanan, D.; Dollár, P.; Zitnick, C.L. Microsoft coco: Common objects in context. In Proceedings of the European Conference on Computer Vision, Zurich, Switzerland, 6–12 September 2014; pp. 740–755.
31. Ministry of Environmental Protection of the People's Republic of China. *GB/T 13271-2014 Emission Standard of Air Pollutants for Boiler*; Environmental Science Press: Beijing, China, 2014.
32. *Tangshan Environment Report*; Tangshan Ecological Environment Bureau: Tangshan, China, 2011.
33. Abadi, M.; Agarwal, A.; Barham, P.; Brevdo, E.; Chen, Z.; Citro, C.; Corrado, G.S.; Davis, A.; Dean, J.; Devin, M. Tensorflow: Large-scale machine learning on heterogeneous distributed systems. *arXiv* **2016**, arXiv:1603.04467.

sensors

MDPI

Article

Transmission Line Vibration Damper Detection Using Deep Neural Networks Based on UAV Remote Sensing Image

Wenxiang Chen [1,2], Yingna Li [1,2,*] and Zhengang Zhao [1,2]

1 Faculty of Information Engineering and Automation, Kunming University of Science and Technology, Kunming 650500, China; wenxiang.chen@stu.kust.edu.cn (W.C.); zhaozhengang@stu.kust.edu.cn (Z.Z.)
2 Computer Technology Application Key Lab of the Yunnan Province, Kunming 650500, China
* Correspondence: liyingna@stu.kust.edu.cn

Abstract: Vibration dampers can greatly eliminate the galloping phenomenon of overhead transmission wires caused by wind. The detection of vibration dampers based on visual technology is an important issue. The current vibration damper detection work is mainly carried out manually. In view of the above situation, this article proposes a vibration damper detection model named DamperYOLO based on the one-stage framework in object detection. DamperYOLO first uses a Canny operator to smooth the overexposed points of the input image and extract edge features, then selectees ResNet101 as the backbone of the framework to improve the detection speed, and finally injects edge features into backbone through an attention mechanism. At the same time, an FPN-based feature fusion network is used to provide feature maps of multiple resolutions. In addition, we built a vibration damper detection dataset named DamperDetSet based on UAV cruise images. Multiple sets of experiments on self-built DamperDetSet dataset prove that our model reaches state-of-the-art level in terms of accuracy and test speed and meets the standard of real-time output of high-accuracy test results.

Keywords: power transmission lines; vibration dampers detection; unmanned aerial vehicle (UAV); deep neural networks; attention mechanism

Citation: Chen, W.; Li, Y.; Zhao, Z. Transmission Line Vibration Damper Detection Using Deep Neural Networks Based on UAV Remote Sensing Image. *Sensors* **2022**, *22*, 1892. https://doi.org/10.3390/s22051892

Academic Editors: Moulay A. Akhloufi and Mozhdeh Shahbazi

Received: 30 January 2022
Accepted: 26 February 2022
Published: 28 February 2022

1. Introduction

The main function of a power line vibration damper is to reduce the vibration of the wire caused by wind galloping. High-voltage transmission towers have large spacing, which makes it easy for the wires to vibrate when subjected to wind. The periodic bending of the suspension caused by the vibration of the wire leads to fatigue damage to the metal wire. In severe cases, accidents such as wire breakage and power tower collapse will be induced. The use of a vibration damper on high voltage transmission lines can reduce the vibration of the wires caused by the wind, thereby reducing the probability of accidents. Therefore, vibration damper detection is an important topic in the inspection of overhead transmission lines [1]. Vibration damper detection refers to obtaining the specific position of the vibration damper in the inspection image. This task is an important prerequisite for the work of vibration damper displacement detection, damage detection, and corrosion detection. At present, vibration damper detection has attracted the attention of researchers in the fields of smart grid and machine vision, with certain progress made [2].

UAV technology has developed rapidly in recent years. UAV has the advantages of convenient operation, easy portability, and low cost [3]. Multi-UAV systems based on wireless sensor networks [4] are used in crop yield estimation [5], object detection [6], and other fields. UAVs have rapidly developed into important auxiliary equipment.

At present, the inspection of overhead transmission lines still mainly relies on visual inspection by staff, which can produce omissions and incorrect judgments for the vibration damper located at a high place; therefore, the use of UAVs for transmission line inspection is an issue of great research value. Researchers have used UAVs for equipment detection and other tasks [7,8]. This article focuses on the issue of vibration damper detection using UAV aerial images.

In early research work, traditional image processing techniques were most widely used in power line inspection scenarios [2,9]. Researchers would select appropriate feature extraction operators according to the actual situation and complete the task of object detection through a threshold setting. Machine learning algorithms were also selected to achieve better detection results [10]. However, such methods are very susceptible to interference from background information, especially when using UAV aerial photography data, as the similar color properties of vibration dampers and power towers can easily cause missed detection.

In recent years, with the exponential growth of machine computing power and data volume, it has become a research hotspot again. Deep learning technologies, especially convolutional neural networks, have opened new research directions in the field of computer vision. There is much research on power components using the state-of-the-art method in the field of object detection [11,12]. However, at present, these works are mainly based on the simple application of the framework, there is no targeted improvement for the characteristics of the vibration damper, and high accuracy of the model requires a large amount of computing resources.

In addition, some studies have used special equipment for imaging or for the physical properties of the device [13,14]. The results of these works are usually excellent, but the extra equipment overhead and high usage cost make such methods unsuitable for power line patrol scenarios.

Aiming at the research status of image-based vibration damper detection, this article proposes a vibration damper detection model based on the one-stage algorithm in target detection. The main contributions of this paper are as follows:

- A proposed vibration damper detection model called DamperYOLO based on the YOLOv4 framework, which is more robust than traditional methods and can achieve a good balance between speed and accuracy, and a vibration damper detection dataset called DamperDetSet based on UAVs aerial images.
- To enhance images, Gaussian filtering is used to smooth the overexposed points in the aerial image and the Canny algorithm is used to extract the contour information in the image.
- Introduction of an attention-based structure in the backbone of DamperYOLO. This module can introduce the edge information extracted by Canny into the forward propagation process of the model and provide semantic guidance for the feature extraction of the network.
- Addressing the problem that the vibration damper is small and difficult to detect in the UAVs aerial image, we used a feature fusion network based on FPN after the backbone. While outputting feature maps of different resolutions, the semantics and underlying feature information of each layer are maintained, which provides a high-quality data basis for the identification of vibration dampers.

The remainder of this article is organized as follows. Section 2 briefly introduces the related work of vibration damper detection. Section 3 introduces the basic framework used in the method proposed in this article. In Section 4, this article introduces the details of DamperYOLO. In Section 5, this article introduces the damper dataset, the experimental details, and a series of comparative experiments. Section 6 provides a brief summary of the work.

2. Related Work

This section focuses on the image-based vibration damper detection research. The existing work is mainly divided into traditional image processing methods, deep learning-based research, and detection methods based on auxiliary equipment.

2.1. Traditional Method

Traditional image processing algorithms use edge detection, color space conversion, and clustering algorithms to extract damper information in images, usually combined with machine learning algorithms for iterative classification tasks.

Wu et al. [2] used the snake model to extract the edge of the vibration damper, but due to the helicopter airborne imaging equipment required, the cruise cost was high. Huang et al. [9] performed corrosion and displacement detection on the vibration damper based on rusty area ratio and color shade index, involving grayscale processing, edge detection, threshold segmentation, morphological processing, and other technologies. Similarly, Song et al. [15] detected the rust problem of the vibration damper based on the histogram. Jin et al. [10] used Harr-like features and a cascade adaboost classifier to classify and detect vibration dampers on overhead lines. Yang et al. [16] performed exponential transformation on the S and V components in the HSV color space to improve the contrast between the front and background. Liu et al. [17] used the canny operator and Hough transform method to detect the displacement of the vibration damper on the high-voltage line. Similarly, Chen et al. [18] used random Hough transformation for the vibration damper detection task. Miao et al. [19] used the wavelet modules maximum method to locate the shock hammer on the transmission line. Pan et al. [20] used a simple extraction operator to monitor the state of the vibration damper. Jin et al. [21] used the Adaboost algorithm to conduct real-time monitoring of the line vibration damper through drones.

Traditional methods use operators and classifiers to identify the vibration damper on the line; the detection accuracy is limited by the complexity of the environmental background, but its advantage lies in its fast detection speed, which is suitable for real-time detection.

2.2. Deep Neural Networks

With the rapid development of deep learning technology, the detection of power line components based on neural networks such as CNNs has gradually become a popular research direction.

Based on YOLOv4, Bao et al. [1] used k-means to analyze the aspect ratio of the anchor to detect damage, corrosion, and displacement faults of the vibration damper. Zhang et al. [11] also used Faster R-CNN to detect damage and corrosion defects of the vibration damper twice, in which the first detection result was used as the second proposal, thereby improving the detection effect. Bao et al. [12] used the Cascade R-CNN framework to locate and detect the damage of the vibration damper. Yang et al. [16] performed the detection task of vibration dampers using Faster R-CNN based on HSV color space transformed images. Guo et al. [22] used YOLOv4 to improve the detection effect of damaged vibration damper. Wang et al. [23] investigated insulator defects in overhead transmission lines, damage to vibration dampers, and foreign objects in bird's nests. Zhang et al. [24] switched to VGG16 as the basic backbone network and performed detection tasks for shockproof hammers and other foreign objects on power towers.

The detection of power line components and foreign objects using deep neural networks has also attracted the attention of researchers. For example, the YOLO framework is used to detect insulators on transmission lines [25] and icing detection [26], change the anchor setting of Faster R-CNN according to the shape characteristics of the insulator [27], using Mask R-CNN to detect line foreign objects [28], defect detection for high-speed rail catenary insulators [29], and detection of wet insulators using infrared images [30]. Usually, these studies are only simple applications of power components datasets, and most of the studies lack targeted transformation for specific environments and scenarios; the solutions provided are mostly trick stacking.

2.3. Auxiliary Equipment

In addition to using common optical images, there is also research that uses other imaging equipment and auxiliary devices to perform detection tasks. For example, a robot is used to reset the vibration damper [14,17,31], and the damage of the vibration damper is detected based on LiDAR data [13]. The damping of the vibration damper is detected based on sensors such as optical ground wire (OPGW) and an all-dielectric self-supporting (ADSS) optical cable [32]. In addition, some researchers [33] designed a rotation-free spacer damper to improve the anti-galloping ability of power lines.

2.4. Researches Summary

There is still room for improvement in the detection of vibration dampers for overhead transmission lines. A summary of these research is as follows:

- Traditional methods based on image processing technology. The detection accuracy is mostly dependent on the quality of the image. If the background in the image is too complex, this leads to the problem that the used feature operator does not cover all situations, which inevitably leads to a decrease in the detection accuracy. The advantage of the traditional method is that it consumes less resources and the calculation speed is fast. Therefore, at present, this type of method is still the most important when the scene is relatively simple, background interference is low, and the real-time requirement is high.
- The method based on deep neural network is the hottest research direction in the field of vibration damper detection. By relying on powerful computing equipment and a large amount of training data, an end-to-end network model can be obtained; on this basis, it is very easy to carry out detection tasks. However, there is currently no public dataset for the vibration damper of overhead transmission lines, and the detection effect of the model is often limited by the lack of computing power of edge devices.
- There is some research work based on auxiliary equipment. Such research uses the characteristics of ultrasonic or infrared imaging equipment to perform the task of vibration damper breakage detection. However, these devices are often inconvenient for use along complex overhead lines, and the maintenance and use costs of the devices are much higher than those of drones.

Combining the characteristics of the abovementioned research work, we not only hope to obtain excellent detection results, but also hope that the model can run in real time on devices lacking computing resources, such as drones. A one-stage method using deep neural network is the most suitable choice. One-stage object detection utilizes the powerful feature extraction capabilities of CNNs to cope with complex application scenarios. At the same time, the detection result does not depend on the proposal, and its calculation speed is fast enough. Therefore, in the following work, based on the one-stage model, we propose a detection method based on the visual characteristics of the vibration damper in the real scene.

3. Basic Knowledge of YOLO

YOLO [34] proposed by Redmon et al. in 2016 is a classic one-stage object detection method. YOLOs [34–37] solves the target detection problem as a regression problem. After an inference of the input image, the positions of all objects in the image, their categories, and corresponding confidence probabilities can be obtained. YOLO divides the input image into SxS grids, and each grid is responsible for detecting objects that fall into the grid. If the coordinates of the center position of an object fall into a grid, then the grid is responsible for detecting the object.

The difference between the backbone in YOLOv4 [37] is that it is based on the Darknet structure in YOLOv3 [36] and borrows the structure of the CSPNet network [38] to propose a network structure called CSPDarknet. The loss function used in training is CIOU [39].

Since the objects to be detected in this paper are only vibration dampers, an overly complex network structure will have a negative impact on feature extraction; therefore, this paper selects the classic ResNet101 [40] as feature extraction network. The objective function of YOLOv4 is as follows:

$$L_{det} = L_{box} + L_{obj} + L_{cls} \tag{1}$$

where L_{box}, L_{obj}, and L_{cls} represent the regression loss, confidence loss, and category loss of the box, respectively. The expression of the box regression loss is as follows:

$$L_{box} = \lambda_{coord} \sum_{i=0}^{S^2} \sum_{j=0}^{B} 1_{i,j}^{obj} \left(1 - \left(IoU - \frac{Distance\_2^2}{Distance\_C^2} - \frac{v^2}{(1 - IoU) + v}\right)\right) \tag{2}$$

where λ_{coord} is the weight of box regression loss, S_i^2 represents the ith grid of $S \times S$ size, B_j represents the jth predicted box of S_i^2, and $1_{i,j}^{obj}$ indicates that there is a target center of the prediction category in the box. IoU is the Intersection-of-Union of the predicted box and ground truth, the calculation formula of IoU is Equation (3), $Distance\_2$ is the Euclidean distance between the center coordinates of Box^p and Box^{gt}, $Distance\_C$ is the diagonal length of the smallest bounding rectangle of Box^p and Box^{gt}, v is a parameter to measure the consistency of the aspect ratio of Box^p and Box^{gt}, and the calculation formula of v is Equation (4).

$$IoU = \frac{|Box^p \cap Box^{gt}|}{|Box^p \cup Box^{gt}|} \tag{3}$$

where Box^p and Box^{gt} represent the predicted box and ground truth, respectively.

$$v = \frac{4}{\pi^2}\left(arctan\frac{w^{gt}}{h^{gt}} - arctan\frac{w^p}{h^p}\right)^2 \tag{4}$$

where w^{gt} and w^p represent the width of the ground truth and predicted box, respectively, while h^{gt} and h^p represent their respective heights.

Similar to the regression loss, the loss function for the target prediction confidence is as follows:

$$L_{obj} = \lambda_{noobj} \sum_{i=0}^{S^2} \sum_{j=0}^{B} 1_{i,j}^{noobj} (c_i - \hat{c}_i)^2 + \lambda_{obj} \sum_{i=0}^{S^2} \sum_{j=0}^{B} 1_{i,j}^{obj} (c_i - \hat{c}_i)^2 \tag{5}$$

where λ_{noobj} and λ_{obj}, respectively, represent the weight of the confidence loss when the object is not included and when it is included. c_i and $\hat{c}_i$, respectively, represent the true value and predicted value of whether there is an object of category i in the current box. The other parameters have the same meaning as in the regression loss.

The category prediction loss uses the classic cross-entropy loss, and its calculation formula is as follows:

$$L_{obj} = \lambda_{noobj} \sum_{i=0}^{S^2} \sum_{j=0}^{B} 1_{i,j}^{noobj} (c_i - \hat{c}_i)^2 + \lambda_{obj} \sum_{i=0}^{S^2} \sum_{j=0}^{B} 1_{i,j}^{obj} (c_i - \hat{c}_i)^2 \tag{6}$$

where λ_{class} represents the weight of the category loss; $\hat{p}_i(c)$ represents the predicted value of the confidence of the current category; and $p_i(c)$ is a conditional probability, which is obtained by obtaining a value of 0 or 1, depending on whether S_i^2 contains the target center, and then multiplying it with IoU.

YOLOv4 uses CSPDarknet53 [38] as its feature extraction network, but CSPDarknet53 has lots of parameters. In addition, the only object to be detected in this paper is the damper. As shown in Table 1, ResNet101 is composed of multiple groups of residual blocks. ResNet has excellent feature extraction ability, which overcomes the problem of low learning efficiency caused by excessive network depth. Therefore, the classic ResNet101 is used as the backbone in this article.

Table 1. Applied kernels of ResNet101 in DamperYOLO.

Layer	Output Size	Kernel Size
conv1	304 × 304	7 × 7, 64
conv2_x	152 × 152	$\begin{bmatrix} 1 \times 1, 64 \\ 3 \times 3, 64 \\ 1 \times 1, 256 \end{bmatrix} \times 3$
conv3_x	76 × 76	$\begin{bmatrix} 1 \times 1, 128 \\ 3 \times 3, 128 \\ 1 \times 1, 512 \end{bmatrix} \times 4$
conv4_x	38 × 38	$\begin{bmatrix} 1 \times 1, 256 \\ 3 \times 3, 256 \\ 1 \times 1, 1024 \end{bmatrix} \times 23$
conv5_x	19 × 19	$\begin{bmatrix} 1 \times 1, 512 \\ 3 \times 3, 512 \\ 1 \times 1, 2048 \end{bmatrix} \times 3$

4. DamperYOLO

In this section, a new framework named DamperYOLO is proposed for the vibration damper detection task of overhead transmission lines based on YOLOv4 [37], Canny algorithm [41], attention mechanism [42] and FPN [43] structure.

4.1. Edge Extraction

The quality of the input image is very important as it is the first step of the whole network detection, which directly affects the subsequent detection process. Although, strong noise immunity is one of the advantages of deep neural networks, no network would want to receive a high-quality input, so that the trained model parameters have more powerful attention to our target. Therefore, we decided to use edge detection techniques to improve the semantic information in images for the purpose of image enhancement, detailed in this subsection.

The canny algorithm is used to extract edge information from UAV aerial images. The canny algorithm is mainly divided into four parts: Gaussian smooth image, gradient magnitude and direction calculation, gradient magnitude nonmaximum suppression, double threshold algorithm detection and edge connection.

Our images are obtained by unmanned aerial photography and are highly susceptible to light reflections to generate exposure points. To reduce the influence of these bright white points, a Gaussian kernel is used to smooth the image.

Compared with the median filter [44] and the mean filter [45], the Gaussian filter assigns different calculation weights to different fields of the current element, which can achieve the purpose of denoising while preserving the gray distribution characteristics of the image. Gaussian filtering is usually implemented by iterative operations on the image with $(2k + 1) \times (2k + 1)$ convolution kernels. The kernel generation equation is shown in Equation (7).

$$H_{ij} = \frac{1}{2\pi\sigma^2} \exp\left(-\frac{(i-(k+1))^2 + (j-(k+1))^2}{2\sigma^2}\right); 1 \le i, j \le (2k+1) \tag{7}$$

where k represents an integer, $(2k + 1)$ represents the size of the convolution kernel, and (i, j) represents the coordinates of one of the points.

The size of the convolution kernel is usually set to an odd number for the convenience of calculation. The larger the kernel, the stronger the processing ability for local noise. In our experiments, kernels with sizes of 3 × 3, 5 × 5, and 9 × 9 were selected for comparison. The experimental results show that the kernel of 5 × 5 has the smallest effect.

After Gaussian smoothing, the background part still contains overexposed points. There is no need to worry about the negative impact this brings to the model, as the network focuses on the ground truth part during training. What must pay attention to is if the feature of the vibration damper is improved, and edge detection is one of the important means of image enhancement. The parts of the image with high gradient variation in the canny algorithm task image represent a higher probability of edges. Therefore, our next step is to extract the gradient information of the image.

Gradients reflect the intensity of local pixel transformations. The greater the gradient change, the greater the change in the corresponding region. The gradient needs to calculate the direction and size of two parts, usually by calculating the gradient of the horizontal and vertical directions to represent a complete gradient. Its calculation formula is shown in Equations (8) and (9).

$$\frac{\partial f}{\partial x} \approx \frac{f(x+1,y) - f(x-1,y)}{2} \tag{8}$$

$$\frac{\partial f}{\partial y} \approx \frac{f(x,y+1) - f(x,y-1)}{2} \tag{9}$$

The direction a and increment b of the gradient can be obtained based on the gradients in the horizontal and vertical directions, as shown in Equations (10) and (11).

$$\theta = \tan^{-1}(\frac{\partial f}{\partial y}/\frac{\partial f}{\partial x}) \tag{10}$$

$$\|\nabla f\| = \sqrt{\left(\frac{\partial f}{\partial x}\right)^2 + \left(\frac{\partial f}{\partial y}\right)^2} \tag{11}$$

Gradient images contain all grayscale variations. Therefore, the canny algorithm uses the nonmaximum suppression method [41] to propose the lower gradient variation in the region.

The nonmaximum suppression algorithm calculates in eight areas around the pixel, retaining the parts with the largest grayscale changes in the horizontal, vertical, and diagonal directions while eliminating other parts with smaller changes by changing the broad-side gradient map to a single pixel width of the side.

The method of the nonmaximum suppression algorithm can only enhance the edge information and cannot guarantee that the remaining part is foreground information. Therefore, the last step of the canny algorithm is to use the double threshold algorithm to separate the foreground and background based on our prior knowledge.

In the double-threshold algorithm, the pixels above the strong edge threshold represent edge information, and the pixels below the weak edge threshold represent background information. The threshold between the two is the pending element, and if there is a strong edge in the eight-neighborhood of these pixels, the pixel is also classified as an edge pixel. Through comparison experiments of 200, 300, and 400 strong edge thresholds, it was found that the threshold of strong edge is best when the threshold is 300, and the weak edge threshold is set to 0.5 times of the strong edge. The formula for classifying gradient map pixels is shown in Equation (12).

$$f(i) = \begin{cases} \text{strong edge}; & \text{i} > 300 \\ \text{weak edge}; & 150 \le \text{i} \le 300 \\ \text{non} - \text{edge}; & \text{i} < 150 \end{cases} \tag{12}$$

To verify the effect of edge detection, we compared the performance of several classical edge detection operators on vibration dampers. As shown in Figure 1, the edge extracted by the Canny operator is the clearest.

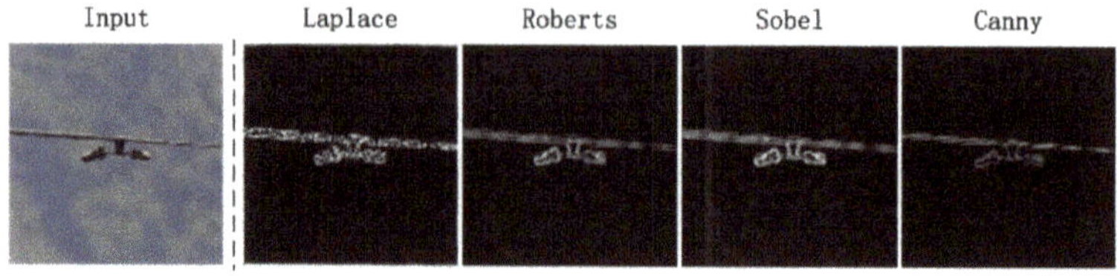

Figure 1. Test examples of edge detection algorithm.

4.2. Attention Mechanism

After obtaining the edge information in the image using the canny algorithm, it can be used to produce positive effects. The attention mechanism [42] originated in the field of NLP and has been introduced into computer vision in recent years. As shown in Figure 2, by introducing additional convolution operations, the attention mechanism can focus on the additional information being added.

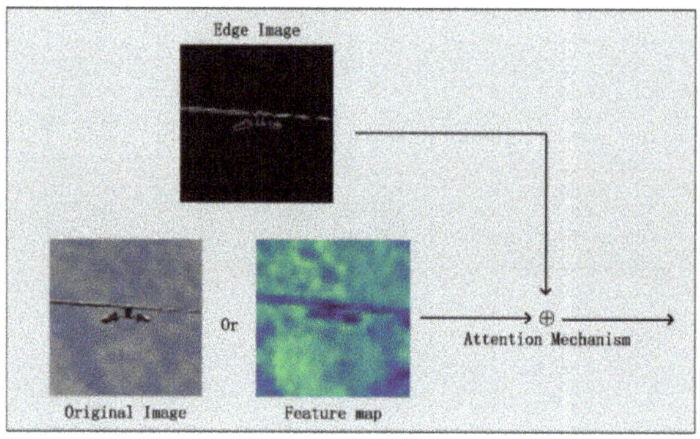

Figure 2. Schematic diagram of the attention mechanism.

The attention mechanism is based on the edge information obtained by the canny algorithm, and performs a convolution operation to obtain the attention weight matrix a. The expression of the convolution operation is shown in Equation (13).

$$I_{\mathrm{A}}^{i} = \mathrm{Softmax}(I^{i} W_{A}^{i} + b_{A}^{i}), \text{for } i = 1,2 \tag{13}$$

where I^i represents the input image, $\{W_A^i, b_A^i\}_{i=1}^{2}$ represents the parameter of the convolution operation, and $\mathrm{Softmax}(\cdot)$ represents the SoftMax function used for normalization.

We multiplied the resulting attention weight matrix with the corresponding input image to obtain the final output:

$$I_A = (I_A^1 \otimes I^1) \oplus (I_A^2 \otimes I^2) \tag{14}$$

where I_A represents the final output result of the attention mechanism, I^1 and I^2 represent the input images, and the symbols $\otimes$ and $\oplus$ represent the multiplication and addition elements of the matrix.

Attention mechanism is used in ResNet101 to send the edge image output by the canny algorithm to the network to enhance the network's ability to focus on the ground truth region during feature extraction. We used an attention mechanism in layers 1, 2, and 3 of ResNet because the network focuses on the low-level features of the input image in the early stage of feature extraction. At the fourth and fifth layers, the output is a feature map with highly abstract semantics. At this time, the introduction of the attention mechanism containing the edge map interferes with the effect of the feature map. A follow-up sensitivity analysis on where the attention mechanism is introduced proves our point.

4.3. Feature Fusion Network

After introducing edge detection and attention mechanisms, our framework improved to a certain extent. However, in the inspection data of overhead transmission lines captured by UAVs, the vibration damper is a small target object. When ResNet101 performs feature extraction, the deep network responds easily to semantic features and the shallow network responds easily to image features. This feature leads to a problem: although the high-level network can respond to semantic features, due to the small size of the Feature Map it does not contain much geometric information, which is not conducive to object detection. This problem is more pronounced for small-sized object detection. The vibration damper easily disappears in the feature map output by the fifth layer of ResNet because the target is small. The disappearance of the vibration damper feature leads to a decrease in detection accuracy.

It is natural to think that a feature map that combines deep and shallow features can be used to meet the needs of small target detection. FPN [43] is a network structure that adopts this idea. FPN uses the idea of image pyramid to solve the problem of difficulty in detecting small-sized objects in object detection scenes. The traditional image pyramid method uses a multiscale image input to construct multiscale features. The biggest problem with this approach is that the recognition time is k times the recognition time of a single image, where k is the number of scaled dimensions.

To improve the detection speed, methods such as Faster R-CNN [46] use a single-scale Feature Map, but the single-scale feature map limits the detection capability of the model, especially for samples with extremely low coverage in the training set (such as larger and smaller samples). Unlike Faster R-CNN, which only uses the top-level Feature Map, SSD [47] uses the hierarchical structure of convolutional networks, starting from conv4_3 of VGG [48], and obtains multiscale Feature Maps through different network layers. Although this method can improve accuracy and does not increase the test time, while it does not use the low-level Feature Map, these low-level features are very helpful for detecting small objects. In response to the above problems, FPN adopts the form of a Feature Map in the pyramid of SSD.

Different from SSD, FPN not only uses deep Feature Map in VGG, but also applies shallow Feature Map. These Feature Maps are efficiently integrated through bottom-up, top-down, and lateral connections, which improve the accuracy without greatly increasing the detection time. Therefore, as shown in Figure 3, this article refers to these practices and introduce a structure composed of FPN and bottom-up after the third, fourth, and fifth layers of ResNet101 so that the semantics and lines of the final output feature maps of the three scales' layer features are more abundant.

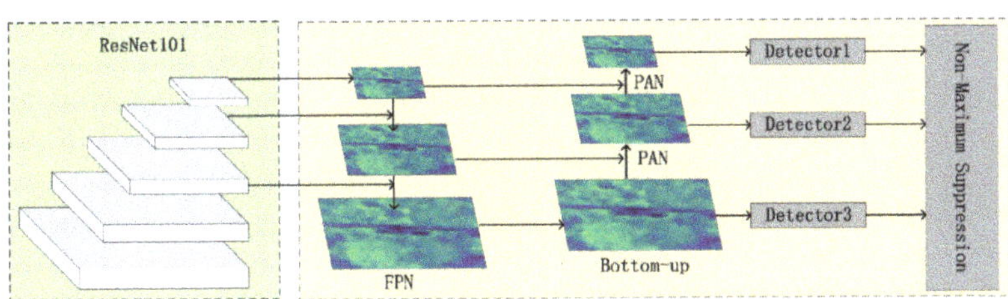

Figure 3. The Feature Fusion Network used for feature transfer containing two parts: the FPN and the Bottom-up module.

DamperYOLO was trained after all framework components were introduced. The training process is as described in Algorithm 1. As shown in Figure 4, the Edge Detection module, the ResnNet101 backbone, Attention Mechanism, the FPN and Bottom-up framework are used to construct the entire vibration damper detection process.

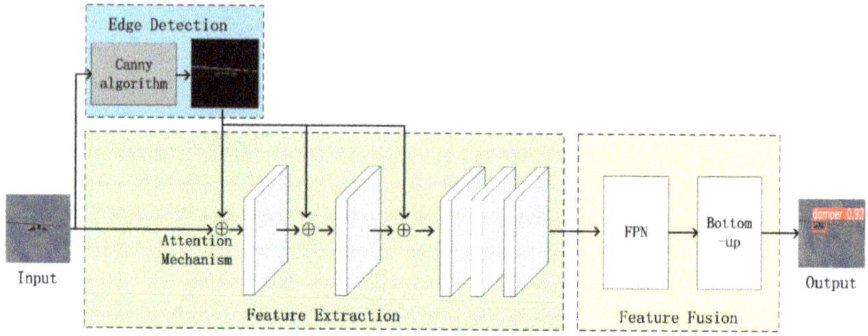

Figure 4. The realization of detection of vibration dampers is divided into three parts: Edge Detection, Feature Extraction, Feature Fusion. First, Edge Detection is used to provide edge information. then Feature Extraction and Feature Fusion are used to obtain feature maps for vibration dampers. Finally, the detection results can be obtained from classifier of YOLOv4.

Algorithm 1: The Training Process of DamperYOLO.

```
Input: Original damper image set I = {I_1, ···, I_N} that each image contains dampers.
Output: DamperYOLO after training.
Initialize DamperYOLO with random weights;
repeat
  for i in 1~epochs do
    for j in 1~N do
      Image augment for I_j;
      Extract feature map using ResNet101;
      Output detection results using YOLO;
      Calculate the penalty value via Formula (2), (5) and (6);
      Minimize Formula (1) to update the parameters of DamperYOLO;
    end for
  end for
until DamperYOLO completes convergence
return
```

5. Experiments and Analysis

5.1. Experiment Description

5.1.1. Dataset

A dataset of vibration dampers for overhead transmission lines is required for the proposed theoretical validation and experimental analysis. At present, although there is a lot of research work on vibration dampers, but there is no completely public vibration damper detection dataset. Moreover, most of the vibration damper data in the article were obtained by geometric transformation methods such as flipping, cutting, and scaling. An insufficient number of vibration dampers would make it difficult to verify the correctness of the proposed theory. Therefore, a dataset was made for vibration damper detection based on the real UAV cruise video of overhead transmission lines, and named DamperDetSet. In the process of making the DamperDetSet dataset, LabelMe was used as a data labeling tool to label the positions of all existing line vibration dampers in the original image. The callout box was kept as close as possible to the smallest enclosing rectangle of the target area.

DamperDetSet contains a total of 3000 images, each of which contains vibration dampers, and the types of vibration dampers are not unique, such as hippocampus antislip vibration dampers, hook wire vibration dampers, etc. We randomly divided all 3000 images into a training set and a test set. The training set contains 2500 images and the test set contains 500 images. The ratio of training set and test set is 5:1. In addition, as the dataset is obtained by shooting with UAVs, the presentation angle of the vibration damper in the image is not unique, which also puts forward higher requirements for the robustness of the model.

5.1.2. Experiment Configuration

In terms of hyperparameter settings in the experiment, we trained DamperYOLO for a total of 200 epochs. The learning rate of the first 100 epochs remained unchanged, and the learning rate of the last 100 epochs gradually decreased to 0. In terms of experimental software settings, all our programs were written in Python language and integrated based on the Pytorch 1.4 platform. In the system environment of the experimental platform, Ubuntu18.04 was used as the operating system. In terms of the hardware environment of the experimental platform, an NVIDIA RTX 2080 GPU was used as the main equipment for training calculation, matched with an AMD R5-3600X CPU and 32 GB RAM.

5.2. *The Baselines*

In the following experiments, we chose one-stage, two-stage, and anchor-free methods as comparison methods.

YOLOv4 [37]: This method is the latest achievement of the YOLO series. After continuing the advantages of the previous work, it introduces the structure of FPN + PAN, which improves the transferability of features in the network; it is also the basis of our proposed model.

Cascade R-CNN [49]: This framework is the latest achievement of the R-CNN series. It creatively introduces a cascade structure. The detection accuracy is state-of-the-art, but its excellent performance consumes a lot of computational resources.

CenterNet [50]: This method is a heatmap-based detection method rather than anchor-based, which has the advantage of fast testing and low space occupancy.

SSD [47]: SSD is another classic one-stage object detection method. It initially utilizes multiple detectors.

RetinaNet [51]: RetinaNet is based on FPN [43], and its contribution is to propose focal loss to solve the problem of category imbalance.

5.3. *Qualitative Evaluation*

To visually compare the difference between the detection effect of DamperYOLO and other baselines, we conducted qualitative analysis and comparison experiments based on the DamperDetSet dataset. The experimental results are shown in Figure 5. As can be seen from Figure 5, under the same test image, the detection effect of CenterNet is not stable enough. This proves that calculation of the heat map will be greatly disturbed by the current anchor-free algorithm in the face of complex scenes such as transmission lines. The performance of the two-stage Cascade R-CNN is very superior. As the latest framework of the R-CNNs series, the results obtained by the second iteration based on the proposal are more accurate. There is also room for improvement in the performance of a single-level SSD. And using VGG16 as a backbone may be weaker than the ResNet-like feature extraction network in feature extraction. RetinaNet and YOLOv4 perform better. Both of them benefit from the latest research results in one-stage. They can obtain high performance with only one calculation, but the edge detection effect of vibration damper still needs to be improved. Finally, DamperYOLO outperforms other one-stages. The detection result image of DamperYOLO proves that our proposed improvement strategy is effective, and its performance is no less than that of Cascade R-CNN.

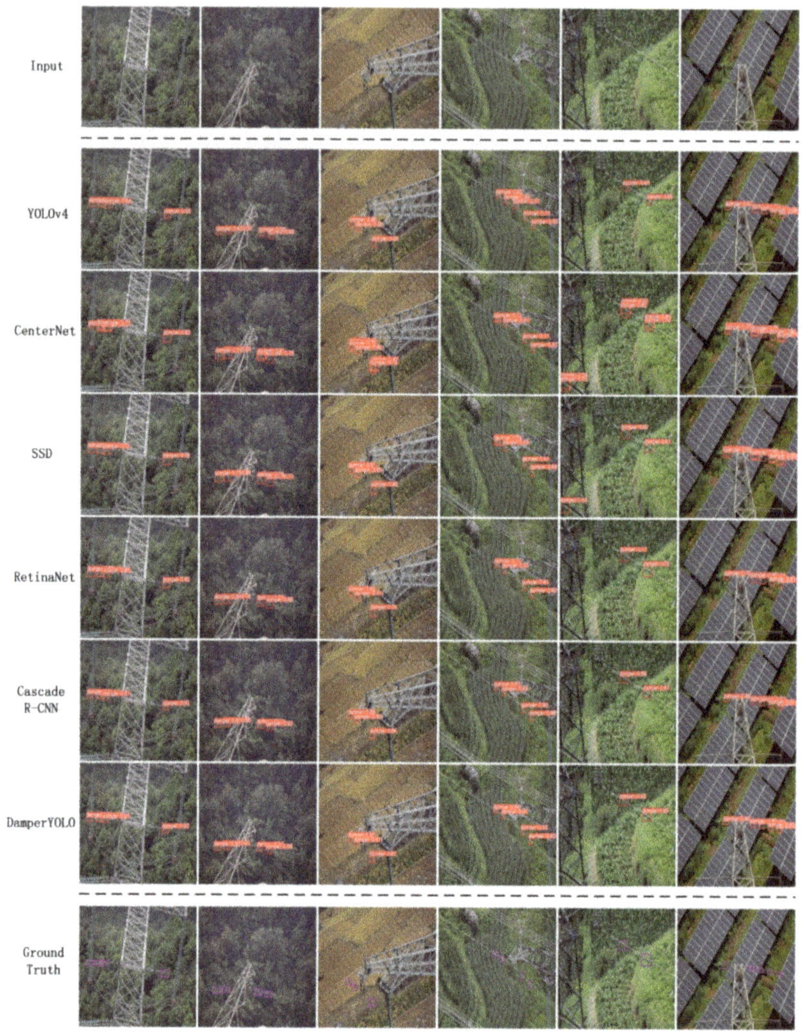

Figure 5. Test examples of each model on the DamperDetSet dataset. Experimental results show that the performance of DamperYOLO is similar to Cascade R-CNN, better than SSD, RetinaNet and YOLOv4 in one-stage class, and CenterNet.

5.4. Quantitative Evaluation

We compared other baselines and performed the quantitative analysis shown in Table 2 with AP in the COCO [52] dataset as the evaluation standard. The calculation of AP is based on the ground truth and the IOU of the prediction result. The calculation formula is shown in Equation (3). The AP calculation results were selected when the IOU was 0.5, 0.7, and 0.9 as the evaluation basis, so that the performance difference of the model under different pressure levels could be more comprehensively evaluated.

Table 2. APs of the different models.

Model	DamperDetSet			FPS
	AP_{50}	AP_{70}	AP_{90}	
YOLOv4	88.23	80.67	73.26	71
SSD	85.71	78.34	71.38	70
RetinaNet	87.18	79.62	72.70	73
CenterNet	84.38	77.25	69.42	118
Cascade R-CNN	92.26	89.52	81.43	31
DamperYOLO	92.62	89.67	81.24	74

As can be seen from Table 2, under the same test picture, thanks to the two-stage detection strategy, the performance of Cascade R-CNN was still stable, and its performance under various AP standards was at the forefront; however, its good score came at the cost of great computation time.

The one-stage RetinaNet and YOLOv4 performed similarly, and YOLOv4 slightly outperformed RetinaNet. Compared with SSD, both of them had a certain degree of lead in terms of indicators, and the latest training tricks available from analysis confers an advantage in accuracy. In addition, the calculation speed of these three methods is faster than Cascade R-CNN, without the intermediate step of proposal, which shortens the calculation time considerably.

The anchor-free based CenterNet had the lowest score; so, it can be concluded that the calculation of the heatmap is very susceptible to interference from similar objects in the background. However, the advantage of the anchor-free class method is that the calculation speed is much faster than other baselines, which is a huge advantage for scenarios with extremely high real-time requirements.

Our proposed DamperYOLO takes the lead on AP, but the score of YOLOv4 is lower than Cascade R-CNN; therefore, the edge extraction, attention mechanism, and feature fusion structure proposed in this paper are better than Cascade R-CNN. The calculation speed of DamperYOLO was similar to other one-stage class methods. Therefore, DamperYOLO is a model with a balance between speed and accuracy.

5.5. Sensitivity Analysis

In this section, multiple sets of sensitivity analysis are performed on each component of DamperYOLO, which includes the choice of backbone, edge extraction, the attention mechanism, the number of training iterations, and the minimum amount of training data.

5.5.1. Backbone

We conducted a sensitivity analysis on the backbone used by DamperYOLO while retaining other improvements. As shown in Table 3, the CSPDarknet53 used by YOLOv4 was improved based on ResNet50, so it performed better. In addition, the only objects need to be detected were dampers. Therefore, we believe that it may be more effective to expand the number of network layers and improve the feature abstraction ability of the backbone. The performance of ResNet101 also supports our idea, but if network layers such as using ResNet152 are added, the improvement is limited, so ResNet101 is used as the backbone.

Table 3. APs of different backbones.

Backbone	DamperDetSet			FPS
	AP_{50}	AP_{70}	AP_{90}	
CSPDarknet53	88.20	80.58	73.28	72
VGG16	82.18	76.54	67.91	71
ResNet50	84.12	77.62	70.42	78
ResNet101(ours)	92.62	89.67	81.24	74
ResNet152	93.25	89.97	82.16	68

5.5.2. Edge Extraction

To verify the effectiveness of preprocessing, a sensitivity analysis was performed on the image denoising, and edge detection used while retaining other improvements constant. Table 4 shows that, compared with not using any preprocessing strategy, using image denoising and edge extraction alone leads to a certain improvement in detection effect. If both are used, the AP50 increase by about five percentage points, which shows that the image augmentation method in this paper is effective.

Table 4. APs of different preprocessing methods.

Preprocessing Method	DamperDetSet			FPS
	AP_{50}	AP_{70}	AP_{90}	
No preprocessing	87.18	79.52	71.83	79
Image denoising	88.92	81.93	73.65	78
Edge extraction	91.25	86.74	79.17	77
Image denoising + Edge extraction	92.62	89.67	81.24	74

5.5.3. Attention Mechanism

The attention mechanism is an important mechanism pioneered in the field of NLP, and was developed in object detection in recent years. In order to verify the effect of adding an attention mechanism in different layers of ResNet101, we conducted a sensitivity analysis for the number of times an attention mechanism is introduced while retaining the other conditions. As shown in Table 5, when the attention mechanism was added to the first three layers of ResNet101, the detection effect improved to a certain extent. However, continuing to introduce attention-blocks containing edge information to the 4th and 5th layers caused a drop in detection accuracy. This is because there is more abstract information in the feature maps extracted by the fourth and fifth layers in ResNet101, and the edge information is the basic feature information. This is counterproductive and reduces the detection performance.

Table 5. APs of different introduction times of the attention mechanism.

Introduced Layer	DamperDetSet			FPS
	AP_{50}	AP_{70}	AP_{90}	
None	86.28	77.36	70.03	81
C1	87.83	80.23	71.37	80
C1, C2	91.38	84.61	77.42	77
C1, C2, C3	92.62	89.67	81.24	74
C1, C2, C3, C4	93.14	90.15	81.92	74
C1, C2, C3, C4, C5	89.27	83.32	73.52	73

5.5.4. Number of Epochs

The number of epochs for experimental training affects the performance of the model. Because the number of training epochs is not enough, the model is under-fitted, and the model has not yet fully learned to identify all the objects to be detected. Excessive training epochs reduce the robustness of the model, the parameters are limited by the existing training data, and the realization of unfamiliar data in the test set is reduced. Therefore, we conducted an evaluation test of the number of training times for the performance of the model, and the test results are shown in Table 6. It can be seen from the table that when the training epoch is 200, the model is the most balanced.

Table 6. APs of different epoch numbers.

Number of Epochs	DamperDetSet			FPS
	AP_{50}	AP_{70}	AP_{90}	
50	71.63	60.62	41.37	79
100	80.51	72.27	65.23	77
150	84.15	80.16	74.38	75
200	92.62	89.67	81.24	74
250	93.31	88.65	80.47	74

5.5.5. Minimum Training Data Experiment

Changes in the amount of training data also affect the final performance of the model. At the same time, by comparing the detection accuracy of the model with different amounts of data, the feature extraction ability of the model can be judged. As shown in Table 7, we conducted experiments with the minimum amount of data. From the results, it can be seen that when the amount of data decreases, the performance of the model has weak performance, which indicates that our data is sufficient. The model performance did not drop significantly until the test set dropped to 1750. Moreover, DamperYOLO had strong robustness and could still learn key feature information on small-scale datasets, which overcame the shortcomings of the previous model's poor generalization ability to a certain extent.

Table 7. Results of minimum training data experimental.

The Amount of Training Set	DamperDetSet			FPS
	AP_{50}	AP_{70}	AP_{90}	
2500 (100%)	92.62	89.67	81.24	74
2250 (90%)	89.51	86.28	78.83	75
2000 (80%)	85.39	81.75	75.41	74
1750 (70%)	82.41	77.40	71.68	74
1500 (60%)	73.97	69.62	64.01	72

5.6. Ablation Analysis

To analyze the functions of the different components of DamperYOLO, an ablative analysis was performed on DamperDetSet. As shown in Table 8, Model B had better indicators than Model A, which indicates that using ResNet101 as the backbone can better extract image features. Model C uses image augmentation for preprocessing, which improves the quality of the input image and provides the model with better training data. Compared with other stages, the performance of Model D has the highest improvement in detection effect. This indicates that the attention mechanism plays a sufficient role, because the attention mechanism allows the model to focus on the edge information of the damper when converging, with the help of the image enhancement model. In addition, it can be seen from other comparative experiments that the additional overhead brought by it is very low, so it is necessary for our task to add an attention mechanism to the backbone.

Table 8. The results of the ablation analysis.

Model	Architecture	AP_{50}	AP_{70}	AP_{90}
A	YOLOv4	86.21	78.45	70.96
B	A + ResNet101	88.57	82.36	73.72
C	B + Edge Extraction	90.82	84.24	76.50
D	C + Attention Mechanism	92.62	89.67	81.24

5.7. Computational Complexity

The network parameters and training time were recorded to evaluate the space and time complexity of the networks. As shown in Table 9, compared with Cascade R-CNN, DamperYOLO has a similar detection effect, but its parameters and training time are greatly reduced. Compared with YOLOv4, the space complexity and the training time are basically unchanged, because we only changed the backbone and added the attention mechanism on its basis, but a higher detection effect was achieved. In addition, CenterNet still consumes the least resources. The computational complexity of SSD is slightly higher than that of RetinaNet, but the detection effect is slightly worse.

Table 9. Network parameters (Param.) and training time of the different models.

Model	Param.	Training Time (h)
YOLOv4	28 M	6.38
SSD	34 M	7.46
RetinaNet	32 M	7.03
CenterNet	14 M	4.05
Cascade R-CNN	184 M	49.84
DamperYOLO	30 M	6.92

6. Conclusions

We propose a power line vibration damper detection model named DamperYOLO based on a deep neural network that can detect the position of the vibration damper in drone inspection aerial images. DamperYOLO first uses the Canny algorithm to obtain the edge information of the original image, then uses the attention mechanism to introduce edge information into ResNet101 to guide feature extraction. Finally, it outputs a feature map that is more conducive to small target detection with the FPN structure. The following conclusions can be drawn through qualitative and quantitative experiments on the power line vibration damper detection dataset built in this paper. Compared with the current baselines in the object detection field, the DamperYOLO proposed in this paper can output state-of-the-art detection accuracy. The results of sensitivity analysis experiments show that edge detection, attention mechanism, and feature pyramid network all significantly improve the detection accuracy. The ablation analysis shows that the attention mechanism and the feature pyramid network improve the accuracy of the output detection results. In addition, DamperYOLO consumes space similar to the computational resources and baselines of other one-stage classes, but the detection accuracy can reach the level of Cascade R-CNN, which shows the superiority of our model. In the future, we will continue to introduce appropriate training tricks for the detection accuracy of DamperYOLO and explore the application of the model to other power line components.

Author Contributions: Conceptualization, W.C. and Y.L.; methodology, W.C., Y.L. and Z.Z.; validation, W.C.; formal analysis, W.C. and Y.L.; investigation, W.C., Y.L. and Z.Z.; resources, W.C., Y.L., and Z.Z.; writing—original draft preparation, W.C.; writing—review and editing, W.C., Y.L. and Z.Z.; visualization, W.C.; supervision, Y.L.; project administration, Y.L.; All authors have read and agreed to the published version of the manuscript.

Funding: This research was supported by the National Natural Science Foundation of China, grant number 61962031, the National Natural Science Foundation of China, grant number 51667011, and the Applied Basic Research Project of Yunnan province, grant number 2018FB095.

Institutional Review Board Statement: Not applicable.

Informed Consent Statement: Not applicable.

Data Availability Statement: The data in this paper are undisclosed due to the confidentiality requirements of the data supplier.

Acknowledgments: We thank the Yunnan Electric Power Research Institute for collecting the transmission line inspection data, which provided a solid foundation for the verification of the model proposed in this paper. The authors thank the reviewers and editors for their constructive comments to improve the quality of this article.

Conflicts of Interest: The authors declare that there are no conflict of interest.

References

1. Bao, W.; Ren, Y.; Wang, N.; Hu, G.; Yang, X. Detection of Abnormal Vibration Dampers on Transmission Lines in UAV Remote Sensing Images with PMA-YOLO. *Remote Sens.* **2021**, *13*, 4134. [CrossRef]
2. Wu, H.; Xi, Y.; Fang, W.; Sun, X.; Jiang, L. Damper detection in helicopter inspection of power transmission line. In Proceedings of the 2014 4th International Conference on Instrumentation and Measurement, Computer, Communication and Control (IMCCC), Harbin, China, 18–20 September 2014; pp. 628–632. [CrossRef]
3. Ma, Y.; Li, Q.; Chu, L.; Zhou, Y.; Xu, C. Real-Time Detection and Spatial Localization of Insulators for UAV Inspection Based on Binocular Stereo Vision. *Remote Sens.* **2021**, *13*, 230. [CrossRef]
4. Hinas, A.; Roberts, J.M.; Gonzalez, F. Vision-Based Target Finding and Inspection of a Ground Target Using a Multirotor UAV System. *Sensors* **2017**, *17*, 2929. [CrossRef]
5. Huang, S.; Han, W.; Chen, H.; Li, G.; Tang, J. Recognizing Zucchinis Intercropped with Sunflowers in UAV Visible Images Using an Improved Method Based on OCRNet. *Remote Sens.* **2021**, *13*, 2706. [CrossRef]
6. Popescu, D.; Stoican, F.; Stamatescu, G.; Chenaru, O.; Ichim, L. A Survey of Collaborative UAV–WSN Systems for Efficient Monitoring. *Sensors* **2019**, *19*, 4690. [CrossRef] [PubMed]
7. Zhang, Y.; Yuan, X.; Li, W.; Chen, S. Automatic Power Line Inspection Using UAV Images. *Remote Sens.* **2017**, *9*, 824. [CrossRef]
8. Liu, Y.; Li, J.X.; Xu, W.; Liu, M.Y. A method on recognizing transmission line structure based on multi-level perception. In Proceedings of the International Conference on Image and Graphics, Shanghai, China, 13–15 September 2017; pp. 512–522.
9. Huang, X.; Zhang, X.; Zhang, Y.; Zhao, L. A method of identifying rust status of dampers based on image processing. *IEEE Trans. Instrum. Meas.* **2020**, *69*, 5407–5417. [CrossRef]
10. Jin, L.J.; Yan, S.J.; Liu, Y. Vibration damper recognition based on Haar-Like features and cascade adaboost classifier. *J. Syst. Simul.* **2012**, *24*, 60–63.
11. Zhang, K.; Hou, Q.; Huang, W. Defect Detection of Anti-vibration Hammer Based on Improved Faster R-CNN. In Proceedings of the 2020 7th International Forum on Electrical Engineering and Automation (IFEEA), Hefei, China, 25–27 September 2020; pp. 889–893. [CrossRef]
12. Bao, W.; Ren, Y.; Liang, D.; Yang, X.; Xu, Q. Defect Detection Algorithm of Anti-vibration Hammer Based on Improved Cascade R-CNN. In Proceedings of the 2020 International Conference on Intelligent Computing and Human-Computer Interaction (ICHCI), Sanya, China, 4–6 December 2020; pp. 294–297. [CrossRef]
13. Hickey, C.; Young, P.; Mayomi, T.; Noctor, J. Fault Investigation and Analysis of an Overhead Transmission Line Vibration Damper. In Proceedings of the 2021 56th International Universities Power Engineering Conference (UPEC), Middlesbrough, UK, 31 August–3 September 2021; pp. 1–6. [CrossRef]
14. Xiao, S.; Wang, H.; Ling, L. Research on a novel maintenance robot for power transmission lines. In Proceedings of the 2016 4th International Conference on Applied Robotics for the Power Industry (CARPI), Jinan, China, 11–13 October 2016; pp. 1–6. [CrossRef]
15. Song, W.; Zuo, D.; Deng, B.; Zhang, H.; Xue, K.; Hu, H. Corrosion defect detection of earthquake hammer for high voltage transmission line. *Chin. J. Sci. Instrum.* **2016**, *37*, 113–117.
16. Yang, H.; Guo, T.; Shen, P.; Chen, F.; Wang, W.; Liu, X. Anti-vibration hammer detection in UAV image. In Proceedings of the 2017 2nd International Conference on Power and Renewable Energy (ICPRE), Chengdu, China, 20–23 September 2017; pp. 204–207. [CrossRef]
17. Liu, Y.; Wen, S.; Chen, Z.; Zhang, D. Research of the Anti-vibration Hammer Resetting Robot Based on Machine Vision. In Proceedings of the 2020 Chinese Control and Decision Conference (CCDC), Hefei, China, 22–24 August 2020; pp. 3708–3712. [CrossRef]
18. Chen, X.; Wu, Y.; Zhao, L. Identification of OPGW vibration damper based on random Hough transformation. Heilongjiang Dianli Jishu Heilongjiang Electric. *Power* **2010**, *32*, 1–2.
19. Miao, S.; Sun, W.; Zhang, H. Intelligent visual method based on wavelet moments for obstacle recognition of high voltage transmission line deicer robot. *Jiqiren* **2010**, *32*, 425–431. [CrossRef]
20. Pan, L.; Xiao, X. Image recognition for on-line vibration monitoring system of transmission line. In Proceedings of the 2009 9th International Conference on Electronic Measurement & Instruments, Beijing, China, 16–19 August 2009; pp. 3–1081. [CrossRef]
21. Liu, Y.; Jin, L. Vibration Damper Recognition of Transmission System Based on Unmanned Aerial Vehicles. In Proceedings of the 2011 Asia-Pacific Power and Energy Engineering Conference, Wuhan, China, 25–28 March 2011; pp. 1–3. [CrossRef]
22. Guo, J.; Xie, J.; Yuan, J.; Jiang, Y.; Lu, S. Fault Identification of Transmission Line Shockproof Hammer Based on Improved YOLO V4. In Proceedings of the 2021 International Conference on Intelligent Computing, Automation and Applications (ICAA), Nanjing, China, 25–27 June 2021; pp. 826–833. [CrossRef]

23. Wang, W.; Wang, Z.; Liu, B.; Yang, Y.; Sun, X. Typical Defect Detection Technology of Transmission Line Based on Deep Learning. In Proceedings of the 2019 Chinese Automation Congress (CAC), Hangzhou, China, 22–24 November 2019; pp. 1185–1189. [CrossRef]
24. Zhang, Z.; Jiang, W.; Yang, J. An Improved Quantization Algorithm for Electric Power Inspection. In Proceedings of the 2021 9th International Electrical Engineering Congress (iEECON), Pattaya, Thailand, 10–12 March 2021; pp. 397–400. [CrossRef]
25. Liu, C.; Wu, Y.; Liu, J.; Sun, Z. Improved YOLOv3 Network for Insulator Detection in Aerial Images with Diverse Background Interference. *Electronics* **2021**, *10*, 771. [CrossRef]
26. Sadykova, D.; Pernebayeva, D.; Bagheri, M.; James, A. IN-YOLO: Real-Time Detection of Outdoor High Voltage Insulators Using UAV Imaging. *IEEE Trans. Power Deliv.* **2020**, *35*, 1599–1601. [CrossRef]
27. Zhao, Z.; Zhen, Z.; Zhang, L.; Qi, Y.; Kong, Y.; Zhang, K. Insulator Detection Method in Inspection Image Based on Improved Faster R-CNN. *Energies* **2019**, *12*, 1204. [CrossRef]
28. Chen, W.; Li, Y.; Li, C. A Visual Detection Method for Foreign Objects in Power Lines Based on Mask R-CNN. *Int. J. Ambient. Comput. Intell. IJACI* **2020**, *11*, 34–47. [CrossRef]
29. Kang, G.; Gao, S.; Yu, L.; Zhang, D. Deep Architecture for High-Speed Railway Insulator Surface Defect Detection: Denoising Autoencoder with Multitask Learning. *IEEE Trans. Instrum. Meas.* **2019**, *68*, 2679–2690. [CrossRef]
30. Cheng, L.; Liao, R.; Yang, L.; Zhang, F. An Optimized Infrared Detection Strategy for Defective Composite Insulators According to the Law of Heat Flux Propagation Considering the Environmental Factors. *IEEE Access* **2018**, *6*, 38137–38146. [CrossRef]
31. Yu, C.; Pan, W.; Lei, X.; Yu, G.; Qin, W.; Zhu, K.; Zheng, H. Simulation of electric field and potential transfer arc during the on-line process of the live working anti-vibration hammer robot. In Proceedings of the 2021 International Conference on Electrical Materials and Power Equipment (ICEMPE), Chongqing, China, 11–15 April 2021; pp. 1–4. [CrossRef]
32. Diana, G.; Falco, M.; Cigada, A.; Manenti, A. On the measurement of overhead transmission lines conductor self-damping. *IEEE Trans. Power Deliv.* **2000**, *15*, 285–292. [CrossRef]
33. Si, J.; Rui, X.; Liu, B.; Zhou, L.; Liu, S. Study on a New Combined Anti-Galloping Device for UHV Overhead Transmission Lines. *IEEE Trans. Power Deliv.* **2019**, *34*, 2070–2078. [CrossRef]
34. Redmon, J.; Divvala, S.; Girshick, R.; Farhadi, A. You only look once: Unified, real-time object detection. In Proceedings of the 2016 IEEE Conference on Computer Vision and Pattern Recognition (CVPR), Las Vegas, NV, USA, 27–30 June 2016; pp. 779–788. [CrossRef]
35. Redmon, J.; Farhadi, A. YOLO9000: Better, faster, stronger. In Proceedings of the IEEE Conference on Computer Vision and Pattern Recognition (CVPR), Honolulu, HI, USA, 21–26 July 2017; pp. 7263–7271. [CrossRef]
36. Redmon, J.; Farhadi, A. YOLOv3: An Incremental Improvement. *arXiv* **2018**, arXiv:1804.02767. Available online: https://arxiv.org/abs/1804.02767 (accessed on 10 January 2022).
37. Bochkovskiy, A.; Wang, C.Y.; Liao, H. YOLOv4: Optimal Speed and Accuracy of Object Detection. *arXiv* **2020**, arXiv:2004.10934. Available online: https://arxiv.org/abs/2004.10934v1 (accessed on 20 January 2022).
38. Wang, C.Y.; Liao, H.M.; Wu, Y.H.; Chen, P.Y.; Hsieh, J.W.; Yeh, I.H. CSPNet: A new backbone that can enhance learning capability of cnn. In Proceedings of the IEEE Conference on Computer Vision and Pattern Recognition Workshop (CVPR Workshop), Seattle, WA, USA, 14–19 June 2020; pp. 1571–1580. [CrossRef]
39. Zheng, Z.; Wang, P.; Liu, W.; Li, J.; Ye, R.; Ren, D. Distance-IoU Loss: Faster and better learning for bounding box regression. In Proceedings of the AAAI Conference on Artificial Intelligence (AAAI), New York, NY, USA, 7–12 February 2020. [CrossRef]
40. He, K.; Zhang, X.; Ren, S.; Sun, J. Deep Residual Learning for Image Recognition. In Proceedings of the 2016 IEEE Conference on Computer Vision and Pattern Recognition (CVPR), Las Vegas, NV, USA, 27–30 June 2016; pp. 770–778. [CrossRef]
41. Canny, J. A Computational Approach to Edge Detection. *IEEE Trans. Pattern Anal. Mach. Intell.* **1986**, *8*, 679–698. [CrossRef]
42. Bahdanau, D.B.; Kyunghyun, C.; Yoshua, B. Neural Machine Translation by Jointly Learning to Align and Translate. *arXiv* **2014**, arXiv:1409.0473. Available online: https://arxiv.org/abs/1409.0473 (accessed on 10 January 2022).
43. Lin, T.Y.; Dollár, P.; Girshick, R.; He, K.; Hariharan, B.; Belongie, S. Feature pyramid networks for object detection. In Proceedings of the IEEE Conference on Computer Vision and Pattern Recognition (CVPR), Honolulu, HI, USA, 21–26 July 2017; pp. 2117–2125. [CrossRef]
44. Tao, C.; Ma, K.K.; Chen, L.H. Tri-state median filter for image denoising. *IEEE Trans. Image Process.* **1999**, *8*, 1834–1838.
45. Zhang, X.M.; Xu, B.S.; Dong, S.Y.; Gan, X.M. Adaptive median-weighted mean hybrid filter. *Opt. Tech.* **2004**, *6*, 652–659.
46. Ren, S.; He, K.; Girshick, R.B.; Sun, J. Faster R-CNN: Towards Real-Time Object Detection with Region Proposal Networks. *IEEE Trans. Pattern Anal. Mach. Intell.* **2017**, *39*, 1137–1149. [CrossRef]
47. Liu, W.; Anguelov, D.; Erhan, D.; Szegedy, C.; Reed, S.; Fu, C.; Berg, A.C. SSD: Single shot multibox detector. In Proceedings of the European Conference on Computer Vision (ECCV), Amsterdam, The Netherlands, 8–16 October 2016; pp. 21–37. [CrossRef]
48. Simonyan, K.; Zisserman, A. Very Deep Convolutional Networks for Large-Scale Image Recognition. *arXiv* **2014**, arXiv:1409.1556. Available online: https://arxiv.org/abs/1409.1556 (accessed on 20 January 2022).
49. Cai, Z.; Vasconcelos, N. Cascade R-CNN: Delving into High Quality Object Detection. In Proceedings of the IEEE Conference on Computer Vision and Pattern Recognition, Salt Lake City, UT, USA, 18–23 June 2018; pp. 6154–6162. [CrossRef]
50. Zhou, X.; Wang, D.; Krähenbühl, P. Objects as Points. *arXiv* **2019**, arXiv:1904.07850. Available online: https://arxiv.org/abs/1904.07850v2 (accessed on 20 January 2022).

51. Lin, T.Y.; Goyal, P.; Girshick, R.; He, K.; Dollár, P. Focal loss for dense object detection. *IEEE Trans. Pattern Anal. Mach. Intell.* **2020**, *42*, 2980–2988. [CrossRef]
52. Lin, T.Y.; Maire, M.; Belongie, S.; Hays, J.; Zitnick, C.L. Microsoft COCO: Common Objects in Context. In Proceedings of the European Conference on Computer Vision-ECCV 2014, Lecture Notes in Computer Science, Zurich, Switzerland, 6–7, 12 September 2014; Springer: Berlin/Heidelberg, Germany, 2014; Volume 8693. [CrossRef]

sensors

MDPI

Article

The Effect of Synergistic Approaches of Features and Ensemble Learning Algorithms on Aboveground Biomass Estimation of Natural Secondary Forests Based on ALS and Landsat 8

Chunyu Du [1,2], Wenyi Fan [1,3], Ye Ma [1], Hung-Il Jin [1,4] and Zhen Zhen [1,3,*]

1 School of Forestry, Northeast Forestry University, Harbin 150040, China; duchunyu402@163.com (C.D.); fanwy@163.com (W.F.); maye799535410@163.com (Y.M.); jhicoco@nefu.edu.cn (H.-I.J.)
2 Jilin Forestry Research Institute, Jilin 132013, China
3 Key Laboratory of Sustainable Forest Ecosystem Management-Ministry of Education, Northeast Forestry University, Harbin 150040, China
4 Faculty of Forest Science, Kim Il Sung University, Pyongyang 999093, Democratic People's Republic of Korea
* Correspondence: zhzhen@syr.edu; Tel.: +86-0451-8219-1219

Abstract: Although the combination of Airborne Laser Scanning (ALS) data and optical imagery and machine learning algorithms were proved to improve the estimation of aboveground biomass (AGB), the synergistic approaches of different data and ensemble learning algorithms have not been fully investigated, especially for natural secondary forests (NSFs) with complex structures. This study aimed to explore the effects of the two factors on AGB estimation of NSFs based on ALS data and Landsat 8 imagery. The synergistic method of extracting novel features (i.e., *COLI*1 and *COLI*2) using optimal Landsat 8 features and the best-performing ALS feature (i.e., elevation mean) yielded higher accuracy of AGB estimation than either optical-only or ALS-only features. However, both of them failed to improve the accuracy compared to the simple combination of the untransformed features that generated them. The convolutional neural networks (CNN) model was much superior to other classic machine learning algorithms no matter of features. The stacked generalization (SG) algorithms, a kind of ensemble learning algorithms, greatly improved the accuracies compared to the corresponding base model, and the SG with the CNN meta-model performed best. This study provides technical support for a wall-to-wall AGB mapping of NSFs of northeastern China using efficient features and algorithms.

Keywords: ensemble learning; machine learning; feature extraction; AGB; NSFs

Citation: Du, C.; Fan, W.; Ma, Y.; Jin, H.-I.; Zhen, Z. The Effect of Synergistic Approaches of Features and Ensemble Learning Algorithms on Aboveground Biomass Estimation of Natural Secondary Forests Based on ALS and Landsat 8. *Sensors* **2021**, *21*, 5974. https://doi.org/10.3390/s21175974

Academic Editors: Moulay A. Akhloufi and Mozhdeh Shahbazi

Received: 8 August 2021
Accepted: 2 September 2021
Published: 6 September 2021

1. Introduction

The Asian temperate mixed forest in northeastern China is one of the three major temperate mixed forests in the world (i.e., northeastern North America, Europe, and East Asia) [1], which is of great strategic importance to the carbon trading of China. The forests of Northeast China have experienced three periods of excessive timber harvesting in the last century, including the period of Russian and Japanese aggression (1896–1945), the period of encouraging excessive harvesting for timber production (1950–1977), and the period of national economic reforms and the broadening of international relations (1978–1998) [2]. The excessive logging and neglected cultivation of forests nearly exhausted exploitable forest reserves in the region [3]. Since the Natural Forest Conservation Program (NFCP) was put into practice in 1998, there was a profound shift in focus from timber production to environmental protection by rehabilitating damaged forest ecosystems, afforesting desertified and degraded areas, and banning logging in natural forests [2]. In this context, natural secondary forests (NSFs) are gradually expanding and gaining importance. NSFs, which account for as much as 70% of the forests of northeastern China, refer to as the natural-regeneration forests after stand-replacing disturbances of primary forests by anthropogenic activities or by extreme natural events [4,5]. Nowadays, the NSFs of northeastern China are gradually recovering from the excessive logging of the 20th century, which led to an extraordinary reduction in the quality of the forest ecosystem. NSFs are of significance to China not only for timber supply, but also for a vital reservoir of biodiversity, potential carbon sequestration, a destination of ecological tourism, and a broad ecological shelter for northeastern China [2].

The accurate estimation of forest aboveground biomass (AGB) has a critical effect on the understanding of forest quality and recovery in the NSFs of northeastern China. AGB is defined as the dry mass of live or dead matter from tree or shrub life forms, typically expressed as a mass per area density (e.g., Mg/ha) [6]. In general, AGB could be obtained by (1) direct harvest method; (2) allometric equation-based method; (3) biomass expansion factor (BEF)-based method; (4) process-based biogeochemical modeling; (5) remote sensing-based estimation method. Although the direct harvest method is the most exact among these methods, it is time-consuming, destructive, and labor-intensive. It is only suitable for AGB estimation of a small area or of individual trees with a small sample size [7] and is usually applied as reference data to establish allometric equations of AGB (e.g., [8,9]). An allometric equation-based method is more flexible and feasible than the direct harvest method to estimate AGB on both individual tree and plot levels. A variety of allometric equations are developed for diverse tree species by modeling the relationship between AGB and various physical parameters of trees, such as diameter at breast height (DBH), tree height, crown diameter, etc. (e.g., [8,10–13]). Similar to an allometric equation-based method, the BEF-based method applied BEF defining as the ratio of all stand biomass to growing stock volume to convert timber volume to biomass [14]. However, the allometric equation-based and BEF-based methods are still time-consuming and expensive because both of them are based on the acquisition of field measurements (such as DBH, tree height), and still limited to plot-level or individual tree-level AGB estimations. Process-based biogeochemical models consider the processes including photosynthesis, absorption, and carbon allocation, and generally couple biology, soil, climate, hydrology, and anthropogenic effects [15]. To some degree, these models could improve the conventional, point-based estimation of biomass over large areas [16]. However, the high uncertainties in biomass estimation due to constraints in data source, spatial resolution, homogeneous assumption, and inaccuracy of models greatly limit the usage of process-based biogeochemical models [15]. The remote sensing-based method is exceedingly appealing for estimating forest biomass on a large scale (e.g., local, regional or global) because of its unique characteristics such as repetitive data acquisition, large coverage, digital format, and so on [15], and of the capability of providing spatially explicit AGB estimates for every pixel location, instead of only the mean or total biomass within a given inventory unit [17,18]. Nowadays, it becomes the most commonly used method for large-scale AGB estimation [19–21].

In the last three decades, researchers have attempted a variety of remotely sensed data sources to estimate AGB. With a relatively long history of data availability, optical satellite imagery (such as Landsat, MODIS, etc.) has become a primary data source for biomass estimation (e.g., [22–25]). In particular, Landsat series satellite imagery is the most commonly used data source for AGB estimation (e.g., [26–30]), mainly because of the continuous, long-term, medium spatial resolution, and cross-calibrated data for global surface observations, and free access policy [31]. However, it is of significance to notice the data saturation in Landsat imagery, which refers to the phenomenon that spectral reflectance values are not sensitive to the change in biomass of mature forest or advanced successional forests even if AGB varies significantly [32,33]. For example, Steininger [34] found that the canopy reflectance in Landsat imagery saturates when the AGB approaches 15 kg/m^2 or over 15 years of age in Brazilian tropical secondary forests. Zhao et al. [33] examined the saturation values in Landsat imagery for different vegetation types in a subtropical region, and found the AGB saturation values for pine forest, mixed forest, Chinese fir forest, broadleaf forest, bamboo forest, and shrub were 159, 152, 143, 123, 75, and 55 Mg/ha, respectively. Data saturation in optical imagery like Landsat significantly lowers the accuracy and increases the uncertainties of AGB estimation [15]. Data saturation still exists in RADAR (Radio Detection and Ranging) data like SAR (Synthetic Aperture Radar) [35]. Generally speaking, saturation values could be higher obtained by longer wavelengths (such as L and P bands) and lower by shorter wavelengths (such as C bands), and also vary for different forest structures [36]. Until now, the data saturation problem caused by remote sensing signals is still one of the biggest obstacles to applying optical imagery and RADAR data for AGB estimation [15,37,38].

Since the 1990s, it has been found that LiDAR (Light Detection and Ranging) is more advantageous than optical imagery for AGB estimation because it is more relative to tree height and produces less estimation error [39]. Meanwhile, LiDAR is unaffected by the data saturation problem, even for high AGB values (>1000 Mg/ha) [40]. Thus, LiDAR data is widely used in AGB estimation in the last two decades. According to the format of return signals, LiDAR can be classified into discrete and continuous LiDAR; according to platforms, LiDAR can be classified into spaceborne, airborne, UAV(Unmanned Aerial Vehicle), terrestrial, backpack/handheld LiDAR; according to the size of the footprint, LiDAR can be classified into small footprint (footprint size <1 m), mid footprint (footprint size: 10–30 m), and large footprint LiDAR (footprint >50 m) [41]. In recent years, Airborne Laser Scanning (ALS) data, a kind of discrete, multiple returns, and small footprint LiDAR data captured from an aerial platform, has received much scientific and operational attention for AGB estimation than any of the other remote sensing data [42]. ALS emits laser pulses towards the ground and receives the pulses reflected from the tree canopy, branches, leaves, trunk, shrub, and then ground to form a three-dimensional profile of forest structure. ALS is far more capable than optical and RADAR sensors in estimating forest parameters and is considered the premier tool for large-scale AGB estimation (e.g., [43–47]). It is beneficial to estimate AGB by capturing both two-dimensional spectral information of the upper canopy and three-dimensional structural information of the canopy. However, the spectral characteristics of vegetation provided by ALS are very limited since most LiDAR systems only work at a single wavelength [48]. Thus, the integration of optical imagery and ALS data has become the most promising approach for large-scale AGB estimation (e.g., [48–53]).

Features are the most direct representation or manifestation of data sources. Feature extraction and selection could greatly influence the accuracy of AGB estimation [54]. A variety of spectral-related features including band combinations, textures, diverse vegetation indices, leaf area index, fraction of vegetation cover, and so on were derived from optical imagery for AGB estimation (e.g., [29,55–58]). Similar, diverse point-based features including height statistics (e.g., mean, maximum, variance, skewness, etc.), canopy-based quantile estimators, canopy relief ratio, laser penetration rates, canopy closure, and so on were extracted from LiDAR data for AGB estimation (e.g., [25,46,52,59,60]). Some researchers directly combined optical imagery and LiDAR features (e.g., [21,61,62]) while a few of them designed novel features derived from optical imagery and LiDAR data to improve AGB estimation. For example, Zhang et al. [48] developed two novel groups of features (i.e., *COLI*1 and *COLI*2) using seven vegetation indices derived from Landsat 8 and the best-performing LiDAR variable (i.e., mean of height). The *COLI*1 and *COLI*2 were generated by the multiplication and ratio combinations of the best-performing LiDAR variable and each vegetation index, respectively. They found that the stacked sparse autoencoder network model with the combination of all *COLI*1, optical, and LiDAR features yielded the highest accuracy of AGB estimation for the coniferous and broadleaf mixed forest of southeast China. However, whether it is more efficient to use novel features extracted from both data than directly combine all features is still needed to be further investigated.

In addition to data sources and features, it is vital to establish a reliable and suitable model to estimate AGB. Currently, most remote sensing-based AGB estimation methods use data-driven empirical models, which can be divided into parametric and non-parametric models [63]. Parametric models explicitly determine parameterized expressions of independent variables (e.g., spectral bands) and the dependent variable of interest (e.g., AGB) assuming the probability distributions of the variables being assessed [63]. Multiple linear regression, a classic parametric model with normality assumption, was the most widely used method in previous AGB studies due to their simplicity and interpretability (e.g., [53,64,65]). Other parametric models, like non-linear regression (e.g., an exponential, power, or polynomial fitting function), were also applied for AGB estimations (e.g., [59,66,67]). Unlike parametric models, nonparametric models are distribution-free methods in which the predictor does not take a predetermined form but is constructed according to information derived from the data. Most machine learning models belongs to non-parametric, such as artificial neural network (ANN), random forest (RF), k-nearest neighbor (KNN), support vector machine (SVM), cubist (CB), classification and regression tree (CART), convolutional neural networks (CNN) and so on. Without the assumption of distribution, the non-parametric machine learning models are extremely flexible and capable of capturing the complex relationships between remote sensing variables and AGB, and widely applied in AGB estimation (e.g., [43,68–73]).

Ensemble learning, a branch of machine learning, is designed to learn tasks by constructing and then integrating multiple learners to produce a strong learner for improving accuracy [74,75]. There are three basic categories of ensemble learning: bagging, boosting, and stacking. RF and adaptive boosting (AdaBoost) algorithms are classic representatives of bagging and boosting algorithms, respectively. RF builds trees using subsamples and a random subset of predictors and can be very effective for estimating AGB due to its robustness to overfitting and noise in the training dataset [43,76,77]. Adaptive boosting is an iterative boosting algorithm that adaptively changes the distribution of the training set based on the performance of previous learners. Another boosting algorithm, called extreme gradient boosting (XGBoost), has been demonstrated to show great advantages in decreasing overestimation of low AGB values and underestimation of high AGB values for a forest type-based biomass estimation using continuous forest inventory data and Landsat 8 imagery [54]. Stacking, first proposed by Wolpert [78], is another method for combining multiple models but is less used than bagging and boosting. Unlike the RF algorithm that the base learner is homogeneous (e.g., regression tree), stacking are heterogeneous ensemble algorithms that could integrate diverse base learners to generate a stronger learner. The stacking algorithm was used to estimate canopy height in forestry (e.g., [79]), however, its potential has not been fully explored in AGB estimation.

Although the synergistic utilization of ALS and optical passive imagery was proved to improve AGB estimation [48], the synergistic approach (i.e., features) has not been fully investigated, especially for NSFs with complex structures. For example, is it more efficient to apply a novel feature extracted from passive imagery and LiDAR data (e.g., *COLI*1 and *COLI*2 in [48]) or directly combine all the features from the two data sources (like [61])? In addition, will ensemble learning algorithms improve the accuracy of AGB estimation for NSFs? Inspired by these questions, this study aimed at exploring the effects of different synergistic approaches of features and ensemble learning algorithms on AGB estimation of NSFs of northeastern China based on ALS and Landsat 8 OLI (Operational Land Imager) imagery. Specifically, the objectives of this study were (1) to investigate the effects of different data sources and classic machine learning algorithms on AGB estimation of a natural secondary forest; (2) to grope for a highly effective approach to combine ALS and Landsat 8 OLI imagery on AGB estimation of a natural secondary forest; (3) to explore the performances of ensemble learning algorithms in estimating AGB of a natural secondary forest; (4) to generate an accurate wall-to-wall AGB map of a natural secondary forest for future forest resources management.

2. Materials and Methods

2.1. Study Area

The study area is located in Maoershan Experimental Forest Farm of Northeast Forestry University (NEFU), Shangzhi, Heilongjiang Province, China, ranging from 127°29′ to 127°44′ E and 45°14′ to 45°29′ (Figure 1). The landform of the forest farm belongs to a low mountain and hilly area. The terrain gradually rises from south to north, with an average elevation of 300 m. The highest mountain is Maoer Mountain, with an elevation of 805 m.

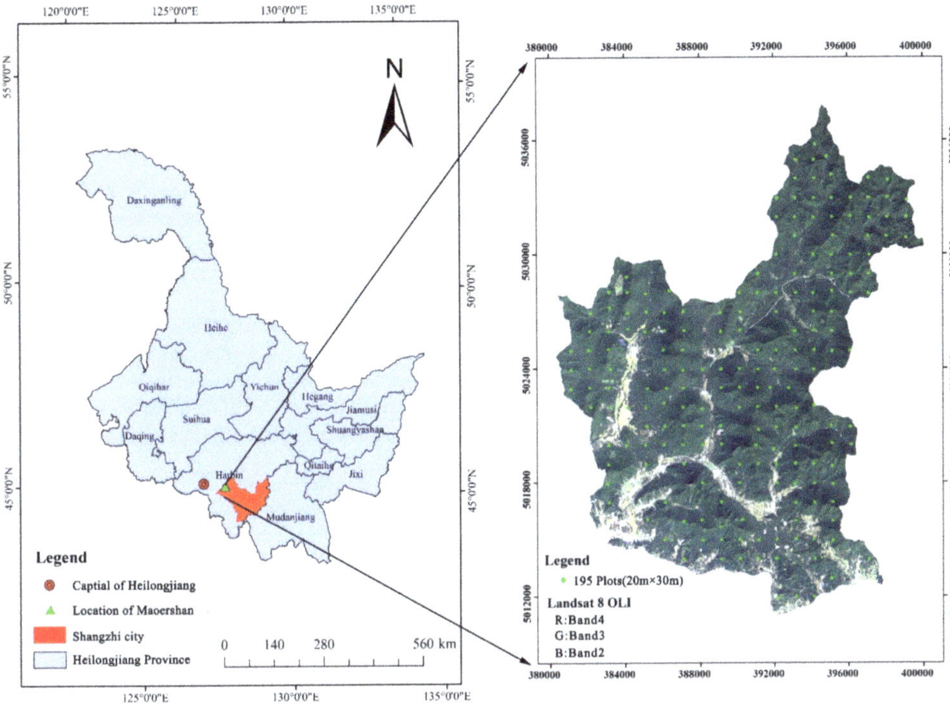

(a) (b)

Figure 1. The location of study area: (**a**) The location of Maoershan Experimental Forest Farm within Heilongjiang Province; (**b**) the locations of 195 plots (20 m × 30 m) within Maoershan (Background: Landsat 8 OLI image).

The total area of the forest farm is 26,496 ha, which belongs to a typical natural secondary forest in northeastern China. The vegetation in the Maoershan area is a part of Changbai plant flora, with the original zonal top-level community of Korean pine broad-leaved forest. Due to the destruction in the last century, the original vegetation has undergone reverse succession. It has formed a forest landscape in which natural secondary forests are dominated by precious broad-leaved forests, poplar and birch forests, oak forests, and so on, and plantations such as red pine and larch are inlaid. The main species include *Betula platyphylla*, *Quercus mongolica*, *Populus davidiana*, *Larix olgensis*, *Pinus sylvestris*, and *Pinus koraiensis*, etc. The average forest coverage rate is 95%, and the total stock is approximately 3.5 million m^3.

2.2. Data Collection

2.2.1. Remotely Sensed Data

The remotely sensed data utilized in this study include ALS data and Landsat 8 OLI imagery. ALS data were obtained in September 2015. It is a secondary product scanned by the LiDAR sensor (Riegl LMS-Q680i) carried by the LiCHY system of the Chinese Academy of Forestry. The maximum frequency of the laser pulse of the LiDAR sensor is 400 kHz, with a wavelength of 1550 nm, a scanning angle of $\pm 30°$, a sampling interval of 1 ns, and vertical accuracy of 0.15 m. The sidelap of this flight strip was designed to be greater than 60%, with an average point cloud density of 3.6 points$\cdot m^{-2}$.

To be consistent with ALS data in time, the Landsat 8 OLI imagery acquired on 13 September 2015 was applied in this study (downloaded from https://earthexplorer.usgs.gov/ (accessed on 1 September 2021)). The scene ID is LC81170282015256LGN01 (L1T-level product), with cloudiness of 1.35%, sun elevation angle of 45.28°, and sun azimuth angle of 154.91°. Seven multispectral bands (band1–band7) of 30 m nominal spatial resolution were utilized in this study. The radiometric resolution of the imagery is 12 bits and the swath width is 185 km × 185 km.

2.2.2. Reference Data

The 195 fixed plots data of continuous forest resources inventory obtained in 2016 was applied as reference data in this study (see Figure 1b). The plot size was 20 m × 30 m and the center of each plot was correctly determined using a GPS (accuracy ±5 m). The diameter at breast height (DBH) of the trees larger than 5 cm and the tree species of each plot were recorded.

The AGB of individual trees was calculated using the species-specific allometric growth equations with DBH. In this study, the allometric growth models developed by [80,81] for the major species of trees and understory in northeastern China were employed to calculate the AGB of individual trees. The allometric growth equation was showed as Equation (1) and the parameters of major species of trees and understory were listed in Table 1.

$$W = a \cdot D^b \tag{1}$$

where W represents aboveground biomass (kg), D represents DBH (cm), a and b are estimated parameters of different species in [80,81]. The AGB of the plot was the cumulative summation of the AGB of individual trees of each plot.

Table 1. Estimated parameters (a and b) of the allometric growth models of different species applied in this study.

Vegetation Types	Latin Names of Species	a	b
Deciduous trees	*Acer mono Maxim.*	0.318	2.081
	Ulmus pumila L.	0.350	1.995
	Populus davidiana Dode	0.078	2.512
	Betula platyphylla	0.313	2.114
	Quercus mongolica Fisch. ex Ledeb.	0.097	2.501
	Tilia mongolica Maxim	0.083	2.422
	Fraxinus mandshurica Rupr./Juglans mandshurica Maxim/Phellodendron amurense Rupr.	0.268	2.118
Coniferous trees	*Larix olgensis Henry*	0.168	2.248
	Pinus koraiensis Sieb.et Zucc.	0.082	2.426
	Picea asperata Mast.	0.067	2.517
	Larix olgensis Henry [1]	0.222	2.174
	Pinus koraiensis Sieb.et Zucc. [1]	0.206	2.117
	Pinus sylvestris var. *mongolica Litv.* [1]	0.080	2.440
Understory	*Acer ginnala*	0.527	2.217
	Syringa reticulata var. *amurensis*	0.395	2.300
	Padus asiatica	0.090	2.696
	Rhamnus yoshinoi	0.169	2.555
	Arbor-like mixed species [2]	0.182	2.487

[1] Represents plantations; otherwise are natural forests. [2] represent arbor-like mixed species of understory that do not have a specific Latin name.

2.3. Methods

To investigate the effects of different synergistic approaches of features and ensemble learning algorithms on AGB estimation of NSFs, a five-step methodology with three experiments of features (Feature experiments I-III) was implemented in this study, including (1) data preprocessing, (2) feature extraction and selection, (3) establishment and evaluation of classic machine learning models, (4) establishment and evaluation of ensemble learning models, (5) wall-to-wall AGB prediction using the most effective algorithm and features. Feature experiment I was designed to explore the effects of features from different data sources (ALS, optical imagery, and combined data) on AGB estimation based on a variety of machine learning algorithms; Feature experiment II was designed to investigate how to efficiently combine the best-performing ALS feature (a unique feature) with several spectral features for AGB estimation, is it better to use novel extracted features or directly combine all the features?; Feature experiment III aims to compare the performance of combining all features for AGB estimation. The feature experiment design and logic of this study were shown in Table 2 and Figure 2, respectively.

Table 2. Feature experiments designed in this study.

Experiment	Data Source	Number of Features [1]	Details
I	ALS	9	Feature 1: Optimal ALS features
	Landsat 8	9	Feature 2: Optimal Landsat 8 features
	ALS + Landsat 8	18	Feature 1 + 2: Optimal ALS and Landsat 8 features
II	ALS + Landsat 8	9	Feature 4: All *COLI1* [2]
		9	Feature 5: All *COLI2* [2]
		10	Feature 2 + 3 [3]: Optimal Landsat 8 features (9) + The best performing ALS feature (1)
III	ALS + Landsat 8	27	Feature 1 + 2 + 4: Optimal ALS features (9) + Optimal Landsat 8 features (9) + All *COLI1* (9)
		27	Feature 1 + 2 + 5: Optimal ALS features (9) + Optimal Landsat 8 features (9) + All *COLI2* (9)

[1] Number of features was determined by the procedure described in Section 2.3.2. [2] *COLI1* and *COLI2* were calculated using Equations (3) and (4) described in Section 2.3.2. [3] Feature 3 is the best performing ALS feature.

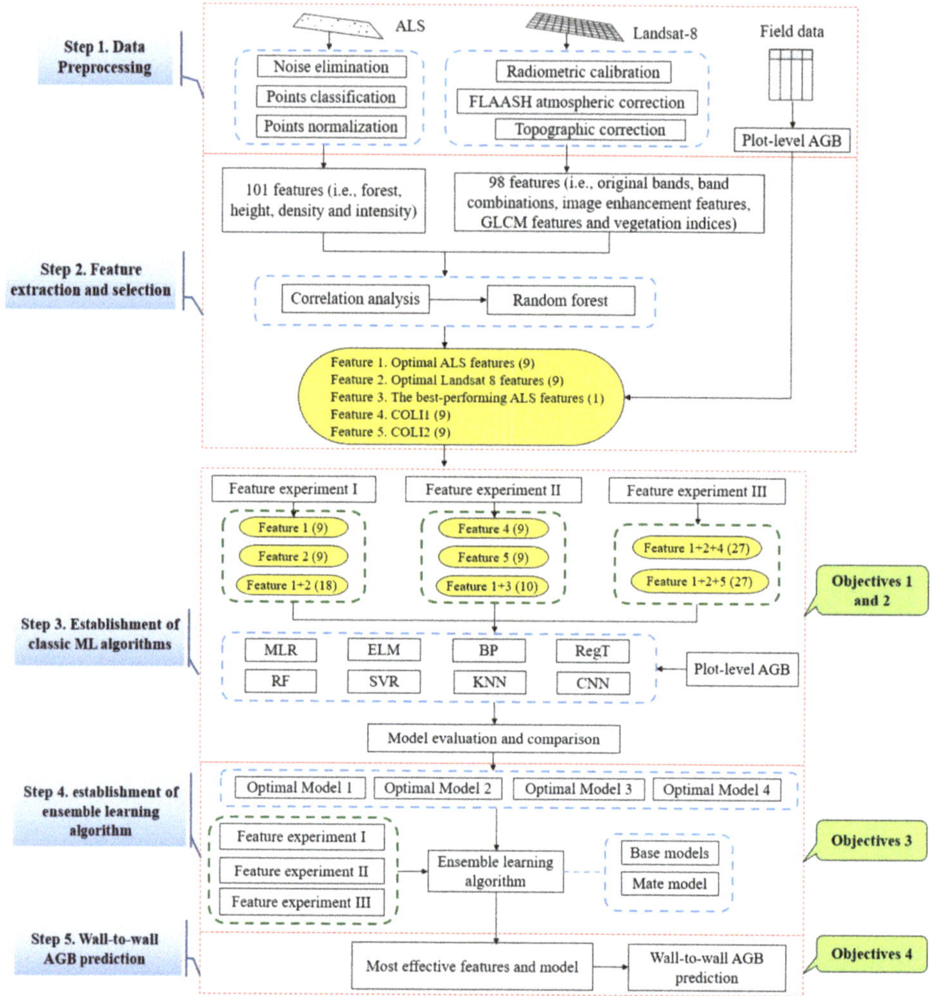

Figure 2. The flowchart of this study. Note: the number in parentheses represents feature number. Feature 1: optimal ALS features; Feature 2: optimal Landsat 8 features; Feature 3: the best performing ALS feature; Feature 4: all *COLI*1s; Feature 5: all *COLI*2s.

2.3.1. Preprocessing of Remotely Sensed Data

The preprocessing of the ALS data includes (1) noise elimination (such as air points, low points, and isolated points). The radius of a fitting plane and the multiples of standard deviation were set to 0.5 m and 1, respectively. The algorithm will automatically calculate the standard deviation of the surrounding fitting plane of a point. If the distance from this point to that plane is less than multiples of standard deviation, this point will be kept. (2) classification of ground and non-ground points. The ground points were classified by improved progressive triangulated irregular network densification (IPTD) filtering algorithm developed in [82]. The maximum building size and maximum terrain angle were set to 20 m and 88°, respectively. (3) normalization of point clouds. A digital terrain model (DTM) with a resolution of 0.5m was generated based on ground points using the inverse distance weighted (IDW) interpolation method. The power of the distance between sampling points and an unknown point was set to 2, and the smallest number of points used for interpolation was 12. Then, the point clouds were normalized by subtracting the DTM value from the elevation of all points. The preprocessing of the ALS data was implemented using LiDAR 360 V3.2 of GreenValley International.

Preprocessing of the Landsat 8 OLI imagery including radiometric calibration, atmospheric correction, and topographic correction was implemented using ENVI 5.3 software. The Fast Line-of-sight Atmospheric Analysis of Spectral Hypercube (FLAASH) radiative transfer model was implemented for atmospheric correction and conversion to surface reflectance in the EVNI environment. The topographic correction was conducted with the well-known Sun Canopy Sensor + C correction (SCS + C) approach using the extension tool of "Topographic Correction_V5.3_4_S1". The SCS + C correction approach reduces overcorrection and is an effective topographic correction method in forested and mountainous terrain [83,84]. The SCS + C topographic correction model can be expressed by Equation (2).

$$L_t = L \cdot \left(\frac{cos\theta \cdot cos\alpha + C}{cosi + C} \right) \quad (2)$$

where L_t is the corrected pixel radiance value of the image; L is the uncorrected pixel radiance value of the image; i is the incidence angles on a horizontal surface; θ is the solar zenith angle; α is the slope angle; C is the semi-empirical parameter. DTM generated from ALS data was applied for topographic correction in this study.

2.3.2. Feature Extraction and Selection

- Feature Extraction

Four categories of 101 features related to forest, height, density, and intensity features were derived from normalized ALS point cloud data. Forest features include canopy cover, leaf area index (LAI), and gap fraction. Canopy cover refers to the proportion of the forest floor covered by the vertical projection of the tree crowns [85]. LAI is one of the most significant variables for representing canopy structure, with the definition of half the total foliage area per unit ground surface area [86]. The gap fraction can be calculated by the ratio of the number of ground points whose elevation is lower than the height threshold (i.e., 2 m in this study) and the total return number. All 101 ALS features, including three forest metrics, 46 elevation metrics, 10 density metrics, and 42 intensity metrics were extracted using LiDAR 360 V3.2 of GreenValley International. The feature details were listed in Table A1 of Appendix A.

A variety of features could be derived from optical imagery. According to previous studies (e.g., [48,54,73,87]), band combinations, vegetation indices, textures (e.g., gray-level co-occurrence matrix (GLCM)) of each band, and image transformations (e.g., principal component analysis, tasseled cap, minimum noise fraction) were extracted as potential predictors for AGB modeling. Therefore, 98 features were selected or extracted from Landsat 8 imagery in this study, including seven original bands (band 1–7), ten band combinations, ten image enhancement features (i.e., three principal components, three tasseled-cap features, and four minimum noise fractions), 56 GLCM features, and 15 vegetation indices. The details of the 98 features derived from Landsat 8 were listed in Table A2 of Appendix A.

- Feature Selection

To avoid the "curse of dimensionality", it is a prerequisite to select the most effective feature for AGB estimation. In this study, the two-step feature selection procedure is implemented, including (1) preliminary selection using Pearson correlation coefficient; and (2) further selection based on variable importance measure using random forest. For the first step, Pearson correlation coefficients of each feature and AGB were calculated and the features with p-value less than 0.05 that significantly correlated with AGB were selected. Then, the selected features were ranked according to variable important measures calculated with random forest. Due to the randomness, the ranking procedure was implemented 10 times to find out the most stable set of features with high ranking.

The two-step feature selection was implemented for ALS and Landsat 8 data, respectively, to select two sets of best-performing features. Among the selected ALS features, the best-performing ALS variable was determined by establishing and evaluating the univariate models of each ALS feature and AGB. The feature selection procedure was implemented using R version 4.0.4 (https://www.r-project.org/ (accessed on 1 September 2021)).

According to [48], two types of indices (*COLI*1 and *COLI*2) incorporating optical imagery and ALS information were established using the best-performing LiDAR variable with each optical spectral vegetation index. The best-performing LiDAR variable was determined by the univariate model of AGB and the LiDAR variable with the highest R^2. The best-performing spectral features of Landsat 8 were selected by the two-step feature selection procedure described above. Then, the generation of *COLI*1 and *COLI*2 based on the best-performing LiDAR variable (only one feature) and the best-performing Landsat 8 features (could be several features) included both feature selection and extraction procedures. For convenience, we still used the notation of [48] but adjusted the equations as follows.

$$COLI1 = SF_i \times BLV \tag{3}$$

$$COLI2 = SF_i\_BLV = \frac{(BLV - SF_i)}{(BLV + SF_i)} \tag{4}$$

where *BLV* is the best-performing LiDAR variable (only one feature), SF_i is a set of best-performing features derived from Landsat 8 imagery (several features). Thus, the number of *COLI*1 or *COLI*2 is identical to the number of best-performing spectral features (SF_i).

2.3.3. Classic Machine Learning Algorithms

In this study, seven classic machine learning algorithms were conducted to estimate the AGB of NSFs, including extreme learning machine (ELM), backpropagation (BP) neural network, regression tree (RegT), RF, support vector regression (SVR), KNN, and CNN. Traditional multiple linear regression (MLR) was applied as a baseline for model comparison.

- ELM

ELM is a class of machine learning methods built on the feedforward neuron network (FNN) for supervised and unsupervised learning problems [88]. ELM is an improvement of FNN and its backpropagation algorithm, which is characterized by random or artificially given weights of the nodes in the hidden layer and does not need to be updated. Compared to single-layer perceptron and SVM, ELM is considered to have possible advantages in terms of learning rate and generalization ability [88].

- BP

BP neural network, proposed by Rumelhart et al. in 1986 [89], is a multilayer feedforward network trained by error backpropagation algorithm and is one of the most widely used neural network models [90]. Its learning rule is to use the fastest descent method to continuously adjust the weights and thresholds of the network by backpropagation to minimize the sum of squared errors of the network. According to error and trials, the BP algorithm was implemented with epochs of 1000 in this study.

- RegT and RF

A regression tree is a basic method built on the principle of minimizing the loss function for a regression problem. The major advantage of the regression tree is the readability of the model and fast computational speed, which make it particularly suitable for integrated learning, such as random forests. RF, proposed by Leo Breiman [76], is based on multiple regression trees, which is capable of capturing the complicated relationship between a response and a set of explanatory variables with the following advantages: robustness to reduce over-fitting, ability to determine variable importance, higher accuracy, fewer parameters that need to be tuned, lower sensitivity to the tuning of the parameters, fast training speed, and anti-noise property. The number of regression trees and the random state of the RF algorithm were set to 1000 and 10, respectively, in this study.

- SVR

SVM is a class of generalized linear algorithms that performs the classification of data in a supervised learning manner, where the decision boundary is the hyperplane of maximum margins solved for the learned samples. SVR is a transformation of SVM designed for regression problems and can perform nonlinear problems by kernel method. Linear kernel and penalty factor of 1 were applied for SVR in this study.

- KNN

The KNN method is a multivariate nonparametric algorithm that uses a set of predictors (Xs) to match each target pixel to a number (K) of most similar (nearest neighbors) reference pixels for which values of response variables (Y) are known. The number of nearest neighbors was set to 5 and uniform weights were utilized in this study.

- CNN

CNN, firstly developed in 1995 for the classification of handwritten images [91], is one of the most representative algorithms of deep learning. CNN interprets spatial data by scanning it using a series of trainable moving windows and has the capability of representation learning in a translation-invariant manner according to its hierarchical structure. In this study, the CNN model had a simple structure with an input layer, a hidden layer, and an output layer, and was implemented using an epoch of 1000 and a batch size of 30.

2.3.4. Ensemble Learning Algorithms

Stacked generalization (SG) which is a layered ensemble learning algorithm [92] was applied in this study. There are two layers designed in the SG algorithm here, including basic models and meta models. The input of the base model is the original training set and the output of the base model is applied as the training set for meta model [93]. The meta model could be a single model or an ensemble model [93,94], like RF. To obtain a better performance of SG, the base models should be accurate and different as much as possible. Thus, the four best-performing machine learning algorithms described in Section 2.3.3 were selected for the base models according to leave-one-out cross-validation and meta models for establishing SG algorithms in this study, which resulted in four SG algorithms. The flowchart of the SG algorithm in this study was presented in Figure 3.

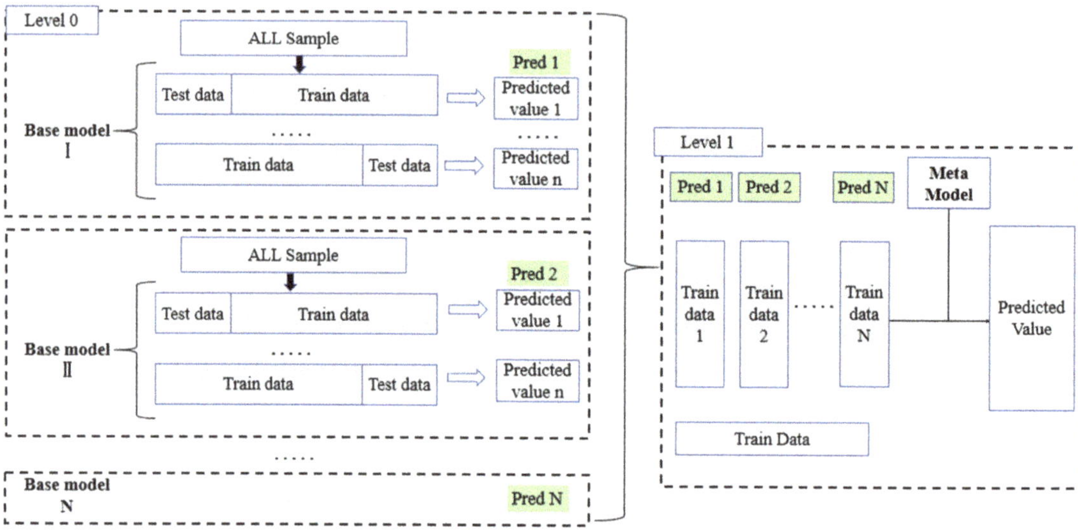

Figure 3. Flowchart of stacked generalization (SG) algorithm in this study. Note: The number of the base model (N) was set to four in this study and 195 iterations were running within each model because of the leave-one-out cross-validation of 195 sample plots.

2.3.5. Model Evaluation

This study adopted a leave-one-out cross-validation method to evaluate the model accuracy. Since 195 sample plots were used in this study, the training and testing data were 194 plots and 1 plot, respectively; and 195 iterations were run for each model. Due to the problems of coefficient of linear determination (R^2) for nonlinear models [95], we avoid applying R^2 of machine learning models established by selected features and AGB. However, R^2 of actual and predicted AGB could be used as an indicator since the relationship of actual and predicted AGB can be described by a simple linear model. Therefore, six indices were applied for model evaluation, including R^2 of actual and predicted AGB, root mean squared error (RMSE), relative root mean squared error (rRMSE), mean absolute error (MAE), mean absolute percentage error (MAPE), and precision measure (PM). The equations were shown as follows:

$$R^2 = \frac{\sum_{i=1}^{n}(\hat{y}_i - \overline{y})^2}{\sum_{i=1}^{n}(y_i - \overline{y})^2} \tag{5}$$

$$RMSE = \sqrt{\frac{1}{n}\sum_{i=1}^{n}(y_i - \hat{y})^2} \tag{6}$$

$$rRMSE = \frac{\sqrt{\frac{1}{n}\sum_{i=1}^{n}(y_i - \hat{y})^2}}{\overline{y}} \tag{7}$$

$$MAE = \frac{1}{n}\sum_{i=1}^{n}|y_i - \hat{y}_i| \tag{8}$$

$$MAPE = \frac{1}{n}\sum_{i=1}^{n}\frac{|y_i - \hat{y}_i|}{y_i} \times 100\% \tag{9}$$

$$PM = \frac{\sum_{i=1}^{n}(y_i - \hat{y}_i)^2}{\sum_{i=1}^{n}(y_i - \overline{y})^2} \tag{10}$$

where n represents the number of observation samples, y_i represents the actual AGB of the ith plot, $\hat{y}_i$ represents the predicted AGB of the ith plot, and $\overline{y}$ represents the mean of the actual AGB. All the model fitting and evaluation procedures in this study were implemented by python 3.7 (https://www.python.org/downloads/ (accessed on 1 September 2021)), TensorFlow 2.2 (https://tensorflow.google.cn/ (accessed on 1 September 2021)) and sklearn (https://scikit-learn.org/stable/ (accessed on 1 September 2021)).

3. Results

3.1. Feature Selection

Due to a large number of extracted features (199 features in total), two-step feature selection was implemented in this study, including preliminary selection using Pearson correlation coefficient; and further selection based on variable importance measure using random forest. Finally, nine ALS features were selected and sorted from highest to lowest variable importance as follows: elev_mean, int_AII_5th, elev_cv, density_7th, int_max, int_AII_40th, int_per_60th, int_per_80th, and int_AII_50th; nine features extracted from Landsat 8 were selected and ranked in descending order of variable importance: MVI5, B1, B76, B65, B53, Entr_B5, B2, ND563, and MVI7. The selected features and their descriptions were listed in Table 3.

Table 3. Feature Selection of ALS and Landsat 8 imagery.

ALS	Feature Descriptions	Landsat 8	Feature Descriptions
elev_mean	Mean value of height	MVI5	$(B_5 + B_4 - B_2)/(B_5 + B_4 + B_2)$
int_AII_5th	The cumulative intensity of 5% points in each pixel	B1	Band 1
elev_cv	Coefficient of variation of height	B76	B_7/B_6
density_7th	The proportion of returns in 7th height interval	B65	B_6/B_5
int_max	Max of intensity	B53	B_5/B_3
int_AII_40th	The cumulative intensity of 40% points in each pixel	Entr_B5	Entropy of band 5
int_per_60th	60% intensity percentile	B2	Band 2
int_per_80th	80% intensity percentile	ND563	$(B_5 + B_6 - B_3)\cdot(B_5 + B_6 + B_3)$
int_AII_50th	The cumulative intensity of 50% points in each pixel	MVI7	$(B_5 - B_7)/(B_5 + B_7)$

To grope for the best-performing ALS feature, simple linear regressions were established to model the relationship between AGB and each ALS feature. The result of univariate models showed that the elevation mean outperformed other ALS features due to higher R^2 and lower *RMSE*, *rRMSE*, *MAE*, *MAPE*, and *PM* (Table 4). Thus, elevation mean was selected as the best-performing ALS feature to generate *COLI*1 and *COLI*2 using Equations (3) and (4).

Table 4. Accuracy assessment of the univariate models with AGB and each ALS feature.

ALS Features	R^2	*RMSE*	*rRMSE*	*MAE*	*MAPE*	*PM*
elev_mean	0.34	60.03	0.40	43.74	0.39	0.67
int_AII_5th	0.13	68.75	0.46	51.70	0.65	0.87
elev_cv	0.08	70.64	0.48	53.99	0.66	0.92
density_7th	0.05	71.89	0.48	53.27	0.87	0.95
int_max	0.19	66.41	0.45	49.03	0.64	0.81
int_AII_40th	0.20	65.97	0.44	49.85	0.63	0.80
int_per_60th	0.17	66.89	0.45	50.68	0.64	0.83
int_per_80th	0.17	66.94	0.45	50.51	0.66	0.83

3.2. *Performance of Classic Machine Learning Algorithms*

3.2.1. Experiment I

The goal of feature experiment I was to explore the effects of features from different data sources (optical imagery, ALS, and combined data) on AGB estimation based on seven classic machine learning algorithms, including ELM, BP, RegT, RF, SVR, KNN, and CNN. MLR was implemented as a baseline for model comparison. Table 5 shows the performance of the eight models using the three sets of features designed in Experiment I.

Table 5. Accuracy assessment of classic machine learning algorithms with three sets of features designed in experiment I.

Features	Algorithm [1]	R^2	*RMSE*	*rRMSE*	*MAE*	*MAPE*	*PM*
Optimal ALS features (Feature 1)	MLR	0.31	52.76	0.37	41.09	38.37	0.67
	ELM	0.31	56.79	0.40	42.61	35.49	0.69
	BP	0.28	61.01	0.42	44.37	36.01	0.71
	RegT	0.21	71.95	0.47	58.55	42.66	1.11
	RF	0.29	61.84	0.41	45.80	37.08	0.72
	SVR	0.40	57.84	0.38	39.32	32.35	0.66
	KNN	0.31	60.95	0.4	45.21	35.36	0.81
	CNN	0.49	51.54	0.34	37.31	30.82	0.41
Optimal Landsat 8 features (Feature 2)	MLR	0.17	66.36	0.47	58.08	44.31	1.05
	ELM	0.12	71.64	0.48	59.73	41.40	1.21
	BP	0.13	68.58	0.49	57.19	42.76	1.04
	RegT	0.14	66.24	0.48	58.59	42.53	0.89
	RF	0.15	67.33	0.44	50.69	43.39	0.92
	SVR	0.07	70.31	0.46	51.65	47.28	1.14
	KNN	0.11	68.95	0.45	52.91	43.31	0.84
	CNN	0.27	62.54	0.41	47.16	43.08	0.72
Optimal ALS and Landsat 8 features (Feature 1 + 2)	MLR	0.25	63.48	0.40	47.21	42.34	0.94
	ELM	0.30	57.49	0.38	42.91	36.42	0.78
	BP	0.29	55.65	0.39	43.4	37.87	0.72
	RegT	0.24	60.86	0.45	55.07	39.18	0.87
	RF	0.28	61.91	0.41	45.36	39.28	0.91
	SVR	0.39	57.8	0.38	39.19	31.3	0.77
	KNN	0.22	65.37	0.43	48.6	34.69	1.07
	CNN	0.97	12.6	0.08	6.43	4.02	0.13

[1] MLR- multiple linear regression; ELM—extreme learning machine; BP—back propagation; RegT—regression tree; RF—random forest; SVR—support vector regression; KNN—k-nearest neighbor regression; CNN—convolutional neural networks

In general, the optimal ALS features (Feature 1) performed significantly better than the optimal Landsat 8 features (Feature 2) for AGB estimation, no matter of algorithms; the combination of the optimal ALS and Landsat 8 features (Feature 1 + 2) performed differently for various algorithms. For each data source, the accuracy of CNN was greatly higher than that of other algorithms, especially for applying both ALS and Landsat 8 features (R^2 = 0.97, *RMSE* = 12.6, *rRMSE* = 0.08, *MAE* = 6.43, *MAPE* = 4.02, *PM* = 0.13). However, it is worth mentioning that the accuracies of other algorithms (except CNN) based on two data sources (Feature 1 + 2) were not significantly improved compared with those based on optimal ALS features (Feature 1), which suggested that the accuracy of AGB estimation not only depends on data sources but also different algorithms. Some algorithms (like RF and SVR) could provide very similar accuracy using both optimal ALS and Landsat 8 features to that using only optimal ALS features, making it meaningless to involve optical imagery. Thus, ALS data are of significance to AGB estimation.

3.2.2. Experiment II

After determining the best-performing ALS feature (i.e., elevation mean), we designed feature experiment II to investigate how to efficiently combine the unique feature with the optimal Landsat 8 features for AGB estimation. Is it better to utilize a novel feature extracted from elevation mean and optimal Landsat 8 features (i.e., *COLI*1 and *COLI*2) or directly combine all the features? A similar feature size in experiment II (i.e., 9 or 10) could avoid the unfair comparison due to the big difference in feature number. Table 6 presented the accuracy assessment of classic machine learning algorithms with three sets of features designed in experiment II. The results showed that the addition of elevation mean significantly improves the accuracies of AGB estimation compared to those using optical features only (Feature 2), no matter how to add it. The models except CNN had very similar performances in AGB estimation for the three feature combinations in experiment II. CNN still showed great advantages like Experiment I, especially for the case of simply combining the optimal Landsat 8 features and elevation mean together (Feature 2 + 3) with the accuracy of R^2 = 0.88, *RMSE* = 24.48, *rRMSE* = 0.16, *MAE* = 10.19, *MAPE* = 7.23, and *PM* = 0.24, followed by the case of all *COLI*2 (Feature 5), and then the case of all *COLI*1 (Feature 4). Thus, it seemed unnecessary to generate the new features (i.e., *COLI*1 or *COLI*2) when CNN was applied for AGB estimation based on the optimal Landsat 8 features and the best-performing ALS feature for NSFs.

Table 6. Accuracy assessment of classic machine learning algorithms with three sets of features designed in experiment II.

Features	Algorithm	R^2	*RMSE*	*rRMSE*	*MAE*	*MAPE*	*PM*
All *COLI*1 (Feature 4)	MLR	0.34	59.50	0.39	45.08	34.07	0.61
	ELM	0.31	59.25	0.41	44.27	37.7	0.66
	BP	0.30	57.34	0.38	45.68	39.39	0.68
	RegT	0.28	62.62	0.43	50.22	45.45	0.72
	RF	0.32	60.14	0.40	43.27	35.55	0.62
	SVR	0.24	69.91	0.46	51.13	43.78	0.85
	KNN	0.26	62.58	0.41	46.3	38.39	0.69
	CNN	0.5	51.06	0.34	38.27	30.48	0.54
All *COLI*2 (Feature 5)	MLR	0.22	61.49	0.48	50.12	39.34	0.72
	ELM	0.25	64.35	0.47	51.07	40.81	0.75
	BP	0.30	62.14	0.47	50.39	38.24	0.78
	RegT	0.24	67.07	0.49	52.41	43.93	0.79
	RF	0.24	63.98	0.42	46.28	39.73	0.74
	SVR	0.26	67.69	0.45	49.05	38.71	0.78
	KNN	0.25	63.51	0.42	47.3	40.05	0.71
	CNN	0.66	42.42	0.28	29.71	22.16	0.45
Optimal Landsat 8 features + The best-performing ASL feature (Feature 2 + 3)	MLR	0.33	60.14	0.40	44.45	40.76	0.70
	ELM	0.29	64.26	0.43	48.39	42.59	0.69
	BP	0.30	63.8	0.41	50.11	44.01	0.70
	RegT	0.25	64.14	0.45	52.34	45.53	0.74
	RF	0.28	62.29	0.41	45.62	41.69	0.71
	SVR	0.29	62.25	0.41	42.00	40.21	0.82
	KNN	0.24	63.38	0.42	46.95	39.24	0.69
	CNN	0.88	24.48	0.16	10.19	7.23	0.24

3.2.3. Experiment III

To investigate the effect of combing optimal ALS and Landsat 8 features and two types of novel features (*COLI*1 or *COLI*2) using classic machine learning algorithms, experiment III was implemented (Table 7). Comparing to the result of applying optimal ALS and Landsat 8 features (Feature 1 + 2) in Table 5, the additions of the novel features, no matter *COLI*1 or *COLI*2, slightly improved the accuracies of most models, like MLR, BP, RegT, RF, and KNN. In addition, the accuracies of all models except RF using optimal ALS and Landsat 8 features and all *COLI*2 (Feature 1 + 2 + 5) were slightly improved compared to those using optimal ALS and Landsat 8 features and all *COLI*1 (Feature 1 + 2 + 4), indicating *COLI*2 were more efficient than *COLI*1 for AGB estimation of NSFs. CNN was still much superior to other algorithms and reached the highest accuracies (R^2 = 0.99, *RMSE* = 6.85, *rRMSE* = 0.04, *MAE* = 2.95, *MAPE* = 1.02, *PM* = 0.03) when optimal ALS and Landsat 8 features and all *COLI*2 (Feature 1 + 2 + 5) was applied.

Table 7. Accuracy assessment of classic machine learning algorithms with two sets of features designed in experiment III.

Features	Algorithm	R^2	*RMSE*	*rRMSE*	*MAE*	*MAPE*	*PM*
Optimal ALS + Landsat 8 features + All *COLI*1 (Feature 1 + 2 + 4)	MLR	0.32	60.50	0.40	45.08	36.07	0.68
	ELM	0.28	63.26	0.42	44.15	37.84	0.81
	BP	0.31	58.71	0.37	40.30	36.98	0.65
	RegT	0.28	62.07	0.42	42.29	38.51	0.79
	RF	0.31	60.32	0.41	43.41	39.26	0.73
	SVR	0.39	57.74	0.39	38.05	35.31	0.66
	KNN	0.29	61.11	0.44	42.87	36.47	0.69
	CNN	0.92	12.02	0.09	11.37	8.3	0.11
Optimal ALS + Landsat 8 features + All *COLI*2 (Feature 1 + 2 + 5)	MLR	0.33	59.38	0.42	44.27	39.50	0.70
	ELM	0.29	61.67	0.43	47.09	40.34	0.81
	BP	0.32	57.74	0.42	48.29	41.60	0.72
	RegT	0.33	65.59	0.42	49.26	42.17	0.83
	RF	0.31	60.61	0.40	44.69	31.08	0.69
	SVR	0.42	56.82	0.37	38.76	29.39	0.68
	KNN	0.32	59.83	0.39	44.34	37.3	0.64
	CNN	0.99	6.85	0.04	2.95	1.02	0.03

3.3. *Performance of Ensemble Learning Algorithms*

3.3.1. Experiment I

To explore the performances of ensemble learning algorithms in estimating AGB based on different feature combinations, Experiment I, II, and III were also implemented using the designed SG algorithms. According to the results of classic machine learning algorithms (Tables 5–7), four best-performing models, that is, RF, SVR, KNN, and CNN, were selected as base models for the SG algorithm. The predictions of base models were applied as the input of the meta model of the SG algorithms, which were also RF, SVR, KNN, and CNN. Thus, there were four SG algorithms due to four meta models, including SG(RF), SG(SVR), SG(KNN), and SG(CNN). Table 8 presented the accuracy assessment of ensemble learning algorithms with three sets of features designed in experiment I. Comparing to the results of base models (Table 5), the SG algorithms greatly improved the accuracy of AGB estimation using the optimal Landsat 8 features (Feature 2) and the combined optimal features (Feature 1 + 2). However, for the case of optimal ALS features (Feature 1), the SG algorithms had slightly lower accuracies than those of base models, except CNN. In general, CNN still performed best as a meta model of SG algorithm, followed by SG algorithm with SVR meta model, and finally with RF meta model as well as KNN model. Although CNN was still an outstanding meta model for all the cases, it was worth noting that the drastic improvements of accuracies brought by SG(SVR), SG(RF), and SG(KNN) compared with their corresponding base model, especially for the Feature 2 and Feature 1 + 2. For example, R^2 of SG(SVR), SG(RF), and SG(KNN) increased approximately 30%–40%

and 60%–70% for Feature 2 and Feature 1 + 2, respectively; alternatively, R^2 of SG(CNN) only increased 49% and 0% for Feature 2 and Feature 1 + 2, respectively. Other indices (*RMSE*, *rRMSE*, *MAE*, *MAPE*, and *PM*) had similar trends, but in the opposite direction. Thus, it had more room for improvement to apply the SG algorithms for relatively weaker learners (like SVR, RF, and KNN) than strong deep learning learners (like CNN).

Table 8. Accuracy assessment of ensemble learning algorithms with three sets of features designed in experiment I.

Features	Algorithm	R^2	*RMSE*	*rRMSE*	*MAE*	*MAPE*	*PM*
Optimal ALS features (Feature 1)	SG(RF)	0.20	65.38	0.43	50.66	42.35	1.03
	SG(SVR)	0.24	63.98	0.42	45.75	41.03	0.92
	SG(KNN)	0.19	66.07	0.44	50.70	42.22	1.24
	SG(CNN)	0.61	45.42	0.30	31.59	24.28	0.37
Optimal Landsat 8 features (Feature 2)	SG(RF)	0.44	54.24	0.36	40.20	32.47	0.57
	SG(SVR)	0.45	54.36	0.36	38.85	34.59	0.65
	SG(KNN)	0.44	54.34	0.36	40.37	32.08	0.53
	SG(CNN)	0.76	35.28	0.23	24.29	18.17	0.26
Optimal ALS and Landsat 8 features (Feature 1 + 2)	SG(RF)	0.93	18.04	0.12	8.78	6.30	0.17
	SG(SVR)	0.97	12.13	0.08	5.70	4.70	0.14
	SG(KNN)	0.9	24.27	0.16	16.76	15.09	0.15
	SG(CNN)	0.97	10.95	0.07	6.58	5.06	0.03

3.3.2. Experiment II

Feature experiment II was also implemented to investigate how to integrate elevation mean and the optimal Landsat 8 features for AGB estimation based on ensemble learning algorithms (Table 9). It showed that the SG algorithms greatly improved the accuracies for all the cases except the SG(CNN) for Feature 5 and Feature 2 + 3, comparing to the accuracies using the corresponding base model (Table 6). When SG algorithms were utilized, the trend that the simple combination of optimal Landsat 8 features and elevation mean (Feature 2 + 3) performed best, followed by all *COLI2* (Feature 5), and finally all *COLI1* (Feature 4) was much more obvious than that using classic machine learning algorithms (Table 6 vs. Table 9). The advantage of applying deep learning algorithm CNN as meta model decreased with the dramatic increase in the accuracies of the other three algorithms (i.e., RF, SVR, and KNN), especially for Feature 5 and Feature 2 + 3. In other words, when the feature set of all *COLI2* or the feature set of optimal Landsat 8 features and elevation mean was applied for AGB estimation, SG(RF), SG(SVR), and SG(KNN) had comparable accuracies to SG(CNN).

Table 9. Accuracy assessment of ensemble learning algorithms with three sets of features designed in experiment II.

Features	Algorithm	R^2	*RMSE*	*rRMSE*	*MAE*	*MAPE*	*PM*
All *COLI1* (Feature 4)	SG(RF)	0.38	57.84	0.38	41.72	33.69	0.68
	SG(SVR)	0.48	52.83	0.35	38.69	32.08	0.62
	SG(KNN)	0.36	58.5	0.39	43.04	34.36	0.71
	SG(CNN)	0.63	43.78	0.29	31.86	25.13	0.49
All *COLI2* (Feature 5)	SG(RF)	0.64	43.13	0.28	30.66	23.11	0.48
	SG(SVR)	0.64	43.28	0.28	31.09	28.28	0.47
	SG(KNN)	0.60	45.85	0.30	32.74	27.00	0.51
	SG(CNN)	0.50	51.31	0.34	36.80	27.66	0.50
Optimal Landsat 8 features + The best-performing ALS feature (Feature 2 + 3)	SG(RF)	0.86	26.94	0.18	14.22	10.66	0.24
	SG(SVR)	0.88	24.61	0.16	10.13	10.25	0.29
	SG(KNN)	0.79	34.06	0.23	22.46	17.88	0.31
	SG(CNN)	0.86	26.45	0.17	14.76	10.35	0.2

3.3.3. Experiment III

The effect of combing optimal ALS and Landsat 8 features and two types of novel features (*COLI*1 or *COLI*2) on AGB estimation using ensemble algorithms was investigated with experiment III (Table 10). Unlike classic machine learning algorithms, the addition of *COLI*1 in ensemble algorithms did not improve the accuracies of AGB estimation, compared to the result of applying optimal ALS and Landsat 8 features (Feature 1 + 2) in Table 8. The SG(SVR) or SG(KNN) with the addition of *COLI*1 even lower R^2 by about 10%–20% than SG(SVR) or SG(KNN) with only Feature 1 + 2 (Table 8). However, the addition of *COLI*2 in ensemble algorithms slightly increased the accuracies of most models except SG(KNN), even though SG algorithms with Feature 1 + 2 had already performed well (Table 8). In general, the SG algorithms with optimal ALS and Landsat 8 features and all *COLI*2 (Feature 1 + 2 + 5) had more stable accuracies than that with optimal ALS and Landsat 8 features and all *COLI*1 (Feature 1 + 2 + 4), no matter which meta model was used, indicating *COLI*2 were more efficient than *COLI*1 for AGB estimation of NSFs. It is still the SG model with CNN meta model that has the highest accuracy (R^2 = 0.99, *RMSE* = 2.02, *rRMSE* = 0.01, *MAE* = 0.87, *MAPE* = 0.73, *PM* = 0.02) when optimal ALS and Landsat 8 features and all *COLI*2 (Feature 1 + 2 + 5) was applied.

Table 10. Accuracy assessment of ensemble learning algorithms with two sets of features designed in experiment III.

Features	Algorithm	R^2	*RMSE*	*rRMSE*	*MAE*	*MAPE*	*PM*
Optimal ALS + Landsat 8 features + All *COLI*1 (Feature 1 + 2 + 4)	SG(RF)	0.95	15.35	0.11	12.44	9.34	0.14
	SG(SVR)	0.71	58.84	0.38	24.02	15.76	0.49
	SG(KNN)	0.86	57.00	0.38	21.38	15.49	0.38
	SG(CNN)	0.97	12.35	0.08	2.02	1.07	0.03
Optimal ALS + Landsat 8 features + All *COLI*2 (Feature 1 + 2 + 5)	SG(RF)	0.98	10.13	0.06	2.48	1.98	0.10
	SG(SVR)	0.95	4.10	0.18	3.20	2.34	0.08
	SG(KNN)	0.96	15.76	0.10	9.04	8.28	0.17
	SG(CNN)	0.99	2.02	0.01	0.87	0.73	0.02

In addition, the ensemble algorithms greatly improved the accuracies of the corresponding features and base model (Table 10 vs. Table 7). For example, if the combination of optimal ALS and Landsat 8 features and all *COLI*1 (Feature 1 + 2 + 4) was utilized, the R^2 of SG(RF) increased more than 60% compared with that of the RF model; *RMSE*, *rRMSE*, *MAE*, *MAPE* and *PM* of SG(RF) decreased by 75%, 73%, 71%, 76%, and 81%, respectively, compared with those of the RF model. Although the CNN base model had already achieved high accuracy, especially when applying the combination of optimal ALS and Landsat 8 features and all *COLI*2 (Feature 1 + 2 + 5 in Table 7), the SG(CNN) still decreased the group of *RMSE*, *rRMSE*, and *MAE* and the group of *MAPE* and *PM* by about 70% and 30%, respectively.

3.4. Wall-to-Wall AGB Predictions

Based on the above results and algorithm efficiency, CNN and the feature set of optimal ALS and Landsat 8 and all *COLI*2 (Feature 1 + 2 + 5) were selected for a wall-to-wall AGB prediction of the entire Maorshan Experimental Forest Farm of NEFU (Figure 4). The predicted AGB varied from 0 to 491.04 Mg/ha, with a mean value of 59.9 Mg/ha and a standard deviation of 48.69 Mg/ha. The area with AGB of 0 or low values was located along rivers, roads, or residential regions, whereas the area with high AGB values was located in the center part (e.g., Zhonglin, Yuejin, Beiling, Donglin, and Xinken working districts) of Maorshan (Figure 4a). However, the embedded pattern of high and low AGB values was obvious for most of the study area, as the enlarged area in Zhonglin working district (Figure 4b).

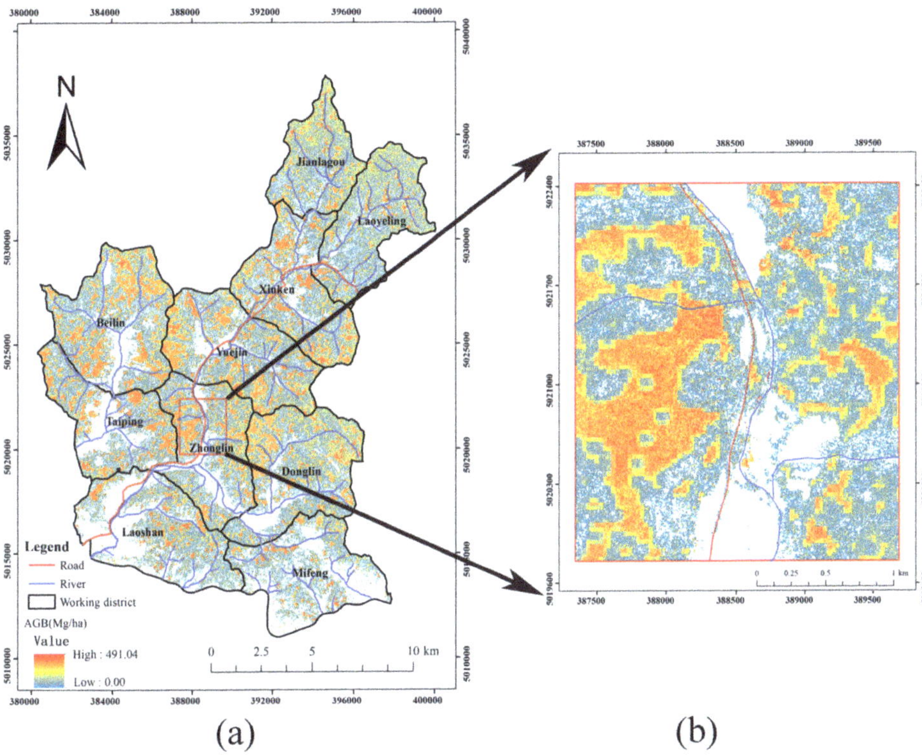

Figure 4. (**a**) The wall-to-wall AGB prediction of the entire study area estimated by the CNN model with optimal ALS features, optimal Landsat 8 features, and all *COLI*2 (Feature 1 + 2 + 5); (**b**) Spatial distribution of AGB for a partial area in Zhonglin working district.

Figure 5 showed the relationship of actual and estimated AGB (Mg/ha) of 195 plots using the CNN algorithm based on different feature sets. For experiment I, it was better to apply ALS than Landsat 8 to predict AGB if only one data source had to be used, which indicated the vertical forest structure was more vital than spectral information for AGB estimation of NSFs. The synergism of optical imagery and ALS markedly increased the accuracy of a single data source (Figure 5c vs. Figure 5a or Figure 5b) since it could effectively alleviate the underestimation of high AGB values. Even only one ALS feature (i.e., elevation mean) was added to the Landsat 8 features (Experiment II), the improvement was obvious and significant. However, it was unnecessary to generate novel features like *COLI*1 or *COLI*2 using the optimal Landsat 8 and elevation mean. It was in evidence that the performance of directly combining them was much better than that of new features (Figure 5f vs. Figure 5d) or Figure 5e), but worse than that of all optimal ALS and Landsat 8 features (Figure 5f vs. Figure 5c) due to the smaller number of features (i.e., 10 vs. 18). The effectiveness of *COLI*1 was very limited because Feature 1 + 2 provided a comparable result to Feature 1 + 2 + 4 (Figure 5c vs. Figure 5g). It is the most efficient to combine all optimal ALS, Landsat 8, and *COLI*2 features, especially for estimating high AGB values (Figure 5h).

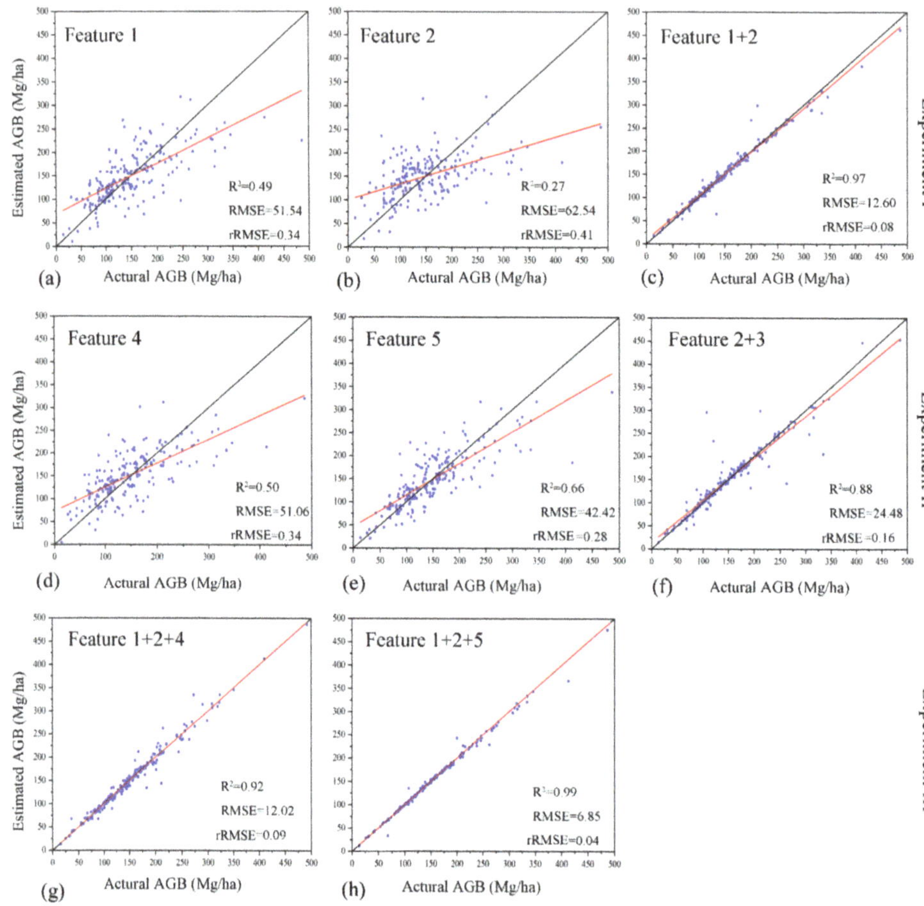

Figure 5. The relationship of actual and estimated AGB (Mg/ha) of 195 plots using CNN algorithm based on (**a**) Feature 1: optimal ALS features; (**b**) Feature 2: Optimal Landsat 8 features; (**c**) Feature 1 + 2: Optimal ALS and Landsat 8 features; (**d**) Feature 4: All *COLI*1; (**e**) Feature 5: All *COLI*2; (**f**) Feature 2 + 3: Optimal Landsat 8 features and the best performing ALS feature; (**g**) Feature 1 + 2 + 4: Optimal ALS features, optimal Landsat 8 features, and all *COLI*1; (**h**) Feature 1 + 2 + 5: Optimal ALS features, optimal Landsat 8 features, and all *COLI*2. Note: The red and black lines represent the fitted regression lines and the line of 45°, respectively.

4. Discussion

4.1. AGB Estimation Using Different Features

The differences in features are responses to the characteristics of different data sources. In this study, we extracted a variety of features and investigated the effects of different synergistic approaches of features derived from ALS and Landsat 8 OLI imagery on AGB estimation of NSFs of northeastern China. For ALS data, besides elevation features, density- (e.g., density_metrics7) and intensity-related (e.g., int_AII_5th, int_max, int_AII_40th, int_per_60th, int_per_80th, and int_AII_50th) metrics also had great potentials in AGB estimation; for Landsat 8 imagery, band combinations and texture are more efficient than vegetation indices, especially MVI5 (i.e., the band combination of band 5, 4 and 2). Unfortunately, some traditional vegetation indices that commonly applied in previous studies [48], for example, the normalized difference vegetation index (NDVI), enhanced vegetation index (EVI), atmospherically resistant vegetation index (ARVI), soil adjusted vegetation index (SAVI), etc., were excluded due to the low correlations with AGB. Only one vegetation index (i.e., ND563) was selected. It might be because that study area is a natural secondary forest with high canopy density which could easily result in the saturation (insensitivity to AGB) of the traditional vegetation indices, which was also confirmed in [96,97]. The low accuracies (e.g., $R^2 < 0.3$) of AGB estimations using the optimal Landsat 8 features (Feature 2), no matter of algorithms, indicating the difficulties of AGB estimation of NSFs as well. Due to the vegetation characteristics, near-infrared and shortwave infrared bands (i.e., band 5, 6, and 7) were more related to AGB estimation than other bands.

Similar to previous studies [48,52,98], it was beneficial to combine ALS data and optical imagery, even only combining one significant feature derived from ALS (like elevation mean). The synergistic method of extracting novel features (i.e., *COLI*1 and *COLI*2) using optimal Landsat 8 features and the best-performing ALS feature (i.e., elevation mean) yielded higher accuracy of AGB estimation than either optical-only or ALS-only features when the same model was implemented. From experiment II and III, it showed that *COLI*2 had more advantages than *COLI*1 in AGB estimations of NSFs, which is different from [48] due to different forest types (NSFs of northeastern China vs. mixed forests of southern China). However, it is surprised to find out that the novel extracted features (*COLI*1 and *COLI*2) were not efficient in improving the accuracy compared to the simple combination of the untransformed features (optimal Landsat 8 features + BLV), which indicated the great convenience and effectiveness brought by just adding the best-performing ALS feature (i.e., elevation mean) to the original set of Landsat 8 features for AGB estimation of NSFs. The number of features was also a vital factor to influence the AGB accuracy. To make sure a fair comparison of synergistic approaches of features, we keep the number of features consistent as much as possible within each experiment. It is a trend that the accuracy of AGB estimation raises with the increase in the number of involved features under the same conditions (e.g., algorithms). Thus, it was not surprising that the combination with 27 features (i.e., Feature 1 + 2 + 4 or Feature 1 + 2 + 5) in experiment III provided the best performances in this study, from a feature size perspective.

4.2. AGB Estimation Using Machine Learning Algorithms

The effect of classic machine learning and ensemble learning algorithms on AGB estimation using different features was explored in this study. The RF algorithm that is one of the most commonly used algorithms in forestry only provided very modest accuracy in this study since it constantly overfits the data, often with poorer predictions [33]. CNN, a deep learning algorithm firstly developed in 1995 for the classification of handwritten images [91], showed absolute advantages compared with other classic algorithms (e.g., ELM, BP, RF, KNN, SVR, etc.). As a representative of deep learning algorithms that is a branch of machine learning, a large and deep CNN (consisting of many-layered convolutions) was further developed in 2012 and achieved a winning top-5 test error rate of 15.3% in the ImageNet ILSVRC-2012 competition [99]. In recent years, the CNN model has been increasingly applied in forestry, for example, for the prediction of forest inventory parameters and identification of different tree species [100,101]. CNN interprets spatial data by scanning it using a series of trainable moving windows and sufficiently complex artificial neural networks and does not require human-derived feature selection in essence [100]. However, to make sure a fair comparison of different models, we keep the feature selection procedure consistent for all models. It means that the CNN model was applied for two-dimensional data of AGB and a set of human-derived features instead of a three-dimensional image. Although the CNN model lost the advantage of automatically extracting and selecting features, it is still sensitive to changes in features and significantly superior to other models (e.g., ELM, BP, RF, KNN, SVR, etc.).

The SG algorithms, a kind of ensemble learning algorithms, applied heterogeneous ensemble methods with different base models and greatly improved the AGB estimation accuracy in this study. RF, KNN, SVR, and CNN were selected as base models since SG algorithms could take advantage of the good and stable predictions from base models.

The good prediction of the CNN base model successfully made the accuracy of the SG algorithms improved and stable no matter of meta-models, which indicated that SG has a stronger generalization ability than base models. In other words, it is more beneficial for weaker learners (e.g., RF, KNN, and SVR) to become stronger learners using SG algorithms than strong learners (e.g., CNN).

However, although the SG algorithm is superior to its corresponding base model, we still recommend employing the CNN model for AGB estimation in practice due to its comparable accuracy and good efficiency. Table 11 summarized the efficiency (i.e., runtime) of all the algorithms with the combination of the optimal ALS and Landsat 8 features, and all COLI2 (Feature 1 + 2 + 5) for AGB estimation of 195 plots on a computer with AMD RX3700x + 16GB + GTX960 4GB. It showed that the runtime of ensemble algorithms (i.e., SG(RF), SG(KNN), SG(SVR), SG(CNN)) was dramatically augmented compared with their corresponding base model (i.e., RF, KNN, SVR, CNN). For example, the efficiency of SG(CNN) is only half of that of the CNN model. Other SG algorithms (i.e., SG(RF), SG(KNN), SG(SVR)) raised the runtime of the corresponding algorithm (i.e., RF, KNN, SVR) even more. The CNN model had the longest runtime but yield the highest accuracy (see Tables 5–7) among classic machine learning algorithms due to the most complex structure. Thus, to balance the workload and accuracy, the wall-to-wall AGB prediction map was generated using the CNN model with the combination of the optimal ALS and Landsat 8 features, and all COLI2 (Feature 1 + 2 + 5) in this study.

Table 11. The runtime of all algorithms with the combination of the optimal ALS and Landsat 8 features, and all COLI2 (Feature 1 + 2 + 5).

Classic Algorithms	Runtime (s)	SG Algorithms	Runtime (s)
MLR	1.2	SG(RF)	8168
ELM	45	SG(SVR)	7798
BP	38	SG(KNN)	7794
RegT	24	SG(CNN)	15170
RF	382		
SVR	12		
KNN	8		
CNN	7384		

4.3. Comparison of Estimated Forest AGB and Current Publications

From the AGB accuracy perspective, the highest accuracy (R^2 = 0.99, *RMSE* = 2.02, *rRMSE* = 0.01, *MAE* = 0.87, *MAPE* = 0.73, *PM* = 0.02) was yielded by SG(CNN) algorithm with the combination of the optimal ALS and Landsat 8 features and all *COLI2* (Feature 1 + 2 + 5) in this study, which was better than other similar AGB studies that applied both LiDAR and optical imagery (e.g., [48,61,69,98,102]). Besides features and algorithms, the high accuracy of this study also benefited from the case of a local study with a relatively small area. It tends to decrease the accuracy for national and global scales. For example, Su et al. [69] provided the R^2 of 0.75 and the RMSE of 42.39 Mg/ha for the AGB estimation of China based on ICESat GLAS laser altimetry data, MODIS, and forest inventory data. Yang et al. [103] produced a global forest AGB map with the R^2 of 0.90 and the *RMSE* of 35.87 Mg/ha using gradient augmented regression trees algorithm based on multiple data sources (e.g., LiDAR-derived forest AGB datasets, field measurements, high-level products from optical satellite imagery, etc.).

Further, we dig into the predicted AGB values of the wall-to-wall map of the entire Maorshan and compared the distributions of AGB values of the wall-to-wall prediction map and 195 sample plots (Figure 6). Although the spatial distribution of AGB values of the wall-to-wall prediction map seemed to be reasonable (Figure 4), it showed that there was still a big difference between the two distributions, especially for the ranges of 0–50 Mg/ha and >200 Mg/ha (Figure 6), indicating the underestimation of high AGB values and overestimation of low AGB values. It suggested that the data saturation in

Landsat imagery was not fully eliminated in this study of natural secondary forests. For Heilongjiang province, the average forest AGB density estimated by [69,104] was 81 Mg/ha and 85 Mg/ha, respectively (using a ratio of 50% for the conversion from forest AGB to AGB carbon stock); for the entire northeastern China, the average forest AGB density estimated by [57,105] was 83.50 Mg/ha and 89.30 Mg/ha, respectively. All these values were significantly higher than the average AGB of 59.9 Mg/ha in this study. The first reason for that could be the different study area: the area of either Heilongjiang province or northeastern China is much larger than Maorshan Experimental Forest Farm and includes the areas with high AGB values, such as Daxing'an Mountains, Xiaoxing'an Mountain, or Changbai Mountains, which results in a higher average AGB value. The second reason could be that the data saturation in this study greatly causes the relatively low average AGB, although the range of predicted AGB (0–491.04 Mg/ha) is reasonable. Thus, how to eliminate data saturation and quantitatively determine saturation for NSFs still need further investigation.

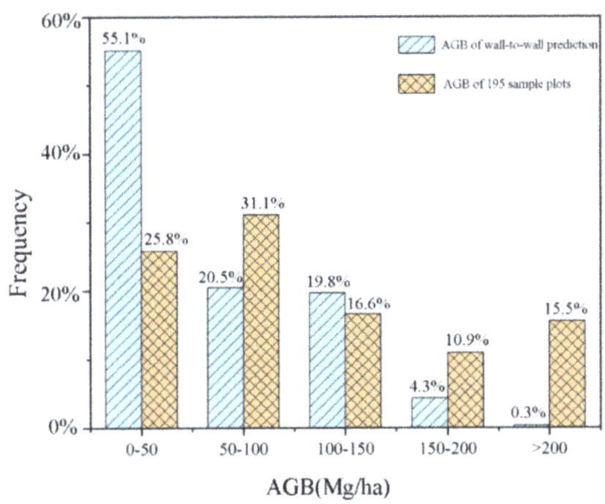

Figure 6. The distributions of AGB values of wall-to-wall prediction map (blue bars with one slash) and 195 sample plots (orange bars with double slashes).

4.4. Limitations and Recommendations

The AGB retrievals with high accuracy from remotely sensed data is not an easy task. Every procedure or factor could greatly influence the accuracy, including data sources, feature extraction and selection, estimation models, and model evaluation, and so on. Although high accuracies of AGB estimation were yielded by the CNN and SG(CNN) models based on the combination of the optimal ALS and Landsat 8 features and all *COLI*2 (Feature 1 + 2 + 5), there were still limitations in this study. First, in this study, we only tested the features (*COLI*1, *COLI*2) proposed by [48] and compared them with the direct combination of these original features that generated them for the AGB estimation of NSFs. It is possible to find a more effective approach to combine ALS and Landsat 8 imagery than *COLI*s for NSFs. Thus, it is still valuable to propose novel features or explore other synergistic approaches based on multiple data sources for various forest types.

The second limitation is that the underestimation of high AGB values and the overestimation of low AGB values were not eliminated from the wall-to-wall prediction map, although the CNN model had good efficiency and high accuracy according to model evaluation results. Data saturation might be responsible for this phenomenon and lead to a much lower average of AGB estimates of the entire study area than those values in similar studies [57,69,104,105]. The high risks of overfitting resulted from the data-driven models could be another possible reason for the big discrepancy between model evaluation results and final wall-to-wall prediction. Thus, the development of models with good generalizability in the estimation of biomass and the interpretation of the physical meaning of models are strongly recommended in further research [17].

In addition, the model evaluation procedure based on leave-one-out cross-validation may be another incentive for the high accuracy of the CNN model using reference data. Leave-one-out cross-validation is a special case of K-fold cross-validation where the number of folds equals the number of records in the data set [106]. Since the evaluated model is applied once for each record, using all other records as a training set and the selected record as a single-item test set, it could tend to yield higher accuracy due to overfitting compared to ten-fold cross-validation, for example, which only uses 90% records to train the model. However, the quantitative effects of different cross-validation procedures on AGB estimations still need to be further investigated. Sometimes, it could be a big difference between the accuracy of the model evaluation procedure using reference data and wall-to-wall prediction values. Thus, besides the traditional model evaluation procedure, we strongly suggest assessing the spatial distribution of AGB estimates based on a wall-to-wall prediction map and distribution of AGB estimates based on histogram compared to existed data.

The AGB estimation in this study was based on an area-based approach (ABA) that develops models to relate AGB with features derived from remotely sensed data at a plot level and apply the models over the whole study area [17]. The fixed plots of continuous forest resources inventory obtained in 2016 had an area of 20 m $\times$30 m with the geolocation error of 5 m, while the pixel size of Landsat 8 was 30 m $\times$ 30 m. Thus, geolocation mismatch between remotely sensed data (i.e., Landsat 8 imagery) and field measurements is another source of uncertainty of AGB estimation [107]. Fortunately, the large plot size (i.e., 195) in this study could greatly decrease the geolocation errors according to [107]: the geolocation errors will be stabilized below 5 m with 20 measurement points and below 3 m with 50 measurement points. Another drawback of this study is the lack of assessing biomass uncertainty based on ABA. It is difficult for AGB estimation using ABA to understand biomass uncertainties at different spatial scales [108]. In recent years, with the development of automatic individual tree crown delineation algorithms in precise forestry (e.g., [109,110]), the AGB estimation based on individual-tree-based approach (ITA) has received more and more attention because field data are needed only for a sample of trees instead of a sample of plots or stands [17]. In addition, ITA allows AGB estimation of tree-level, plot-level, and propagation of errors in an up-scaling framework [108]. Thus, it is appealing and worth estimating AGB based on ITA for a large-scale forest and quantifying its uncertainty from tree-level to plot-level then to stand-level in an up-scaling framework in subsequent research.

5. Conclusions

Accurate quantification of AGB plays a vital role in forest carbon sequestration in the context of climate change. In this study, we investigated the effects of different synergistic approaches of features and ensemble learning algorithms on AGB estimation of natural secondary forests of northeastern China based on ALS and Landsat 8 OLI imagery. It is conducive to combine active and passive data to improve the accuracy of AGB estimation. Unlike the previous study implemented in southeastern China [48], we found that *COLI*2 features are more effective in AGB estimation than *COLI*1 features for the NSFs. Sometimes, it might be more convenient and efficient to adopt the simple combination of the untransformed features (e.g., the optimal Landsat 8 features + BLV) than the novel features (i.e., *COLI*1 or *COLI*2), especially for NSFs of northeastern China. The CNN model was much superior to multiple linear regression and other classic machine learning algorithms (i.e., ELM, BP, RegT, RF, SVR, KNN) no matter of feature sets, and reached the highest accuracies (R^2 = 0.99, $RMSE$ = 6.85, $rRMSE$ = 0.04, MAE = 2.95, $MAPE$ = 1.02, PM = 0.03) when optimal ALS and Landsat 8 features and all *COLI*2 (Feature 1 + 2 + 5) was applied. Ensemble learning algorithms (SG(RF), SG(SVR), SG(KNN), SG(CNN)) that took advantage of the good and stable predictions from the base models (i.e., RF, SVR, KNN, CNN) greatly improved the accuracy of AGB and had stronger generalization ability compared to its corresponding base model. The ensemble learning algorithm is exceedingly adept to train weaker learners to strong learners, especially when applying heterogeneous ensemble strategy. The SG model with CNN meta-model performed best (R^2 = 0.99, *RMSE* = 2.02, *rRMSE* = 0.01, *MAE* = 0.87, *MAPE* = 0.73, *PM* = 0.02) with the feature combination of the optimal ALS and Landsat 8 features and all *COLI*2 (Feature 1 + 2 + 5) in this study. However, considering both the efficiency (i.e., runtime) and accuracy, a wall-to-wall AGB prediction map of Maoershan was generated using the CNN model and Feature 1 + 2 + 5, instead of the SG(CNN) model. The average and standard deviation of the estimated AGB of Maoershan Experimental Forest Farm in 2015 was 59.9 Mg/ha and 48.69 Mg/ha, respectively, ranging from 0 to 491.04 Mg/ha. The lower average value than that of similar studies for northeastern China maybe because of the different study areas, data saturation, overfitting of the algorithm, and leave-one-out cross-validation. Estimating data saturation, developing advanced algorithms, understanding the effects of the different cross-validation procedures, and quantifying the sources of error are still fundamental and significant to AGB estimation at all levels.

Author Contributions: Conceptualization, W.F. and Z.Z.; methodology, C.D.; software, Y.M.; validation, C.D., Y.M., and Z.Z.; formal analysis, C.D. and H.-I.J.; data curation and preprocessing, W.F. and H.-I.J.; writing—original draft preparation, C.D.; writing—review and editing, Z.Z.; visualization, Y.M.; supervision, Z.Z.; project administration, Z.Z.; funding acquisition, Z.Z. All authors have read and agreed to the published version of the manuscript.

Funding: This research was funded by National Natural Science Foundation of China "Multi-scale forest aboveground biomass estimation and its spatial uncertainty analysis based on individual tree detection techniques", 32071677; "The Fundamental Research Funds for the Central Universities", 2572019CP15,2572020BA05.

Institutional Review Board Statement: Not applicable.

Informed Consent Statement: Not applicable.

Data Availability Statement: The data presented in this study are available on request from the corresponding author.

Conflicts of Interest: The authors declare no conflict of interest.

Appendix A

Table A1. The 101 features extracted from ALS data in this study.

Feature Group	Feature Name	Feature Descriptions [111]
Forest features [1] (3 features)	CC	Canopy cover: $CC = N_{\text{veg}}/N$
	G	Gap fraction: $G = N'/N$
	LAI	Leaf area index: $LAI = -\cos(A)\cdot\ln(G)/k$
Elevation features (46 features) [2]	elev_AAD	Average absolute deviation of elevation: $\sum_{i=1}^{n}(\lvert Z_i - \overline{Z}\rvert)/n$
	elev_CRR	Canopy relief ratio of elevation: $(\overline{Z} - Z_{\min})/(Z_{\max} + Z_{\min})$
	elev_AIH_*i*th	The cumulative height of *i*% points in each pixel is the AIH of the pixel, *i* = 1%, 5%, 10%, 20%, 25%, 30%, 40%, 50%, 60%, 70%, 75%, 80%, 90%, 95%, 99%
	elev_AIH_IQ	AIH interquartile distance: AIH75%–AIH25%
	elev_GM_2	Generalized means for the 2nd power: $\sqrt[2]{\sum_{i=1}^{n} Z_i^3/n}$

Table A1. *Cont.*

Feature group	Feature Name	Feature Descriptions [111]
	elev_GM_3	Generalized means for the 3rd power: $\sqrt[3]{\sum_{i=1}^{n} Z_i^3/n}$
	elev_cv	Coefficient of variation of elevation: $Z_{std}/\overline{Z} \times 100\%$
	elev_IQ	Elevation percentile interquartile distance: Elev75%–Elev25%
	elev_kurt	Kurtosis of elevation
	elev_MMAD	Median of median absolute deviation of elevation
	elev_max	Maximum of elevation
	elev_min	Minimum of elevation
	elev_mean	Mean of elevation
	elev_med	Median of elevation
	elev_per_*i*th	*i*th elevation percentiles, i = 1%, 5%, 10%, 20%, 25%, 30%, 40%, 50%, 60%, 70%, 75%, 80%, 90%, 95%, 99%
	elev_skew	Skewness of elevation
	elev_std	Standard deviation of elevation
	elev_var	Variance of elevation
Density features (10 features)	density_*i*th	The proportion of returns in *i*th height interval, i = 1–10
Intensity features (42 features) [3]	int_AAD	Average absolute deviation of intensity: $\sum_{i=1}^{n} (\lvert I_i - \overline{I} \rvert)/n$
	int_cv	Coefficient of variation of intensity: $I_{std}/\overline{I} \times 100\%$
	int_AII_*i*th	The cumulative intensity of X% points in each pixel is the AII of the pixel, i = 1%, 5%, 10%, 20%, 25%, 30%, 40%, 50%, 60%, 70%, 75%, 80%, 90%, 95%, 99%
	int_kurt	Kurtosis of intensity
	int_MMAD	Median of median absolute deviation of intensity
	int_max	Maximum of intensity
	int_min	Minimum of intensity
	int_mean	Mean of intensity
	int_med	Median of intensity
	int_per_*i*th	*i*th intensity percentiles, i = 1%, 5%, 10%, 20%, 25%, 30%, 40%, 50%, 60%, 70%, 75%, 80%, 90%, 95%, 99%
	int_skew	Skewness of intensity
	int_std	Standard deviation of intensity
	int_var	Variance of intensity
	Int_IQ	Intensity percentile interquartile distance: Int75%–Int25%

[1] N_{veg}: point number of vegetation; N: the total return number; N': the number of ground points whose elevation is lower than the height threshold of 2m for separating ground and tree points; A: average scanning angle; k: extinction coefficient, which is closely related to the leaf inclination angle distribution of the canopy. [2] n is the number of points in a pixel; Z_i: the elevation of i point within a pixel, $\overline{Z}$, Z_{min}, Z_{max}, Z_{std} are the average, minimum, maximum, and standard deviation of elevation of all points within a pixel, respectively; AIH75% and AIH25% represents the 75% and 25% AIH statistical layer, respectively. [3] I_i: the elevation of i point within a pixel, $\overline{I}$, I_{min}, I_{max}, I_{std} are the average, minimum, maximum, and standard deviation of intensity of all points within a pixel, respectively; Int75% and Int25% are 75% and 25% intensity statistical layer, respectively.

Table A2. The 98 spectral features extracted from Landsat 8 OLI imagery in this study.

Feature Group	Feature Name	Feature Descriptions
Original bands (7 features)	Bi [1]	Band1–7 of Landsat 8 OLI image
Band combination (10 features)	Albedo	$0.246\ B_2 + 0.146\ B_3 + 0.191 \cdot B_4 + 0.304 \cdot B_5 + 0.105 \cdot B_6 + 0.008 \cdot B_7$ [112]
	B4/Albedo	$B_4/(0.246 \cdot B_2 + 0.146 \cdot B_3 + 0.191 \cdot B_4 + 0.304 \cdot B_5 + 0.105 \cdot B_6 + 0.008 \cdot B_7)$ [112,113]
	B24	B_2/B_4 [113]
	B74	B_7/B_4 [113]
	B76	B_7/B_6 [113]
	B547	$B_5 \cdot B_4/B_7$ [113]
	B65	B_6/B_5 [113]
	B345	$B_3 \cdot B_4/B_5$ [113]
	B53	B_5/B_3 [113]
	VIS234	$B_2 + B_3 + B_4$ [113]
GLCM features [2] (56 features)	Mean_B_i	Mean of each band
	Var_B_i	Variance of each band
	Hom_B_i	Homogeneity of each band
	Cont_B_i	Contrast of each band
	Diss_B_i	Dissimilarity of each band
	Entr_B_i	Entropy of each band
	Sec_B_i	Second moment of each band
	Corr_B_i	Correlation of each band
Image enhancement features (10 features)	Bright	Brightness from tasseled cap transformation: $0.3521 \cdot B_2 + 0.3899 \cdot B_3 + 0.3825 \cdot B_4 + 0.6985 \cdot B_5 + 0.2343 \cdot B_6 + 0.1867 \cdot B_7$ [114]
	Green	Greenness from tasseled cap transformation: $-0.3301 \cdot B_2 - 0.3455 \cdot B_3 - 0.4508 \cdot B_4 + 0.6970 \cdot B_5 - 0.0448 \cdot B_6 - 0.2840 \cdot B_7$ [114]
	Wet	Wetness from tasseled cap transformation: $0.2651 \cdot B_2 + 0.2367 \cdot B_3 + 0.1296 \cdot B_4 + 0.059 \cdot B_5 - 0.7506 \cdot B_6 - 0.5386 \cdot B_7$ [114]
	PC1	The first principal component from principal component analysis (PCA): $0.111 \cdot B_3 + 0.870 \cdot B_5 + 0.423 \cdot B_6 + 0.192 \cdot B_7$
	PC2	The second principal component from PCA: $0.198 \cdot B_1 + 0.217 \cdot B_2 + 0.267 \cdot B_3 + 0.376 \cdot B_4 - 0.436 \cdot B_5 + 0.430 \cdot B_6 + 0.571 \cdot B_7$
	PC3	The third principal component from PCA: $0.295 \cdot B_1 + 0.324 \cdot B_2 + 0.398 \cdot B_3 + 0.473 \cdot B_4 + 0.183 \cdot B_5 - 0.615 \cdot B_6 - 0.12 \cdot B_7$
	MNF1	The first band of minimum noise fraction rotation (MNF): $-0.2632 \cdot B_1 - 0.3528 \cdot B_2 - 0.0737 \cdot B_3 - 0.0618 \cdot B_4 - 0.7457 \cdot B_5 - 0.4898 \cdot B_6 + 0.031 \cdot B_7$
	MNF2	The second band of MNF: $-0.0441 \cdot B_1 - 0.0781 \cdot B_2 - 0.1869 \cdot B_3 - 0.0389 \cdot B_4 - 0.7523 \cdot B_5 - 0.4280 \cdot B_6 - 0.4542 \cdot B_7$
	MNF3	The third band of MNF: $-0.2387 \cdot B_1 - 0.2230 \cdot B_2 + 0.0947 \cdot B_3 - 0.0195 \cdot B_4 + 0.5277 \cdot B_5 + 0.7731 \cdot B_6 - 0.0885 \cdot B_7$
	MNF4	The fourth band of MNF: $0.0199 \cdot B_1 - 0.00013 \cdot B_2 - 0.01021 \cdot B_3 - 0.1027 \cdot B_4 - 0.4377 \cdot B_5 - 0.69145 \cdot B_6 - 0.565 \cdot B_7$
Vegetation indices (15 features)	NDVI	Normalized vegetation index 1: $(B_5 - B_4)/(B_5 + B_4)$ [113]
	RVI	Ratio vegetation index: B_5/B_4 [113]
	DVI	Difference vegetation index: $B_5 - B_4$ [113]
	EVI	Enhanced vegetation index: $2.5 \cdot (B_5 - B_4)/(B_5 + 6 \cdot B_4 - 7.5 \cdot B_2 + 1)$ [113]
	MSAVI	Modified soil-adjusted vegetation index: $[(B_5 - B_4)/(B_5 + B_4 + L)] \cdot (1 + L)$ [3] [115]
	ARVI	Atmospherically resistant vegetation index: $(B_5 - 2 \cdot B_4 + B_2)/(B_5 + 2 \cdot B_4 - B_2)$ [113]
	TVI	Triangular vegetation index: $\sqrt{(B_5 - B_4)/(B_5 + B_4) + 0.5}$ [113]

Table A2. *Cont.*

Feature Group	Feature Name	Feature Descriptions
	PVI	Perpendicular vegetation index: $\sqrt{(0.355 \cdot B_5 - 0.149 \cdot B_4)^2 + (0.355 \cdot B_4 - 0.852 \cdot B_5)^2}$ [113]
	MSR	Modified simple ratio vegetation index : $(B_5/B_4 - 1)/\sqrt{B_5/B_4 + 1}$ [113]
	SLAVI	Specific leaf area vegetation index: $B_5/(B_4 + B_7)$ [113]
	MVI5	Moisture vegetation index 1: $(B_5 + B_4 - B_2)/(B_5 + B_4 + B_2)$ [116]
	MVI7	Moisture vegetation index 2: $(B_5 - B_7)/(B_5 + B_7)$ [116]
	NLI	Nonlinear index : $\left(B_5^2 - B_4\right)/\left(B_5^2 + B_4\right)$ [113]
	RDVI	Renormalized difference vegetation index : $(B_5 - B_4)/\sqrt{B_5 + B_4}$ [113]
	ND563	Normalized difference vegetation index 2: $(B_5 + B_6 - B_3)/(B_5 + B_6 + B_3)$ [113]

[1] The index i represents the band index (1–7). [2] GLCM: gray-level co-occurrence matrix. [3] $L = 2 \cdot s \cdot (B_5 - B_4) \cdot (B_5 - s \cdot B_4)/(B_5 + B_4)$ where s is the slope of the soil line from a plot of red versus near infrared brightness values.

References

1. Wang, C. Biomass allometric equations for 10 co-occurring tree species in Chinese temperate forests. *For. Ecol. Manag.* **2006**, *222*, 9–16. [CrossRef]
2. Yu, D.; Zhou, L.; Zhou, W.; Ding, H.; Wang, Q.; Wang, Y.; Wu, X.; Dai, L. Forest management in northeast China: History, problems, and challenges. *Environ. Manag.* **2011**, *48*, 1122–1135. [CrossRef] [PubMed]
3. Zhang, P.; Shao, G.; Zhao, G.; Le Master, D.C.; Parker, G.R.; Dunning, J.B., Jr.; Li, Q. China's forest policy for the 21st century. *Science* **2000**, *288*, 2135–2136. [CrossRef]
4. Zhu, J.; Liu, S. Conception of secondary forest and its relation to ecological disturbance degree. *Chin. J. Ecol.* **2007**, *26*, 1085–1093. (In Chinese)
5. Yang, K.; Zhu, J.; Zhang, M.; Yan, Q.; Sun, O. Soil microbial biomass carbon and nitrogen in forest ecosystems of northeast China: A comparison between natural secondary forest and larch plantation. *J. Plant. Ecol.* **2010**, *3*, 175–182. [CrossRef]
6. CEOS Land Product Validation Subgroup. Available online: https://lpvs.gsfc.nasa.gov/AGB/AGB_home.html (accessed on 29 July 2021).
7. Vashum, K.T.; Jayakumar, S. Methods to estimate above-ground biomass and carbon stock in natural forests—A review. *J. Ecosyst. Ecography* **2012**, *2*, 116. [CrossRef]
8. Dong, L.; Zhang, L.; Li, F. Developing two additive biomass equations for three coniferous plantation species in northeast China. *Forests* **2016**, *7*, 136. [CrossRef]
9. Bond-Lamberty, B.; Wang, C.; Gower, S.T. Aboveground and belowground biomass and sapwood area allometric equations for six boreal tree species of northern Manitoba. *Can. J. For. Res.* **2002**, *32*, 1441–1450. [CrossRef]
10. Brown, S.; Gillespie, A.R.; Lugo, A.E. Biomass estimation methods for tropical forests with applications to forest inventory data. *For. Sci.* **1989**, *35*, 881–902.
11. Nelson, B.W.; Mesquita, R.; Pereira, J.L.; de Souza, S.G.A.; Batista, G.T.; Couto, L.B. Allometric regressions for improved estimate of secondary forest biomass in the central Amazon. *For. Ecol. Manag.* **1999**, *117*, 149–167. [CrossRef]
12. Chung-Wang, X.; Ceulemans, R. Allometric relationships for below- and above-ground biomass of young Scots pines. *For. Ecol. Manag.* **2004**, *203*, 177–186.
13. Chave, J.; Riéra, B.; Dubois, M. Estimation of biomass in a neotropical forest of French Guiana: Spatial and temporal variability. *J. Trop. Ecol.* **2001**, *17*, 79–96. [CrossRef]
14. Fang, J.; Chen, A.; Peng, C.; Zhao, S.; Ci, L. Changes in forest biomass carbon storage in china between 1949 and 1998. *Science* **2001**, *292*, 2320–2322. [CrossRef]
15. Lu, D.; Chen, Q.; Wang, G.; Liu, L.; Li, G.; Moran, E. A survey of remote sensing-based aboveground biomass estimation methods in forest ecosystems. *Int. J. Digit. Earth* **2014**, *9*, 63–105. [CrossRef]
16. White, J.; Coops, N.; Scott, N. Estimates of New Zealand forest and scrub biomass from the 3-PG model. *Ecol. Model.* **2000**, *131*, 175–190. [CrossRef]
17. Chen, Q. LiDAR remote sensing of vegetation biomass. In *Remote Sensing of Natural Resources*; CRC PRESS: Boca Raton, FL, USA, 2014.
18. Jenkins, J.C.; Birdsey, R.A.; Pan, Y. Biomass and NPP estimation for the mid-Atlantic region (USA) using plot-level forest in-ventory data. *Ecol. Appl.* **2001**, *11*, 1174–1193. [CrossRef]
19. Cao, L.; Pan, J.; Li, R.; Li, J.; Li, Z. Integrating airborne LiDAR and optical data to estimate forest aboveground biomass in arid and semi-arid regions of China. *Remote Sens.* **2018**, *10*, 532. [CrossRef]

20. Endres, A.; Mountrakis, G.; Jin, H.; Zhuang, W.; Manakos, I.; Wiley, J.J.; Beier, C.M. Relative importance analysis of Landsat, waveform LIDAR and PALSAR inputs for deciduous biomass estimation. *Eur. J. Remote Sens.* **2016**, *49*, 795–807. [CrossRef]
21. Laurin, G.V.; Chen, Q.; Lindsell, J.; Coomes, D.A.; Del Frate, F.; Guerriero, L.; Pirotti, F.; Valentini, R. Above ground biomass estimation in an African tropical forest with lidar and hyperspectral data. *ISPRS J. Photogramm. Remote Sens.* **2014**, *89*, 49–58. [CrossRef]
22. Foody, G.M.; Boyd, D.; Cutler, M. Predictive relations of tropical forest biomass from Landsat TM data and their transferability between regions. *Remote Sens. Environ.* **2003**, *85*, 463–474. [CrossRef]
23. Myneni, R.B.; Dong, J.; Tucker, C.J.; Kaufmann, R.K.; Kauppi, P.E.; Liski, J.; Zhou, L.; Alexeyev, V.; Hughes, M.K. A large carbon sink in the woody biomass of Northern forests. *Proc. Natl. Acad. Sci. USA* **2001**, *98*, 14784–14789. [CrossRef]
24. Thenkabail, P.S.; Enclona, E.A.; Ashton, M.S.; Legg, C.; De Dieu, M.J. Hyperion, IKONOS, ALI and ETM plus sensors in the study of African rainforests. *Remote Sens. Environ.* **2004**, *90*, 23–43. [CrossRef]
25. Clark, D.B.; Read, J.M.; Clark, M.L.; Cruz, A.M.; Dotti, M.F.; Clark, D.A. Application of 1-m and 4-m resolution satellite data to ecological studies of tropical rain forests. *Ecol. Appl.* **2004**, *14*, 61–74. [CrossRef]
26. Gasparri, N.I.; Parmuchi, M.G.; Bono, J.; Karszenbaum, H.; Montenegro, C.L. Assessing multi-temporal Landsat 7 ETM + images for estimating above-ground biomass in subtropical dry forests of Argentina. *J. Arid. Environ.* **2010**, *74*, 1262–1270. [CrossRef]
27. Gömez, C.; White, J.C.; Wulder, M.A.; Alejandro, P. Historical forest biomass dynamics modelled with Landsat spectral trajectories. *ISPRS J. Photogramm. Remote Sens.* **2014**, *93*, 14–28. [CrossRef]
28. Dube, T.; Mutanga, O. Investigating the robustness of the new Landsat-8 Operational Land Imager derived texture metrics in estimating plantation forest aboveground biomass in resource constrained areas. *ISPRS J. Photogramm. Remote Sens.* **2015**, *108*, 12–32. [CrossRef]
29. Kelsey, K.C.; Neff, J.C. Estimates of aboveground biomass from texture analysis of landsat imagery. *Remote Sens.* **2014**, *6*, 6407–6422. [CrossRef]
30. Dube, T.; Mutanga, O. Evaluating the utility of the medium-spatial resolution Landsat 8 multispectral sensorin quantifying aboveground biomass in uMgeni catchment, South Africa. *ISPRS J. Photogramm. Remote Sens.* **2015**, *101*, 36–46. [CrossRef]
31. Loveland, T.R.; Irons, J.R. Landsat 8: The plans, the reality, and the legacy. *Remote Sens Environ.* **2016**, *185*, 1–6. [CrossRef]
32. Lu, D. Aboveground biomass estimation using Landsat TM data in the Brazilian Amazon. *Int. J. Remote Sens.* **2005**, *26*, 2509–2525. [CrossRef]
33. Zhao, P.; Lu, D.; Wang, G.; Wu, C.; Huang, Y.; Yu, S. Examining spectral reflectance saturation in Landsat imagery and cor-responding solutions to improve forest aboveground biomass estimation. *Remote Sens.* **2016**, *8*, 469. [CrossRef]
34. Steininger, M.K. Satellite estimation of tropical secondary forest above-ground biomass: Data from Brazil and Bolivia. *Int. J. Remote Sens.* **2000**, *21*, 1139–1157. [CrossRef]
35. Lucas, R.M.; Held, A.A.; Phinn, S.R.; Saatchi, S. Tropical forests. In *Remote Sensing for Natural Resource Management and Environmental Monitoring*, 3rd ed.; Ustin, S.D., Ed.; John Wiley & Sons, Inc.: Hoboken, NJ, USA, 2004; Volume 3, pp. 239–315.
36. Le Toan, T.; Quegan, S.; Woodward, I.; Lomas, M.; Delbart, N.; Picard, G. Relating radar remote sensing of biomass to mod-elling of forest carbon budgets. *Clim. Chang.* **2004**, *67*, 379–402. [CrossRef]
37. Waring, R.H.; Way, J.; Hunt, E.R.; Morrissey, L.; Ranson, K.J.; Weishampel, J.F.; Oren, R.; Franklin, S.E. Imaging radar for ecosystem studies. *BioScience* **1995**, *45*, 715–723. [CrossRef]
38. Zolkos, S.; Goetz, S.; Dubayah, R. A meta-analysis of terrestrial aboveground biomass estimation using lidar remote sensing. *Remote Sens. Environ.* **2012**, *128*, 289–298. [CrossRef]
39. Gonzalez, P.; Asner, G.P.; Battles, J.J.; Lefsky, M.A.; Waring, K.M.; Palace, M. Forest carbon densities and uncertainties from Lidar, QuickBird, and field measurements in California. *Remote Sens. Environ.* **2010**, *114*, 1561–1575. [CrossRef]
40. Means, J.E.; Acker, S.A.; Harding, D.J.; Blair, J.B.; Lefsky, M.A.; Cohen, W.B.; Harmon, M.E.; McKee, W.A. Use of large-footprint scanning airborne lidar to estimate forest stand characteristics in the western cascades of Oregon. *Remote Sens. Environ.* **1999**, *67*, 298–308. [CrossRef]
41. Lu, D.; Chen, Q.; Wang, G.; Moran, E.; Batistella, M.; Zhang, M.; Laurin, G.V.; Saah, D. Aboveground forest biomass estima-tion with Landsat and Lidar data and uncertainty analysis of the estimates. *Int. J. For. Res.* **2012**, *2012*, 250–265.
42. Mauya, E.W.; Ene, L.T.; Bollandsås, O.M.; Gobakken, T.; Naesset, E.; Malimbwi, R.E.; Zahabu, E. Modelling aboveground forest biomass using airborne laser scanner data in the miombo woodlands of Tanzania. *Carbon Balance Manag.* **2015**, *10*, 28. [CrossRef]
43. Gleason, C.J.; Im, J. Forest biomass estimation from airborne LiDAR data using machine learning approaches. *Remote Sens. Environ.* **2012**, *125*, 80–91. [CrossRef]
44. Ioki, K.; Tsuyuki, S.; Hirata, Y.; Phua, M.H.; Wong, W.V.C.; Ling, Z.Y.; Saito, H.; Takao, G. Estimating above-ground bio-mass of tropical rainforest of different degradation levels in Northern Borneo using airborne LiDAR. *For. Ecol. Manag.* **2014**, *328*, 335–341. [CrossRef]
45. Hansen, E.H.; Gobakken, T.; Bollandsås, O.M.; Zahabu, E.; Næsset, E. Modeling aboveground biomass in dense tropical submontane rainforest using airborne laser scanner data. *Remote Sens.* **2015**, *7*, 788–807. [CrossRef]
46. Magdon, P.; González-Ferreiro, E.; Pérez-Cruzado, C.; Purnama, E.S.; Sarodja, D.; Kleinn, C. Evaluating the potential of ALS data to increase the efficiency of aboveground biomass estimates in tropical peat–swamp forests. *Remote Sens.* **2018**, *10*, 1344. [CrossRef]

47. Adhikari, H.; Heiskanen, J.; Siljander, M.; Maeda, E.; Heikinheimo, V.; Pellikka, P.K.E. Determinants of aboveground bio-mass across an Afromontane landscape mosaic in Kenya. *Remote Sens.* **2017**, *9*, 827. [CrossRef]
48. Zhang, L.; Shao, Z.; Liu, J.; Cheng, Q. Deep learning based retrieval of forest aboveground biomass from combined LiDAR and landsat 8 data. *Remote Sens.* **2019**, *11*, 1459. [CrossRef]
49. Clark, M.L.; Roberts, D.A.; Ewel, J.J.; Clark, D.B. Estimation of tropical rain forest aboveground biomass with small-footprint lidar and hyperspectral sensors. *Remote Sens. Environ.* **2011**, *115*, 2931–2942. [CrossRef]
50. Egberth, M.; Nyberg, G.; Næsset, E.; Gobakken, T.; Mauya, E.; Malimbwi, R.; Katani, J.; Chamuya, N.; Bulenga, G.; Olsson, H. Combining airborne laser scanning and Landsat data for statistical modeling of soil carbon and tree biomass in Tanzanian Miombo woodlands. *Carbon Balance Manag.* **2017**, *12*, 8. [CrossRef] [PubMed]
51. Heiskanen, J.; Adhikari, H.; Piiroinen, R.; Packalen, P.; Pellikka, P.K. Do airborne laser scanning biomass prediction models benefit from Landsat time series, hyperspectral data or forest classification in tropical mosaic landscapes? *Int. J. Appl. Earth Obs. Geoinf.* **2019**, *81*, 176–185. [CrossRef]
52. Phua, M.H.; Johari, S.A.; Wong, O.C.; Ioki, K.; Mahali, M.; Nilus, R.; Coomes, D.A.; Maycock, C.R.; Hashim, M. Synergistic use of Landsat 8 OLI image and airborne LiDAR data for above-ground biomass estimation in tropical lowland rainforests. *For. Ecol. Manag.* **2017**, *406*, 163–171. [CrossRef]
53. Li, S.; Quackenbush, L.J.; Im, J. Airborne lidar sampling strategies to enhance forest aboveground biomass estimation from landsat imagery. *Remote Sens.* **2019**, *11*, 1906. [CrossRef]
54. Li, Y.; Li, C.; Li, M.; Liu, Z. Influence of variable selection and forest type on forest aboveground biomass estimation using machine learning algorithms. *Forests* **2019**, *10*, 1073. [CrossRef]
55. Blackard, J.A.; Finco, M.V.; Helmer, E.H.; Holden, G.R.; Hoppous, M.L.; Jacobs, D.M.; Lister, A.J.; Moisen, G.G.; Nelson, M.D.; Riemann, R.; et al. Mapping US forest biomass using nationwide forest inventory data and moderate resolution information. *Remote Sens. Environ.* **2008**, *112*, 1658–1677. [CrossRef]
56. Houghton, R.A.; Lawrence, K.T.; Hackler, J.L.; Brown, S. The spatial distribution of forest biomass in the Brazilian amazon: A comparison of estimates. *Glob. Chang. Biol.* **2001**, *7*, 731–746. [CrossRef]
57. Tan, K.; Piao, S.; Peng, C.; Fang, J. Satellite-based estimation of biomass carbon stocks for northeast China's forests between 1982 and 1999. *For. Ecol. Manag.* **2007**, *240*, 114–121. [CrossRef]
58. Chen, L.; Ren, C.; Zhang, B.; Wang, Z.; Xi, Y. Estimation of forest above-ground biomass by geographically weighted regres-sion and machine learning with sentinel imagery. *Forests* **2018**, *9*, 582. [CrossRef]
59. Lim, K.S.; Treitz, P.M. Estimation of above ground forest biomass from airborne discrete return laser scanner data using canopy-based quantile estimators. *Scand. J. For. Res.* **2004**, *19*, 558–570. [CrossRef]
60. Zhao, K.; Popescu, S.; Nelson, R. Lidar remote sensing of forest biomass: A scale-invariant estimation approach using air-borne lasers. *Remote Sens. Environ.* **2009**, *113*, 182–196. [CrossRef]
61. Kulawardhana, R.W.; Popescu, S.; Feagin, R. Fusion of lidar and multispectral data to quantify salt marsh carbon stocks. *Remote Sens. Environ.* **2014**, *154*, 345–357. [CrossRef]
62. Li, W.; Niu, Z.; Wang, C.; Huang, W.; Chen, H.; Gao, S.; Li, D.; Muhammad, S. Combined use of airborne LiDAR and satellite GF-1 data to estimate leaf area index, height, and aboveground biomass of maize during peak growing season. *IEEE J. Sel. Top. Appl. Earth Obs. Remote Sens.* **2015**, *8*, 4489–4501. [CrossRef]
63. Verrelst, J.; Camps-Valls, G.; Muñoz-Marí, J.; Rivera, J.P.; Veroustraete, F.; Clevers, J.G.P.W.; Moreno, J. Optical remote sensing and the retrieval of terrestrial vegetation bio-geophysical properties—A review. *ISPRS J. Photogramm.* **2015**, *108*, 273–290. [CrossRef]
64. Asner, G.P.; Hughes, R.F.; Varga, T.A.; Knapp, D.E.; Kennedy-Bowdoin, T. Environmental and biotic controls over above-ground biomass throughout a tropical rain forest. *Ecosystems* **2009**, *12*, 261–278. [CrossRef]
65. Lucas, R.M.; Cronin, N.; Lee, A.; Moghaddam, M.; Witte, C.; Tickle, P. Empirical relationships between AIRSAR backscatter and LiDAR-derived forest biomass, Queensland, Australia. *Remote Sens. Environ.* **2006**, *100*, 407–425. [CrossRef]
66. Patenaude, G.; Hill, R.; Milne, R.; Gaveau, D.; Briggs, B.; Dawson, T. Quantifying forest above ground carbon content using LiDAR remote sensing. *Remote Sens. Environ.* **2004**, *93*, 368–380. [CrossRef]
67. St-Onge, B.; Hu, Y.; Vega, C. Mapping the height and above-ground biomass of a mixed forest using lidar and stereo Ikonos images. *Int. J. Remote Sens.* **2008**, *29*, 1277–1294. [CrossRef]
68. Xie, Y.; Sha, Z.; Yu, M.; Bai, Y.; Zhang, L. A comparison of two models with Landsat data for estimating above ground grassland biomass in Inner Mongolia, China. *Ecol. Model.* **2009**, *220*, 1810–1818. [CrossRef]
69. Su, Y.; Guo, Q.; Xue, B.; Hu, T.; Alvarez, O.; Tao, S.; Fang, J. Spatial distribution of forest aboveground biomass in China: Estimation through combination of spaceborne lidar, optical imagery, and forest inventory data. *Remote Sens. Environ.* **2016**, *173*, 187–199. [CrossRef]
70. Li, M.; Im, J.; Quackenbush, L.J.; Liu, T. Forest biomass and carbon stock quantification using airborne LiDAR data: A case study over huntington wildlife forest in the adirondack park. *IEEE J. Sel. Top. Appl. Earth Obs. Remote Sens.* **2014**, *7*, 3143–3156. [CrossRef]
71. Ou, G.; Li, C.; Lv, Y.; Wei, A.; Xiong, H.; Xu, H.; Wang, G. Improving aboveground biomass estimation of pinus densata forests in yunnan using landsat 8 imagery by incorporating age dummy variable and method comparison. *Remote Sens.* **2019**, *11*, 738. [CrossRef]

72. Serrano, P.M.L.; López-Sánchez, C.A.; Álvarez-González, J.G.; García-Gutiérrez, J. A Comparison of machine learning techniques applied to landsat-5 tm spectral data for biomass estimation. *Can. J. Remote Sens.* **2016**, *42*, 690–705. [CrossRef]
73. Dong, L.; Du, H.; Han, N.; Li, X.; Zhu, D.; Mao, F.; Zhang, M.; Zheng, J.; Liu, H.; Huang, Z.; et al. Application of convolutional neural network on lei bamboo Above-Ground-Biomass (AGB) estimation using worldview-2. *Remote Sens.* **2020**, *12*, 958. [CrossRef]
74. Luo, M.; Wang, Y.; Xie, Y.; Zhou, L.; Qiao, J.; Qiu, S.; Sun, Y. Combination of feature selection and catboost for prediction: The first application to the estimation of aboveground biomass. *Forests* **2021**, *12*, 216. [CrossRef]
75. Sonobe, R.; Yamaya, Y.; Tani, H.; Wang, X.; Kobayashi, N.; Mochizuki, K. Crop classification from Sentinel-2-derived vege-tation indices using ensemble learning. *J. Appl. Remote Sens.* **2018**, *12*, 26019. [CrossRef]
76. Breiman, L. Random Forests. *Mach Learn.* **2001**, *45*, 5–23. [CrossRef]
77. Zeng, N.; Ren, X.; He, H.; Zhang, L.; Zhao, D.; Ge, R.; Li, P.; Niu, Z. Estimating grassland aboveground biomass on the Ti-betan Plateau using a random forest algorithm. *Ecol. Indic.* **2019**, *102*, 479–487. [CrossRef]
78. Wolpert, D.H. Stacked generalization. *Neural Netw.* **1992**, *5*, 241–259. [CrossRef]
79. Jiang, F.; Zhao, F.; Ma, K.; Li, D.; Sun, H. Mapping the forest canopy height in northern china by synergizing ICESat-2 with sentinel-2 using a stacking algorithm. *Remote Sens.* **2021**, *13*, 1535. [CrossRef]
80. Dong, L. Study on the Compatible Modles of Tree Biomass for Main Species in Heilongjiang Province. Master's Thesis, Northeast Forestry University, Harbin, Heilongjiang, China, 2012. (In Chinese).
81. Li, X.; Guo, Q.; Wang, X.; Zheng, H. Allometry of understory tree species in a natural secondary forest in northeast China. *Sci. Silvae Sin.* **2010**, *46*, 22–32. (In Chinese)
82. Zhao, X.; Guo, Q.; Su, Y.; Xue, B. Improved progressive TIN densification filtering algorithm for airborne LiDAR data in for-ested areas. *ISPRS J. Photogramm.* **2016**, *117*, 79–91. [CrossRef]
83. Soenen, S.A.; Peddle, D.R.; Coburn, C.A. SCS + C: A modified Sun-canopy-sensor topographic correction in forested terrain. *IEEE T. Geosci. Remote* **2005**, *43*, 2148–2159. [CrossRef]
84. Soenen, S.A.; Peddle, D.R.; Hall, R.J.; Coburn, C.A.; Hall, F.G. Estimating aboveground forest biomass from canopy reflectance model inversion in mountainous terrain. *Remote Sens Environ.* **2010**, *114*, 1325–1337. [CrossRef]
85. Jennings, S.; Brown, N.; Sheil, D. Assessing forest canopies and understorey illumination: Canopy closure, canopy cover and other measures. *Forestry* **1999**, *72*, 59–74. [CrossRef]
86. Chen, J.; Black, T. Measuring leaf area index of plant canopies with branch architecture. *Agric. For. Meteorol.* **1991**, *57*, 1–12. [CrossRef]
87. Ou, G.; Lv, Y.; Xu, H.; Wang, G. Improving forest aboveground biomass estimation of pinus densata forest in yunnan of southwest china by spatial regression using Landsat 8 images. *Remote Sens.* **2019**, *11*, 2750. [CrossRef]
88. Huang, G.-B.; Zhu, Q.-Y.; Siew, C.-K. Extreme learning machine: Theory and applications. *Neurocomputing* **2006**, *70*, 489–501. [CrossRef]
89. Rumelhart, D.E.; Hinton, G.E.; Williams, R.J. Learning representations by back-propagating errors. *Nature* **1986**, *323*, 533–536. [CrossRef]
90. Zhu, Y.; Liu, K.; Liu, L.; Wang, S.; Liu, H. Retrieval of mangrove aboveground biomass at the individual species level with worldview-2 images. *Remote Sens.* **2015**, *7*, 12192–12214. [CrossRef]
91. LeCun, Y.; Bengio, Y. Convolutional networks for images, speech, and time series. In *The Handbook of Brain Theory and Neural Network*; MIT Press: Cambridge, MA, USA, 1995; Volume 3361, pp. 1–14.
92. Chen, T.; Guestrin, C. XGBoost: A scalable tree boosting system. In Proceedings of the 22nd ACM SIGKDD International Conference on Knowledge Discovery and Data Mining, San Francisco, CA, USA, 13–17 August 2016; Association for Computing Machinery: New York, NY, USA, 2016; pp. 785–794.
93. Feng, L.; Li, Y.; Wang, Y.; Du, Q. Estimating hourly and continuous ground-level PM2.5 concentrations using an ensemble learning algorithm: The ST-stacking model. *Atmos. Environ.* **2019**, *223*, 117242. [CrossRef]
94. Wen, L.; Hughes, M. Coastal wetland mapping using ensemble learning algorithms: A comparative study of bagging, boosting and stacking techniques. *Remote Sens.* **2020**, *12*, 1683. [CrossRef]
95. Book, S.A.; Yong, P.H. The trouble with R2. *J. Parametr.* **2006**, *25*, 87–114. [CrossRef]
96. Van Der Meer, F.; Bakker, W.; Scholte, K.; Skidmore, A.; De Jong, S.; Clevers, E.A.; Epema, G. Spatial scale variations in veg-etation indices and above-ground biomass estimates: Implications for MERIS. *Int. J. Remote Sens.* **2001**, *22*, 3381–3396. [CrossRef]
97. Huang, S.; Tang, L.; Hupy, J.P.; Wang, Y.; Shao, G. A commentary review on the use of normalized difference vegetation index (NDVI) in the era of popular remote sensing. *J. For. Res.* **2020**, *32*, 1–6. [CrossRef]
98. Wu, Z.; Dye, D.; Vogel, J.; Middleton, B. Estimating forest and woodland aboveground biomass using active and passive re-mote sensing. *Photogramm. Eng. Rem. S.* **2016**, *82*, 271–281. [CrossRef]
99. Krizhevsky, A.; Sutskever, I.; Hinton, G.E. ImageNet classification with deep convolutional neural networks. *Commun. ACM* **2017**, *60*, 84–90. [CrossRef]
100. Ayrey, E.; Hayes, D.J. The use of three-dimensional convolutional neural networks to interpret LiDAR for forest inventory. *Remote Sens.* **2018**, *10*, 649. [CrossRef]
101. Fricker, G.A.; Ventura, J.D.; Wolf, J.A.; North, M.P.; Davis, F.W.; Franklin, J. A convolutional neural network classifier iden-tifies tree species in mixed-conifer forest from hyperspectral imagery. *Remote Sens.* **2019**, *11*, 2326. [CrossRef]

102. Fassnacht, F.; Hartig, F.; Latifi, H.; Berger, C.; Hernández, J.; Corvalán, P.; Koch, B. Importance of sample size, data type and prediction method for remote sensing-based estimations of aboveground forest biomass. *Remote Sens. Environ.* **2014**, *154*, 102–114. [CrossRef]
103. Yang, L.; Liang, S.; Zhang, Y. A new method for generating a global forest aboveground biomass map from multiple high-level satellite products and ancillary information. *IEEE J. Sel. Top. Appl. Earth Obs. Remote Sens.* **2020**, *13*, 2587–2597. [CrossRef]
104. Guo, Z.; Hu, H.; Li, P.; Li, N.; Fang, J. Spatio-temporal changes in biomass carbon sinks in China's forests from 1977 to 2008. *Sci. China Life Sci.* **2013**, *56*, 661–671. [CrossRef]
105. Zhang, Y.; Liang, S.; Sun, G. Forest biomass mapping of northeastern china using GLAS and MODIS Data. *IEEE J. Sel. Top. Appl. Earth Obs. Remote Sens.* **2013**, *7*, 140–152. [CrossRef]
106. Sammut, C.; Webb, G.I. (Eds.) Leave-one-out cross-validation. In *Encyclopedia of Machine Learning*, 2020 ed.; Springer: Boston, MA, USA, 2011.
107. Réjou-Méchain, M.; Barbier, N.; Couteron, P.; Ploton, P.; Vincent, G.; Herold, M.; Mermoz, S.; Saatchi, S.; Chave, J.; de Bois-sieu, F.; et al. Upscaling forest biomass from field to satellite measurements: Sources of errors and ways to reduce them. *Surv. Geophys.* **2019**, *40*, 881–911. [CrossRef]
108. Xu, Q.; Man, A.; Fredrickson, M.; Hou, Z.; Pitkänen, J.; Wing, B. Quantification of uncertainty in aboveground biomass esti-mates derived from small-footprint airborne LiDAR. *Remote Sens. Environ.* **2018**, *216*, 514–528. [CrossRef]
109. Zhen, Z.; Quackenbush, L.J.; Stehman, S.V.; Zhang, L. Agent-based region growing for individual tree crown delineation from airborne laser scanning (ALS) data. *Int. J. Remote Sens.* **2015**, *36*, 1965–1993. [CrossRef]
110. Zhao, Y.; Hao, Y.; Zhen, Z.; Quan, Y. A region-based hierarchical cross-section analysis for individual tree crown delineation using ALS Data. *Remote Sens.* **2017**, *9*, 1084. [CrossRef]
111. GreenValley International. *LiDAR360 V3.2 User Guide*; GreenValley International, Ltd.: Beijing, China, 2019.
112. Olmedo, G.F.; Ortega-Farías, S.; de la Fuente-Sáiz, D.; Fonseca-Luego, D.; Fuentes-Peñailillo, F. water: Tools and functions to es-timate actual evapotranspiration using land surface energy balance models in R. *R J.* **2016**, *8*, 352–369. [CrossRef]
113. Xu, T.; Cao, L.; Shen, X.; She, G. Estimates of subtropical forest biomass based on airborne LiDAR and Landsat 8 OLI data. *Chin. J. Plant Ecol.* **2015**, *39*, 309–321. (In Chinese)
114. Li, B.; Di, C.; Yan, X. Study of derivation of tasseled cap transformation for Landsat 8 OLI images. *Sci. Surv. Mapp.* **2016**, *41*, 102–107. (In Chinese)
115. Qi, J.; Chehbouni, A.; Huete, A.R.; Kerr, Y.H.; Sorooshian, S. A modified soil adjusted vegetation index. *Remote Sens. Environ.* **1994**, *48*, 119–126. [CrossRef]
116. Zhou, L.; Ou, G.; Wang, J.; Xu, H. Light saturation point determination and biomass remote sensing estimation of *Pinus kesiya var. langbianensis* forest based on spatial regression models. *Sci. Silvae Sin.* **2020**, *56*, 38–46. (In Chinese)

sensors

MDPI

Article

Crop Disease Classification on Inadequate Low-Resolution Target Images

Juan Wen, Yangjing Shi, Xiaoshi Zhou and Yiming Xue *

College of Information and Electrical Engineering, China Agricultural University, Beijing 100083, China; wenjuan@cau.edu.cn (J.W.); sy20183081433@cau.edu.cn (Y.S.); zhouxiaoshi0713@163.com (X.Z.)

* Correspondence: xueym@cau.edu.cn

Received: 1 July 2020; Accepted: 13 August 2020; Published: 16 August 2020

Abstract: Currently, various agricultural image classification tasks are carried out on high-resolution images. However, in some cases, we cannot get enough high-resolution images for classification, which significantly affects classification performance. In this paper, we design a crop disease classification network based on Enhanced Super-Resolution Generative adversarial networks (ESRGAN) when only an insufficient number of low-resolution target images are available. First, ESRGAN is used to recover super-resolution crop images from low-resolution images. Transfer learning is applied in model training to compensate for the lack of training samples. Then, we test the performance of the generated super-resolution images in crop disease classification task. Extensive experiments show that using the fine-tuned ESRGAN model can recover realistic crop information and improve the accuracy of crop disease classification, compared with the other four image super-resolution methods.

Keywords: super-resolution; Generative Adversarial Networks; Convolutional Neural Networks; disease classification

1. Introduction

Crop diseases are generally caused by the environment, soil, pests and pathogens. They pose a severe threat to the quality and security of agricultural production [1,2]. At the same time, crop diseases also cause losses to farmers. Taking prompt action could reduce losses. However, it is hard to detect the diseases in time through manual work.

With the development of computer science, it has become a hot topic to identify crop diseases based on computer vision and machine learning techniques. Earlier studies were based on feature extraction techniques. Alsuwaidi et al. [3] applied adaptive feature selection and ensemble learning for crop disease classification. Pantazi et al. [4] employed Local Binary Patterns (LBPs) for feature extraction and a one-class classifier to classify leaf diseases in various crop species. In recent years, image analysis methods based on deep learning have been used for crop disease identification and other purposes in agriculture, such as plant phenotypic analysis. Jia et al. [5] used transfer learning to classify tomato pests and diseases on leaf images based on VGG16 network. Zhang et al. [6] proposed global pooling dilated convolutional neural network (GPDCNN), which integrated the advantages of global pooling and dilated convolution to identify cucumber leaf diseases. Meanwhile, in order to construct a cost-effective system to diagnose diseases and symptoms of mango leaves, a multi-layer convolutional neural network (MCNN) [7] was proposed to classify mango leaves infected by anthracnose disease. It surpassed other approaches on a real-time dataset that includes 1070 images of the Mango tree leaves. Furthermore, based on the open dataset Plant Village [8], Too et al. [9] conducted a comparative study on the fine-tuned convolutional neural network (CNN) models for crop disease identification, including VGG16 [10], Inception V4 [11], ResNet with 50, 101 and 152 layers [12], and DenseNets [13] with 121 layers.

Since unmanned aerial vehicles (UAVs) have become increasingly popular in the agriculture industry in the past few years, some attempts have been made to identify crop diseases based on UAV images. Su et al. [14] collected UAV multispectral images by low-altitude UAVs and low-cost multispectral cameras. They then applied machine learning algorithms to monitor wheat yellow rust, making a significant contribution to yellow rust monitoring at farmland scales. Similar to their work, Cao et al. applied low-altitude remote sensing UAV images to detect Sclerotinia sclerotiorum on oilseed rape leaves [15]. Additionally, Kerkech et al. [16] utilized color information of UAV images to detect vine diseases based on a CNN model.

No matter what device was used to obtain the experimental images, one thing in common among these previous work was that high-resolution (HR) images were required for model training to ensure classification accuracy. In order to obtain HR images, high-quality cameras or sensors are required [17], which are costly and inefficient. In particular, if a UAV is used to capture HR images, it has to fly at a low altitude [14]. However, the drone propellers' spinning motion will create turbulence and shake the leaves, which makes pictures blurry and unclear. According to Torres-Sánchez et al. [18], the ideal application scenario for UAVs is to fly at a high altitude to capture as many plants as possible. However, in such a case, the resolution of images will not be high enough for disease recognition. To solve this problem, Yamamoto et al. [19] first utilized the super-resolution (SR) method to transform low-resolution (LR) images to HR images for crop disease recognition. They applied a super-resolution convolution neural network (SRCNN) [20] to recover tomato leaf details and achieved better performance comparing with the results obtained from the original LR images. Cap et al. [21] used SRCNN and a Generative Adversarial Network (GAN) [22] to generate high-resolution images for detecting cucumber diseases, largely boosting the classification performance.

Because GAN has shown excellent ability in image SR tasks, in this article, we train a crop image super-resolution model based on GAN. Then we conduct crop disease classification on the generated SR images. Specifically, an enhanced super-resolution GAN (ESRGAN) [23] is trained to generate SR images on the Plant Village dataset [8], which is an open-source dataset with multiple plants and diseases. One major problem in our work is that it can be challenging to train a stable GAN model with insufficient labeled datasets. To address this issue, we use data augmentation to increase training samples. Furthermore, a base model pre-trained on ImageNet [24] is adopted to set the initial parameters of ESRGAN, and then transfer learning is applied to fine-tune the model twice in different learning rates to achieve a better quality of the SR images. Since tomato samples have more disease categories than other plants in the Plant Village dataset, tomato is chosen as the target crop in this paper. A VGG16 network is trained by transfer-learning and utilized to identify different types of tomato diseases, in order to verify the classification performance on the generated SR images. Extensive experiments are conducted to show the superiority of the proposed method compared with SRCNN and three conventional image scaling methods: bilinear, cubic, and lanczos4.

Our main contributions are mainly—(1) to handle low-resolution crop images, an ESRGAN model is built and trained to generate the HR images which are comparable to the original images. (2) To make the model work appropriately in case of inadequate crop data, we apply the transfer learning strategy to fine-tune the parameters of the ESRGAN in two separate steps. (3) Using the fine-tuned ESRGAN, which is one of the most potential SR algorithms, we can recover more realistic crop images and further improve the accuracy of crop disease classification.

The remainder of this article is as follows. Section 2 introduces the effective architecture of ESRGAN. Section 3 describes proposed method in details. Experimental details and results are covered in Section 4. Finally, the conclusion is provided in Section 5.

2. Related Work

2.1. Image Super-Resolution Methods

Image SR methods aim to recover detailed and spatial HR images from the corresponding LR images [25]. Recently, deep learning-based SR methods have become a persistent hot topic. SRCNN proposed by Dong et al. [20] established a mapping between low- and high-resolution images, which became a pioneer work of deep learning-based methods. After that, different network architectures and other strategies were put forward to improve the SR performance, mainly evaluated by Peak Signal-to-Noise Ratio (PSNR) [26–32]. In recent years, Shamsolmoali et al. introduced a progressive dilated convolution network which used progressive dilated densely connections and nonlinear learnable activation function to obtain complex features. Consequently, the network achieved satisfying performance in image SR tasks with few layers [33]. Yamamoto et al. [19] applied SRCNN to recover SR tomato leaf images and showed that the accuracy obtained on SR images was better by a large margin than those on LR images. However, images reconstructed via PSNR-oriented approaches can only capture limited perceptually relevant differences, that is, higher PSNR does not necessarily reflect a better perceptual result [34].

To improve the visual quality of SR images, some researchers proposed perceptual-driven methods. Perceptual loss [35] was applied to optimize SR model in feature space rather than pixel space. Furthermore, some researchers introduced GAN to generate SR images resembling realistic images. One of the milestones of GAN-based methods was SRGAN [34], which was constructed by residual blocks [12] and optimized with perceptual loss. Experiments showed that SRGAN significantly enhanced the visual quality of reconstruction over the PSNR-oriented methods. Based on SRGAN, Wang et al. proposed ESRGAN [23]. They improved the generator by designing the Residual-in-Residual Dense Block (RRDB), which had high capacity and low training complexity. Moreover, they improved the discriminator by utilizing Relativistic average GAN (RaGAN) [36]. Benefit from the adversarial structure and perceptual-driven SR strategies, ESRGAN can generate SR images with excellent visual effect.

GAN-based SR models are used in various image SR tasks. In scene recognition tasks, Wang et al. [37] proposed a text-attentional Conditional Generative Adversarial Network (CGAN) for text image SR in natural scene. The proposed model introduced effective channel and spatial attention mechanisms to enhance the original CGAN. It performed well on the public text image dataset. In handwriting recognition tasks, an end-to-end trainable framework was proposed by jointing GAN, deep back projection network (DBPN), and bidirectional long short term memory (BLSTM) [38]. The framework achieved state-of-the-art performances on both printed and handwritten document enhancement and recognition. In object recognition tasks, Xi et al. [39] proposed a Representation Learning Generative Adversarial Network (RLGAN) to generate SR image representation for tiny object recognition. RLGAN significantly improved the classification results on the challenging task of LR object recognition.

2.2. Transfer Learning

At present, more and more machine learning application scenarios have appeared. The existing supervised methods with better performance require a large amount of labeled data. Labeling data is a tedious and costly task. As one of the solutions, transfer learning has attracted more and more attention. Recently, many transfer learning approaches have emerged. Chen et al. [40] proposed a novel subspace alignment method for domain adaptation (DA). The method generated source subspace close to the target subspace by re-weighting the source samples. To match the source domain and target domain, data transformation and mapping are often used. In Reference [41], Xiao et al. proposed a projection-based feature transformation method for feature adaption between source and target domain.

In classification tasks, transfer learning allows us to learn a general classifier using a large amount of labeled data from the source domain and a small amount of labeled data from the target domain. A robust information-theoretic transfer learning framework was proposed in Reference [42] for classifier adaptation. The framework compensated for the loss of generalization performance caused by insufficient data through prior knowledge modeling. Furthermore, a novel deep transfer learning (DTL) model was proposed by applying sparse auto-encoder (SAE) and the maximum mean discrepancy term (MMDT) [43]. SAE extracted raw data features, and MMDT minimized the discrepancy penalty between training and testing data. The prediction accuracy of DTL on the famous motor bearing dataset was as high as 99.82%. Based on transfer learning, it is easier to achieve domain-invariant representation and domain transformation for GANs. A novel transfer learning framework with GAN architecture was proposed in Reference [44]. The model contains three parts: an encoder, a generator, and a duplex adversarial discriminators. It achieved state-of-the-art performance on unsupervised domain adaptation of digital classification and target recognition.

3. Materials and Methods

3.1. Proposed Method

In this paper, our task is to conduct crop disease classification based on inadequate low-resolution target images. To ensure the classification performance, we apply image super-resolution methods to transform the low-resolution crop images into HR images, trying to see how the performance can be improved by using these HR images instead. ESRGAN is chosen in our experiments due to its powerful ability in image SR tasks. Like most GAN-based models, ESRGAN can easily lead to non-convergence or over-fitting under insufficient data. One of the biggest challenges of our work is that there are not enough crop images to train our ESRGAN. In this paper, data augmentation and transfer learning are used to train ESRGAN under insufficient target images. First, we apply a basic model pre-trained on a public dataset ImageNet [24], which contains 1000 different classes. Then, the model parameters are fine-tuned with small-scale target images from the Plant Village dataset [8] to improve SR performance. Figure 1 shows the three-step process of our work.

(1) Data processing: as shown in Figure 1a, to build the classification model, it is necessary to prepare the LR and HR image pairs for model training. Images from the Plant Village dataset can be considered as HR images because these images themselves are of high quality. So we denote the cropped images with size of 128 × 128 pixels from Plant Village dataset as I^{HR}. Then I^{HR} are flipped and rotated to enlarge the number of training samples. We obtain the HR images by bicubic interpolation with downsampling factor $r = 4$. In this way, I^{HR} can be converted to LR image I^{LR} and the pair (I^{LR}, I^{HR}) can be used as the training sample of our GAN model.

(2) Model training: the process is shown in Figure 1b. Firstly we get a pre-trained generator G of ESRGAN, which is trained on ImageNet and saved as RRDB_ESRGAN_x4.pth, available on the website: https://github.com/xinntao/ESRGAN. Then we fine-tune this ESRGAN model using the crop dataset. We iteratively train the generator and the discriminator with adversarial training strategy. We end up with a well-trained G, which can be used to transfer the LR target images into HR ones. Details can be seen in Section 4.2.

(3) Evaluation: the evaluation is depicted in Figure 1c. Four other SR methods will be used for comparison. We first evaluate the quality of generated images I^{SR} by PSNR, structural similarity index (SSIM) [45], and perceptual index (PI) [46]. Then the classification results based on VGG16 [10] through different SR methods are compared and analyzed.

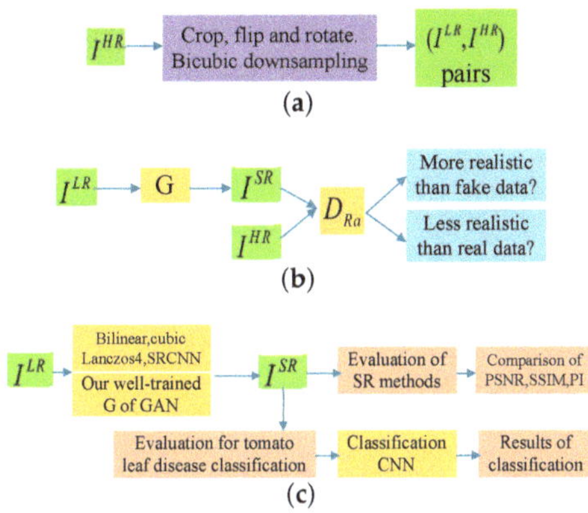

Figure 1. The three steps of our work. (**a–c**) represent the process of data processing, Generative Adversarial Network (GAN) model training and model evaluation, respectively.

3.2. Network Architecture

Our model adopts the training strategy of the original GAN, which optimizes the generator and discriminator in an alternating manner. The task of the generator G is to fool the discriminator by generating SR images similar to HR images. Conversely, the discriminator (denoted as D_{Ra}) is trained to distinguish the generated images from the real ones. In contrast to PSNR-oriented SR methods, ESRGAN applies perceptual loss in G to get natural and high-quality images.

(1) The Generator: The generator is depicted in Figure 2. The input LR image I^{LR} is fed to a convolutional layer with 3×3 filter kernels followed by LeaklyReLU as the activation function. 23 RRDBs, each of which is composed of dense blocks [13] and a multi-level residual network with five convolutional layers, are connected to the first convolutional layer [12] (See in Figure 3). In general, the RRDBs can magnify network capacity. Another convolutional layer with 3×3 kernels and 64 feature maps is added after the RRDB group to integrate features and match the data dimension. The scale factors of two upsampling layers are set to 2 to achieve image SR for $4\times$ upscaling factors. Other convolutional layers are the same as the first one except that the final convolutional layer has three feature maps.

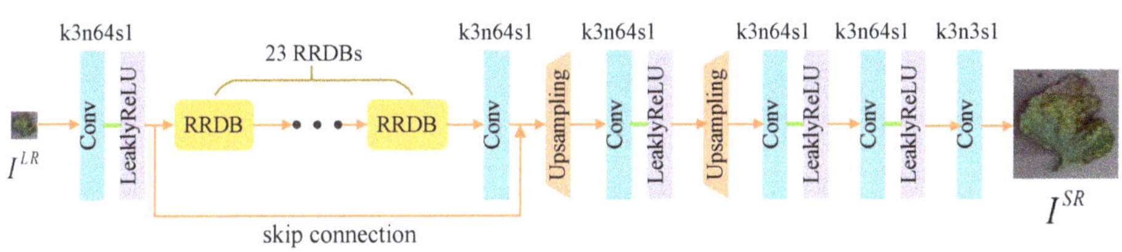

Figure 2. Architecture of generator network G. In each convolutional layer, k, n, and s represent kernel size, number of feature maps, and stride.

(2) The Discriminator: The discriminator is based on RaGAN [36]. It learns to determine which of the two input images is more realistic. The architecture of D_{Ra} is depicted in Figure 4. It contains ten convolutional layers with 3×3 and 4×4 filter kernels appearing in an alternating way. Specifically, the kernel size k, the number of feature maps n, and stride s in each convolutional layer are showed in Figure 4. Batch-normalization (BN) layers [47] are connected behind convolutional layers to counteract the internal co-variate shift. I^{HR} denotes the real HR crop image, and I^{SR} is the fake HR image generated by the generator from the LR image I^{LR}. I^{HR} has the same size as I^{SR}. Two dense layers

and a final sigmoid activation function are used to predict the probability that an original real image I^{HR} is relatively more realistic than a generated fake image I^{SR}.

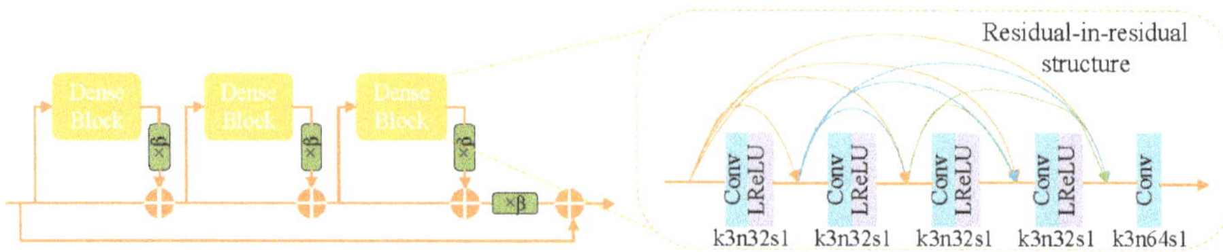

Figure 3. Residual-in-Residual Dense Block (RRDB) with residual scaling parameter β. In each convolutional layer, k, n, and s represent kernel size, number of feature maps, and stride.

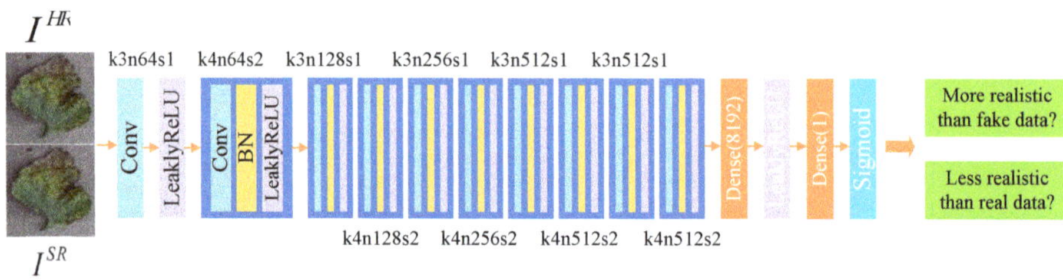

Figure 4. Architecture of discriminator network D_{Ra}. In each convolutional layer, k, n, and s represent kernel size, number of feature maps, and stride.

(3) Loss Functions: D_{Ra} has two outputs, denoted by D_{real} and D_{fake}, respectively. D_{real} is the average probability that the predicted result of the discriminator is an original HR image, and D_{fake} is the average probability that the predicted result of the discriminator is the generated SR image. They can be expressed as Equations (1) and (2).

$$D_{real} = C(I^{HR}) - E(C(I^{SR})) \tag{1}$$

$$D_{fake} = C(I^{SR}) - E(C(I^{HR})), \tag{2}$$

where $C(I)$ means discriminator output. $E(\cdot)$ means taking the average in the mini-batch data.

The loss of the discriminator D_{Ra} is denoted by L_D^{Ra}. It can be divided into two parts: $L_{D_{real}^{Ra}}$ and $L_{D_{fake}^{Ra}}$. Formulas of L_D^{Ra}, $L_{D_{real}^{Ra}}$ and $L_{D_{fake}^{Ra}}$ can be expressed as Equations (3)–(5), respectively.

$$L_D^{Ra} = L_{D_{real}^{Ra}} + L_{D_{fake}^{Ra}} \tag{3}$$

$$L_{D_{real}^{Ra}} = -E_{I^{HR}}[\log(D_{Ra}(I^{HR}, I^{SR}))] \tag{4}$$

$$L_{D_{fake}^{Ra}} = -E_{I^{SR}}[\log(1 - D_{Ra}(I^{SR}, I^{HR}))], \tag{5}$$

where $D_{Ra}(I^{HR}, I^{SR}) = \sigma(C(I^{HR}) - E_{I^{SR}}[C(I^{SR})])$, σ means sigmoid function.

The adversarial loss for generator G can be expressed as a symmetrical form as Equation (6).

$$\begin{aligned} L_G^{Ra} = & - E_{I^{HR}}[\log(1 - D_{Ra}(I^{HR}, I^{SR}))] \\ & - E_{I^{SR}}[\log(D_{Ra}(I^{SR}, I^{HR}))] \end{aligned} \tag{6}$$

Furthermore, the total loss of G is shown in Equation (7):

$$L_G = L_{perceptual} + \alpha L_G^{Ra} + \beta L_1, \tag{7}$$

where $L_1 = E_{I^{SR}}||I^{SR} - I^{HR}||_1$ is the content loss which is used to evaluate the 1-norm distance between the recovered image I^{SR} and the ground-truth I^{HR}. L_G^{Ra} is an adversarial loss for generator, and we choose SR-MINC loss [46] as an appropriate perceptual loss $L_{perceptual}$, which is based on a fine-tuned VGG model for objection recognition and focuses on textures instead of object [48]. α, β are the coefficients to balance different loss terms.

3.3. Datasets and Metrics

The crop disease images used in our experiments are obtained from Plant Village dataset [8], which includes 54,309 images of 14 kinds of crops, such as tomato, corn, grape, apple, and soybean (available at: https://github.com/spMohanty/PlantVillage-Dataset/tree/master/raw/color). Since tomato is one of the most produced crops and has the largest number of diseases in the Plant Village dataset, it is chosen as the target crop in this paper. The size of each image in Plant Village is 256 × 256 pixels (denoted as original HR images). The number of tomato images is up to 18,160 in this dataset. There are 9 kinds of tomato disease classes, as well as the healthy class, shown in Table 1.

Table 1. The number of each category tomato leaf images in Plant Village dataset.

No.	Name of Category	Number of Pictures
0	bacterial spot	2027
1	early blight	1000
2	late blight	1909
3	mold leaf	952
4	septoria leaf spot	1771
5	spider mites	1676
6	target spot	1404
7	tomato yellow curl virus	5357
8	tomato mosaic virus	373
9	healthy	1591

PSNR and SSIM [45] are two common metrics for evaluating the quality of images. They are frequently used to evaluate SR algorithms. PSNR between two images f and g with $m \times n$ pixels is defined as below Equation (8). A higher PSNR indicates better quality of generated images.

$$PSNR = 10 \cdot \log_{10}(\frac{255^2}{MSE}) \tag{8}$$

where

$$MSE = \frac{1}{mn}\sum_{i=0}^{m-1}\sum_{j=0}^{n-1}[f(i,j) - g(i,j)]^2. \tag{9}$$

And SSIM is calculated in Equation (10). Higher value of SSIM indicates better image quality.

$$SSIM = \frac{(2\mu_x\mu_y + C_1)(2\sigma_{xy} + C_2)}{(\mu_x^2\mu_y^2 + C_1)(\sigma_x^2 + \sigma_y^2 + C_2)}, \tag{10}$$

where x and y represent the 7 × 7 windows in image f and g, μ_x and μ_y represent the average value of x and y, σ_x^2 and σ_y^2 represent the variance of x and y, and σ_{xy} represents the covariance of x and y. C_1 and C_2 are variables to stabilize the division with weak denominators. Since we use RGB multi-channel images, these indices are calculated for each channel and then the average values of the channels are calculated.

However, several studies indicate that PSNR and SSIM cannot thoroughly evaluate perceptual-driven SR methods, such as SRGAN [34] and ESRGAN [23]. For this reason, Ledig et al. [34] proposed the mean opinion score (MOS) testing. In addition, Wang et al. [23] suggested applying PI in PIRM-SR Challenge [46] as an evaluation metric (more details in https://www.pirm2018.org/PIRM-SR.html). To better measure model performance, we also use PI for quantitative evaluation. Calculation of PI value depends on Ma's score [49] and NIQE [50]. The expression is shown below in Equation (11). A lower PI value represents better perceptual quality. In other words, the image is more real and natural. We use the MATLAB program provided by sponsors of the competition to calculate PI values.

$$PI = \frac{1}{2}((10 - Ma) + NIQE). \tag{11}$$

3.4. Crop Disease Classification

Since VGG16 [10] is a standard and straightforward image classification model, which performs well in the balance between training time and classification accuracy, it is chosen as the classifier in our experiments. We apply the classic VGG16 model, which consists of 13 convolution layers and 3 dense layers. The size of input and output layers of VGG16 is variable and adaptable. When the size of the input images changes, we need to change the setting of the width and height of the input layer of VGG16. In other words, the width and height of the input layer should be equal to the width and height of the input images. Similarly, the number of output classes should be equal to the number of neurons of the output layer. Specifically, if we perform a 6-class classification experiment with image size 64×64 pixels, the width and height of the input layer should be set to 64, and the number of neurons of the output layer should be set to 6. If we perform a 10-class classification experiment with image size 128×128 pixels, the width and height of the input layer are modified to 128, and the number of neurons in the output layer is modified to 10. Each layer is followed by ReLU activation function, which increases the non-linearity. Moreover, the MaxPooling layers are added to the second, fourth, seventh, tenth, and twelfth convolutional layers to reduce the dimension. Small filters with size 3×3 are applied to reduce the numbers of parameters and improve computational efficiency. Meanwhile, we fine-tune the VGG16 classification models trained on ImageNet with the Plant Village dataset, to achieve better classification performance and save computing resources.

4. Experiments

4.1. Experiment Setup

Most computations are conducted using python 3.5 on Ubuntu 16.04 system in our experiments. We implement the models with the PyTorch framework (version 1.1.0) and train them using a NVIDIA GeForce GTX 1070 GPU. A small part of image processing and PI calculation are carried out by MATLAB 2018a. We divide 18160 tomato leaf images from Plant Village database as training, validation, and testing sets, accounting for 80%, 10%, and 10%, respectively. All experiments apply a scaling factor of $\times 4$ between LR and HR images. The size of the original HR images is 256×256 pixels. Since a larger patch size requires more computing resources and training time, the cropped HR patch size is 128×128 pixels. Furthermore, cropped HR images are flipped and rotated for data augmentation. Since GPU memory is an issue, the batch size is set to 16. In future work, we will consider accumulating gradients across batches to optimize the training process and improve efficiency. SRCNN [20] consists of three convolutional layers, and the size of the kernel is 9×9, 1×1, and 5×5. Mean-square error (MSE) is used as the loss function of the model. We trained SRCNN on the Plant Village dataset for comparison.

4.2. Train with Transfer Learning

We use a pre-trained ESRGAN model provided by Wang et al. [23] to initialize the parameters for better quality and faster convergence (available on: https://github.com/xinntao/ESRGAN). This model is trained on ImageNet and does not work well in crop images. However, Wang only released the pre-trained generator G (denoted as G_{pre}) and did not release the pre-trained discriminator D. We fine-tune our model twice to compare the training performance in different training conditions. In the first fine-tuning, we use the pre-trained generator model G_{pre} as the initialization of our G and initialize D_{Ra} randomly. This causes an imbalance between the abilities of D_{Ra} and G. In other words, G's generation ability is strong, and D_{Ra}'s discriminative ability is poor. When the first fine-tuning finished, we got the trained G (denoted as G_1) and the trained D_{Ra} (denoted as D_{Ra1}).The turbulent orange training curves in Figure 5 indicates insufficient training of the first fine-tuning step. So we consider carrying out the second fine-tune training with different hyperparameters settings. In the second fine-tuning, we utilize G_1 model and D_{Ra1} as initialization of G and D_{Ra}. Because G_1 and D_{Ra1} have learned certain feature distribution, the discriminator becomes more powerful, and the abilities of G and D_{Ra} become relatively balanced. Thus, we get the G_2 and D_{Ra2} at the end of the second fine-tuning.

To be specific, in the first fine-tuning step, we train the generator G using the loss function in Equation (7) with $\alpha = 5 \times 10^{-3}$ and $\beta = 1 \times 10^{-2}$, where learning rate is set to 1×10^{-4} and halved at $[50k, 100k, 200k, 300k, 400k]$ iterations (learning rate decay factor $\gamma = 0.5$). The learning rate setting for discriminator is the same as the generator. We use Adam [51] with $\beta_1 = 0.9$ and $\beta_2 = 0.99$ as the optimizer of generator and discriminator. The maximum number of iterations is set to $500k$, and checkpoint is saved every $5k$ steps (Settings are referred to Reference [23]). It took about six days for the first fine-tuning. In the second fine-tuning process, we used the trained model G_1 as initialization for G and the corresponding D_{Ra1} as initialization for D_{Ra}. The learning rate of G and D_{Ra} is set to 5×10^{-5}, which is smaller than previous settings. Moreover, the learning rate is adjusted dynamically to help the model converge. The learning rate is halved at $[50k, 125k, 200k, 300k]$ iterations. Loss function coefficients are also modified: $\alpha = 1 \times 10^{-4}$ and $\beta = 5 \times 10^{-3}$. These settings emphasize the perceptual loss term. The maximum number of iterations is set to $400k$. Other settings remain unchanged. It took around five days for second fine-tuning.

Furthermore, since BN layers are removed to make training stable, training such a deep network becomes a problem. When the weights are updated, the distribution of the inputs in deep layers may change after each mini-batch, making the algorithm difficult to converge. To solve this problem, we use residual scaling strategy [11], which scales down the residuals by multiplying a constant between 0 and 1 before adding them to the main path to prevent instability. Using smaller initialization parameters in the residual structure can make training easier to converge.

The comparison of two fine-tuning steps is shown in Figure 5. The orange curves show the first fine-tuning process, and the blue ones show the second fine-tuning process. We can see that the blue curves are smoother than the orange curves, revealing that the second fine-tuning is more stable and reliable. Figure 5a,b represent the two average relativistic output of D_{Ra}: D_{real} and D_{fake}. In the second fine-tuning process, the value of D_{real} and D_{fake} finally stabilized at 30 and -30, respectively. And this indicates good training of D_{Ra}. l_g_per, l_g_gan, and l_g_con in Figure 5c–e, represent perceptual loss, adversarial loss, and content loss of G, respectively. It can be seen that the loss of the second training has decreased. PSNR is one of the metrics for evaluating SR methods. As shown in Figure 5f, compared to the first fine-tuning, the PSNR of the second fine-tuning is higher, which also reflects the good performance of the second fine-tuning. However, we can see that in the second training step, the average PSNR gradually decreases as the number of iterations increases. That is because the optimization goal of perceptual-driven SR methods is to minimize perceptual loss instead of mean squared reconstruction error (MSRE). This type of method sacrifices the PSNR performance in exchange for better image visual perception.

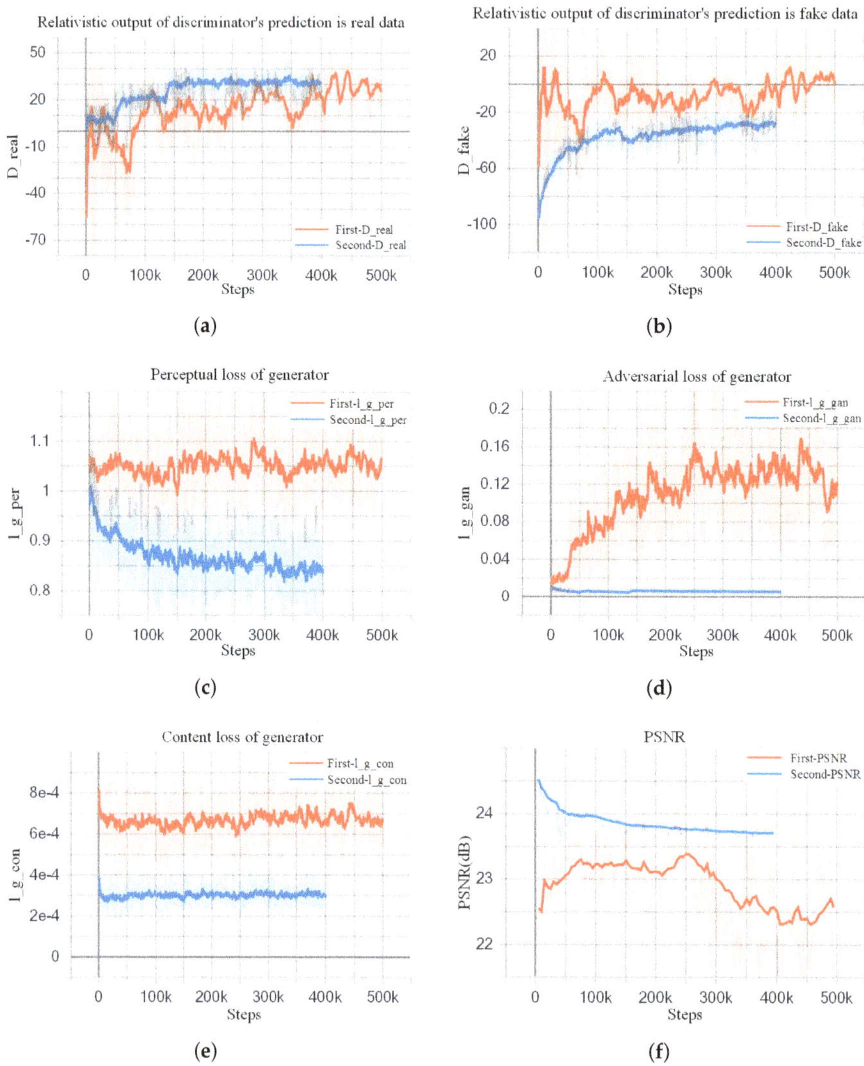

Figure 5. Comparison results of the first and second fine-tune training. In (**a**,**b**), The blue curves are smoother and have a larger mean absolute value of the difference between D_{real} and D_{fake} than orange ones, which indicates better training of D_{Ra} in second fine-tuning process. The sudden change of the discriminator output at 50k and 125k should be caused by the changes in the learning rate. In (**c**–**e**), blue curves are smoother with smaller absolute losses. In (**f**), compared to the first fine-tuning, the PSNR of the second fine-tuning is higher, which also reflects the good performance of the second fine-tuning.

An example of SR images generated by the pre-trained G_{pre}, first fine-tuned G_1, and the second fine-tuned G_2 can be seen in Figure 6. It can be observed that the image in Figure 6a only contains basic leaf shape and color information but lacks detailed information on lesions. After the first fine-tuning process, the image in Figure 6b is clearer and has sharper edges. However, it still lacks detailed information due to the different initialization strategies for the generator and the discriminator. The generated SR image from G_2 is realistic and natural, as shown in Figure 6c.

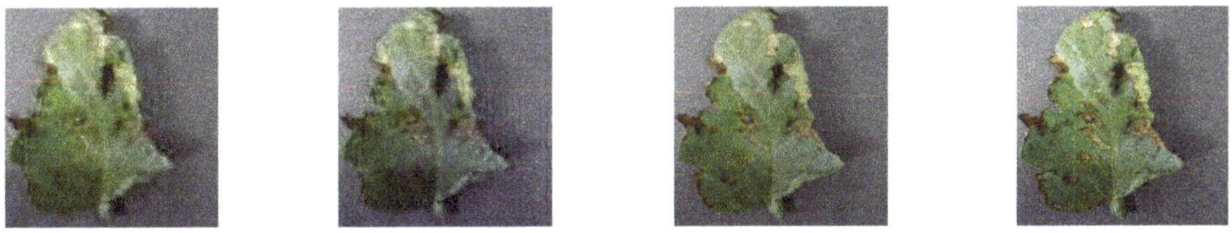

(a) (b) (c) (d)

Figure 6. Visual comparison of super-resolution (SR) images generated from three training stages. (**a**) is from the pre-trained G_{pre} based on ImageNet, (**b**) is from G_1 after the first fine-tuning and (**c**) is from G_2 after the second fine-tuning. (**d**) is the original high-resolution (HR) image.

4.3. Evaluation of the Generated SR Images

To evaluate the quality of the generated SR images, we display some test image results in Figure 7, in which PSNR (evaluated on all RGB channels), SSIM, and PI (evaluation index for PRIM-SR Challenge) are compared. Among them, "ESRGAN without ft" in the sixth column means the results for ESRGAN without fine-tuning. It can be seen in Figure 7 that the PI values of three generated SR images by our second fine-tuned ESRGAN (denoted as ft_ESRGAN) are the lowest. However, their PNSR and SSIM values are not the highest. That is because, unlike these PSNR-oriented approaches, ESRGAN is mainly minimizing perceptual loss to enhance visual quality instead of minimizing MSRE. Besides, our ft_ESRGAN achieves better visual performance with more natural and authentic textures than the other four approaches.

Figure 7. Examples of generated SR images. Our ft_ESRGAN produces sharper and more natural texture with richer visual information. "ESRGAN without ft" in the sixth column means ESRGAN without fine-tuning. And "ft_ESRGAN" in the seventh column means ESRGAN with second fine-tuning. [$\times 4$ upscaling].

We also calculate the average PSNR and SSIM of SR images generated by different SR methods from the test set (including 1812 images). The PI calculation is time-consuming, it takes about a minute to calculate PI value for one image. So we randomly choose 100 images from the test set (10 images are randomly chosen per category). The results are shown in Table 2. The average PNSR and SSIM of PSNR-oriented SRCNN are the highest, and the average PI of our perceptual-driven ft_ESRGAN is the lowest, which indicates that ft_ESRGAN could generate more realistic SR images with more comprehensive crop lesion details.

Table 2. The average Peak Signal-to-Noise Ratio (PSNR), structural similarity index (SSIM) and perceptual index (PI) of SR images generated by different SR methods.

Evaluation Index	Bilinear	Cubic	Lanczos4	SRCNN	ESRGAN without ft	ft_ESRGAN
PSNR(dB)	24.45	24.49	24.50	**25.42**	21.72	23.70
SSIM	0.4776	0.4710	0.4734	**0.5126**	0.3886	0.4678
PI	7.23	7.07	7.00	7.16	7.12	**6.10**

4.4. Classification Results

To verify whether the generated SR images by ft_ESRGAN contain rich information for classification, we conduct crop disease classification experiments on tomato leaves. Then we compare our model with the bilinear, cubic, lanczos4, and SRCNN. Considering the problem of data balance, we first choose 6 categories of tomato leaf images, each of which has a similar amount of samples. These 6 categories are bacterial spot (2027 images), late blight (1909), septoria leaf spot (1771), spider mites (1676), target spot (1404), and healthy (1591), respectively. The total number is 10,478 (see Table 1). Based on these original images, we conduct comparative experiments with different image sizes. By down-sampling HR images through bicubic kernel, we get two groups of LR images with 16×16 and 32×32 pixels. Then we reconstruct SR images using bilinear, cubic, lanczos4, SRCNN, and our ft_ESRGAN with a magnification scaling factor of $\times 4$. After reconstruction, we generate two groups of SR images with 64×64 and 128×128 pixels for each SR method. We also show the classification results on HR and LR images as the upper and lower bounds of the experiment.

In these classification experiments, the image samples are randomly divided to form the training, validation, and testing sets with a ratio of 0.8, 0.1, and 0.1. We use a VGG16 model trained on ImageNet as the initialization for our classifier. We modify the setting of the width and height of the input layer and the number of output classes of the output layer to fit our image sizes of this 6-class classification task. Stochastic Gradient Descent (SGD) is used for optimization, and the learning rate is set to be 1×10^{-4}. The maximum number of iterations is set to be 1×10^4. The 6-class classification results on the test set are shown in Table 3.

From Table 3, we can see that the classification accuracies through SR images are much higher than the ones through LR images. The proposed ft_ESRGAN achieves the highest accuracies, reaching 93.59% and 95.60% for SR images with the sizes 64×64 and 128×128 pixels, respectively. Moreover, classification performance based on deep learning methods (SRCNN and ft_ESRGAN) is better than the conventional image scaling methods (Bilinear, Cubic, and Lanczos4).

To further evaluate the classification performance of the proposed model on an unbalanced dataset, we also conduct a comparative experiment in all 10 categories (see Table 1) using a similar process. The learning rate is set to 5×10^{-5}, and the maximum of iterations is 1.5×10^4. The number of neurons in the output layer is modified to 10. Other settings are the same as the 6-class classification experiments. The 10-class classification results are shown in Table 4.

From Table 4, it can be observed that the classification accuracies on SR images are much higher compared with those on LR images under both image sizes. Moreover, the classification performance on the generated SR images obtained by our ft_ESRGAN model is better than other methods. The above experiments show that the proposed ft_ESRGAN model can generate images with useful and specific information for classification tasks.

Table 3. Comparison of classification results for low-resolution (LR) and SR images based on 6 categories.

	Sample Size	Total Numbers	Accuracy
LR	16 × 16	1045	56.94%
bilinear	64 × 64	1045	88.42%
cubic	64 × 64	1045	88.61%
lanczos4	64 × 64	1045	89.36%
SRCNN	64 × 64	1045	92.54%
ft_ESRGAN	64 × 64	**1045**	**93.59%**
HR	64 × 64	**1045**	**95.41%**
LR	32 × 32	1045	80.48%
bilinear	128 × 128	1045	93.30%
cubic	128 × 128	1045	92.82%
lanczos4	128 × 128	1045	93.78%
SRCNN	128 × 128	1045	94.26%
ft_ESRGAN	128 × 128	**1045**	**95.60%**
HR	128 × 128	**1045**	**97.80%**

Table 4. Comparison of classification results for LR and SR images based on 10 categories.

	Sample Size	Total Numbers	Accuracy
LR	16 × 16	1812	52.65%
bilinear	64 × 64	1812	71.14%
cubic	64 × 64	1812	73.01%
lanczos4	64 × 64	1812	72.79%
SRCNN	64 × 64	1812	80.13%
ft_ESRGAN	64 × 64	**1812**	**85.38%**
HR	64 × 64	**1812**	**89.96%**
LR	32 × 32	1812	72.74%
bilinear	128 × 128	1812	82.40%
cubic	128 × 128	1812	81.02%
lanczos4	128 × 128	1812	83.44%
SRCNN	128 × 128	1812	86.09%
ft_ESRGAN	128 × 128	**1812**	**90.78%**
HR	128 × 128	**1812**	**95.14%**

From Tables 3 and 4, we can see that classification accuracy on LR images is the lowest. It reveals that LR images contain less useful information that can be captured by VGG16 for classification than SR or HR ones. Besides, because the size of the LR images is smaller than the size of SR and HR images, VGG16 may not be well trained for LR images due to its large amount of parameters, resulting in low classification accuracy. That is to say, VGG16 may not be a good tool for classifying the LR images. In this paper, the LR image accuracy is considered as a lower bound for classification, helping us to study the impact of SR methods for the classification tasks.

To study the classification accuracy on each category, we show the confusion matrix for the second group of 10-class classification experiment (LR images: 32 × 32 pixels, SR and HR images: 128 × 128 pixels) in Figure 8. The results are normalized to 0–1 by the number of elements in each category. From Figure 8, We can see the classification accuracy gradually increases from LR to SR to HR. Among the chosen SR methods, our ft_ESRGAN performance is closest to the upper bound—the classification performance on HR images. Healthy class is the easiest category to identify. Furthermore, class 1 (early blight) and class 2 (late blight) are quite confounding. Similarly, class 0 (bacterial spot), class 4 (septoria leaf spot), and class 6 (target spot) are hard to distinguish from each other, too.

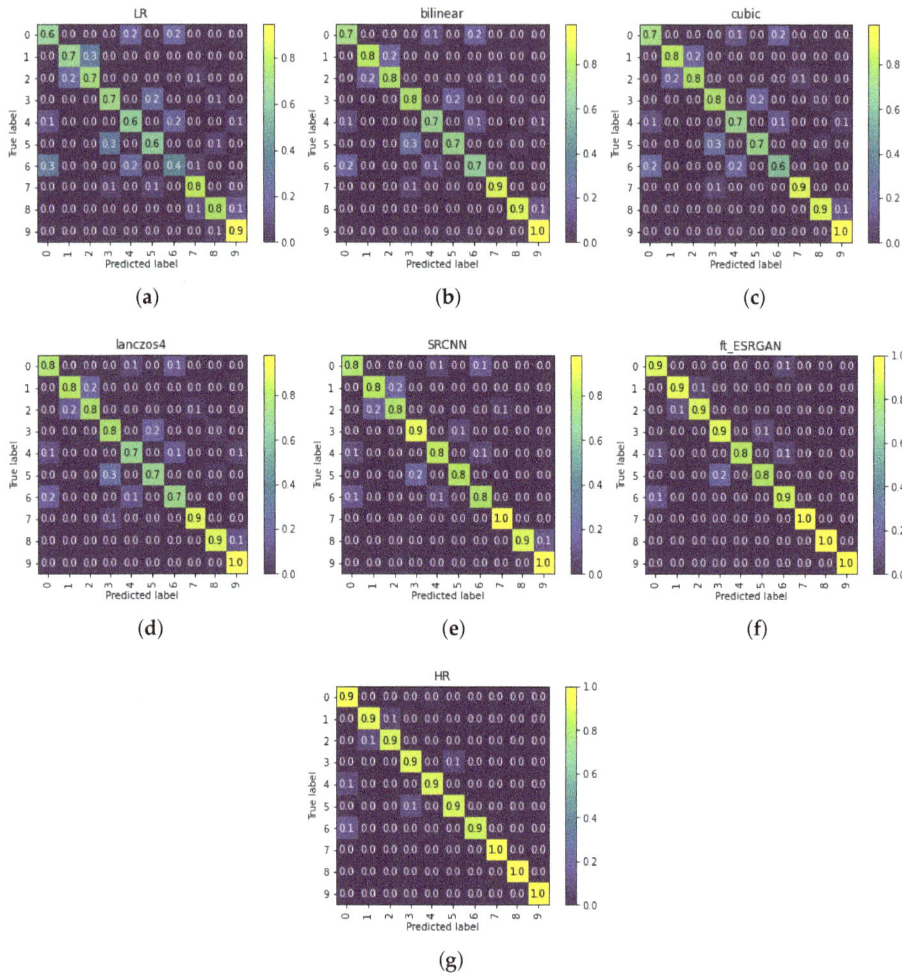

Figure 8. Confusion matrix of disease classification using using LR (32 × 32 pixels), SR (128 × 128 pixels) and HR images (128 × 128 pixels). Numbers on x and y axes indicate the ID of diseases in Table 1. (**a**) LR; (**b**) Bilinear; (**c**) Cubic; (**d**) Lanczos4; (**e**) SRCNN; (**f**) ft_ESRGAN; (**g**) HR.

5. Conclusions

In this paper, we have proposed a method for crop disease identification on LR images by transferring LR images to SR images based on GAN. First, we employ ESRGAN on LR images to generate the corresponding SR images. Due to insufficient crop data, we apply transfer learning to fine-tune the model trained on ImageNet. After two fine-tuning steps, our SR model reaches a stable state, and the generated images achieve an excellent visual effect. Then we conduct disease classification experiments using the generated SR images. Experimental results show that the classification accuracy can be significantly improved by applying the proposed SR model, indicating that our SR model can reconstruct the useful information for identifying crop diseases. Due to the powerful reconstruction ability of ESRGAN, the performance achieved by the proposed model is better than those achieved by the other four methods. In our research, we utilized disease images taken by ground cameras rather than UAV cameras. Although our approach should be effective on UAV images, it is still necessary to verify our approach to images from UAV cameras for practical application in future works. Besides, The training efficiency and generalization ability of the model can be further improved. Furthermore, we can apply the SR model in object detection tasks. In this way, we can detect multiple diseases on one crop images.

Author Contributions: Conceptualization, J.W. and Y.X.; Data curation, X.Z.; Formal analysis, J.W. and X.Z.; Funding acquisition, J.W.; Investigation, Y.S.; Methodology, J.W. and Y.S.; Project administration, Y.X.; Resources, J.W. and Y.X.; Software, Y.S. and X.Z.; Supervision, Y.X.; Validation, Y.S. and X.Z.; Visualization, Y.S. and X.Z.; Writing—original draft, Y.S.; Writing—review & editing, J.W. and Y.X. All authors have read and agreed to the published version of the manuscript.

Funding: This work was supported by the National Natural Science Foundation of China (No.61802410), and the Chinese Universities Scientific Fund (2018XD002 & 2018QC024).

Conflicts of Interest: The authors declare no conflict of interest.

References

1. Harvey, C.A.; Zo Lalaina, R.; Rao, N.S.; Radhika, D.; Hery, R.; Rivo Hasinandrianina, R.; Haingo, R.; Mackinnon, J.L. Extreme vulnerability of smallholder farmers to agricultural risks and climate change in Madagascar. *Philos. Trans. R. Soc. Lond. Ser. A* **2014**, *369*, 20130089. [CrossRef]
2. Tai, A.P.K.; Martin, M.V.; Heald, C.L. Threat to future global food security from climate change and ozone air pollution. *Nat. Clim. Chang.* **2014**, *4*, 817–821. [CrossRef]
3. Alsuwaidi, A.; Grieve, B.; Yin, H. Feature-Ensemble-Based Novelty Detection for Analyzing Plant Hyperspectral Datasets. *IEEE J. STARS* **2018**, *PP*, 1–15. [CrossRef]
4. Pantazi, X.; Moshou, D.; Tamouridou, A.A. Automated leaf disease detection in different crop species through image features analysis and One Class Classifiers. *Comput. Electron. Agric.* **2019**, *156*, 96–104. [CrossRef]
5. Jia. S.; Jia, P.; Hu, S.; Liu, H. Automatic detection of tomato diseases and pests based on leaf images. In Proceedings of the 2017 Chinese Automation Congress (CAC), Jinan, China, 20–22 October 2017; pp. 2537–2510. [CrossRef]
6. Zhang, S.; Zhang, S.; Zhang, C.; Wang, X.; Shi, Y. Cucumber leaf disease identification with global pooling dilated convolutional neural network. *Comput. Electron. Agric.* **2019**, *162*, 422–430. [CrossRef]
7. Singh, U.P.; Chouhan, S.S.; Jain, S.; Jain, S. Multilayer Convolution Neural Network for the Classification of Mango Leaves Infected by Anthracnose Disease. *IEEE Access* **2019**, *7*, 43721–43729. [CrossRef]
8. Hughes, D.P.; Salathe, M. An open access repository of images on plant health to enable the development of mobile disease diagnostics through machine learning and crowdsourcing. *arXiv* **2015**, arXiv:1511.08060v2.
9. Too, E.C.; L.Y.; Njuki, S.; Liu, Y. A comparative study of fine-tuning deep learning models for plant disease identification. *Comput. Electron. Agric.* **2018**, 272–279. [CrossRef]
10. Simonyan, K.; Zisserman, A. Very Deep Convolutional Networks for Large-Scale Image Recognition. *arXiv* **2014**, arXiv 1409.1556.
11. Szegedy, C.; Ioffe, S.; Vanhoucke, V.; Alemi, A. Inception-v4, Inception-ResNet and the Impact of Residual Connections on Learning. *arXiv* **2016**, arXiv:1602.07261.
12. He, K.; Zhang, X.; Ren, S.; Sun, J. Deep Residual Learning for Image Recognition. In Proceedings of the IEEE Conference on Computer Vision and Pattern Recognition (CVPR) 2016, Las Vegas, NV, USA, 27–30 June 2016; pp. 770–778. [CrossRef]
13. Huang, G.; Liu, Z.; van der Maaten, L.; Weinberger, K.Q. Densely Connected Convolutional Networks. In Proceedings of the 2017 IEEE Conference on Computer Vision and Pattern Recognition (CVPR), Honolulu, HI, USA, 21–26 July 2016; pp. 2261–2269. [CrossRef]
14. Su, J.; Liu, C.; Coombes, M.; Hu, X.; Wang, C.; Xu, X.; Li, Q.; Guo, L.; Chen, W.-H. Wheat yellow rust monitoring by learning from multispectral UAV aerial imagery. *Comput. Electron. Agric.* **2018**, *155*, 157–166. [CrossRef]
15. Cao, F.; Liu, F.; Guo, H.; Kong, W.; Zhang, C.; Yong, H. Fast Detection of Sclerotinia Sclerotiorum on Oilseed Rape Leaves Using Low-Altitude Remote Sensing Technology. *Sensors* **2018**, *18*, 4464. [CrossRef] [PubMed]
16. Kerkech, M.; Hafiane, A.; Canals, R. Deep learning approach with colorimetric spaces and vegetation indices for vine diseases detection in UAV images. *Comput. Electron. Agric.* **2018**, *155*, 237–243. [CrossRef]

17. Abdulridha, J.; Batuman, O.; Ampatzidis, Y. UAV-Based Remote Sensing Technique to Detect Citrus Canker Disease Utilizing Hyperspectral Imaging and Machine Learning. *Remote Sens.* **2019**, *11*, 1373. [CrossRef]
18. Torres-Sánchez, J.; López-Granados, F.; De Castro, A.; Peña-Barragán, J.M. Configuration and Specifications of an Unmanned Aerial Vehicle (UAV) for Early Site Specific Weed Management. *PLoS ONE* **2013**, *8*, e58210. [CrossRef]
19. Yamamoto, K.; Togami, T.; Yamaguchi, N. Super-Resolution of Plant Disease Images for the Acceleration of Image-based Phenotyping and Vigor Diagnosis in Agriculture. *Sensors* **2017**, *17*, 2557. [CrossRef]
20. Chao, D.; Chen, C.L.; He, K.; Tang, X. Learning a Deep Convolutional Network for Image Super-resolution. In Proceedings of the European Conference on Computer Vision 2014—ECCV 2014, Zurich, Switzerland, 6–12 September 2014; pp. 184–199. [CrossRef]
21. Cap, Q.H.; Tani, H.; Uga, H.; Kagiwada, S.; Iyatomi, H. Super-Resolution for Practical Automated Plant Disease Diagnosis System. In Proceedings of the 2019 53rd Annual Conference on Information Sciences and Systems (CISS), Baltimore, MD, USA, 20–22 March 2019; pp. 184–199. [CrossRef]
22. Goodfellow, I.J.; Pouget-Abadie, J.; Mirza, M.; Bing, X.; Warde-Farley, D.; Ozair, S.; Courville, A.; Bengio, Y. Generative Adversarial Nets. In Proceedings of the Advances in Neural Information Processing Systems 27 (NIPS 2014), Montreal, QC, Canada, 8–13 December 2014; pp. 2672–2680.
23. Wang, X.; Yu, K.; Wu, S.; Gu, J.; Liu, Y.; Dong, C.; Qiao, Y.; Loy, C.C. ESRGAN: Enhanced Super-Resolution Generative Adversarial Networks. In Proceedings of the European Conference on Computer Vision (ECCV) 2018, Munich Germany, 8–14 September 2018; pp. 63–79. [CrossRef]
24. Deng, J.; Dong, W.; Socher, R.; Li, L.J.; Li, K.; Li, F.F. ImageNet: A Large-Scale Hierarchical Image Database. In Proceedings of the IEEE Conference on Computer Vision and Pattern Recognition (CVPR), Miami, FL, USA, 20–25 June 2009; pp. 248–255. [CrossRef]
25. Glasner, D.; Bagon, S.; Irani, M. Super-resolution from a single image. In Proceedings of the IEEE International Conference on Computer Vision (ICCV), Kyoto, Japan, 29 September–2 October 2009; pp. 349–356. [CrossRef]
26. Lai, W.S.; Huang, J.B.; Ahuja, N.; Yang, M.H. Deep Laplacian Pyramid Networks for Fast and Accurate Super-Resolution. In Proceedings of the IEEE Conference on Computer Vision and Pattern Recognition (CVPR), Honolulu, HI, USA, 21–26 July 2017; pp. 624–632. [CrossRef]
27. Kim, J.; Lee, J.K.; Lee, K.M. Deeply-Recursive Convolutional Network for Image Super-Resolution. In Proceedings of the IEEE Conference on Computer Vision and Pattern Recognition (CVPR), Las Vegas, NV, USA, 27–30 June 2016; pp. 1637–1645. [CrossRef]
28. Tai, Y.; Yang, J.; Liu, X. Image Super-Resolution via Deep Recursive Residual Network. In Proceedings of the IEEE Computer Vision and Pattern Recognition (CVPR), Honolulu, HI, USA, 21–26 July 2017; pp. 2790–2798. [CrossRef]
29. Tai, Y.; Yang, J.; Liu, X.; Xu, C. MemNet: A Persistent Memory Network for Image Restoration. In Proceedings of the 2017 IEEE International Conference on Computer Vision (ICCV), Venice, Italy, 22–29 October 2017; pp. 4549–4557. [CrossRef]
30. Haris, M.; Shakhnarovich, G.; Ukita, N. Deep Back-Projection Networks For Super-Resolution. In Proceedings of the IEEE Conference on Computer Vision and Pattern Recognition (CVPR), Salt Lake City, UT, USA, 18–22 June 2018; pp. 1664–1673. [CrossRef]
31. Zhang, Y.; Tian, Y.; Kong, Y.; Zhong, B.; Fu, Y. Residual Dense Network for Image Super-Resolution. In Proceedings of the IEEE Conference on Computer Vision and Pattern Recognition (CVPR), Salt Lake City, UT, USA, 18–22 June 2018; pp. 2472–2481. [CrossRef]
32. Zhang, Y.; Li, K.; Li, K.; Wang, L.; Zhong, B.; Fu, Y. Image Super-Resolution Using Very Deep Residual Channel Attention Networks. In Proceedings of the European Conference on Computer Vision (ECCV), Munich, Germany, 8–14 September 2018; pp. 2–16.
33. Pourya Shamsolmoali, X.L.; Wang, R. Single image resolution enhancement by efficient dilated densely connected residual network. *Signal Process. Image Commun.* **2019**, *79*, 13–23. [CrossRef]
34. Ledig, C.; Theis, L.; Huszar, F.; Caballero, J.; Aitken, A.; Tejani, A.; Totz, J.; Wang, Z.; Shi, W. Photo-Realistic Single Image Super-Resolution Using a Generative Adversarial Network. In Proceedings of the IEEE Computer Vision and Pattern Recognition (CVPR), Las Vegas, NV, USA, 26 June–1 July 2016; pp. 105–114. [CrossRef]

35. Johnson, J.; Alahi, A.; Li, F.F. Perceptual losses for real-time style transfer and super-resolution. In Proceedings of the European Conference on Computer Vision (ECCV), Amsterdam, The Netherlands, 11–14 October 2016; pp. 694–711. [CrossRef]
36. Jolicoeur-Martineau, A. The relativistic discriminator: a key element missing from standard GAN. *arXiv* **2018**, arXiv:1807.00734.
37. Wang, Y.; Su, F.; Qian, Y. Text-Attentional Conditional Generative Adversarial Network for Super-Resolution of Text Images. In Proceedings of the 2019 IEEE International Conference on Multimedia and Expo (ICME), Shanghai, China, 8–12 July 2019; pp. 1024–1029. [CrossRef]
38. Ray, A.; Sharma, M.; Upadhyay, A.; Makwana, M.; Chaudhury, S.; Trivedi, A.; Singh, A.; Saini, A. An End-to-End Trainable Framework for Joint Optimization of Document Enhancement and Recognition. In Proceedings of the 2019 International Conference on Document Analysis and Recognition (ICDAR), Sydney, Australia, 22–25 September 2019.
39. Xi, Y.; Zheng, J.; Jia, W.; He, X.; Li, H.; Ren, Z.; Lam, K.M. See Clearly in the Distance: Representation Learning GAN for Low Resolution Object Recognition. *IEEE Access* **2020**, [CrossRef]
40. Chen, S.; Zhou, F.; Liao, Q. Visual domain adaptation using weighted subspace alignment. In Proceedings of the 2016 Visual Communications and Image Processing (VCIP), Chengdu, China, 27–30 November 2016; pp. 1–4. [CrossRef]
41. Xiao, M.; Guo, Y. Feature Space Independent Semi-Supervised Domain Adaptation via Kernel Matching. *IEEE Trans. Pattern Anal. Mach. Intell.* **2015**, *37*, 54–66. [CrossRef]
42. Ramachandran, A.; Gupta, S.; Rana, S.; Venkatesh, S. Information-theoretic transfer learning framework for Bayesian optimisation. In Proceedings of the 2018 European Conference on Machine Learning (ECML), Dublin, Ireland, 10–14 September 2018; pp. 827–842.
43. Wen, L.; Gao, L.; Li, X. A New Deep Transfer Learning Based on Sparse Auto-Encoder for Fault Diagnosis. *IEEE Trans. Syst. Man Cybern. Syst.* **2017**, *49*, 136–144. [CrossRef]
44. Hu, L.; Kan, M.; Shan, S.; Chen, X. Duplex Generative Adversarial Network for Unsupervised Domain Adaptation. In Proceedings of the IEEE Conference on Computer Vision and Pattern Recognition (CVPR), Salt Lake City, UT, USA, 18–22 June 2018; pp. 1498–1507. [CrossRef]
45. Wang, Z.; Bovik, A.C.; Sheikh, H.R.; Simoncelli, E.P. Image quality assessment: from error visibility to structural similarity. *IEEE Trans. Image Process.* **2004**, *13*, 600–612. [CrossRef] [PubMed]
46. Blau, Y.; Mechrez, R.; Timofte, R.; Michaeli, T.; Zelnik-Manor, L. 2018 PIRM Challenge on Perceptual Image Super-resolution. In Proceedings of the European Conference on Computer Vision (ECCV), Munich, Germany, 8–14 September 2018.
47. He, K.; Zhang, X.; Ren, S.; Sun, J. Identity Mappings in Deep Residual Networks. In Proceedings of the European Conference on Computer Vision (ECCV), Amsterdam, The Netherlands, 11–14 October 2016; pp. 630–645.
48. Bell, S.; Upchurch, P.; Snavely, N.; Bala, K. Material Recognition in the Wild with the Materials in Context Database. In Proceedings of the IEEE Conference on Computer Vision and Pattern Recognition (CVPR), Boston, MA, USA, 7–12 June 2015; pp. 3479–3487.
49. Ma, C.; Yang, C.Y.; Yang, X.; Yang, M.H. Learning a No-Reference Quality Metric for Single-Image Super-Resolution. *Comput. Vis. Image Underst.* **2016**, *158*, 1–16. [CrossRef]
50. Mittal, A.; Soundararajan, R.; Bovik, A. Making a "Completely Blind" Image Quality Analyzer. *IEEE Signal Process Lett.* **2013**, *20*, 209–212. [CrossRef]
51. Kingma, D.; Ba, J. Adam: A Method for Stochastic Optimization. *arXiv* **2014**, arXiv:1412.6980.

sensors

MDPI

Article

Flash-Flood Potential Mapping Using Deep Learning, Alternating Decision Trees and Data Provided by Remote Sensing Sensors

Romulus Costache [1,2], Alireza Arabameri [3,*], Thomas Blaschke [4], Quoc Bao Pham [5,6,*], Binh Thai Pham [7], Manish Pandey [8,9], Aman Arora [10], Nguyen Thi Thuy Linh [11,12] and Iulia Costache [13]

1 Research Institute of the University of Bucharest, 90-92 Sos. Panduri, 5th District, 050663 Bucharest, Romania; romulus.costache@icub.unibuc.ro
2 National Institute of Hydrology and Water Management, București-Ploiești Road, 97E, 1st District, 013686 Bucharest, Romania
3 Department of Geomorphology, Tarbiat Modares University, Tehran 36581-17994, Iranmailto
4 Department of Geoinformatics–Z_GIS, University of Salzburg, 5020 Salzburg, Austria; Thomas.Blaschke@sbg.ac.at
5 Environmental Quality, Atmospheric Science and Climate Change Research Group, Ton Duc Thang University, Ho Chi Minh City 700000, Vietnam
6 Faculty of Environment and Labour Safety, Ton Duc Thang University, Ho Chi Minh City 700000, Vietnam
7 Geotechnical Engineering Deparment, University of Transport Technology, Hanoi 100000, Vietnam; binhpt@utt.edu.vn
8 University Center for Research & Development (UCRD), Chandigarh University, Punjab 140413, India; manish07sep@gmail.com
9 Department of Civil Engineering, University Institute of Engineering, Chandigarh University, Punjab 140413, India
10 Department of Geography, Faculty of Natural Sciences, Jamia Millia Islamia, New Delhi 110025, India; aman.jmi01@gmail.com
11 Institute of Research and Development, Duy Tan University, Danang 550000, Vietnam; nguyentthuylinh58@duytan.edu.vn
12 Faculty of Environmental and Chemical Engineering, Duy Tan University, Danang 550000, Vietnam
13 Faculty of Geography, University of Bucharest, Bd. Nicolae Bălcescu No 1, 1st District, 010041 Bucharest, Romania; iulia.elena.costache@gmail.com
* Correspondence: a.arabameri@modares.ac.ir (A.A.); phambaoquoc@tdtu.edu.vn (Q.B.P.)

Citation: Costache, R.; Arabameri, A.; Blaschke, T.; Pham, Q.B.; Pham, B.T.; Pandey, M.; Arora, A.; Linh, N.T.T.; Costache, I. Flash-Flood Potential Mapping Using Deep Learning, Alternating Decision Trees and Data Provided by Remote Sensing Sensors. *Sensors* **2021**, *21*, 280. https://doi.org/10.3390/s21010280

Received: 6 November 2020
Accepted: 22 December 2020
Published: 4 January 2021

Abstract: There is an evident increase in the importance that remote sensing sensors play in the monitoring and evaluation of natural hazards susceptibility and risk. The present study aims to assess the flash-flood potential values, in a small catchment from Romania, using information provided remote sensing sensors and Geographic Informational Systems (GIS) databases which were involved as input data into a number of four ensemble models. In a first phase, with the help of high-resolution satellite images from the Google Earth application, 481 points affected by torrential processes were acquired, another 481 points being randomly positioned in areas without torrential processes. Seventy percent of the dataset was kept as training data, while the other 30% was assigned to validating sample. Further, in order to train the machine learning models, information regarding the 10 flash-flood predictors was extracted in the training sample locations. Finally, the following four ensembles were used to calculate the Flash-Flood Potential Index across the Bâsca Chiojdului river basin: Deep Learning Neural Network–Frequency Ratio (DLNN-FR), Deep Learning Neural Network–Weights of Evidence (DLNN-WOE), Alternating Decision Trees–Frequency Ratio (ADT-FR) and Alternating Decision Trees–Weights of Evidence (ADT-WOE). The model's performances were assessed using several statistical metrics. Thus, in terms of Sensitivity, the highest value of 0.985 was achieved by the DLNN-FR model, meanwhile the lowest one (0.866) was assigned to ADT-FR ensemble. Moreover, the specificity analysis shows that the highest value (0.991) was attributed to DLNN-WOE algorithm, while the lowest value (0.892) was achieved by ADT-FR. During the training procedure, the models achieved overall accuracies between 0.878 (ADT-FR) and 0.985 (DLNN-WOE). K-index shows again that the most performant model was DLNN-WOE (0.97). The Flash-Flood Potential Index (FFPI) values revealed that the surfaces with high and very high flash-flood susceptibility cover between 46.57% (DLNN-FR) and 59.38% (ADT-FR) of the study zone. The use of the Receiver Operating Characteristic (ROC) curve for results validation highlights the fact that $FFPI_{DLNN\text{-}WOE}$ is characterized by the most precise results with an Area Under Curve of 0.96.

Keywords: flash-flood potential index; remote sensing sensors; bivariate statistics; deep learning neural network; alternating decision trees; ensemble models

1. Introduction

In recent decades, climate change and its related phenomena, e.g., flash floods, have had significant negative effects worldwide for both human society and environment [1]. The extreme rainfalls, extreme river discharge values, and therefore the flash-flood risk are characterized by a continuous increasing trend [2]. This trend is also validated by the huge amount of damages that flash floods generate worldwide. Therefore, an increasing number of studies in the literature approaching the subject of flash-flood susceptibility can be also observed [3–6]. Moreover, the estimation of flood risk and vulnerability became an essential and mandatory procedure which should be included in the Flood Risk Management strategy [7]. In this regard, the Geographic Informational Systems (GIS) and Remote Sensing (RS) techniques represent the necessary tools, which facilitate the spatial modelling and mapping of flash-flood susceptible areas. It is worth emphasizing the crucial role of Remote Sensing sensors in the observation's campaigns conducted for the identification of areas already affected by flash-flood processes [8]. Thus, without the RS sensors, the correct inventory of the torrential areas, which favor the occurrence of flash flood, will be impossible. Consideration of the previously affected areas and their involvement as input data in more advanced techniques such as machine learning or bivariate statistics, is of a real help to estimate as accurate as possible the flash-flood susceptibility within a specific catchment [9].

In recent years, new techniques and models have been developed by researchers worldwide [10–35]. During the last 6 years, several studies have been individualized regarding the flash-flood susceptibility investigations, which were carried out through the integration of GIS techniques with bivariate statistical models such as: frequency ratio [36], weights of evidence [37], statistical index [38], evidential belief function [39], certainty factor [40], or index of entropy [41]. Another category of methods successfully used in this type of study are those included in Multicriteria Decision Making such as: Technique for Order Preference by Similarity to Ideal Solution (TOPSIS) [42], Analytical Hierarchy Process (AHP) [43], Analytical Network Process (ANP) [44] or Vlse Kriterijuska Optamizacija I Komoromisno Resenje (VIKOR) [45]. Promising results in terms of flash-flood susceptibility were also provided by machine learning models such as: logistic regression [46], naïve bayes [47], artificial neural network [48], random forest [49,50], support vector machine [51], neuro-fuzzy inference system [52], *k*-nearest neighbor [53] or deep learning neural network [54]. The attempts of researchers to combine models from the same category or from different categories to generate ensemble algorithms that are considered much more accurate than the stand-alone ones should also be noted [55]. In this regard, the following examples can be provided: Fuzzy Unordered Rules Induction Algorithm (FURIA) [3], Bayesian-based machine learning models [9], machine learning and multicriteria decision making ensembles [7], machine learning and bivariate statistics ensembles [56].

Taking into account the previously presented aspects, the main purpose of the proposed research work is to estimate the susceptibility to flash floods in the basin of the Bâsca Chiojdului river from Romania. Estimation of flash-flood exposure will be based on the data collected using Remote Sensing sensors and the GIS database and their use in a number of four ensemble models generated by combining bivariate statistics with deep learning neural networks and alternating decision trees. Thus, on the one hand, the Frequency Ratio and Weights of Evidence bivariate statistical models will be used; these being combined with deep learning neural network and alternating decision trees. The construction of Receiver Operating Characteristic (ROC) curve and the calculation of several statistical metrics will ensure the validation of the results and the evaluation of the models' performances. It is worthwhile to note that the present study is intended to enrich the scientific literature regarding the flash-flood susceptibility assessment by proposing, for the first time in the literature, the combination above mentioned of four machine learning ensemble models with the GIS and remote sensing techniques.

2. Study Area

The Bâsca Chiojdului river basin from Romania, on which the present research is focused, has a total area of 340 km^2. The basin has an elevation which varies from 242 m to 1463 m, and a slope angle with an average value of 12.3°. It should be noted that a percentage of 79% of the total area is characterized by slope angles higher than 7° [57]. The circularity ratio, that is another important feature with a high influence on flash-flood susceptibility, has a value of 0.46, while the river basin concentration time is 7.27 h [36]. The low concentration time highlights a high predisposition of the study area to the flash-flood events. The forest vegetation covers a total percentage of 50%, while in terms of the soil component, the hydrological group B accounts for approximately 41% of the total research area.

The lithology consists mainly of the sedimentary rocks included in the Paleogene and Cretaceous flysch. The climate is characterized by a high continentalism degree, and especially in the warm season, the heavy rainfalls often lead to severe flash-flood phenomena. Due to the geographical characteristics of the Bâsca Chiojdului river basin, the socio-economic elements located across its territory suffered material losses following the flash-flood propagation. The most important flash-flood event occurred in 1975, when the maximum discharge value (300 m^3/s) of the Bâsca Chiojdului river reached the historical maximum [57]. More information regarding the main flash floods occurred across the study area, as well as the damages caused by these phenomena, can be found in the research works carried out by: Costache and Zaharia [10], Prăvălie and Costache [57], Costache et al. [38], Zarea and Gheorghe [58], Prăvălie and Costache [59].

3. Data

In order to carry out the present study, data consisting of torrential areas polygons and flash-flood predictors were gathered.

3.1. Torrential Area Inventory and Sampling

The inventory of surfaces previously affected by a specific process is essential for an accurate prediction of the areas where that phenomenon can occur in the future [60]. In the present research work, we consider the torrential surfaces as the spatial indicator for the areas with a high susceptibility for flash-flood genesis. In order to identify, as accurate as possible, the areas affected by torrential phenomena, analysis of the images provided by the Remote Sensing sensors was mandatory. This fact highlights the crucial role that this type of sensor has in the analysis of natural hazard susceptibility. Thus, using the Google Earth imagery a total area of 34 km^2 was delimited. These surfaces were created by the accelerated surface runoff occurring on the slopes. The manner in which these surfaces are delineated is described in the study carried out by Costache [61]. According to Costache and Zaharia [8], the torrential areas are defined as the areas characterized by the unified presence of torrential microform of relief such as ravines and gullies, which are generated by surface runoff. They are located in the upper part of the river basin, where the absence of vegetation and the high slopes favor the production of such phenomena. In order to be taken into account in the present study, a sample of 481 points representing locations where the torrential runoff took place was extracted from the entire delimited area. Moreover, another sample of 481 points was placed within the study area, representing points without torrential processes (Figure 1). Both torrential pixels and non-torrential pixels were divided into training (70%) and validating (30%) samples. This division was necessary in order to train the models and then to validate the results regarding the susceptibility to flash floods.

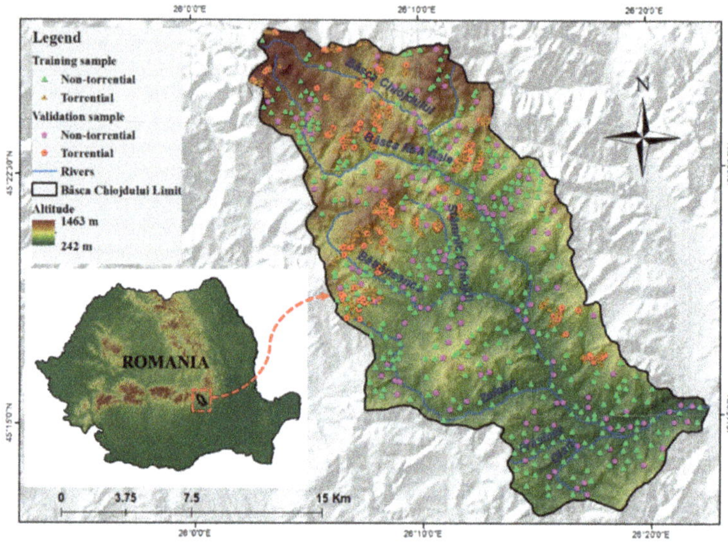

Figure 1. Study area location.

3.2. Flash-Flood Predictors

For the realization of this study, a number of 10 flash-flood conditioning factors were taken into account. Their main properties are described in the following lines. **Slope angle** was calculated using the Digital Elevation Model (DEM) taken from Shuttle Radar Topographic Mission (SRTM) 30 m database and processed in ArcGIS 10 software. A high value of slope angle will influence in a positive water runoff velocity, while the low values of the same parameter will be restrictive for the surface runoff occurrence [56]. For the study area, the map of slope angle was designed by splitting its range of values into five classes as following [12]: <3°; 3°–7°; 7.1°–15°; 15.1°–25°; >25° (Figure 2a). Another water surface runoff predictor is represented by the **Topographic Wetness Index (TWI)** calculated by the DEM processing in SAGA GIS 2.1.0. The algorithm used to calculate this index requires the use of the area upslope to each pixel and the tangent value of the slope value recorded in the same pixel [53]. The generation of TWI map was possible following the partition of its values into the next five classes using *Natural Breaks* method: 3.15–6.1, 6.11–7.78, 7.79–10.21, 10.22–14.5, 14.51–24.59 (Figure 2b). **Topographic Position Index (TPI)** is a mandatory flash-flood predictor which should be involved in the susceptibility related studies because its values emphasize the altitude difference between the location of a specific point and its neighboring area [62]. This important morphometric indicator was achieved at a spatial resolution of 30 m and its values ranging from −20 to 20 were divided into the next five classes using *Natural Breaks* method: (−20)–(−3.8), (−3.7)–(−1.1), (−1.1)–1.3, 1.4–4.5, 4.6–20 (Figure 2c). **Profile curvature** is mainly used to delineate the surfaces on which an accelerated surface runoff is manifested from those on which a decelerated surface runoff occurs [63]. According to the literature [38], positive profile curvature is characteristic for areas with a decelerated water runoff, while the negative values show the surfaces that increase the water runoff velocity. Across the study area, the profile curvature was classified into the following three intervals: (−3)–0, 0.1–0.9, 1–2 (Figure 2d). The ability of **convergence index** morphometric factor consists of the differentiation of the areas belonging the river valleys from those which are situated along the interfluvial lines. This index, achieved by DEM processing in SAGA GIS 2.1.0, was classified according to the literature: (−99)–(−3), (−2.9)–(−2), (−1.9)–(−1), (−0.9)–0, 0–99 (Figure 2e). **Stream Power Index (SPI)** is another morphometric factor that is generated in SAGA GIS 2.1.0 based on the values of upslope region that drains into a pixel and the tangent applied to the slope angle [64]. This predictor, which shows the capacity of the river for sediment transport, was mapped using the following classes values: <50, 50–500, 501–2000, 2001–5000, >5000 (Figure 2f). **Slope aspect** (Figure 3a) is the seventh morphometric index taken into account for the present research. The slope orientation has a big influence in the surface runoff process because the humidity condition will vary due to the different quantity of solar radiation [65]. The slope aspect predictor was derived from the DEM.

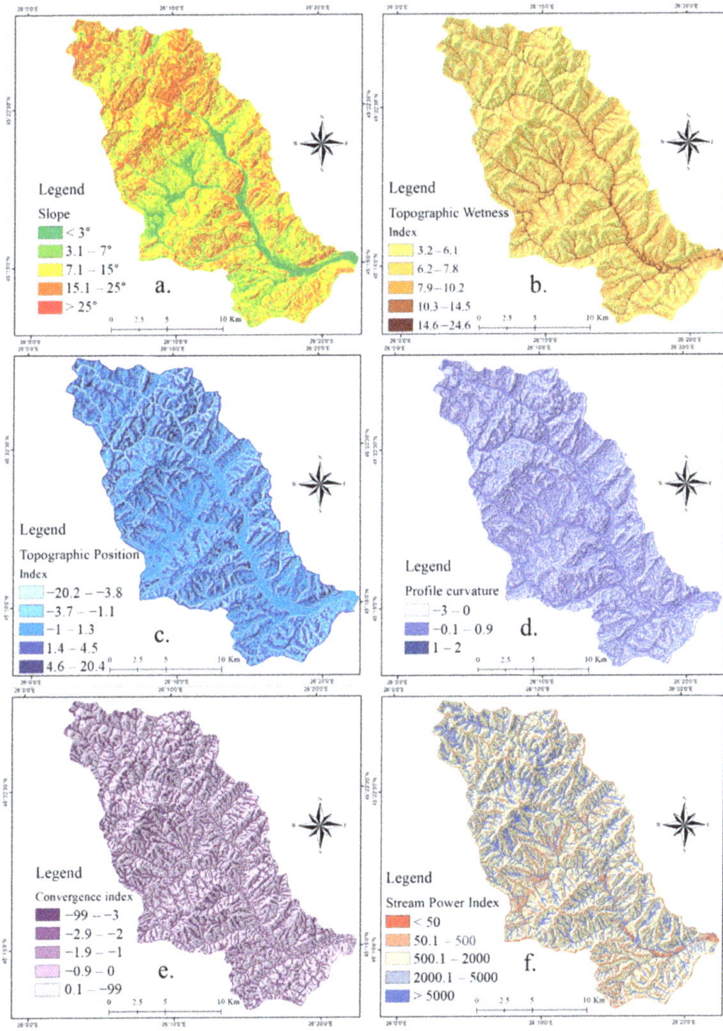

Figure 2. Flash-flood predictors: (**a**) Slope; (**b**) Topographic Wetness Index (TWI); (**c**) Topographic Position Index (TPI); (**d**) Profile curvature; (**e**) Convergence index; (**f**) Stream Power Index (SPI).

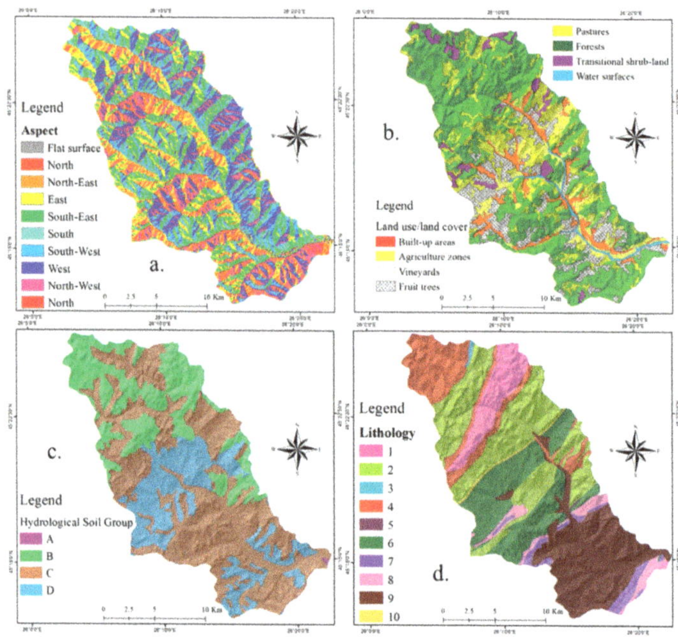

Figure 3. Flash-flood predictors: (**a**) Aspect; (**b**) Land use; (**c**) Hydrological soil groups; (**d**) Lithology.

Land use, which is the main interface between the torrential rainfalls and the ground surface, has an important influence on the runoff velocity [66]. For the present study, the land use layer was taken from the **Corine Land Cover** 2018 database. According to Figure 3b, a number of eight land use categories were delineated within the study area perimeter. **Hydrological soil group** was considered as a flash-flood predictor in the present research due to its incontestable influence on vertical infiltration of water in the ground [67]. Within the Bâsca Chiojdului cathcment, all of the four hydrological soil groups are present (Figure 3c). A similar contribution, as soil groups, in flash-flood genesis is held by the **lithological groups**. In the area of the Bâsca Chiojdului catchment, a total of 10 lithological groups can be found (Figure 3d).

4. Methods

The main steps of the methodological workflow are synthetically described in Figure 4.

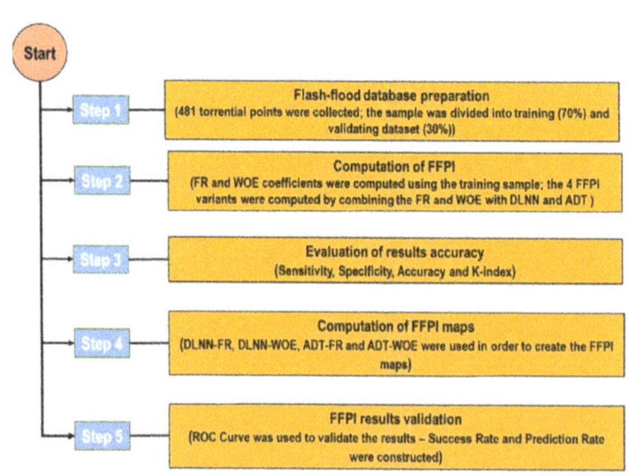

Figure 4. Flowchart of the methodological steps applied in this research.

4.1. Linear Support Vector Machine (LSVM) for Feature Selection

In a study that aims to estimate the qualitative flash-flood susceptibility, it is imperative to analyze the predictive ability of flash-flood conditioning factors in order to see if they all manage to contribute to some extent to the genesis of flash floods. In the present research paper, the evaluation of the prediction ability of flash-flood predictors was determined using Linear Support Vector Machine (LSVM). This method is widely used because it is able to remove redundant and irrelevant information from input data [68]. The following equation is used to compute the predictive ability through LSVM algorithm [69]:

$$f(x) = sign\left(C^T * i + j\right) \quad (1)$$

where C^T is equal to the inverse of weight matrix attributed to each flash-flood predictor, $i = (i1, i2, \ldots, i11)$ is the vector containing the ten flash-flood predictors, j is equal to the offset value calculated from the hyper-plane origin [5].

This algorithm was applied with the help of Weka 9.3 software.

4.2. Weights of Evidence (WOE)

The bivariate statistics model represented by Weights of Evidence (WOE) is a very frequently used algorithm involved in the studies focused on natural hazards predisposition evaluation [40]. In this study, the WOE model is used to calculate the weight that each factor class/category has in relation to the genesis of the flash-flood process. In order to derive the WOE coefficients, first, computing the positive (W^+) and negative (W^-) weights is required. The positive weight highlights the association between a factor class/category and the torrential points, while the negative weight indicates the absence of this spatial association [36]. The following relations should be employed in the weights computation [70]:

$$W^+ = \ln \frac{P\{B|S\}}{P\{B|\overline{S}\}} \quad (2)$$

$$W^- = \ln \frac{P\{\overline{B}|S\}}{P\{\overline{B}|\overline{S}\}} \quad (3)$$

where: W^+—positive weight, W^-—negative weight, P—the probability, B—the presence of flash-flood predictor, $\overline{B}$—the absence of flash-flood predictor, S—the presence of torrential pixels, $\overline{S}$—the absence of torrential pixels.

The final WOE coefficients can be derived using the next equation [71]:

$$Wf = Wplus + \mathrm{Wmin}total - \mathrm{Wmin} \quad (4)$$

where: *Wplus*—positive weight of a class factor, Wmin—negative weight of a class factor, W*mintotal*—the total of all negative weights in a multiclass map.

The final WOE values will be used as input data into the Deep Learning and Alternating Decision Tree models through which the flash-flood susceptibility will be determined.

4.3. Frequency Ratio (FR)

Frequency Ratio (FR) is the second bivariate statistical model which will be employed in order to prepare the input data in the Deep Learning and Alternating Decision Tree algorithms. The FR model consists of the calculation of the ratio between the sum of torrential pixels within a specific category of predictor, and the sum of torrential pixels within the entire study zone. The following relation can be used to estimate the FR coefficients [72]:

$$FR = \frac{\frac{Np(LXi)}{\sum_{i=1}^{m} Np(LXi)}}{\frac{Np(Xj)}{\sum_{j=1}^{n} Np(Xj)}} \quad (5)$$

where: *FR*—the frequency ratio of class *i* of factor *j*; *Np(LXi)*—the number of pixels with torrentiality within class *i* of factor variable *X*; *Np(Xj)*—the number of pixels within factor variable *Xj*; m—the number of classes in the factor variable *Xi*; *n*—the number of factors in the study area.

4.4. Deep Learning Neural Network (DLNN)

Besides one hidden layer neural networks, the Deep Learning Neural Network (DLNN) is characterized by a feed-forward architecture which contains more than one hidden layer [73]. Due to this fact, DLNN model is considered better than the simple neural network in terms of complex classification problems [74]. In the DLNN structure, the information from the input layer will be transmitted to the hidden layers where it is processed and then forwarded to the output layer. Further, the backpropagation algorithm will be employed to send back the error from the output layer to the input layer [75]. The training procedure of DLNN, which is a type of fee-forward neural network, is ensured by the application of Rectified Linear Unit (ReLU) activation function [76]. This function, which is able to reduce the vanishing gradient, is expressed as follows:

$$r(x) = \begin{cases} |x \; if \; x > 0 \\ |0 \; if \; x \leq 0 \end{cases} = \max(0, x) \tag{6}$$

where x is the input signal transmitted to neuron, while r is the ReLU function.

The derivate associated to the ReLU function, which are required by the back-propagation algorithm, can be calculated using the following relation:

$$r'(x) = \begin{cases} |1, \; x > 0 \\ |0, \; x \leq 0 \end{cases} \tag{7}$$

It should be remarked that the cross-entropy function is also involved in the training procedure because it helps the DLNN to achieve a higher degree of accuracy [77]. The cross-entropy is mathematically described using the next equation:

$$E = -\frac{1}{N}\sum_{n=1}^{N} M\ln(P) + (1-M)\ln(1-P) \tag{8}$$

where N is the total number of records in training sample; M is the predictor values, while P is the predicted values.

The adaptive momentum (Adam) prediction model, implied in the stochastic optimization process, is used to complete the training process of DLNN. Through the Adam model, the first and second moments could be computed via the exponential moving averages highlighted through the next relations [78]:

$$m_t = \beta_1 m_{t-1} + (1-\beta_1) g_t \tag{9}$$

$$v_t = \beta_2 v_{t-1} + (1-\beta_2) g_t^2 \tag{10}$$

where m and v are the values of the moving averages, g represents current mini-batch gradient, β is new hyper-parameters computed via the algorithm.

In order to apply the DLNN-FR and DLNN-WOE ensembles, the specific lines of code were written in R programming language. More specifically, the Keras and Lime package from R Studio were used in this regard.

4.5. Alternating Decision Tree

Alternating Decision Tree (ADT) model is an ensemble of the decision tree and boosting method [79]. ADT structure has a lower complexity than decision tree models such as Rotation Forest, Classification and Regression Tree or Random Forest [80]. ADT model uses a natural extension of decision tree and voted stumps and is formed by prediction alternate layers and nodes of decision [81]. Within the ADT algorithm, the decision nodes will specify the predicate condition; meanwhile the prediction nodes will be characterized by a single number [80].

Let c_1 be the value of a precondition, c_2 the value of a base condition, and a and b the values of two real numbers; then a and b will be computed using the relations [82]:

$$a = 0.5 * \ln \frac{W_+(c_1 \cap c_2)}{W_-(c_1 \cap c_2)},\ b = 0.5 * \ln \frac{W_+(c_1 \cap \overline{c_2})}{W_-(c_1 \cap \overline{c_2})} \tag{11}$$

where W denotes the sum of the values from any prediction node, and the best c_1 and c_2 are estimated by minimizing the Z_t (c_1, c_2), determined as follows:

$$z_t(c_1c_2) = 2\sqrt{W_+(c_1 \cap c_2) * W_-(c_1 \cap c_2)} + \sqrt{W_+(c_1 \cap \overline{c_2}) * W_-(c_1 \cap \overline{c_2})} + W(\overline{c_2}) \tag{12}$$

The ADT-FR and ADT-WOE ensembles were run and implemented in Weka software.

4.6. Model Performance and Results Validation

4.6.1. Statistical Measures

At the first stage, the models' performance assessment will consist of the computation of the next statistical metrics: specificity, sensitivity, accuracy, kappa index. The aforementioned indices will be computed using the next mathematical relations:

$$k = \frac{p_o - p_e}{1 - p_e} \tag{13}$$

$$Sensitivity = \frac{TP}{TP + FN} \tag{14}$$

$$Specificity = \frac{TN}{FP + TN} \tag{15}$$

$$Accuracy = \frac{TP + TN}{TP + FP + TN + FN} \tag{16}$$

where *TP* (True Positive) and *TN* (True Negative) are the sum of points that will be correctly classified, *FP* (False Positive) and *FN* (False Negative) are the sum of points erroneously classified; k is kappa coefficient, p_o is the sum of initially established torrential pixels, and p_e is the sum of predicted torrential pixels.

4.6.2. ROC Curve

The second stage of results validation implied the application of the ROC curve and Area Under Curve (AUC) to measure the model performance. An AUC closer to 1 will highlight a performant model, while the values near to 0 will indicate a weak prediction ability of the models [83,84]. The Success Rate will represent a first form of ROC curve which will be constructed with the training samples, while the Prediction Rate is the second variant of ROC curve which will be designed with the help of validation sample. The AUC values will be determined using the next formula:

$$AUC = \frac{(\sum TP + \sum TN)}{(P + N)} \tag{17}$$

where P is the sum of points having torrential phenomena and N is the sum of non-torrential points.

5. Results

5.1. Feature Selection Using LSVM

According to the results achieved through Weka software, the application of LSVM provided the next scores: slope (0.659), profile curvature (0.476), land use (0.429), tpi (0.394), twi (0.362), convergence index (0.338), hydrological soil group (0.283), spi (0.253), lithology (0.231) and aspect (0.162) (Figure 5).

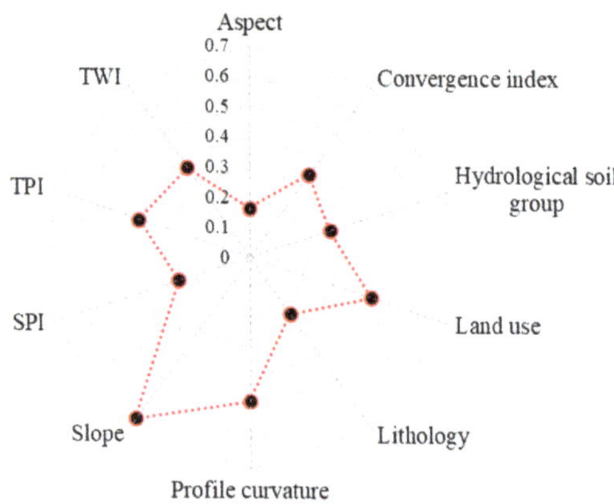

Figure 5. Linear Support Vector Machine (LSVM) scores assigned to flash-flood predictors.

5.2. FR and WOE Coefficients

The values of FR and WOE coefficients are inserted in Table 1. The largest value of FR coefficients (7.295) was achieved by TWI class between 14.6 and 24.6, followed by slope class between 15 and 25° (3.925), SPI values lower than 50 (3.205), built-up areas land use category (2.715) and TPI class between −1 and 1.3 (1.695) (Figure 6). In terms of WOE weights, the highest score was assigned to built-up areas land use category (3.96), followed by TWI class between 14.6 and 24.6 (2.67), slope class between 15 and 25° (2.48), SPI values lower than 50 (1.88) and TPI class between −1 and 1.3 (1.39).

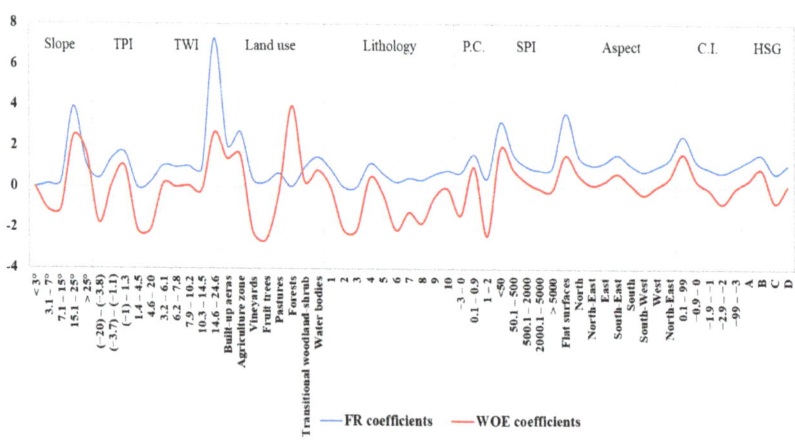

Figure 6. Distribution of FR and WOE coefficients within the classes of flash-flood predictors.

In order to be used as input in ADT and DLNN models, the FR and WOE values were normalized between 0 and 1.

Table 1. FR and WOE coefficients.

Factor	Class	FR	FR Standardized Coefficients	WoE Coefficients	WoE Standardized Coefficients
Slope	<3°	0.000	0.000	0.000	0.307
	3.1–7°	0.152	0.039	−1.100	0.000
	7.1–15°	0.245	0.062	−1.100	0.000
	15.1–25°	3.925	1.000	2.480	1.000
	>25°	1.125	0.287	1.720	0.788
TPI	(−20)–(−3.8)	0.435	0.257	−1.740	0.116
	(−3.7)–(−1.1)	1.415	0.835	0.160	0.727
	(−1)–1.3	1.695	1.000	1.010	1.000
	1.4–4.5	0.000	0.000	−2.100	0.000
	4.6–20	0.245	0.145	−2.100	0.000
TWI	3.2–6.1	1.055	0.030	0.160	0.116
	6.2–7.8	0.975	0.017	0.010	0.063
	7.9–10.2	1.025	0.025	0.080	0.088
	10.3–14.5	0.865	0.000	−0.170	0.000
	14.6–24.6	7.295	1.000	2.670	1.000
Land use	Built-up areas	2.035	0.750	3.960	1.000
	Agriculture zone	2.715	1.000	1.610	0.642
	Vineyards	0.365	0.134	−2.190	0.063
	Fruit trees	0.245	0.090	1.390	0.608
	Pastures	0.675	0.249	−0.300	0.351
	Forests	0.000	0.000	−2.600	0.000
	Transitional woodland-shrub	0.965	0.355	0.270	0.438
	Water bodies	1.485	0.547	0.840	0.524
Lithology	1	0.895	0.768	0.000	0.745
	2	0.000	0.000	−2.100	0.000
	3	0.000	0.000	−2.100	0.000
	4	1.165	1.000	0.450	0.904
	5	0.665	0.571	−0.350	0.621
	6	0.245	0.210	−2.100	0.000
	7	0.435	0.373	−1.230	0.309
	8	0.355	0.305	−1.770	0.117
	9	0.635	0.545	−0.490	0.571
	10	0.815	0.700	−0.070	0.720
Profile curvature	−3–0	0.705	0.237	−1.370	0.299
	0.1–0.9	1.605	1.000	0.980	1.000
	1–2	0.425	0.000	−2.370	0.000
SPI	<50	3.205	1.000	1.880	1.000
	50.1–500	1.615	0.329	0.870	0.498
	500.1–2000	1.025	0.080	0.270	0.199
	2000.1–5000	0.835	0.000	−0.060	0.035
	>5000	0.975	0.059	−0.130	0.000
Aspect	Flat surfaces	3.645	1.000	1.560	1.000
	North	1.535	0.262	0.610	0.503
	North-East	1.095	0.108	0.130	0.251
	East	1.205	0.147	0.280	0.330
	South-East	1.605	0.287	0.690	0.545
	South	1.115	0.115	0.170	0.272
	South-West	0.785	0.000	−0.350	0.000
	West	1.015	0.080	0.040	0.204
	North-East	1.405	0.217	0.490	0.440
Convergence index	0.1–99	2.515	1.000	1.650	1.000
	−0.9–0	1.285	0.317	0.360	0.469
	−1.9–−1	0.915	0.111	−0.110	0.276
	−2.9–−2	0.715	0.000	−0.780	0.000
	−99–−3	0.985	0.150	−0.040	0.305
HSG	A	1.325	0.697	0.360	0.669
	B	1.595	1.000	0.890	1.000
	C	0.705	0.000	−0.710	0.000
	D	1.105	0.449	0.090	0.500

5.3. Models Performance Assessment

The configuration, in terms of the hardware and software environments, that was required for the computational modelling, is presented in Table 2.

Table 2. Hardware and software environmental configuration used for modelling.

Configuration	Parameter
CPU	Intel(R) Core(TM) i7–7500@2.70 GHz
RAM	16.0 GB DDR4
GPU	NVIDIA GeForce MX330
Hard disk	SSD 512 GB M.2 PCIe
Operating system	Windows 10 Pro

It is mandatory that before the final mapping of flash-flood potential, the model's performance must be evaluated in order to verify its reliability in the methodological process. Thus, in terms of the training dataset, the DLNN-WOE ensemble achieved the highest accuracy (0.985), followed by DLNN-FR (0.982), ADT-FR (0.923) and ADT-WOE (0.92). In terms of the validating sample, the highest accuracy was achieved by DLNN-WOE (0.92), followed by DLNN-FR (0.903), ADT-WOE (0.896) and ADT-FR (0.878) (Table 3).

Table 3. Statistical metrics used to evaluate model's performance.

	Models	TP	TN	FP	FN	Sensitivity	Specificity	Accuracy	k-Index
Training	DLNN-FR	330	332	7	5	0.985	0.979	0.982	0.964
	DLNN-WOE	334	330	3	7	0.979	0.991	0.985	0.970
	ADT-FR	312	310	25	27	0.920	0.925	0.923	0.846
	ADT-WOE	309	311	28	26	0.922	0.917	0.920	0.840
Validating	DLNN-FR	132	128	12	16	0.892	0.914	0.903	0.806
	DLNN-WOE	137	128	7	16	0.895	0.948	0.920	0.840
	ADT-FR	129	124	15	20	0.866	0.892	0.878	0.757
	ADT-WOE	132	126	12	18	0.880	0.913	0.896	0.792

5.4. Results of Machine Learning Ensembles

5.4.1. DLNN-FR and DLNN-WOE Results

The DLNN based ensembles were trained by establishing the maximum number of epochs to 100 (Figure 7).

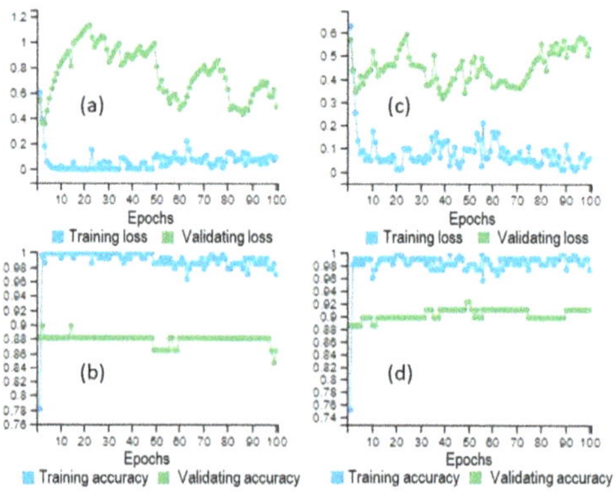

Figure 7. DLNN based ensemble running outputs (**a**) Training and Validating loss of DLNN-FR; (**b**) Training and Validating accuracy of DLNN-FR; (**c**) Training and Validating loss of DLNN-WOE; (**d**) Training and Validating accuracy of DLNN-WOE).

Figure 7 highlights the variability of loss and model accuracy according to the epochs number and also for both training and validating samples. Particularly, in the case of the DLNN-FR model, the best performances were achieved with the following model parameters: a number of two hidden layers; a maximum number of 100 hidden neurons in each hidden layer; a dropout rate of 0.3; a batch size value of 5 and a validation split of 0.3. The same number of hidden layers and neurons was used also in the case of DLNN-WOE, while the other parameters have the following value: a dropout rate of 0.4; a batch size of 4 and a validation rate of 0.2. The architecture of the DLNN-based ensembles are represented in Figure 8.

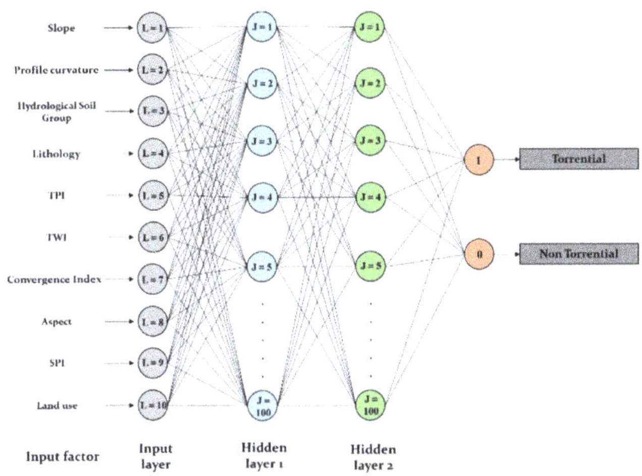

Figure 8. Deep Learning Neural Network architecture.

The next step in the flash-flood susceptibility computation process is the derivation of the flash-flood predictor's importance. In terms of DLNN-FR, the highest importance was assigned to slope factor (0.2). On the second-place rank, land use (0.143), followed by profile curvature (0.12), TWI (0.109), hydrological soil group (0.097), lithology (0.094), TPI (0.08), SPI (0.067), convergence index (0.061) and aspect (0.029) (Figure 9). The application of DLNN-WOE revealed that the most important factor was slope (0.235), and is followed by land use (0.149), SPI (0.089), hydrological soil group (0.086), TPI (0.086), TWI (0.082), lithology (0.074), convergence index (0.072), profile curvature (0.064) and aspect (0.063).

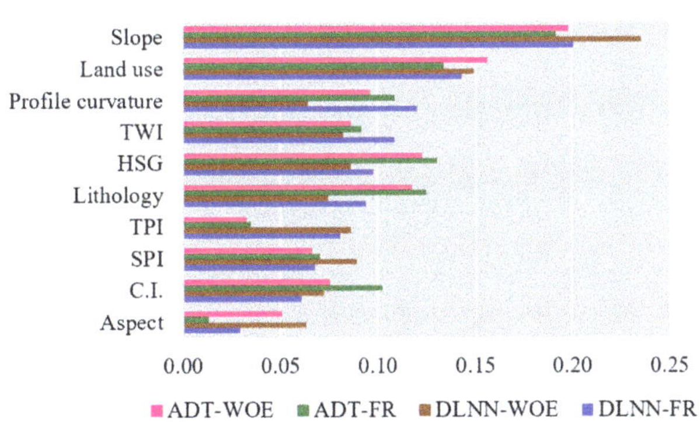

Figure 9. Flash-flood predictors importance.

The weights of flash-flood predictors were used in ArcGIS map algebra in order to derive the flash-flood potential index values. All the Flash-Flood Potential Index (FFPI) results, with values between 0 and 1, were reclassified in five classes using Natural Breaks method. In terms $FFPI_{DLNN\text{-}FR}$, the very low flash-flood potential values cover around 7.5% of the study area and range between 0 and 0.42 (Figure 10a). The low flash-flood potential appears on around 15.6% of Bâsca Chiojdului river catchment and has values ranging from 0.43 and 0.55. It should be remarked that these values are mainly spread on the southern half of the area. The medium flash-flood potential has a span of 30.28% of the entire territory (Figure 11) and is characterized by $FFPI_{DLNN\text{-}FR}$ between 0.56 and 0.66. These values are uniformly distributed across the study zone. The high and very high flash-flood potential appears on areas with $FFPI_{DLNN\text{-}FR}$ higher than 0.67 and covers approximately 46.57% of the research area. This potential degree is mainly present in the northern half of Bâsca Chiojdului river basin.

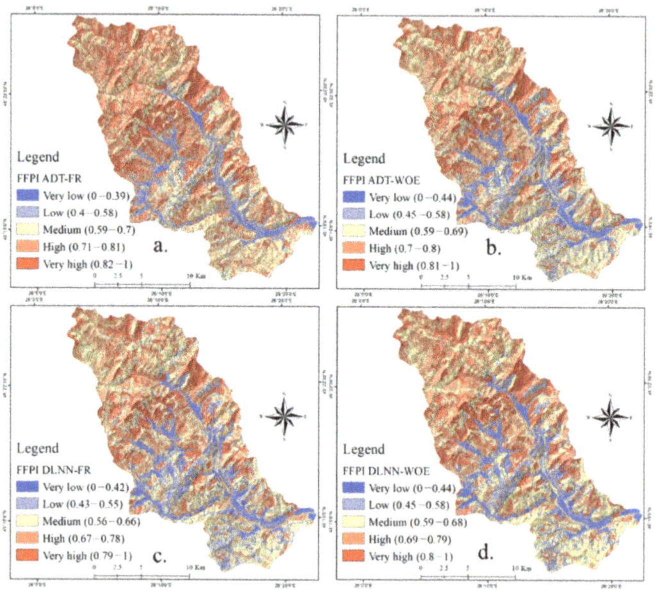

Figure 10. Flash Flood Potential Index (**a**) DLNN-FR; (**b**) DLNN-WOE; (**c**) ADT-FR; (**d**) ADT-WOE.

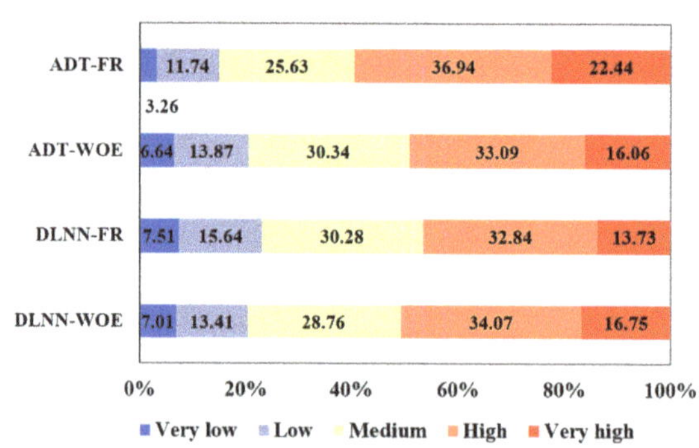

Figure 11. Flash-Flood Potential Index (FFPI) classes weights.

In terms of $FFPI_{DLNN\text{-}WOE}$, the very low flash-flood potential is characteristic for a percentage of 7% from the entire study perimeter, while the low values of the same indicator cover an area of 13.41% of the total territory. Ranging from 0.59 to 0.68 (Figure 10b), the medium flash-flood potential spans accross approximately 28.76% of the Bâsca Chiojdului river catchment. High and very high flash-flood susceptibility has values of $FFPI_{DLNN\text{-}WOE}$ higher than 0.69 and is spread over more than 50% of the research zone. It should be noted that the areas delineated through DLNN-WOE have a lower degree of fragmentation than the areas delineated by DLNN-FR.

5.4.2. ADT-FR and ADT-WOE Results

A trial procedure was applied in order to determine the best parameter associated with the highest accuracy of ADT-FR and ADT-WOE for both training and validating samples. Thus, in terms of ADT-FR, the highest accuracies (0.923 for training and 0.878 for validating) were achieved after 23 iterations, while in terms of ADT-WOE the best accuracies (0.92 for training and 0.896 for validating) were determined after a number of 28 iterations (Table 4). Once the best parameters were determined, the optimally pruned decision trees were constructed (Figure 12a,b) and the flash-flood predictors importance were calculated.

Table 4. The optimal parameters of the ADT based ensembles.

Models	No. of Iterations	Seed	Training Accuracy	Validating Accuracy
ADT-FR	23	6	0.923	0.878
ADT-WOE	28	8	0.920	0.896

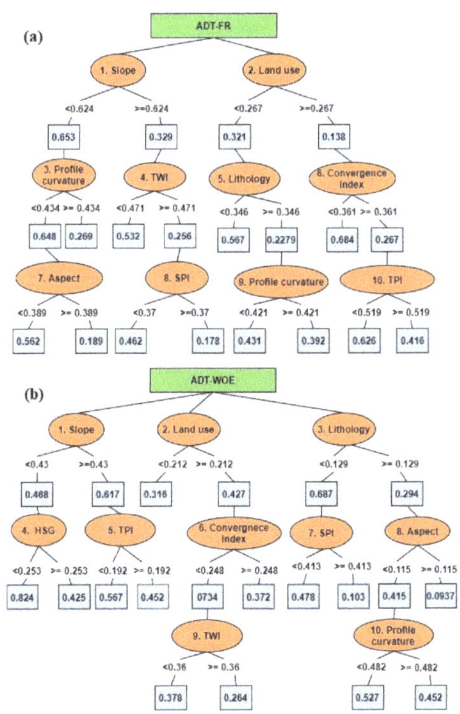

Figure 12. Optimally pruned Decision Tree Structure for ADT based ensembles ((**a**) ADT-FR and (**b**) ADT-WOE ensembles).

Therefore, in terms of ADT-FR, the highest importance was assigned to slope factor (0.191). On the second-place rank land use (0.134), followed by hydrological soil group

(0.131), lithology (0.125), profile curvature (0.108), convergence index (0.102), TWI (0.091), SPI (0.07), TPI (0.034) and aspect (0.013) (Figure 9). The application of ADT-WOE revealed that the most important factor was slope (0.198), and is followed by land use (0.156), hydrological soil group (0.123), lithology (0.117), profile curvature (0.096), TWI (0.086), convergence index (0.075), SPI (0.066), aspect (0.051) and TWI (0.032).

As in the case of the previous two ensembles, the $FFPI_{ADT\text{-}FR}$ and $FFPI_{ADT\text{-}WOE}$ were calculated. In terms $FFPI_{ADT\text{-}FR}$, the very low flash-flood potential spans around 3.26% of the study area and has values between 0 and 0.39 (Figure 10c). The low flash-flood potential is distributed on around 11.74% of the Bâsca Chiojdului river catchment and has values ranging from 0.4 to 0.58. The medium flash-flood potential spans 25.63% of the entire territory and has values between 0.59 and 0.7 (Figure 10c). The high and very high flash-flood potentials appear on areas with $FFPI_{ADT\text{-}FR}$ higher than 0.71 and cover approximately 59.38% of the research area. In terms of $FFPI_{ADT\text{-}WOE}$, the very high flash-flood potential covers 6.64% of the entire study perimeter, while the low values are spread over 13.87% of the total territory. With values from 0.59 to 0.69 (Figure 10d), the medium flash-flood potential occurs over 30.34% of the Bâsca Chiojdului river catchment. The high and very high flash-flood potential indices have values higher than 0.7 and account for almost 50% of the study zone.

5.5. Results Validation Using ROC Curve

The validation of the FFPI results provided by each ensemble model was carried out using the ROC curve method. Thus, in the case of the Success Rate, the highest performance was achieved by $FFPI_{DLNN\text{-}WOE}$ with an AUC of 0.96, being followed by $FFPI_{DLNN\text{-}FR}$ (AUC = 0.942), $FFPI_{ADT\text{-}WOE}$ (AUC = 0.94) and $FFPI_{ADT\text{-}FR}$ (AUC = 0.919) (Figure 13a). If we analyze the Prediction Rate outcomes, it can be seen that the same $FFPI_{DLNN\text{-}WOE}$ indicator achieved the highest performance (AUC = 0.921), followed by $FFPI_{DLNN\text{-}FR}$ (0.92), $FFPI_{ADT\text{-}WOE}$ (0.909) and $FFPI_{ADT\text{-}FR}$ (AUC = 0.879).

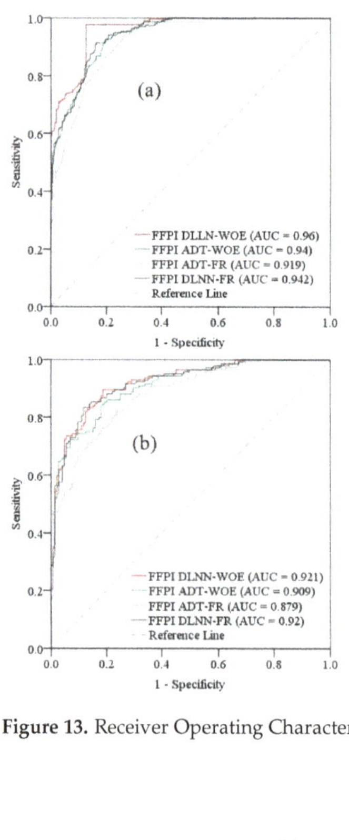

Figure 13. Receiver Operating Characteristic (ROC) curve (**a**) Success Rate; (**b**) Prediction Rate.

6. Discussions

With the undeniable advancement of technology, there are more and more possibilities to monitor the dangerous phenomena that occur on the Earth's surface. In this regard, it is worth remembering the rapid advance of observation techniques of the terrestrial surface by means of remote sensing sensors—with the help of which, the surfaces affected by natural hazards can be observed.

Thus, the present paper used images taken with the help of these sensors to identify the areas already affected by the torrential runoff from the Earth's surface. It should be mentioned that the most accurate identification of these areas is essential in obtaining results with high accuracy and which can be further used by the competent authorities in risk assessment and in adopting the most appropriate measures to reduce future damage caused by these hazards. Thus, by analyzing the images provided by remote sensing sensors, on the river basin of the river Bâsca Chiojdului, areas affected by torrential runoff totaling a total area of 34 km^2, representing about 10% of the entire study area, were identified. Furthermore, in order to capitalize on the delimited surfaces, a sample of about 481 was generated, taking a sample of points affected by torrential phenomena transposed into relief microforms such as ravines. In order to ensure the correctness of the modelling results, another sample of 481 points was generated from the areas where the torrential phenomena did not take place; the entire data set being then divided into training and validating data. The values of 10 flash-flood predictors were also used as input data. It should be noted that Remote Sensing sensors also played a crucial role in generating 8 of the 10 flash-flood predictors. Thus, all morphometric parameters were derived from the digital terrain model taken from the SRTM database, 30 m which was acquired using radar techniques. In addition, the land use, taken from the Corine Land Cover 2018 database, was generated by the supervised classification of the images provided by the Remote Sensing sensors.

Data on the presence of phenomena and the values of the main predictors of flash-flood genesis were included in two of the state-of-the-art machine learning models represented by Deep Learning Neural Networks and Alternating Decision Trees. These two models are recommended due to the very good results they provided following their application in previous studies on the estimation of susceptibility to natural hazards [79,80]. For a higher degree of results objectivity, it was decided to process the training sample by assigning some coefficients using the bivariate methods statistics, Frequency Ratio and Weights of Evidence. This method has proven to be very useful in previous studies [46,56] where the initial data were processed with bivariate statistics algorithms.

The combination of DLNN with WOE proved to be the most efficient because the accuracy achieved during the training process exceeded 98%, while ROC curve applied to the final product $FFPI_{DLNN\text{-}WOE}$ showed a maximum AUC of 0.96. This value of AUC exceeds the value obtained by Costache et al. [38], when, by applying the hybrid combination between Multilayer Perceptron (MLP) and Statistical Index, for the same study area and for the FFPI calculation, a maximum AUC value of 0.94 was obtained. These results confirm the findings from the literature according to which DLNN, whose architecture includes several hidden layers, is able to surpass the MLP performances whose architecture includes a single hidden layer [57]. Moreover, the MLP performance from the previous study was exceeded by the DLNN-FR ensemble model, characterized by an AUC of 0.942. Overall, in the Bâsca Chiojdului basin, the models showed a percentage of the high and very high flash-flood potential between 46.57% (DLNN-FR) and 59.38% (ADT-FR).

7. Conclusions

In light of the continuous increase in the flash-flood events' frequency, the present research work proposed a workflow through which the areas susceptible to flash floods are identified based on remote sensing and GIS data included in Deep Learning and Alternating Decision Trees ensembles. Thus, using 418 torrential and 481 non-torrential locations along with 10 flash-flood predictors, the Flash-Flood Potential Index was determined across the

Bâsca Chiojdului river basin. Using as input data the FR and WOE coefficients, the FFPI was computed using the following four ensembles: DLNN-FR, DLNN-WOE, ADT-FR and ADT-WOE. As was expected, the slope angle and land use resulted in being the most important flash-flood predictors. The highest results accuracy was achieved by the DLNN-WOE model which is characterized by an AUC–ROC curve of 0.985. The percentage (59.38%) of high and very high FFPI classes was revealed by the application of ADT-FR ensemble.

The main novelty of this study is represented by the application for the first time in the literature of the four ensemble models for determining flash-flood potential index values.

This work is of real importance for the governmental authorities which can use the results in order to improve the measures taken to mitigate the negative effects of flash-flood hazards within the study area.

Author Contributions: Conceptualization, R.C., A.A. (Alireza Arabameri) and Q.B.P.; data curation, R.C., A.A. (Alireza Arabameri), Q.B.P., B.T.P., M.P. and A.A. (Aman Arora); methodology, R.C., A.A. (Alireza Arabameri), Q.B.P., B.T.P., M.P., A.A. (Aman Arora), N.T.T.L. and I.C.; writing—original draft. R.C., A.A. (Alireza Arabameri), Q.B.P., B.T.P., M.P. and A.A. (Aman Arora); writing—review and editing, R.C., A.A. (Alireza Arabameri), Q.B.P., B.T.P., M.P., A.A. (Aman Arora), T.B., N.T.T.L. and I.C. All authors have read and agreed to the published version of the manuscript.

Funding: This research was partly funded by the Austrian Science Fund (FWF) through the Doctoral College GIScience (DK W 1237-N23) at the University of Salzburg.

Institutional Review Board Statement: Not applicable.

Informed Consent Statement: Not applicable.

Data Availability Statement: Data available on request.

Conflicts of Interest: There is no coflict of interest.

References

1. Halkos, G.; Skouloudis, A. Investigating resilience barriers of small and medium-sized enterprises to flash floods: A quantile regression of determining factors. *Clim. Dev.* **2020**, *12*, 57–66. [CrossRef]
2. Bezak, N.; Mikoš, M. Investigation of Trends, Temporal Changes in Intensity-Duration-Frequency (IDF) Curves and Extreme Rainfall Events Clustering at Regional Scale Using 5 min Rainfall Data. *Water* **2019**, *11*, 2167. [CrossRef]
3. Bui, D.T.; Tsangaratos, P.; Ngo, P.-T.T.; Pham, T.D.; Pham, B.T. Flash flood susceptibility modeling using an optimized fuzzy rule based feature selection technique and tree based ensemble methods. *Sci. Total Environ.* **2019**, *668*, 1038–1054. [CrossRef] [PubMed]
4. Cao, C.; Xu, P.; Wang, Y.; Chen, J.; Zheng, L.; Niu, C. Flash flood hazard susceptibility mapping using frequency ratio and statistical index methods in coalmine subsidence areas. *Sustainability* **2016**, *8*, 948. [CrossRef]
5. Costache, R. Flash-Flood Potential assessment in the upper and middle sector of Prahova river catchment (Romania). A comparative approach between four hybrid models. *Sci. Total Environ.* **2019**, *659*, 1115–1134. [CrossRef]
6. Elkhrachy, I. Flash flood hazard mapping using satellite images and GIS tools: A case study of Najran City, Kingdom of Saudi Arabia (KSA). *Egypt. J. Remote Sens. Space Sci.* **2015**, *18*, 261–278. [CrossRef]
7. Costache, R.; Bui, D.T. Identification of areas prone to flash-flood phenomena using multiple-criteria decision-making, bivariate statistics, machine learning and their ensembles. *Sci. Total Environ.* **2020**, *712*, 136492. [CrossRef]
8. Janizadeh, S.; Avand, M.; Jaafari, A.; Phong, T.V.; Bayat, M.; Ahmadisharaf, E.; Prakash, I.; Pham, B.T.; Lee, S. Prediction Success of Machine Learning Methods for Flash Flood Susceptibility Mapping in the Tafresh Watershed, Iran. *Sustainability* **2019**, *11*, 5426. [CrossRef]
9. Hosseini, F.S.; Choubin, B.; Mosavi, A.; Nabipour, N.; Shamshirband, S.; Darabi, H.; Haghighi, A.T. Flash-flood hazard assessment using ensembles and Bayesian-based machine learning models: Application of the simulated annealing feature selection method. *Sci. Total Environ.* **2020**, *711*, 135161. [CrossRef]
10. Liu, Y.-X.; Yang, C.-N.; Sun, Q.-D.; Wu, S.-Y.; Lin, S.-S.; Chou, Y.-S. Enhanced embedding capacity for the SMSD-based data-hiding method. *Signal Process. Image Commun.* **2019**, *78*, 216–222. [CrossRef]
11. Zhao, C.; Li, J. Equilibrium Selection under the Bayes-Based Strategy Updating Rules. *Symmetry* **2020**, *12*, 739. [CrossRef]
12. Xiong, Q.; Zhang, X.; Wang, W.-F.; Gu, Y. A Parallel Algorithm Framework for Feature Extraction of EEG Signals on MPI. *Comput. Math. Methods Med.* **2020**, *2020*, 9812019. [CrossRef] [PubMed]
13. Zhu, Q. Research on Road Traffic Situation Awareness System Based on Image Big Data. *IEEE Intell. Syst.* **2019**, *35*, 18–26. [CrossRef]
14. Fu, X.; Yang, Y. Modeling and analysis of cascading node-link failures in multi-sink wireless sensor networks. *Reliab. Eng. Syst. Saf.* **2020**, *197*, 106815. [CrossRef]

15. Fu, X.; Pace, P.; Aloi, G.; Yang, L.; Fortino, G. Topology Optimization Against Cascading Failures on Wireless Sensor Networks Using a Memetic Algorithm. *Comput. Netw.* **2020**, *177*, 107327. [CrossRef]
16. Zenggang, X.; Zhiwen, T.; Xiaowen, C.; Xue-min, Z.; Kaibin, Z.; Conghuan, Y. Research on Image Retrieval Algorithm Based on Combination of Color and Shape Features. *J. Signal Process. Syst.* **2019**, 1–8. [CrossRef]
17. Zuo, C.; Sun, J.; Li, J.; Asundi, A.; Chen, Q. Wide-field high-resolution 3d microscopy with fourier ptychographic diffraction tomography. *Opt. Lasers Eng.* **2020**, *128*, 106003. [CrossRef]
18. Long, Q.; Wu, C.; Wang, X. A system of nonsmooth equations solver based upon subgradient method. *Appl. Math. Comput.* **2015**, *251*, 284–299. [CrossRef]
19. Zhu, J.; Shi, Q.; Wu, P.; Sheng, Z.; Wang, X. Complexity analysis of prefabrication contractors' dynamic price competition in mega projects with different competition strategies. *Complexity* **2018**, *2018*, 5928235. [CrossRef]
20. Xiong, L.; Zhang, H.; Li, Y.; Liu, Z. Improved stability and H∞ performance for neutral systems with uncertain Markovian jump. *Nonlinear Anal. Hybrid Systems* **2016**, *19*, 13–25.
21. Wu, T.; Cao, J.; Xiong, L.; Zhang, H. New Stabilization Results for Semi-Markov Chaotic Systems with Fuzzy Sampled-Data Control. *Complexity* **2019**, *2019*, 7875305. [CrossRef]
22. Wu, T.; Xiong, L.; Cheng, J.; Xie, X. New results on stabilization analysis for fuzzy semi-Markov jump chaotic systems with state quantized sampled-data controller. *Inf. Sci.* **2020**, *521*, 231–250. [CrossRef]
23. Shi, K.; Wang, J.; Tang, Y.; Zhong, S. Reliable asynchronous sampled-data filtering of T–S fuzzy uncertain delayed neural networks with stochastic switched topologies. *Fuzzy Sets Syst.* **2020**, *381*, 1–25. [CrossRef]
24. Shi, K.; Wang, J.; Zhong, S.; Tang, Y.; Cheng, J. Non-fragile memory filtering of TS fuzzy delayed neural networks based on switched fuzzy sampled-data control. *Fuzzy Sets Syst.* **2020**, *394*, 40–64. [CrossRef]
25. Xu, M.; Li, T.; Wang, Z.; Deng, X.; Yang, R.; Guan, Z. Reducing complexity of HEVC: A deep learning approach. *IEEE Trans. Image Process.* **2018**, *27*, 5044–5059. [CrossRef]
26. Lv, Z.; Qiao, L. Deep belief network and linear perceptron based cognitive computing for collaborative robots. *Appl. Soft Comput.* **2020**, *92*, 106300. [CrossRef]
27. Lv, Z.; Xiu, W. Interaction of edge-cloud computing based on SDN and NFV for next generation IoT. *IEEE Internet Things J.* **2019**, *7*, 5706–5712. [CrossRef]
28. Chen, H.; Chen, A.; Xu, L.; Xie, H.; Qiao, H.; Lin, Q.; Cai, K. A deep learning CNN architecture applied in smart near-infrared analysis of water pollution for agricultural irrigation resources. *Agric. Water Manag.* **2020**, *240*, 106303. [CrossRef]
29. Chen, H.; Qiao, H.; Xu, L.; Feng, Q.; Cai, K. A Fuzzy Optimization Strategy for the Implementation of RBF LSSVR Model in Vis–NIR Analysis of Pomelo Maturity. *IEEE Trans. Ind. Inform.* **2019**, *15*, 5971–5979. [CrossRef]
30. Qian, J.; Feng, S.; Tao, T.; Hu, Y.; Li, Y.; Chen, Q.; Zuo, C. Deep-learning-enabled geometric constraints and phase unwrapping for single-shot absolute 3D shape measurement. *APL Photonics* **2020**, *5*, 046105. [CrossRef]
31. Qian, J.; Feng, S.; Li, Y.; Tao, T.; Han, J.; Chen, Q.; Zuo, C. Single-shot absolute 3D shape measurement with deep-learning-based color fringe projection profilometry. *Opt. Lett.* **2020**, *45*, 1842–1845. [CrossRef] [PubMed]
32. Chao, L.; Zhang, K.; Li, Z.; Zhu, Y.; Wang, J.; Yu, Z. Geographically weighted regression based methods for merging satellite and gauge precipitation. *J. Hydrol.* **2018**, *558*, 275–289. [CrossRef]
33. Zhang, S.; Pak, R.Y.; Zhang, J. Vertical time-harmonic coupling vibration of an impermeable, rigid, circular plate resting on a finite, poroelastic soil layer. *Acta Geotech.* **2020**, 1–25.
34. Yang, S.; Deng, B.; Wang, J.; Li, H.; Lu, M.; Che, Y.; Wei, X.; Loparo, K.A. Scalable Digital Neuromorphic Architecture for Large-Scale Biophysically Meaningful Neural Network with Multi-Compartment Neurons. *IEEE Trans. Neural Netw. Learn. Syst.* **2019**, 1–15. [CrossRef] [PubMed]
35. Tsai, Y.H.; Wang, J.; Chien, W.T.; Wei, C.Y.; Wang, X.; Hsieh, S.H. A BIM-based approach for predicting corrosion under insulation. *Autom. Constr.* **2019**, *107*, 102923. [CrossRef]
36. Costache, R.; Zaharia, L. Flash-flood potential assessment and mapping by integrating the weights-of-evidence and frequency ratio statistical methods in GIS environment–case study: Bâsca Chiojdului River catchment (Romania). *J. Earth Syst. Sci.* **2017**, *126*, 59. [CrossRef]
37. Khosravi, K.; Nohani, E.; Maroufinia, E.; Pourghasemi, H.R. A GIS-based flood susceptibility assessment and its mapping in Iran: A comparison between frequency ratio and weights-of-evidence bivariate statistical models with multi-criteria decision-making technique. *Nat. Hazards* **2016**, *83*, 947–987. [CrossRef]
38. Costache, R.; Hong, H.; Pham, Q.B. Comparative assessment of the flash-flood potential within small mountain catchments using bivariate statistics and their novel hybrid integration with machine learning models. *Sci. Total Environ.* **2020**, *711*, 134514. [CrossRef]
39. Tien Bui, D.; Khosravi, K.; Shahabi, H.; Daggupati, P.; Adamowski, J.F.; Melesse, A.M.; Thai Pham, B.; Pourghasemi, H.R.; Mahmoudi, M.; Bahrami, S. Flood spatial modeling in northern Iran using remote sensing and gis: A comparison between evidential belief functions and its ensemble with a multivariate logistic regression model. *Remote Sens.* **2019**, *11*, 1589. [CrossRef]
40. Razandi, Y.; Pourghasemi, H.R.; Neisani, N.S.; Rahmati, O. Application of analytical hierarchy process, frequency ratio, and certainty factor models for groundwater potential mapping using GIS. *Earth Sci. Inform.* **2015**, *8*, 867–883. [CrossRef]
41. Siahkamari, S.; Haghizadeh, A.; Zeinivand, H.; Tahmasebipour, N.; Rahmati, O. Spatial prediction of flood-susceptible areas using frequency ratio and maximum entropy models. *Geocarto Int.* **2018**, *33*, 927–941. [CrossRef]

42. Yang, W.; Xu, K.; Lian, J.; Ma, C.; Bin, L. Integrated flood vulnerability assessment approach based on TOPSIS and Shannon entropy methods. *Ecol. Indic.* **2018**, *89*, 269–280. [CrossRef]
43. Razavi Termeh, S.V.; Pourghasemi, H.R.; Alidadganfard, F. Flood Inundation Susceptibility Mapping using Analytical Hierarchy Process (AHP) and TOPSIS Decision Making Methods and Weight of Evidence Statistical Model (Case Study: Jahrom Township, Fars Province). *J. Watershed Manag. Res.* **2018**, *9*, 67–81. [CrossRef]
44. Dano, U.L.; Balogun, A.-L.; Matori, A.-N.; Wan Yusouf, K.; Abubakar, I.R.; Said Mohamed, M.A.; Aina, Y.A.; Pradhan, B. Flood susceptibility mapping using GIS-based analytic network process: A case study of Perlis, Malaysia. *Water* **2019**, *11*, 615. [CrossRef]
45. Khosravi, K.; Shahabi, H.; Pham, B.T.; Adamowski, J.; Shirzadi, A.; Pradhan, B.; Dou, J.; Ly, H.-B.; Gróf, G.; Ho, H.L. A comparative assessment of flood susceptibility modeling using Multi-Criteria Decision-Making Analysis and Machine Learning Methods. *J. Hydrol.* **2019**, *573*, 311–323. [CrossRef]
46. Ali, S.A.; Parvin, F.; Pham, Q.B.; Vojtek, M.; Vojteková, J.; Costache, R.; Linh, N.T.T.; Nguyen, H.Q.; Ahmad, A.; Ghorbani, M.A. GIS-based comparative assessment of flood susceptibility mapping using hybrid multi-criteria decision-making approach, naïve Bayes tree, bivariate statistics and logistic regression: A case of Topl'a basin, Slovakia. *Ecol. Indic.* **2020**, *117*, 106620. [CrossRef]
47. Chen, W.; Li, Y.; Xue, W.; Shahabi, H.; Li, S.; Hong, H.; Wang, X.; Bian, H.; Zhang, S.; Pradhan, B. Modeling flood susceptibility using data-driven approaches of naïve bayes tree, alternating decision tree, and random forest methods. *Sci. Total Environ.* **2020**, *701*, 134979. [CrossRef]
48. Chapi, K.; Singh, V.P.; Shirzadi, A.; Shahabi, H.; Bui, D.T.; Pham, B.T.; Khosravi, K. A novel hybrid artificial intelligence approach for flood susceptibility assessment. *Environ. Model. Softw.* **2017**, *95*, 229–245. [CrossRef]
49. Avand, M.; Janizadeh, S.; Naghibi, S.A.; Pourghasemi, H.R.; Khosrobeigi Bozchaloei, S.; Blaschke, T. A Comparative Assessment of Random Forest and k-Nearest Neighbor Classifiers for Gully Erosion Susceptibility Mapping. *Water* **2019**, *11*, 2076. [CrossRef]
50. Pham, B.T.; Prakash, I.; Bui, D.T. Spatial prediction of landslides using a hybrid machine learning approach based on random subspace and classification and regression trees. *Geomorphology* **2018**, *303*, 256–270. [CrossRef]
51. Choubin, B.; Moradi, E.; Golshan, M.; Adamowski, J.; Sajedi-Hosseini, F.; Mosavi, A. An Ensemble prediction of flood susceptibility using multivariate discriminant analysis, classification and regression trees, and support vector machines. *Sci. Total Environ.* **2019**, *651*, 2087–2096. [CrossRef] [PubMed]
52. Wang, Y.; Hong, H.; Chen, W.; Li, S.; Panahi, M.; Khosravi, K.; Shirzadi, A.; Shahabi, H.; Panahi, S.; Costache, R. Flood susceptibility mapping in Dingnan County (China) using adaptive neuro-fuzzy inference system with biogeography based optimization and imperialistic competitive algorithm. *J. Environ. Manag.* **2019**, *247*, 712–729. [CrossRef] [PubMed]
53. Costache, R.; Pham, Q.B.; Sharifi, E.; Linh, N.T.T.; Abba, S.; Vojtek, M.; Vojteková, J.; Nhi, P.T.T.; Khoi, D.N. Flash-Flood Susceptibility Assessment Using Multi-Criteria Decision Making and Machine Learning Supported by Remote Sensing and GIS Techniques. *Remote Sens.* **2020**, *12*, 106. [CrossRef]
54. Bui, D.T.; Hoang, N.-D.; Martínez-Álvarez, F.; Ngo, P.-T.T.; Hoa, P.V.; Pham, T.D.; Samui, P.; Costache, R. A novel deep learning neural network approach for predicting flash flood susceptibility: A case study at a high frequency tropical storm area. *Sci. Total Environ.* **2020**, *701*, 134413.
55. Arabameri, A.; Saha, S.; Chen, W.; Roy, J.; Pradhan, B.; Bui, D.T. Flash flood susceptibility modelling using functional tree and hybrid ensemble techniques. *J. Hydrol.* **2020**, *587*, 125007. [CrossRef]
56. Tehrany, M.S.; Pradhan, B.; Jebur, M.N. Flood susceptibility mapping using a novel ensemble weights-of-evidence and support vector machine models in GIS. *J. Hydrol.* **2014**, *512*, 332–343. [CrossRef]
57. Prăvălie, R.; Costache, R. The analysis of the susceptibility of the flash-floods' genesis in the area of the hydrographical basin of Bāsca Chiojdului river/Analiza susceptibilitatii genezei viiturilor īn aria bazinului hidrografic al rāului Bāsca Chiojdului. *Forum Geogr.* **2014**, *13*, 39–49. [CrossRef]
58. Zarea, R.; Gheorghe, M. Dangerous hydrological phenomena on the Hydrographic Basin Bâsca Chiojdului. In *Buletinul Institutului Politehnic Din Iaşi*; Universitatea Tehnică "Gheorghe Asachi" din Iaşi Tomul LVI (LX), Fasc: Iași, Romania, 2010; pp. 37–48.
59. Prăvălie, R.; Costache, R. Assessment of socioeconomic vulnerability to floods in the Bâsca Chiojdului catchment area. *Rom. Rev. Reg. Stud.* **2014**, *10*, 2.
60. Chen, W.; Li, W.; Chai, H.; Hou, E.; Li, X.; Ding, X. GIS-based landslide susceptibility mapping using analytical hierarchy process (AHP) and certainty factor (CF) models for the Baozhong region of Baoji City, China. *Environ. Earth Sci.* **2016**, *75*, 63. [CrossRef]
61. Costache, R. Flash-flood Potential Index mapping using weights of evidence, decision Trees models and their novel hybrid integration. *Stoch. Environ. Res. Risk Assess.* **2019**, *33*, 1375–1402. [CrossRef]
62. Skentos, A. Topographic Position Index based landform analysis of Messaria (Ikaria Island, Greece). *Acta Geobalcanica* **2018**, *4*, 7–15. [CrossRef]
63. Bui, D.T.; Pradhan, B.; Revhaug, I.; Tran, C.T. A comparative assessment between the application of fuzzy unordered rules induction algorithm and J48 decision tree models in spatial prediction of shallow landslides at Lang Son City, Vietnam. In *Remote Sensing Applications in Environmental Research*; Springer: Berlin/Heidelberg, Germany, 2014; pp. 87–111.
64. De Rosa, P.; Fredduzzi, A.; Cencetti, C. Stream Power Determination in GIS: An Index to Evaluate the Most'Sensitive'Points of a River. *Water* **2019**, *11*, 1145. [CrossRef]
65. Corrao, M.V.; Link, T.E.; Heinse, R.; Eitel, J.U. Modeling of terracette-hillslope soil moisture as a function of aspect, slope and vegetation in a semi-arid environment. *Earth Surf. Process. Landf.* **2017**, *42*, 1560–1572. [CrossRef]

66. Zhang, K.; Ruben, G.B.; Li, X.; Li, Z.; Yu, Z.; Xia, J.; Dong, Z. A comprehensive assessment framework for quantifying climatic and anthropogenic contributions to streamflow changes: A case study in a typical semi-arid North China basin. *Environ. Model. Softw.* **2020**, 104704. [CrossRef]
67. Zhang, K.; Wang, Q.; Chao, L.; Ye, J.; Li, Z.; Yu, Z.; Yang, T.; Ju, Q. Ground Observation-based Analysis of Soil Moisture Spatiotemporal Variability Across A Humid to Semi-Humid Transitional Zone in China. *J. Hydrol.* **2019**, *574*, 903–914. [CrossRef]
68. Singh, C.; Walia, E.; Kaur, K.P. Enhancing color image retrieval performance with feature fusion and non-linear support vector machine classifier. *Optik* **2018**, *158*, 127–141. [CrossRef]
69. Lin, S.-W.; Lee, Z.-J.; Chen, S.-C.; Tseng, T.-Y. Parameter determination of support vector machine and feature selection using simulated annealing approach. *Appl. Soft Comput.* **2008**, *8*, 1505–1512. [CrossRef]
70. Lee, S.; Kim, Y.-S.; Oh, H.-J. Application of a weights-of-evidence method and GIS to regional groundwater productivity potential mapping. *J. Environ. Manag.* **2012**, *96*, 91–105. [CrossRef]
71. Van Westen, C. *Statistical Landslide Hazards Analysis, ILWIS 2.1 for Windows Application Guide;* International Institute for Aerospace Survey and Earth Sciences (ITC) Publication: Enschede, The Netherlands, 1997.
72. Lee, S.; Pradhan, B. Landslide hazard mapping at Selangor, Malaysia using frequency ratio and logistic regression models. *Landslides* **2007**, *4*, 33–41. [CrossRef]
73. Nielsen, M.A. *Neural Networks and Deep Learning;* Determination Press: San Francisco, CA, USA, 2015; Volume 25.
74. Schmidhuber, J. Deep learning in neural networks: An overview. *Neural Netw.* **2015**, *61*, 85–117. [CrossRef]
75. Agarap, A.F. Deep learning using rectified linear units (relu). *arXiv* **2018**, arXiv:180308375.
76. Bui, Q.-T.; Nguyen, Q.-H.; Nguyen, X.L.; Pham, V.D.; Nguyen, H.D.; Pham, V.-M. Verification of novel integrations of swarm intelligence algorithms into deep learning neural network for flood susceptibility mapping. *J. Hydrol.* **2020**, *581*, 124379. [CrossRef]
77. Huang, Z.; Li, J.; Weng, C.; Lee, C.-H. Beyond Cross-Entropy: Towards Better Frame-Level Objective Functions for Deep Neural Network Training in Automatic Speech Recognition. In Proceedings of the INTERSPEECH—15th Annual Conference of the International Speech Communication Association, Singapore, 14–18 September 2014.
78. Goodfellow, I.; Bengio, Y.; Courville, A. *Deep Learning;* MIT Press: Cambridge, MA, USA, 2016; ISBN 0-262-33737-1.
79. Wu, Y.; Ke, Y.; Chen, Z.; Liang, S.; Zhao, H.; Hong, H. Application of alternating decision tree with AdaBoost and bagging ensembles for landslide susceptibility mapping. *Catena* **2020**, *187*, 104396. [CrossRef]
80. Hong, H.; Pradhan, B.; Xu, C.; Bui, D.T. Spatial prediction of landslide hazard at the Yihuang area (China) using two-class kernel logistic regression, alternating decision tree and support vector machines. *Catena* **2015**, *133*, 266–281. [CrossRef]
81. Freund, Y.; Mason, L. The Alternating Decision Tree Learning Algorithm. *ICML* **1999**, *99*, 124–133.
82. Khosravi, K.; Pham, B.T.; Chapi, K.; Shirzadi, A.; Shahabi, H.; Revhaug, I.; Prakash, I.; Bui, D.T. A comparative assessment of decision trees algorithms for flash flood susceptibility modeling at Haraz watershed, northern Iran. *Sci. Total Environ.* **2018**, *627*, 744–755. [CrossRef] [PubMed]
83. Sahana, M.; Pham, B.T.; Shukla, M.; Costache, R.; Thu, D.X.; Chakrabortty, R.; Satyam, N.; Nguyen, H.D.; Phong, T.V.; Le, H.V. Rainfall Induced Landslide Susceptibility Mapping Using Novel Hybrid Soft Computing Methods Based on Multi-layer Perceptron Neural Network Classifier. *Geocarto Int.* **2020**, 1–25. [CrossRef]
84. Wang, S.; Zhang, K.; van Beek, L.P.; Tian, X.; Bogaard, T.A. Physically-based landslide prediction over a large region: Scaling low-resolution hydrological model results for high-resolution slope stability assessment. *Environ. Model. Softw.* **2020**, *124*, 104607. [CrossRef]

sensors

MDPI

Article

Modular Neural Networks with Fully Convolutional Networks for Typhoon-Induced Short-Term Rainfall Predictions

Chih-Chiang Wei * and Tzu-Heng Huang

Department of Marine Environmental Informatics and Center of Excellence for Ocean Engineering, National Taiwan Ocean University, Keelung 20224, Taiwan; Azeros@proguide.com.tw

* Correspondence: ccwei@ntou.edu.tw

Abstract: Taiwan is located at the edge of the northwestern Pacific Ocean and within a typhoon zone. After typhoons are generated, strong winds and heavy rains come to Taiwan and cause major natural disasters. This study employed fully convolutional networks (FCNs) to establish a forecast model for predicting the hourly rainfall data during the arrival of a typhoon. An FCN is an advanced technology that can be used to perform the deep learning of image recognition through semantic segmentation. FCNs deepen the neural net layers and perform upsampling on the feature map of the final convolution layer. This process enables FCN models to restore the size of the output results to that of the raw input image. In this manner, the classification of each raw pixel becomes feasible. The study data were radar echo images and ground station rainfall information for typhoon periods during 2013–2019 in southern Taiwan. Two model cases were designed. The ground rainfall image-based FCN (GRI_FCN) involved the use of the ground rain images to directly forecast the ground rainfall. The GRI combined with rain retrieval image-based modular convolutional neural network (GRI-RRI_MCNN) involved the use of radar echo images to determine the ground rainfall before the prediction of future ground rainfall. Moreover, the RMMLP, a conventional multilayer perceptron neural network, was used to a benchmark model. Forecast horizons varying from 1 to 6 h were evaluated. The results revealed that the GRI-RRI_MCNN model enabled a complete understanding of the future rainfall variation in southern Taiwan during typhoons and effectively improved the accuracy of rainfall forecasting during typhoons.

Keywords: typhoon; rainfall; convolutional networks; image segmentation; prediction

Citation: Wei, C.-C.; Huang, T.-H. Modular Neural Networks with Fully Convolutional Networks for Typhoon-Induced Short-Term Rainfall Predictions. *Sensors* **2021**, *21*, 4200. https://doi.org/10.3390/s21124200

Academic Editor: Moulay A. Akhloufi

Received: 18 April 2021
Accepted: 16 June 2021
Published: 18 June 2021

1. Introduction

Taiwan is located in the northwestern Pacific Ocean within an area frequently hit by typhoons. After their formation, typhoons often move along the west Pacific Ocean and strike Taiwan with strong winds and torrential rain. On average, three to four typhoons land in Taiwan each year [1]. Southern Taiwan lies in a subtropical zone. The main rainy season in southern Taiwan is the typhoon season between May and October. Nearly no rainfall occurs in the other months. Therefore, the main water source in southern Taiwan is the rainfall caused by typhoons. However, the short-duration heavy rainfall of typhoons not only provides abundant water but also causes disasters, such as debris flows, river water surges, and downstream flooding [2,3]. Typhoons commonly strike southern Taiwan, for example Typhoon Fung-Wong in 2014 and Typhoons Nepartak, Meranti, and Typhoon Megi in 2016, which caused severe disasters and property losses [4,5]. Therefore, an accurate rainfall forecasting model is urgently required for southern Taiwan to accurately predict the real-time rainfall during typhoon periods and prevent the disasters resulting from heavy rainfall in local areas.

In recent years, considerable developments have occurred in machine learning (ML). Scholars have used various ML-based algorithms along with ground observation data, namely one-dimensional (1-D) data, for precipitation estimation and prediction; for example, artificial neural networks [6–10] and support vector machines [11,12] have been employed to predict rainfall using 1-D ground rainfall data. Although rain gauges provide relatively accurate point rainfall estimates near the ground surface, they cannot effectively capture the spatial variability of rainfall [13,14].

Remote sensing has attracted increasing attention in weather analysis and forecasting. Various types of image data have been collected for remote sensing applications. The development of weather surveillance radars has enabled quantitative precipitation estimation with extremely high spatial resolutions. Weather radars, which have the advantages of wide coverage and round-the-clock observation, are critical devices for meteorological observation [15]. Accordingly, the application of two-dimensional (2-D) radar images compensates for the insufficient 1-D spatial rainfall data collected from land-based observation stations. Many studies have used the statistical relationships between the radar reflectivity and the rain rate or nonlinear regression to establish rainfall estimation models. These studies have achieved favorable outcomes [16–25]. However, the interpretation of these image data is a crucial emerging topic.

Deep learning (DL) is a prominent branch of ML. DL mainly involves using neural-network-based ML algorithms to develop advanced computational technology that can be applied in image recognition. A DL neural network structure is a multilayer neural network architecture that uses two-dimensional matrices to calculate images. Therefore, advanced computer processing units (i.e., graphics cards) are required to execute DL tasks successfully [26]. The convolutional neural networks (CNNs) developed by LeCun et al. [27] is a basic DL image recognition technology. The structure of the CNN model comprises a convolutional layer and pooling layer. A complete CNN model is established using a fully connected layer, which converts two-dimensional images into one-dimensional arrays, and multilayer perceptron network model structures [28]. Such a network structure enables the CNN model to achieve favorable image recognition accuracy [29–32]. CNN algorithms have also been successfully applied to rainfall estimation and hydrological problems. For example, Pan et al. [33] used CNN model stacks with several convolution and pooling operators to extract intricate and valuable circulation features for precipitation estimation. Sadeghi et al. [34] estimated the precipitation rate by processing images in the infrared (IR) and water vapor bands (obtained from geostationary satellites) by using CNNs. Wang et al. [35] proposed the dilated causal CNN model to predict the water level changes during typhoons. Wei [36] proposed a regional extreme precipitation and construction suspension estimation system and used a deep CNN model to enhance the extreme rainfall forecasting capability of this system. The aforementioned studies developed CNNs for precipitation susceptibility mapping by using various 2-D remote images.

Newly emerging DL skills were employed in the study case. First, a fully convolutional network (FCN) developed by Long et al. [37] was employed to conduct image recognition. The FCN was developed as an extension of the CNN for semantic segmentation to address the shortcomings of CNN and increase the prediction accuracy for the rapid recognition of various object representations. To facilitate the pixel classification of images, upsampling was conducted in the FCN model for classifying every pixel on the feature map of the final convolutional layer. The FCN model used deconvolution to match the class of every pixel in a feature map with the corresponding class in the original image and thus solved the problem of semantic segmentation. To the best of our knowledge, few studies have used FCNs for rainfall estimation and prediction. Moreover, Eppel [38] proposed modular convolutional neural networks (MCNNs) that apply FCNs to segment an image into vessel and background area; in that study, the vessel region was used as an input for a second net that recognized the contents of a glass vessel.

The current study developed a DL-based rainfall prediction model, for which the source data are both 1-D ground observation data and 2-D remote sensing imageries, to predict precipitation during typhoons. Southern Taiwan was selected as the research area. This study used the hourly rainfall data of ground stations and radar echo images in southern Taiwan to establish an hourly rainfall forecast model. Toward the aforementioned goal, this study has the following features:

(1) This study employed an FCN, which employs the convolutional and pooling layers for extracting image features, to predict the precipitation during typhoons.
(2) To address the input–output patterns in the FCN modeling process using 2-D array data, this study converted the rainfall data of ground stations into 2-D images.
(3) This study employed the net architecture of MCNN with FCNs, which enabled the integration of the radar echo image and ground observation data as model inputs for enhancing the accuracy of rainfall intensity prediction.

2. Experimental Area and Data

2.1. Region and Gauges

The longitude and latitude ranges of southern Taiwan are 120.11–121.59° E and 22.00–23.34° N, respectively (Figure 1). The area of southern Taiwan is 11,434 km^2, which accounts for 31.59% of the total area of Taiwan. As displayed in the right part of Figure 1, southern Taiwan has 51 weather stations, comprising six Central Weather Bureau (CWB) weather stations (red dots) and 45 automatic detection stations (blue dots). The CWB weather stations are located at Tainan, Kaohsiung, Hengchun, Taitung, Dawu, and Lanyu (coordinates are provided in Table 1). This study used the six CWB weather stations as the experimental sites.

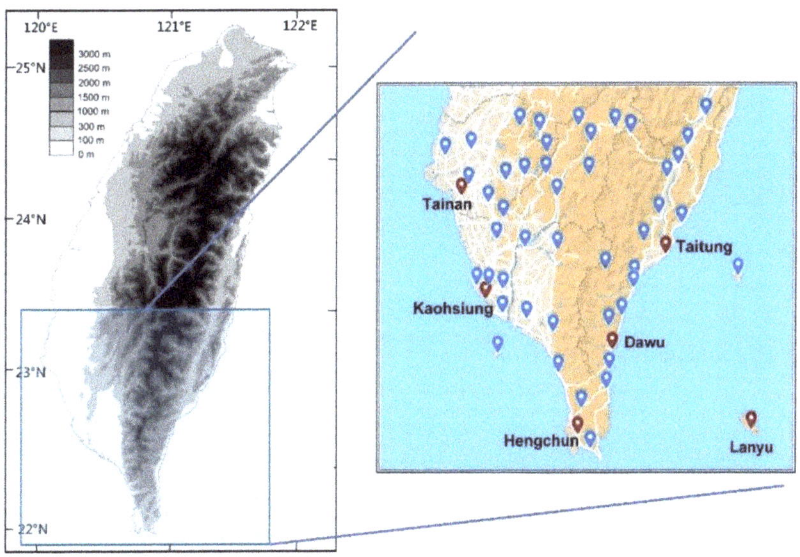

Figure 1. Map of the research area.

Table 1. Weather station information.

Station	Elevation (m)	Longitude (° E)	Latitude (° N)
Tainan	40.8	120.2047	22.9932
Kaohsiung	2.3	120.3157	22.5660
Hengchun	22.1	120.7463	22.0038
Dawu	8.1	120.9038	22.3557
Taitung	9.0	121.1546	22.7522
Lanyu	324.0	121.5583	22.0370

The left part of Figure 1 indicates that the Central Mountain Range (CMR) runs south–north and divides Taiwan into the eastern and western regions. The total length of the CMR is approximately 340 km, and its width from east to west is approximately 80 km.

The average altitude of the range is approximately 2500 m [39]. The Tainan and Kaohsiung stations are located to the west of the CMR, the Hengchun station is located to the south of the CMR, the Taitung and Dawu stations are located to the east of the CMR, and the Lanyu station is located in an outlying island (bottom right of Figure 1).

2.2. Typhoons and Radar Mosaics

In Taiwan, the CWB creates radar echo images (REIs) by using different colors to represent the spatial echo intensity of the reflected signals received by radars from rain particles [40]. REIs are used to reflect the variations of water vapor during typhoon circulation. Wu and Kuo [41] indicated that useful typhoon-related data can be obtained when a typhoon affects Taiwan by setting up an around-the-island Doppler radar network, enhanced surface rain gauge network, and integrated sounding system. This study collected radar images starting from 2013 because the resolution and color appearance of these images were different from those of the radar images captured before 2013. According to the CWB's Typhoon Database [42], 22 typhoon events occurred in Taiwan from 2013 to 2019 (Table 2).

Table 2. Typhoon events in Taiwan from 2013 to 2019.

Typhoon	Periods	Intensity	Pressure at Typhoon Center (hPa)	Maximum Wind Speed of Typhoon Center (m/s)
Soulik	2013/07/11–13	Severe	925	51
Cimaron	2013/07/17–18	Mild	998	18
Trami	2013/08/20–22	Mild	970	30
Kong-Rey	2013/08/27–29	Mild	980	25
Usagi	2013/09/19–22	Severe	910	55
Habigis	2014/06/14–15	Mild	992	20
Matmo	2014/07/21–23	Moderate	960	38
Fung-Wong	2014/09/19–22	Mild	985	25
Noul	2015/05/10–11	Severe	925	51
Linfa	2015/07/06–09	Mild	975	30
Chanhom	2015/07/09–11	Moderate	935	48
Soudelor	2015/08/06–09	Moderate	930	48
Goni	2015/08/20–23	Severe	925	51
Dujuan	2015/09/27–29	Severe	925	51
Nepartak	2016/07/06–09	Severe	905	58
Meranti	2016/09/12–15	Severe	900	60
Megi	2016/09/25–28	Moderate	940	45
Nesat	2017/07/28–30	Moderate	955	40
Hatitang	2017/07/29–31	Mild	990	20
Hato	2017/08/20–22	Moderate	965	33
Guchol	2017/09/06–07	Mild	998	18
Bailu	2019/08/24–25	Mild	975	30

According to the CWB, the maximum wind speeds of mild, moderate, and severe typhoons are 17.2–32.6, 32.7–50.9, and >51 m/s, respectively. Seven severe, six moderate, and nine mild typhoons occurred in southern Taiwan during the study period.

Figure 2 displays the accumulated precipitation of each typhoon in descending order. The top nine typhoons in terms of precipitation, namely Typhoons Trami, Kong-Rey, Usagi, Habigis, Fung-Wong, Nepartak, Meranti, Megi, and Hato, had relatively high precipitation (accumulated precipitation > 100 mm), whereas the others had relatively low precipitation.

This study collected 1412 radar mosaic images with a resolution of 1024 × 1024 pixels. Here, one pixel corresponded to an actual distance of 0.7 × 0.7 km. Figure 3 displays the REIs of nine typhoons approaching the study region. These typhoons all resulted in accumulated precipitation >100 mm.

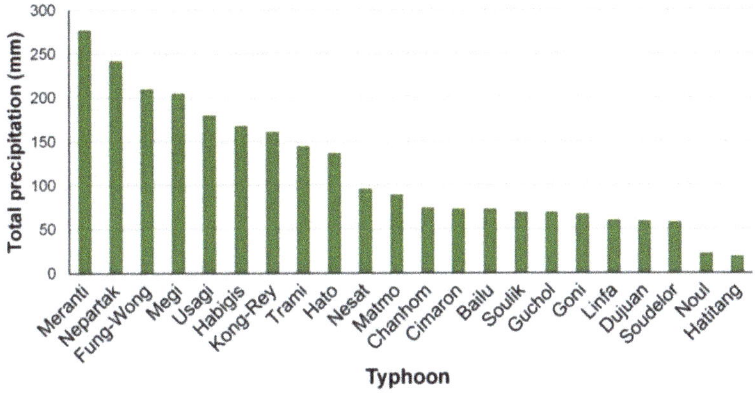

Figure 2. Total precipitation of typhoons between 2013 and 2019.

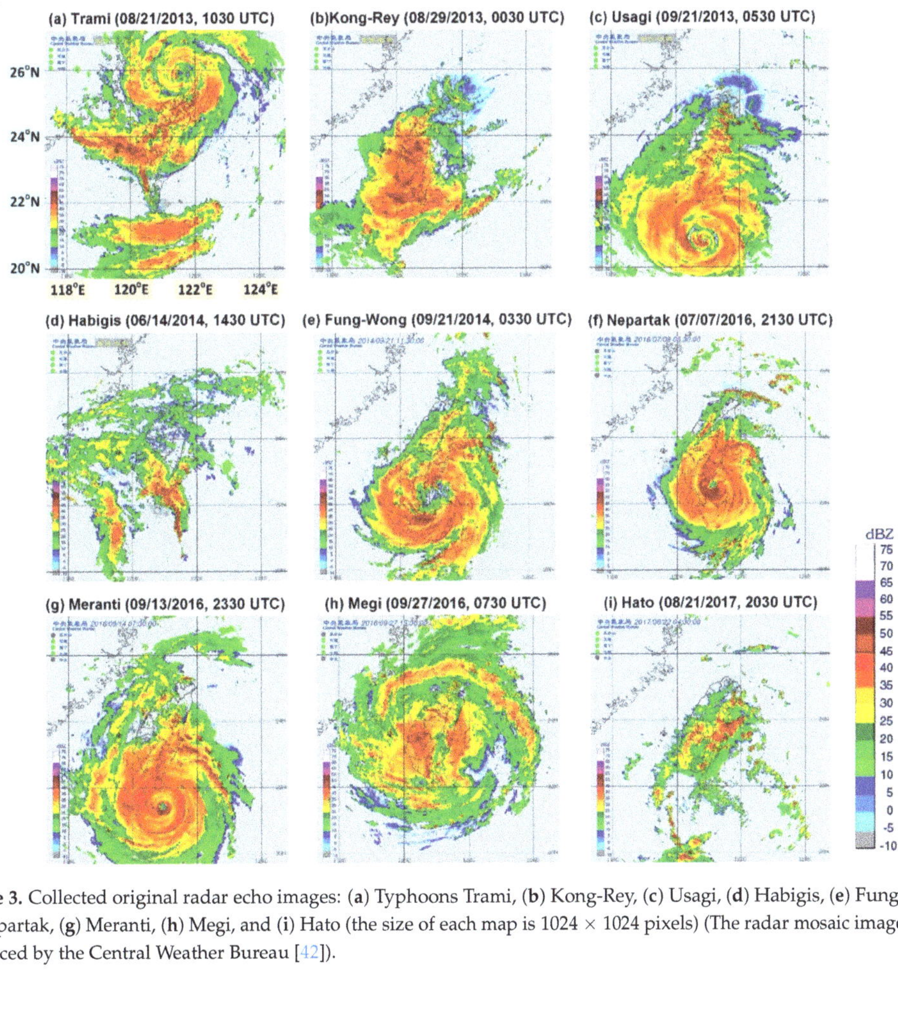

Figure 3. Collected original radar echo images: (**a**) Typhoons Trami, (**b**) Kong-Rey, (**c**) Usagi, (**d**) Habigis, (**e**) Fung-Wong, (**f**) Nepartak, (**g**) Meranti, (**h**) Megi, and (**i**) Hato (the size of each map is 1024 × 1024 pixels) (The radar mosaic images were produced by the Central Weather Bureau [42]).

3. Model Development

This study used the Python programming language to establish models. The Tensorflow (version 2.1) and Keras libraries of Python were used for ML computation. The model computation environment was an ASUS-TS300E9 computer (ASUSTek Computer Inc., Taipei City, Taiwan). The computer clock rate was 3.5 GHz. The computer included 16 GB RAM (DDR4-2400) and a GeForce GTX 1080 Ti X 11G graphics card (Micro-Star International Co., Ltd., New Taipei City, Taiwan).

3.1. Data Division

This study divided the data of typhoon events into training, validation, and testing sets. The training sets were used to tune the model parameters, and the validation sets were used to verify the trained model. To avoid the data leakage and bias problem in the rainfall prediction model, this study randomly split the typhoons ranked 1 to 9 in terms of precipitation into training, validation, and testing sets; that is, rank = 2, 6 and 9 for training set (Nepartak, Habigis, and Hato), rank = 3, 5 and 8 for validation set (Fung-Wong, Usagi, and Trami), and rank = 1, 4 and 7 for testing set (Meranti, Megi, and Kong-Rey). In addition, the remaining typhoons (relative low precipitation) were added for training set. In total, the training, validation, and testing sets comprised 926, 240 and 246 hourly records, respectively.

3.2. Image Preprocessing

In the study, all the inputs and outputs in the modeling process in this study were two-dimensional images. First, when labeling the REI images, the latitudinal and longitudinal range of the original radar images was 117.32–124.79° E and 21.70–27.17° N (Figure 3). Because the original images had a wide geographical range, cropping was required to obtain the image size of study area (120.11–121.59° E and 22.00–23.34° N). Therefore, the raw REIs were cut to a size of 192 × 192 pixels to completely cover the study area. According to the legend of dBZ (Figure 3), there are 17 colors (where dBZ ranging from −10 to 75 dBZ, divided by 5 dBZ). Therefore, the number of categories was 17. These REI images were then encoded into RGB channels (i.e., red, green, and blue) and pixel values at each channel are integer values between 0 and 255. Here, a one-hot encoding was applied to the RGB representation of an REI image when pixel-based images were used as the model inputs.

Second, the rainfall data of ground stations had to be converted into two-dimensional ground rainfall images (namely GRIs). The inverse distance weighting method proposed by Shepard [43] was employed. In this method, an interpolating function is used to identify an interpolated value at a given point based on samples by using the inverse distance weighting method as follows:

$$u(\mathbf{x}) = \begin{cases} \frac{\sum_{i=1}^{N} w_i(\mathbf{x}) u_i}{\sum_{i=1}^{N} w_i(\mathbf{x})}, \text{ if } d(\mathbf{x}, \mathbf{x}_i) \neq 0 \text{ for all } i \\ u_i, \text{ if } d(\mathbf{x}, \mathbf{x}_i) = 0 \text{ for some } i \end{cases} \quad (1)$$

where $w_i(\mathbf{x}) = \frac{1}{d(\mathbf{x},\mathbf{x}_i)^p}$ is a weighting function; $\mathbf{x}$ denotes an interpolated (unknown) point; $\mathbf{x}_i$ is an interpolating (known) point; d is a given distance from $\mathbf{x}_i$ to $\mathbf{x}$; N is the total number of known points used in interpolation; and p is a positive real number, called the power parameter.

This study employed the commonly used p = 2 and subsequently identified the suitable N value. This study found that when $N \leq 4$, the GRIs were varied; however, when $N \geq 5$, the GRIs were more stable and invariant. Figure 4 depicts the GRIs of Typhoons Trami, Kong-Rey, Usagi, Habigis, Fung-Wong, Nepartak, Meranti, Megi, and Hato using the inverse distance weighting method when p = 2 and N = 5. Here, the size of GRI maps is the same as the cropped REI maps (i.e., 192 × 192 pixels). Subsequently, when labeling the GRI images, this study partitioned the precipitation scale into several intervals to label the categorical values. According to the collected typhoons, the range of rain rate from 0 to 76 mm/h. This study divides the rain rate by 5 mm/h. Here, we let the no rain as a special case, as class "0". Thus, the total number of rain intensity categories was 17. For example, if the rain rate was 13 mm/h, it was labeled as class "3". Then, each pixel of the GRI images can be labeled by classes 0 to 16. Finally, these GRI images were encoded into RGB channels when the GRI images were used as the model targets.

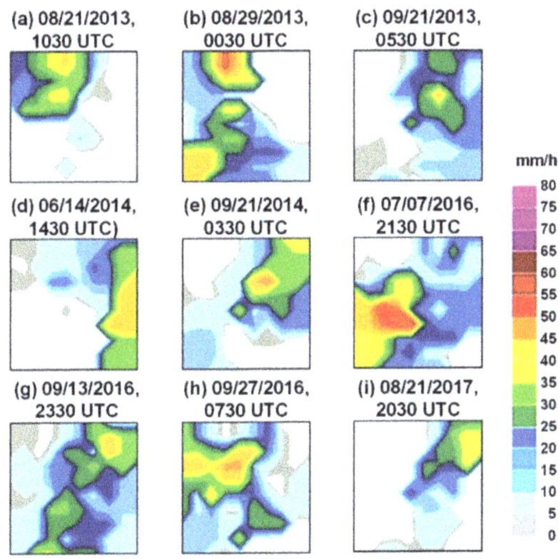

Figure 4. Generated GRIs of Typhoons (**a**) Trami, (**b**) Kong-Rey, (**c**) Usagi, (**d**) Habigis, (**e**) Fung-Wong, (**f**) Nepartak, (**g**) Meranti, (**h**) Megi, and (**i**) Hato. (the size of each map is 192 × 192 pixels).

3.3. Designed Model Cases

In this study, two rainfall prediction models were developed on the basis of two types of neural networks: The GRI-based FCNs (GRI_FCNs) and GRI combined with rain retrieval image (RRI)-based MCNNs (GRI-RRI_MCNNs). The developed GRI_FCN (Figure 5) adopted segmentation steps using a standard FCN, which segmented the image into objects by classifying every pixel in the image into one of a given set of categories. The framework of the GRI_FCN included input, downsampling, upsampling, and output layers. Before FCN modeling was conducted, the 1-D rainfall data of ground stations were converted into 2-D GRIs. In the GRI_FCN model, the GRIs were adopted to predict the ground rainfall directly, and the output results were the predicted GRIs.

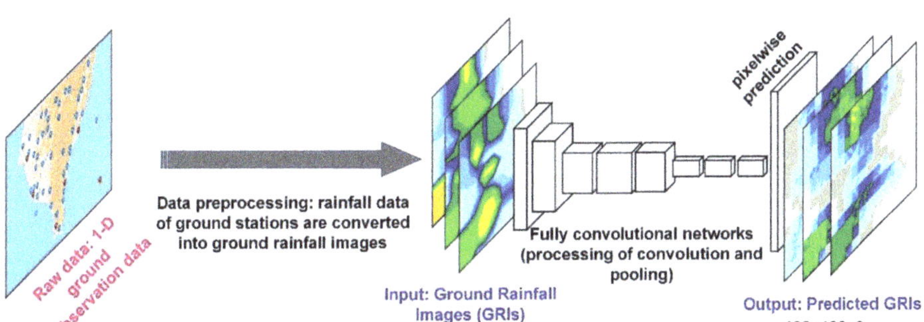

Figure 5. Architecture of the GRI-based fully convolutional networks. (an image of GRI contains a three-dimensional array of size $h \times w \times d$, where h = 192 and w = 192 are spatial dimensions, and d = 3 is the color channel dimension).

The GRI-RRI_MCNN model employed a modular semantic segmentation approach using serially connected FCN networks. The first FCN net identified current ground precipitation, and the output of this net (i.e., rain retrievals) was used by a second FCN net to identify and segment the future ground precipitation (i.e., rain predictions; Figure 6). The GRI-RRI_MCNN involved two steps: in step 1, REIs were used to retrieve the ground rainfall (the GRIs are the model learning targets). The outputs were RRIs. Step 2 involved the fusion (using a summation method) of the RRIs and GRIs obtained in step 1 to create new images. These new images were subsequently used as the input to predict the ground rainfall, and the output results were the predicted GRI images.

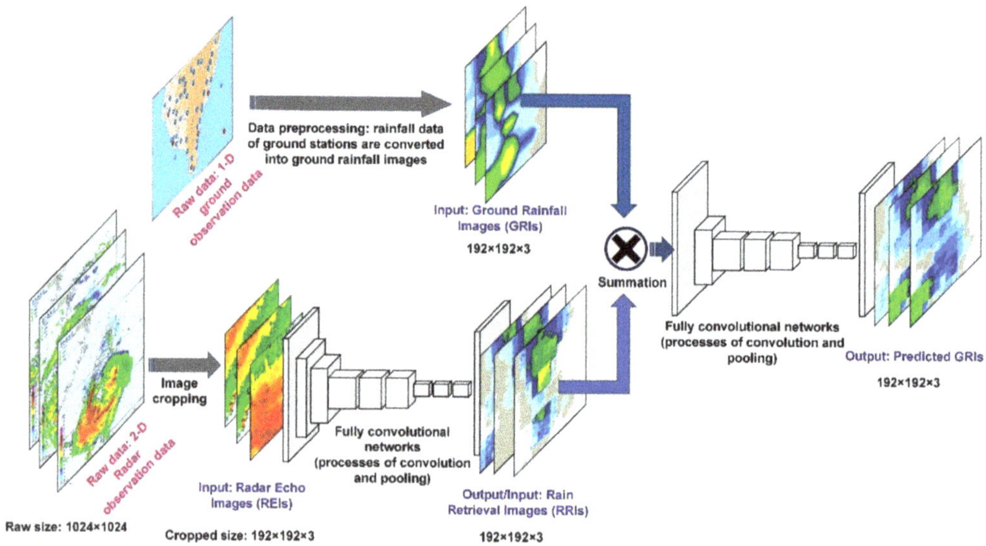

Figure 6. Architecture of blending GRI-RRI-based modular convolutional neural networks. (the images of GRI, REI and RRI contain a three-dimensional array of size 192 × 192 × 3).

The convolution and pooling processes of the FCN in GRI_FCNs and GRI-RRI_MCNNs were identical to those of the CNN. The net architecture of the CNN has been described by [27,44]. In general, CNNs are constructed by stacking two types of interweaved layers: convolutional and pooling (subsampling) layers [45]. The convolutional layer is the core component of a CNN. This layer outputs feature maps by computing the dot product between the local region in the input feature maps and a filter. The pooling layer performs downsampling on feature maps by computing the maximum or average value of a sub-region [46]. An FCN has more neural net layers than a CNN does. An FCN conducts upsampling on the feature map of the final convolution layer. This design enables FCN models to restore the size of the output results to that of the raw input images. Therefore, the classification is performed for every raw image pixel [37]. An FCN can theoretically accept an input image of any size and produce output images of the same size because an FCN is trained end-to-end for pixel-to-pixel semantic segmentation (or pixel-wise prediction).

When running the GRI_FCN and GRI-RRI_MCNN models, the parameter settings of the convolutional and pooling layers were as follows: kernel size = (2, 2), padding method = same, maxpooling with filter size = (2, 2), strides = (2, 2), and the activation function = rectified linear unit function. Moreover, the settings of output layers were as follows: kernel size = (8, 8), strides = (8, 8), and the activation function = softmax function. The loss function was categorical cross entropy. The number of intermediate layers in the FCNs can be seen in the following section.

3.4. Modeling

Two types of neural network models (i.e., GRI_FCN and GRI-RRI_MCNN models) were established to examine the suitable network structures and image size. First, this study evaluated the accuracies of the FCN-32s, FCN-16s, and FCN-8s architectures by using the GRI_FCN model. Figure 7 reveals the intermediate layers (involving convolution layers and pooling layers) in these FCNs. These FCN-type architectures contained the processes of conv1–conv7 and pool1–pool5. In the figure, FCN-32s upsampled stride 32 predictions back to pixels in a single step. Subsequently, FCN-16s combined stride 16 predictions from both the final layer and the pool4 layer, at stride 16, while retaining high-level semantic information. Finally, FCN-8s used additional predictions from pool3, at stride 8, to enhance precision. The FCN employed the upsampling method to increase the pixel accuracy of the output results. Table 3 lists the total numbers of trainable variables in the FCN-32s, FCN-16s, and FCN-8s for GRI_FCN and GRI-RRI_MCNN models.

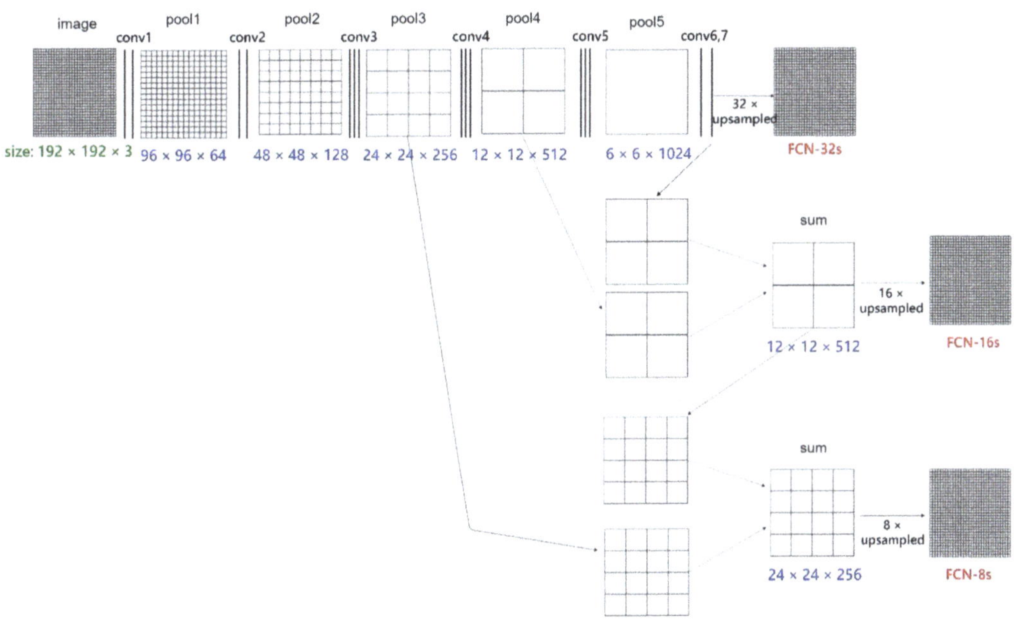

Figure 7. Architecture of GRI_FCN-based FCN-32s, FCN-16s, and FCN-8s and the size information of input images and feature maps in each conv-pool stage. (these FCN-type architectures contain the processes of conv1–conv7 and pool1–pool5; the architecture was referred to [37] and modified for modeling the model cases in the work).

Table 3. Total numbers of trainable variables in the GRI_FCN and GRI-RRI_MCNN models.

Model	FCN-32s	FCN-16s	FCN-8s
GRI_FCN	1.175×10^8	1.343×10^8	1.351×10^8
GRI-RRI_MCNN	2.350×10^8	2.685×10^8	2.701×10^8

Figure 8 depicts the learning curves of GRI_FCNs and GRI-RRI_MCNNs for a FCN-8s network architecture using training set and validation set for a forecast horizon of 1 h. For the training set, the accuracy increased as the epoch number increased for both models (Figure 8a,c). In contrast, the accuracy for the validation set stops increasing after about 80 and 60 epochs for GRI_FCNs and GRI-RRI_MCNNs, respectively. Nonetheless, the categorical cross entropy loss decreased when the epoch number increased for both models (Figure 8b,d). In contrast, the loss values for the validation set began increasing after about

80 and 60 epochs for both models. In order to prevent overfitting, this study stopped training the models at around 80 and 60 epochs respectively for both models.

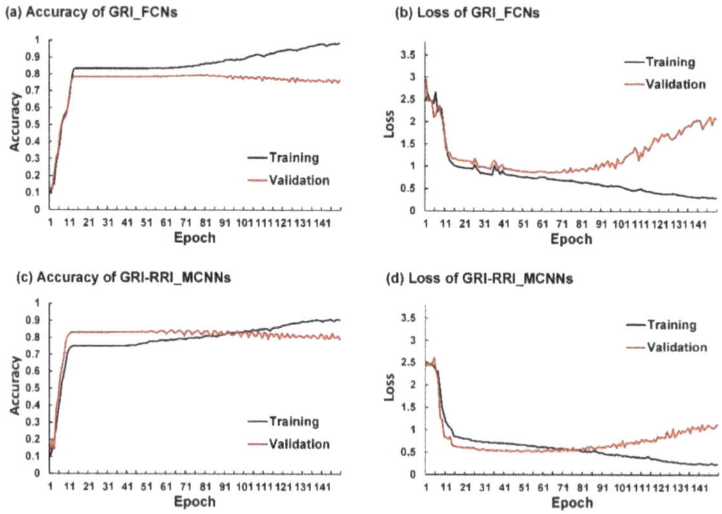

Figure 8. Learning curves for FCN-8s network architecture using training set (black line) and validation set (red line): (**a**) accuracy of GRI_FCNs; (**b**) loss of GRI_FCNs; (**c**) accuracy of GRI-RRI_MCNNs; (**d**) loss of GRI-RRI_MCNNs.

According to [47], the probability of detection (POD) is equal to the number of hits divided by the total number of rain observations; thus it gives a measure of the proportion of rain events successfully forecast. Here, the POD measure was employed to the evaluate the accuracy of per-rain-intensity-category. Figure 9 plots the diagram for POD scores for GRI_FCNs and GRI-RRI_MCNNs as FCN-based architectures were applied. In the figure, the POD scores decreased when the rain-intensity category number increased using GRI_FCNs and GRI-RRI_MCNNs. This trend implies that these cases might correctly predict light rain but misclassify for heavier rain.

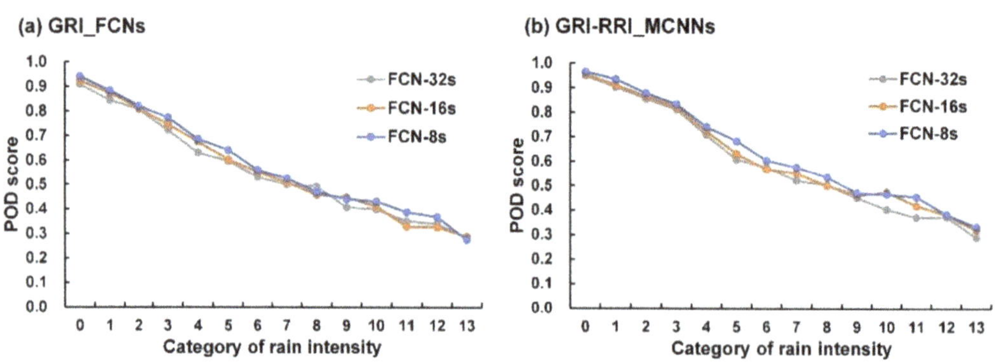

Figure 9. POD scores for FCN-based network architectures using validation set: (**a**) GRI_FCNs; (**b**) GRI-RRI_MCNNs.

Moreover, to evaluate overall accuracy, this study adopted two commonly used categorical metrics in semantic segmentation: pixel accuracy (PA) and mean intersection over union (MIoU). The PA represents the percentage of image pixels classified correctly. The

MIoU first computes the intersection over union for each semantic class and then computes the average over classes. Using the same processing, this study performed the weights training for a forecast horizon of 2–6 h. Table 4 lists the PA and MIoU performance metrics of FCN-32s, FCN-16s, and FCN-8s for forecasted horizons of 1–6 h. The results revealed that FCN-8s exhibited optimal performance in terms of the PA and MIoU. Therefore, this study used FCN-8s as the model structure.

Table 4. Accuracy performance of various network structures using the validation set.

Model	Network Structures		Forecasted Horizons (h)			
			t + 1	t + 2	t + 3	t + 6
GRI_FCN	FCN-32s	PA	77.8%	73.2%	68.9%	51.2%
		MIoU	55.2%	50.3%	42.9%	30.2%
	FCN-16s	PA	78.5%	74.6%	70.8%	52.7%
		MIoU	55.1%	52.4%	47.5%	32.2%
	FCN-8s	PA	79.4%	76.9%	72.7%	56.9%
		MIoU	56.7%	53.8%	48.8%	33.5%
GRI-RRI_MCNN	FCN-32s	PA	81.9%	76.7%	71.6%	55.6%
		MIoU	57.0%	53.7%	45.6%	31.9%
	FCN-16s	PA	82.8%	77.8%	74.3%	58.5%
		MIoU	57.5%	55.2%	48.6%	34.2%
	FCN-8s	PA	83.6%	79.2%	75.3%	60.9%
		MIoU	58.7%	56.8%	50.3%	36.5%

4. Simulation of Typhoons

4.1. Accuracy Results of the Testing Set

Rainfall prediction was performed for three typhoons (i.e., Kong-Rey, Meranti, and Megi) to evaluate the effectiveness of the designed GRI_FCN and GRI-RRI_MCNN models. Figure 10 displays the predicted GRI images when using the testing set. To examine the accuracy of model performance, this study also calculated the PA and MIoU metrics. Figure 11 reveals that the GRI-RRI_MCNN model outperformed the GRI_FCN model for all lead times.

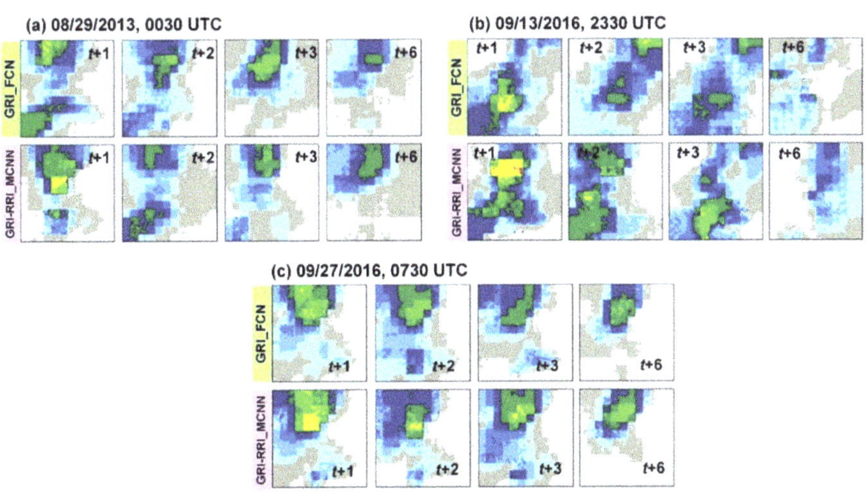

Figure 10. Predicted GRIs using the testing set: (**a**) Typhoons Kong-Rey, (**b**) Meranti, and (**c**) Megi.

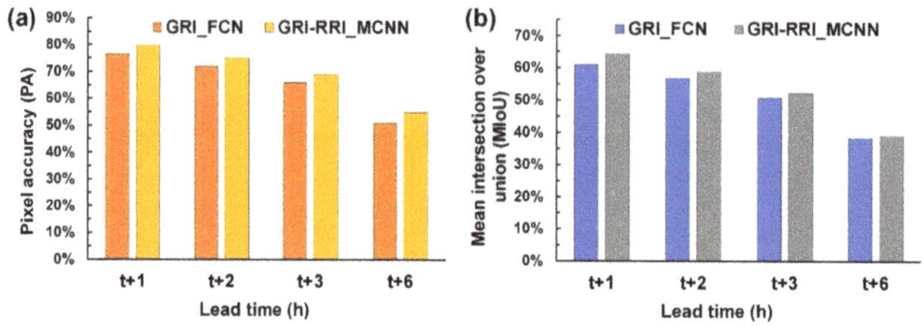

Figure 11. Accuracy performance of the GRI_FCN and GRI-RRI_MCNN in the testing set.

4.2. Evaluation of Rainfall Amounts at Weather Stations

The classified outputs of every pixel in the predicted GRIs in GRI_FCN and GRI-RRI_MCNN were subsequently transformed into original rain amounts (i.e., mm/h). The research region contained 51 weather stations, comprising six CWB weather stations and 45 automatic detection stations. This study selected six CWB weather stations (i.e., Tainan, Kaohsiung, Hengchun, Dawu, Taitung, and Lanyu), which are located in various parts of southern Taiwan, to evaluate the predicted rainfall amounts.

Wei and Hsieh [44] presented a radar mosaic-based multilayer perceptron (RMMLP) model, which is a conventional type of artificial neural networks that includes input, hidden, and output layers. The additional fully connected layer directly receives the cropped radar mosaic images to be flattened to a 1-D array. Here, the RMMLP model was used to a benchmark model and compared with those results made by GRI_FCN and GRI-RRI_MCNN in the six weather stations. Figures 12–14 depict the rainfall prediction results of the six weather stations during Typhoons Kong-Rey, Meranti, and Megi.

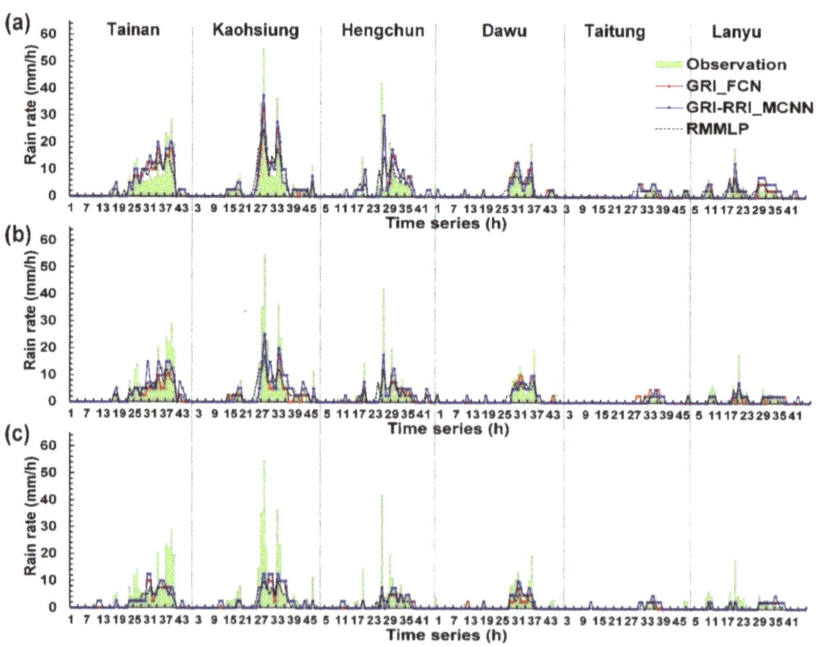

Figure 12. Station prediction results for Typhoon Kong-Rey at lead times of (**a**) 1 h, (**b**) 3 h, and (**c**) 6 h.

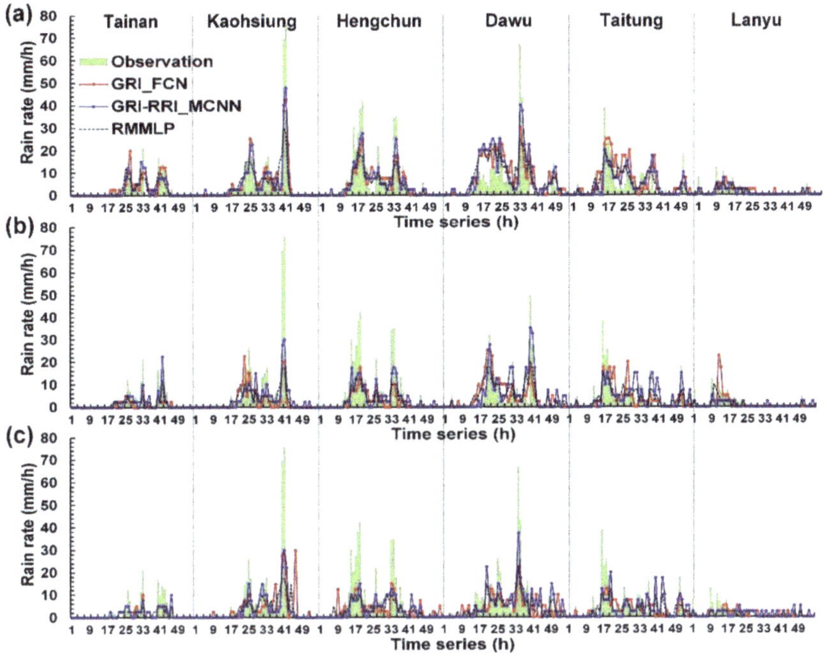

Figure 13. Station prediction results for Typhoon Meranti at lead times of (a) 1 h, (b) 3 h, and (c) 6 h.

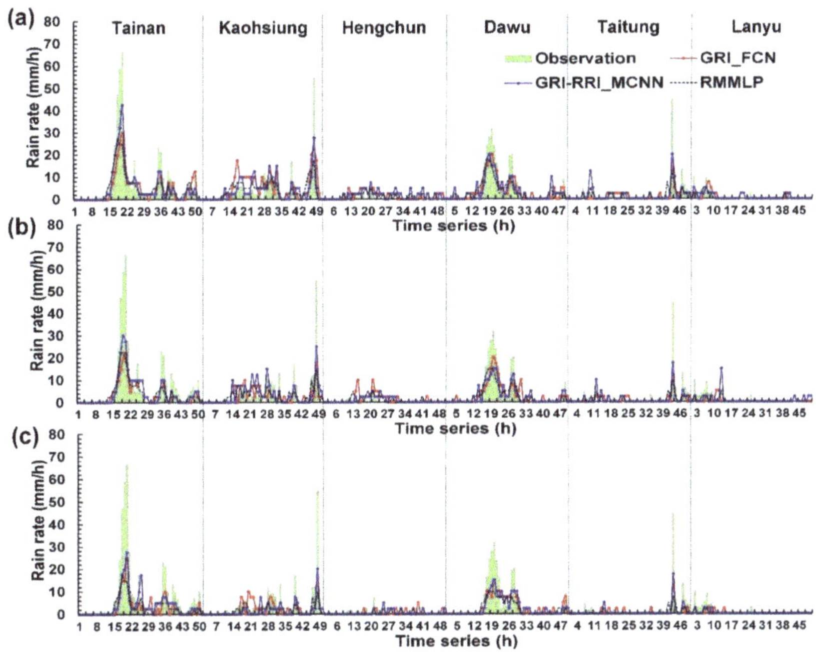

Figure 14. Station prediction results for Typhoon Megi at lead times of (a) 1 h, (b) 3 h, and (c) 6 h.

The tracks of Typhoons Kong-Rey, Meranti, and Megi are illustrated in Figure 13. First, the center of Typhoon Kong-Rey (Figure 15a) moved northward along the eastern coast of Taiwan. Although Typhoon Kong-Rey did not land in Taiwan, its circulation caused heavy rainfall in Taiwan. The highest maximum hourly rainfall data for Typhoon Kong-Rey were observed at the Kaohsiung station (55 mm/h), followed by the Hengchun (42.5 mm/h) station. The results of the prediction models indicated that when the lead time was 1 h (Figure 12a), the trends in the predicted and observed rainfall values for the stations were consistent; however, the peak rainfall was underestimated in the prediction models. When the lead times were 3 and 6 h (Figure 12b,c), more accurate prediction results were obtained in GRI-RRI_MCNN than in GRI_FCN and RMMLP.

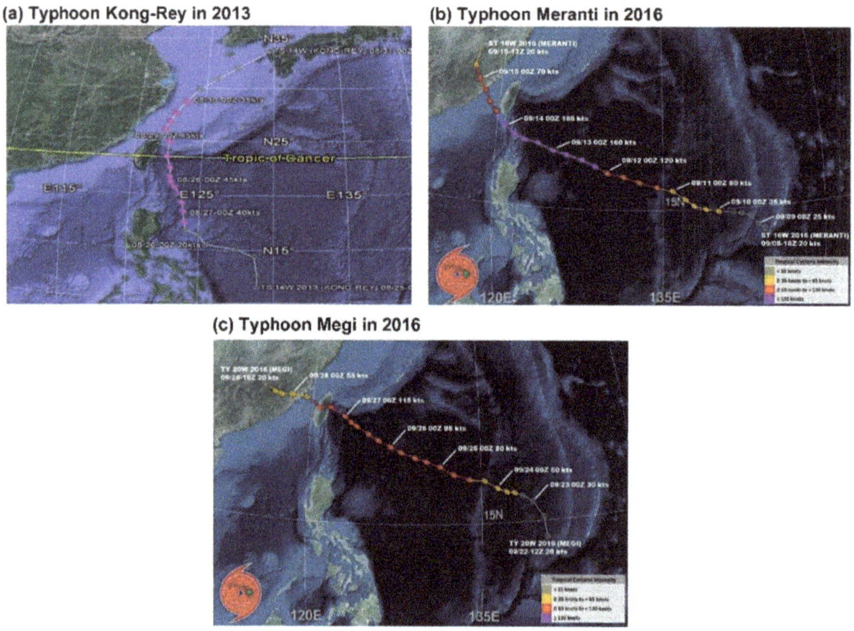

Figure 15. Paths of (**a**) Typhoon Kong-Rey, (**b**) Typhoon Meranti, and (**c**) Typhoon Megi (the maps were obtained from the website of the Joint Typhoon Warning Center [48].

The center of Typhoon Meranti (Figure 15b) passed through the Bashi Channel (near the Hengchun station) and moved northwestward toward Mainland China through the Taiwan Strait. Although Typhoon Meranti did not land in Taiwan, its circulation caused heavy rainfall in Taiwan. The highest maximum hourly rainfall in the western part of the study area was observed at the Kaohsiung station (76.0 mm/h) and that in the eastern part of the study area was observed at the Dawu station (67.0 mm/h). The prediction results in Figure 13 indicate that the rainfall tendencies of each station were accurately predicted by the models. The peak rainfall and volume of underestimation increased with the prediction time.

The center of Typhoon Megi (Figure 15c) moved eastward, landed in Taiwan, and subsequently passed through central Taiwan. After landing, the typhoon circulation covered almost all of Taiwan. When the typhoon center passed through the CMR, the circulation formed a windward slope in the western side of Taiwan, which resulted in heavy rainfall in this region. The highest maximum hourly rainfall was observed at the Tainan station (67.0 mm/h), followed by the Kaohsiung station (55.0 mm/h). The prediction results in Figure 14 indicate that the models accurately predicted the rainfall trends of each station.

4.3. Performance Levels for Predicted Rainfall Amounts

This study employed the mean absolute error (MAE), root mean square error (RMSE), relative MAE (rMAE), relative RMSE (rRMSE), and coefficient efficiency (CE) to calculate model performance for the predicted rainfall amounts. These criteria are defined as follows:

$$\mathrm{MAE} = \frac{1}{N}\sum_{t=1}^{N}\left|R_{t,\mathrm{pre}} - R_{t,\mathrm{obs}}\right| \tag{2}$$

$$\mathrm{RMSE} = \sqrt{\frac{\sum_{t=1}^{N}\left(R_{t,\mathrm{pre}} - R_{t,\mathrm{obs}}\right)^2}{N}} \tag{3}$$

$$\mathrm{CE} = 1 - \frac{\sum_{t=1}^{N}\left(R_{t,\mathrm{obs}} - R_{t,\mathrm{pre}}\right)^2}{\sum_{t=1}^{N}\left(R_{\mathrm{obs}} - \overline{R}_{\mathrm{obs}}\right)^2} \tag{4}$$

where N is the total number of observations, $R_{t,\mathrm{pre}}$ is the predicted rain rate at time t, $R_{t,\mathrm{obs}}$ is the observed rain rate at time t, $\bar{R}_{\mathrm{pre}}$ is the average of predicted rain rates, and $\bar{R}_{\mathrm{obs}}$ is the average of observed rain rates.

Figure 16 depicts the MAE, rMAE, RMSE, rRMSE, and CE of the results obtained at the six CWB stations. First, the absolute errors (i.e., the MAE and RMSE) were used to evaluate the obtained results (Figure 16a,c). The evaluation indicated that the absolute errors of GRI-RRI_MCNN were smaller than those of GRI_FCN and RMMLP. The values of the aforementioned parameters for the six stations in GRI-RRI_MCNN were compared. The results revealed that the Lanyu station had the lowest absolute errors among the six stations because this station was located at the sea and experienced limited rainfall and terrain effects. Among the remaining land stations, the largest absolute errors were observed at the Dawu station, followed by the Hengchun, Taitung, Kaohsiung, and Tainan stations.

Because the precipitation data of the typhoons differed among the stations, we used relative errors (i.e., the rMAE and rRMSE) to evaluate the quality of prediction. Figure 16b indicates that rMAE values of the different stations were not considerably different. Figure 16d indicates that the rRMSE exhibited greater differences among stations than the rMAE did. A comparison of the stations in mainland Taiwan revealed that the rRMSE variations at the Kaohsiung and Tainan stations were higher than those at the Dawu, Hengchun, and Taitung stations.

The overall CE was evaluated using the metric values for GRI-RRI_MCNN. As displayed in Figure 16e, the greatest CE was obtained for the Hengchun station, followed by the Tainan, Kaohsiung stations, Dawu, Taitung, and Lanyu stations. A higher prediction efficiency was obtained for the stations to the west of the CMR (i.e., the Hengchun, Tainan, and Kaohsiung stations) than for the stations to the east of the CMR (i.e., the Dawu, Taitung, and Lanyu stations).

To determine the model performance for each station for different lead times, the RMSE and CE curves of each station were plotted (Figure 17). Figure 17a displays the RMSE–CE–lead time curves for the Tainan station. The RMSE–CE–lead time curves for the other stations are displayed in Figure 17b–f. The curves in Figure 17 indicate that the case model errors increased, and the CE gradually decreased as the prediction time increased.

To understand the improved percentage of the predictions using GRI-RRI_MCNN and GRI_FCN models compared to the benchmark (i.e., RMMLP), we defined the improvement metric $\mathrm{IMP}_{\mathrm{CE}}$, as

$$\mathrm{IMP}_{\mathrm{CE}}\ (\%) = \left(\mathrm{CE}_i - \mathrm{CE}_{\mathrm{RMMLP}}\right) \times 100 \tag{5}$$

where CE_i is the CE value at a specific model, and $\mathrm{CE}_{\mathrm{RMMLP}}$ is the CE value at the benchmark.

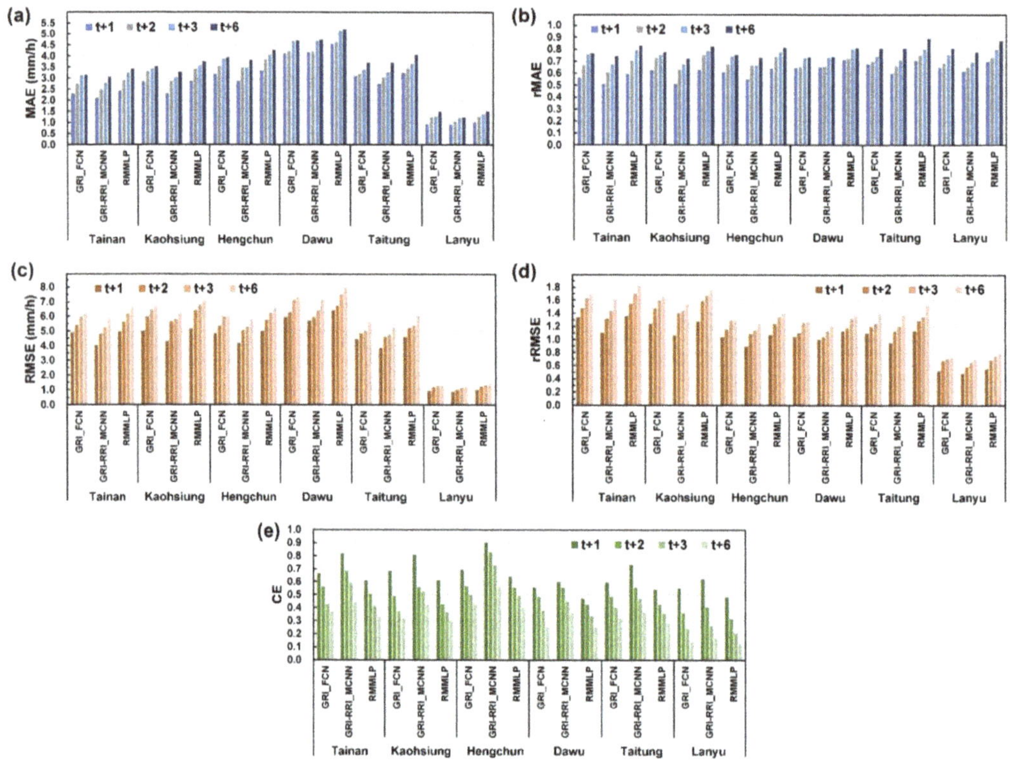

Figure 16. Performance levels of six stations in future (1–6 h) predictions: (**a**) MAE, (**b**) rMAE, (**c**) RMSE, (**d**) rRMSE, and (**e**) CE.

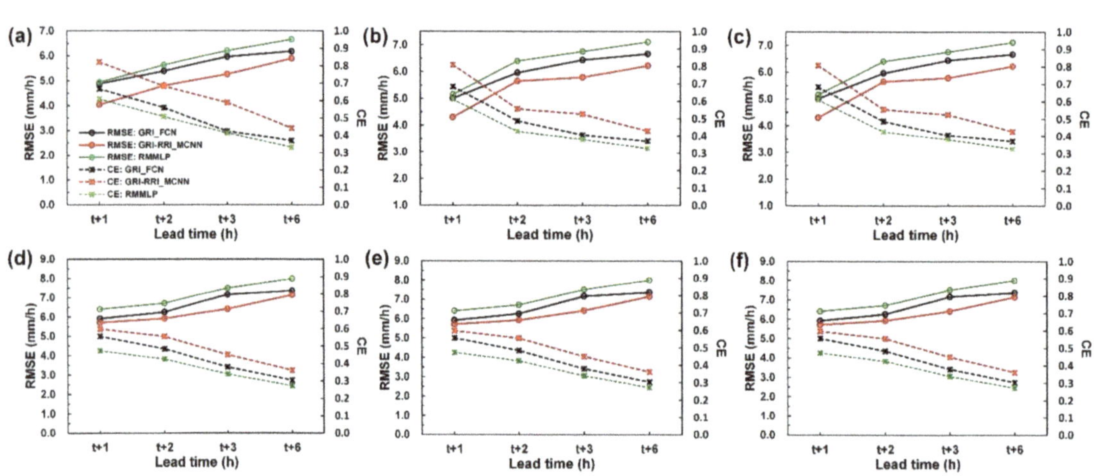

Figure 17. Performance levels in terms of RMSE and CE at (**a**) Tainan station, (**b**) Kaohsiung station, (**c**) Hengchun station, (**d**) Dawu station, (**e**) Taitung station, and (**f**) Lanyu station.

We calculated the average IMP_{CE} measures of six stations for 1–6 h predictions using GRI-RRI_MCNN and GRI_FCN. After calculation, the average IMP_{CE} of GRI-RRI_MCNN and GRI_FCN were respective values of 18.9% and 6.5% for 1 h predictions, 14.9% and 5.5% for 2 h, 13.6% and 4.7% for 3 h, and 9.7% and 3.7% for 6 h. Therefore, we determined that the improvement metric resulting from GRI-RRI_MCNN was higher than that from GRI_FCN.

4.4. Discussion

The hyetograph error indicator performance in GRI-RRI_MCNN was superior to that in GRI_FCN and RMMLP. Better prediction of the peak rainfall time was achieved in GRI-RRI_MCNN than in GRI_FCN and RMMLP. These indicated the GRI-RRI_MCNN effectively predicted the typhoon rainfalls. However, the peak values were underestimated in these models probably because the typhoon circulation structures changed rapidly, especially under the effect of the CMR, which increased the uncertainty and difficulty in predicting transient changes in the typhoon rainfall in real time.

The movement of the typhoons affected the rainfall at each ground station. Under the effect of the CMR, if a station was windward of typhoon circulations, the rainfall was heavy; otherwise, the rainfall was relatively low. The prediction efficiency was higher for the stations to the west of the CMR (i.e., the Hengchun, Tainan, and Kaohsiung stations) than for the stations to the east of the CMR (i.e., the Dawu, Taitung, and Lanyu stations).

5. Conclusions

Typhoons cause severe disasters and damage in southern Taiwan. Accurate prediction of the hourly rainfall caused by typhoons can reduce life and property losses and damages. This study used the FCN model for DL image recognition to analyze the REIs and ground rain data. The collected data were analyzed for predicting the future (1–6-h) rainfall caused by typhoons in the study area. FCNs, which are extensions of CNNs, improve the defects of CNN and solve semantic segmentation problems. An FCN comprises neural net layers and performs upsampling on the feature map of the final convolution layer; thus, the FCN model can restore the size of the output results to that of the raw input images. Therefore, classification is performed for every pixel to address semantic segmentation problems.

This study collected data related to 22 typhoons that affected southern Taiwan from 2013 to 2019. Two model cases were designed. The GRI_FCN involved the use of GRIs to directly predict ground rainfall. The GRI-RRI_MCNN involved the use of REIs to retrieve the ground rainfall before the prediction of the future ground rainfall. Moreover, the RMMLP, a conventional multilayer perceptron neural networks, was used to a benchmark model. The performance of the GRI_FCN, GRI-RRI_MCNN, and RMMLP models was compared for three typhoons, namely Typhoons Kong-Rey in 2013, Meranti in 2016, and Megi in 2016. The rainfall prediction results were obtained for six ground stations in southern Taiwan (i.e., the Tainan, Kaohsiung, Hengchun, Taitung, Dawu, and Lanyu stations). This study used the GRI_FCN and GRI-RRI_MCNN models to establish a rainfall prediction model for generating the predicted GRIs of southern Taiwan. These predicted GRIs were used to assess the predicted rainfall of each station. Overall, the GRI-RRI_MCNN model enabled the typhoon rainfall in southern Taiwan to be predicted with high accuracy.

This study used the inverse distance weighting method to convert the rainfall data of ground stations into two-dimensional rainfall maps. However, the inverse distance interpolation may introduce significant artifacts such as color discrepancy and blurriness in regions where ground measurements are sparse, such as mountain area. Therefore, in the future this study suggests that remote regions could be masked in the interpolated rainfall maps where no sites are nearby and performed partial convolution [49], instead of standard convolution in the presented work.

Author Contributions: C.-C.W. conceived and designed the experiments and wrote the manuscript; T.-H.H. and C.-C.W. carried out this experiment and analysis of the data and discussed the results. All authors have read and agreed to the published version of the manuscript.

Funding: Support for this study provided by the Ministry of Science and Technology, Taiwan under Grant No. MOST110-2622-M-019-001 is greatly appreciated.

Informed Consent Statement: Not applicable.

Data Availability Statement: The typhoon information and radar reflectivity image were obtained from the Central Weather Bureau of Taiwan, which are available at https://rdc28.cwb.gov.tw/ (accessed on 10 January 2021) and https://e-service.cwb.gov.tw/HistoryDataQuery/index.jsp (accessed on 10 January 2021).

Acknowledgments: The authors acknowledge data provided by Taiwan's Central Weather Bureau.

Conflicts of Interest: The authors declare no conflict of interest.

References

1. Huang, W.K.; Wang, J.J. Typhoon damage assessment model and analysis in Taiwan. *Nat. Hazards* **2015**, *79*, 497–510. [CrossRef]
2. Cheung, K.; Yu, Z.; Elsberry, R.L.; Bell, M.; Jiang, H.; Lee, T.C.; Lu, K.C.; Oikawa, Y.; Qi, L.; Rogers, R.F.; et al. Recent advances in research and forecasting of tropical cyclone rainfall. *Trop. Cyclone Res. Rev.* **2018**, *7*, 106–127. [CrossRef]
3. Teng, H.; Done, J.M.; Lee, C.; Kuo, Y. Dependence of probabilistic quantitative precipitation forecast performance on typhoon characteristics and forecast track error in Taiwan. *Weather Forecast.* **2020**, *35*, 585–607. [CrossRef]
4. Chung, K.S.; Yao, I.A. Improving radar echo Lagrangian extrapolation nowcasting by blending numerical model wind information: Statistical performance of 16 typhoon cases. *Mon. Weather Rev.* **2020**, *148*, 1099–1120. [CrossRef]
5. Huang, C.Y.; Chou, C.W.; Chen, S.H.; Xie, J.H. Topographic rainfall of tropical cyclones past a mountain range as categorized by idealized simulations. *Weather Forecast.* **2020**, *35*, 25–49. [CrossRef]
6. Chiang, Y.M.; Chang, F.J.; Jou, B.J.D.; Lin, P.F. Dynamic ANN for precipitation estimation and forecasting from radar observations. *J. Hydrol.* **2007**, *334*, 250–261. [CrossRef]
7. Jin, L.; Yao, C.; Huang, X.Y. A nonlinear artificial intelligence ensemble prediction model for typhoon intensity. *Mon. Weather Rev.* **2008**, *136*, 4541–4554. [CrossRef]
8. Kashiwao, T.; Nakayama, K.; Ando, S.; Ikeda, K.; Lee, M.; Bahadori, A. A neural network-based local rainfall prediction system using meteorological data on the Internet: A case study using data from the Japan Meteorological Agency. *Appl. Soft Comput.* **2017**, *56*, 317–330. [CrossRef]
9. Lin, F.R.; Wu, N.J.; Tsay, T.K. Applications of cluster analysis and pattern recognition for typhoon hourly rainfall forecast. *Adv. Meteorol.* **2017**, 5019646. [CrossRef]
10. Wei, C.C. Examining El Niño–Southern Oscillation effects in the subtropical zone to forecast long-distance total rainfall from typhoons: A case study in Taiwan. *J. Atmos. Ocean. Technol.* **2017**, *34*, 2141–2161. [CrossRef]
11. Lin, G.F.; Chen, G.R.; Wu, M.C.; Chou, Y.C. Effective forecasting of hourly typhoon rainfall using support vector machines. *Water Resour. Res.* **2009**, *45*, 8. [CrossRef]
12. Wei, C.C.; Roan, J. Retrievals for the rainfall rate over land using Special Sensor Microwave/Imager data during tropical cyclones: Comparisons of scattering index, regression, and support vector regression. *J. Hydrometeorol.* **2012**, *13*, 1567–1578. [CrossRef]
13. Gires, A.; Onof, C.; Maksimović, Č.; Schertzer, D.; Tchiguirinskaia, I.; Simoes, N. Quantifying the impact of small scale unmeasured rainfall variability on urban runoff through multifractal downscaling: A case study. *J. Hydrol.* **2012**, *442*, 117–128. [CrossRef]
14. Wei, C.C. Comparison of river basin water level forecasting methods: Sequential neural networks and multiple-input functional neural networks. *Remote Sens.* **2020**, *12*, 4172. [CrossRef]
15. Wei, C.C.; Hsu, C.C. Real-time rainfall forecasts based on radar reflectivity during typhoons: Case study in southeastern Taiwan. *Sensors* **2021**, *21*, 1421. [CrossRef] [PubMed]
16. Abdourahamane, Z.S.; Acar, R.; Serkan, S. Wavelet-copula-based mutual information for rainfall forecasting applications. *Hydrol. Process.* **2019**, *33*, 1780. [CrossRef]
17. Berne, A.; Krajewski, W.F. Radar for hydrology: Unfulfilled promise or unrecognized potential? *Adv. Water Resour.* **2013**, *51*, 357–366. [CrossRef]
18. Biswas, S.K.; Chandrasekar, V. Cross-validation of observations between the GPM dual-frequency precipitation radar and ground based dual-polarization radars. *Remote Sens.* **2018**, *10*, 1773. [CrossRef]
19. Bordoy, R.; Bech, J.; Rigo, T.; Pineda, N. Analysis of a method for radar rainfall estimation considering the freezing level height. *J. Mediterr. Meteorol. Climatol.* **2010**, *7*, 25–39. [CrossRef]
20. Bringi, V.N.; Rico-Ramirez, M.A.; Thurai, M. Rainfall estimation with an operational polarimetric C-band radar in the United Kingdom: Comparison with a gauge network and error analysis. *J. Hydrometeorol.* **2011**, *12*, 935–954. [CrossRef]
21. He, X.; Refsgaard, J.C.; Sonnenborg, T.O.; Vejen, F.; Jensen, K.H. Statistical analysis of the impact of radar rainfall uncertainties on water resources modeling. *Water Resour. Res.* **2011**, *47*, W09526. [CrossRef]
22. Prat, O.P.; Barros, A.P. Exploring the transient behavior of Z–R relationships: Implications for radar rainfall estimation. *J. Appl. Meteorol. Climatol.* **2009**, *48*, 2127–2143. [CrossRef]
23. Qiu, Q.; Liu, J.; Tian, J.; Jiao, Y.; Li, C.; Wang, W.; Yu, F. Evaluation of the radar QPE and rain gauge data merging methods in Northern China. *Remote Sens.* **2020**, *12*, 363. [CrossRef]

24. Sahlaoui, Z.; Mordane, S. Radar rainfall estimation in Morocco: Quality control and gauge adjustment. *Hydrology* **2019**, *6*, 41. [CrossRef]
25. Smith, J.A.; Baeck, M.L.; Meierdiercks, K.L.; Miller, A.J.; Krajewski, W.F. Radar rainfall estimation for flash flood forecasting in small urban watersheds. *Adv. Water Resour.* **2007**, *30*, 2087–2097. [CrossRef]
26. Hossain, S.; Lee, D. Deep learning-based real-time multiple-object detection and tracking from aerial imagery via a flying robot with GPU-based embedded devices. *Sensors* **2019**, *19*, 3371. [CrossRef]
27. LeCun, Y.; Bottou, L.; Bengio, Y.; Haffner, P. Gradient-based learning applied to document recognition. *Proc. IEEE* **1998**, *86*, 2278–2323. [CrossRef]
28. Zhao, Z.; Zheng, P.; Xu, S.; Wu, X. Object detection with deep learning: A review. *IEEE Trans. Neural Netw. Learn. Syst.* **2019**, *30*, 3212–3232. [CrossRef]
29. Krizhevsky, A.; Sutskever, I.; Hinton, G.E. ImageNet classification with deep convolutional neural networks. In Proceedings of the 25th International Conference on Neural Information Processing Systems, Lake Tahoe, NV, USA, 5–8 December 2012; pp. 1097–1105.
30. Song, T.; Wang, Z.; Xie, P.; Han, N.; Jiang, J.; Xu, D. A novel dual path gated recurrent unit model for sea surface salinity prediction. *J. Atmos. Ocean. Technol.* **2019**, *37*, 317–325. [CrossRef]
31. Szegedy, C.; Liu, W.; Jia, Y.; Sermanet, P.; Reed, S.; Anguelov, D.; Erhan, D.; Vanhoucke, V.; Rabinovich, A. Going deeper with convolutions. *arXiv* **2014**, arXiv:1409.4842.
32. Kim, J.H.; Batchuluun, G.; Park, K.R. Pedestrian detection based on faster R-CNN in nighttime by fusing deep convolutional features of successive images. *Expert Syst. Appl.* **2018**, *114*, 15–33. [CrossRef]
33. Pan, B.; Hsu, K.; AghaKouchak, A.; Sorooshian, S. Improving precipitation estimation using convolutional neural network. *Water Resour. Res.* **2019**, *55*. [CrossRef]
34. Sadeghi, M.; Asanjan, A.A.; Faridzad, M.; Nguyen, P.; Hsu, K.; Sorooshian, S.; Braithwaite, D. PERSIANN-CNN: Precipitation estimation from remotely sensed information using artificial neural networks–convolutional neural networks. *J. Hydrometeorol.* **2019**, *20*, 2273–2289. [CrossRef]
35. Wang, J.H.; Lin, G.F.; Chang, M.J.; Huang, I.H.; Chen, Y.R. Real-time water-level forecasting using dilated causal convolutional neural networks. *Water Resour. Manag.* **2019**, *33*, 3759–3780. [CrossRef]
36. Wei, C.C. Real-time extreme rainfall evaluation system for the construction industry using deep convolutional neural networks. *Water Resour. Manag.* **2020**, *34*, 2787–2805. [CrossRef]
37. Long, J.; Shelhamer, E.; Darrell, T. Fully convolutional networks for semantic segmentation. In Proceedings of the IEEE Conference on Computer Vision and Pattern Recognition (CVPR), Boston, MA, USA, 7–12 June 2015; pp. 3431–3440.
38. Eppel, S. Hierarchical semantic segmentation using modular convolutional neural networks. *arXiv* **2017**, arXiv:1710.05126.
39. Wei, C.C. Simulation of operational typhoon rainfall nowcasting using radar reflectivity combined with meteorological data. *J. Geophys. Res. Atmos.* **2014**, *119*, 6578–6595. [CrossRef]
40. Central Weather Bureau (CWB). 2020. Available online: http://www.cwb.gov.tw/V7/index.htm (accessed on 1 December 2020).
41. Wu, C.C.; Kuo, Y.H. Typhoons affecting Taiwan: Current understanding and future challenges. *Bull. Am. Meteorol. Soc.* **1999**, *80*, 67–80. [CrossRef]
42. Central Weather Bureau (CWB). Typhoon Database. 2021. Available online: https://rdc28.cwb.gov.tw/ (accessed on 10 January 2021).
43. Shepard, D. A two-dimensional interpolation function for irregularly-spaced data. In Proceedings of the 1968 23rd ACM National Conference, New York, NY, USA, 27–29 August 1968; pp. 517–524.
44. Wei, C.C.; Hsieh, P.Y. Estimation of hourly rainfall during typhoons using radar mosaic-based convolutional neural networks. *Remote Sens.* **2020**, *12*, 896. [CrossRef]
45. Sun, M.; Song, Z.; Jiang, X.; Pan, J.; Pang, Y. Learning pooling for convolutional neural network. *Neurocomputing* **2017**, *224*, 96–104. [CrossRef]
46. Zhang, W.; Tang, P.; Zhao, L. Remote sensing image scene classification using CNN-CapsNet. *Remote Sens.* **2019**, *11*, 494. [CrossRef]
47. McBride, J.L.; Ebert, E.E. Verification of quantitative precipitation forecasts from operational numerical weather prediction models over Australia. *Weather Forecast.* **2000**, *15*, 103–121. [CrossRef]
48. Joint Typhoon Warning Center (JTWC). 2021. Available online: https://www.metoc.navy.mil/jtwc/jtwc.html. (accessed on 20 January 2021).
49. Liu, G.; Reda, F.A.; Shih, K.J.; Wang, T.C.; Tao, A.; Catanzaro, B. Image inpainting for irregular holes using partial convolutions. *arXiv* **2018**, arXiv:1804.07723.

sensors

MDPI

Article

Novel Ensemble Approach of Deep Learning Neural Network (DLNN) Model and Particle Swarm Optimization (PSO) Algorithm for Prediction of Gully Erosion Susceptibility

Shahab S. Band [1,2,*], Saeid Janizadeh [3], Subodh Chandra Pal [4], Asish Saha [4], Rabin Chakrabortty [4], Manouchehr Shokri [5] and Amirhosein Mosavi [6,7]

1 Institute of Research and Development, Duy Tan University, Da Nang 550000, Vietnam
2 Future Technology Research Center, National Yunlin University of Science and Technology, 123 University Road, Section 3, Douliou, Yunlin 64002, Taiwan
3 Department of Watershed Management Engineering and Sciences, Faculty in Natural Resources and Marine Science, Tarbiat Modares University, 14115-111 Tehran, Iran; janizadehsaeid@modares.ac.ir
4 Department of Geography, The University of Burdwan, West Bengal, Burdwan 713104, India; scpal@geo.buruniv.ac.in (S.C.P.); asishsaha01@gmail.com (A.S.); rabingeo8@gmail.com (R.C.)
5 Institute of Structural Mechanics, Bauhaus Universität Weimar, 99423 Weimar, Germany; manouchehr.shokri@uni-weimar.de
6 Environmental Quality, Atmospheric Science and Climate Change Research Group, Ton Duc ThangUniversity, Ho Chi Minh City 700000, Vietnam; amirhosein.mosavi@tdtu.edu.vn
7 Faculty of Environment and Labour Safety, Ton Duc Thang University, Ho Chi Minh City 700000, Vietnam
* Correspondence: shamshirbandshahaboddin@duytan.edu.vn

Received: 11 August 2020; Accepted: 24 September 2020; Published: 30 September 2020

Abstract: This study aims to evaluate a new approach in modeling gully erosion susceptibility (GES) based on a deep learning neural network (DLNN) model and an ensemble particle swarm optimization (PSO) algorithm with DLNN (PSO-DLNN), comparing these approaches with common artificial neural network (ANN) and support vector machine (SVM) models in Shirahan watershed, Iran. For this purpose, 13 independent variables affecting GES in the study area, namely, altitude, slope, aspect, plan curvature, profile curvature, drainage density, distance from a river, land use, soil, lithology, rainfall, stream power index (SPI), and topographic wetness index (TWI), were prepared. A total of 132 gully erosion locations were identified during field visits. To implement the proposed model, the dataset was divided into the two categories of training (70%) and testing (30%). The results indicate that the area under the curve (AUC) value from receiver operating characteristic (ROC) considering the testing datasets of PSO-DLNN is 0.89, which indicates superb accuracy. The rest of the models are associated with optimal accuracy and have similar results to the PSO-DLNN model; the AUC values from ROC of DLNN, SVM, and ANN for the testing datasets are 0.87, 0.85, and 0.84, respectively. The efficiency of the proposed model in terms of prediction of GES was increased. Therefore, it can be concluded that the DLNN model and its ensemble with the PSO algorithm can be used as a novel and practical method to predict gully erosion susceptibility, which can help planners and managers to manage and reduce the risk of this phenomenon.

Keywords: gully erosion susceptibility; deep learning neural network; DLNN; particle swarm optimization; PSO; geohazard; geoinformatics; ensemble model; erosion; hazard map; spatial model; deep learning; natural hazard; extreme events

1. Introduction

Biodiversity in a given area depends on, to a large extent, and supports the most vital natural resources in the soil, which also contribute to the provision of basic human needs such as food, fresh air, and clean water [1]. Therefore, human survival largely depends on the soil component. Soil erosion in the form of gully erosion is a serious global problem, and it continues to pose a threat to soil and water resources, particularly in arid and semi-arid regions of Iran [2,3]. Among the several types of water-induced erosion, gully erosion is a more intense form of soil erosion [4] and is one of the most complex geomorphic phenomena on the Earth's surface [5]. Such erosional activities also change the shape of the Earth's landform and produce a rugged topography, which is not suitable for production activities, construction of communication networks, etc. Thus, water-induced soil erosion is the main cause of the destruction of agricultural land, vegetation, and ecosystems, and is ultimately responsible for a devastating land degradation phenomenon. It has been estimated that the annual rate of global soil erosion is approximately 75 billion tons [6]. From an international perspective, Iran ranks second in terms of land losses, and the annual rate of soil erosion is close to 2 to 2.5 billion tons [7]. It has also been predicted that Iran's average soil erosion rate is 30–32 tons/ha/year, which is 4.3 times the world average (Food and Agriculture Organization of the United Nations (FAO), 1984). In Iran, soil erosion has been estimated to have caused more than USD 1 billion in economic losses (FAO, 2015) and is a national threat [8]. Thus, it is necessary to protect the soil from erosion and to avoid the phenomenon of land degradation worldwide. The main cause of intensive water-related gully erosion and its development is a long hot/dry season followed by an extremely wet period. Therefore, extreme rainfall causes a large amount of surface runoff over the infiltration capacity and easily transports loose soil particles onto the downward slope. Thus, soil erosion related to water in Iran is a major barrier to sustainable development in the areas of agriculture, watershed management, and other activities related to resource development [9]. Hence, the preparation of a gully erosion susceptibility (GES) map is essential for sustainable management, development, and protection of the most vital natural resource on the Earth's surface, i.e., soil, from intense gully formation and development.

Before preparing a GES map, it is necessary to understand the definition of a gully, its morphological characteristics, causes of occurrences, conditioning factors, and its ultimate impact on the land surface. A gully can be defined as a deep, narrow channel with a depth of more than 30 cm, usually produced by surface and subsurface runoff after a heavy downpour with a temporary flow of water within that channel [8]. Gullies generally transport a large amount of sediment from the high slope or plateau of the unprotected soil surface, i.e., areas with less vegetation, to the down-slope areas of a watershed. It is also a fact that within 5% of the area of a watershed, between 10% and 94% of sediment moves downwards due to gully erosion [10]. According to Poesen [11], different factors affect gully erosion, and these factors are classified into two categories: (a) anthropogenic activities such as excessive use of farm land, overgrazing, unplanned manner of road construction, deforestation, etc., and (b) physical conditions such as topography, climate, vegetation cover, mineral composition in the soil, etc. Depending upon the depth, gullies are classified into three types, i.e., if the depth is <0.3 m then it is called a grove, if the depth is between 0.3 and 2 m it is called a shallow gully, and if the depth is >2 m it is known as a deep gully [12]. Intensive gully erosion causes many environmental problems, such as accumulation of sediment in rivers and devastating floods, as it removes fertile soils, which has a serious impact on agricultural fields, minimizes soil water storage capacity, destroys roads, and ultimately produces badlands [13–15]. It is also a well-known fact that similar factors are not responsible for the occurrence of gullies in several places in the world. Gullies are generally formed and developed based on the local topographical, climatological, and hydrological characteristics. Therefore, different gully-prone areas and associated factors need to be identified by mapping the gully erosion susceptibility. Not only this, but a suitable prediction model along with the identification of respective favorable gully erosion conditioning factors (GECFs) are also essential for an unbiased prediction result. Several methods such as statistical, machine learning (ML), and ensemble algorithms have been used for mapping GES, with the combination of remote sensing and geographic information systems. Thus, GES mapping, using the aforementioned newly developed methods, can help land use planners to maintain soil and water resources sustainably and accurately. Furthermore, the potential of the respective region will ultimately increase when suitable measures are taken.

In recent times, ML algorithms have been widely used for the spatial prediction of several natural hazards such as flooding, landslides [16], wildfires [17], etc. Several researchers throughout the world have carried out GES mapping by using statistical as well as ML algorithms. Some of the widely used statistical methods to predict GES mapping are frequency ratio [7], logistic regression [18], weight of evidence (WoE) [19], index of entropy (IoE) [5], etc. Besides statistical methods, different ML algorithms have also been widely used to predict GES mapping such as artificial neural network (ANN) [20], support vector machine (SVM) [20], random forest (RF) (Hosseinalizadeh et al. 2019), multi-layer perception (MLPC) approaches [21], classification and regression tree (CART) [22], boosted regression tree (BRT) [7], particle swarm optimization (PSO) [23], multi-variate adaptive regression spline (MARS) [5], and maximum entropy [24]. Ensemble models have also been widely used for their novelties and capabilities in the comprehensive analysis of GES mapping [25]. Ensemble models are applied for high precision and predictive analysis of any kind of natural hazard susceptibility mapping. In other words, the presentation of an ML model is significantly enhanced by using an ensemble model. Along with machine learning models, different ensemble models have also been used for gully erosion modeling [20].

In very recent times, the deep neural learning network (DLNN) is a striking ML algorithm and has been widely used by several research groups. This method was proposed for the first time in 2006 and includes different key features of ML as well as artificial intelligence (AI). The DLNN algorithm consists of fully convolutional neural networks (CNNs), deep belief networks (DBNs), stacked auto-encoder (SAE) networks, etc. [26]. In addition to this, the Adaptive moment estimation (Adam) and Rectified Linear Unit (ReLU) algorithms were used for training and activation purposes in every learning unit of a DLNN model [27]. Generally, the DLNN algorithm has been used in different fields such as feature extraction and transformation through supervised and unsupervised processes, recognition of patterns, and classification [28]. On the other hand, the particle swarm optimization (PSO) algorithm is an extended part of AI and an amalgamation of the conventional ML techniques. The PSO algorithm is based on swarm intelligence, and it is straightforward with efficient universal optimization techniques [26]. PSO is used for the feature selection of a dataset through optimization techniques.

Deep learning (DL) and traditional ML algorithms have some basic differences, namely that the DL algorithm needs a big data size to perform and analyze successfully, and in the case of ML algorithms, they are performed in a certain way according to established rules. The DL algorithm requires a lot more matrix operation functions than the ML algorithm does to perform well [29]. In the case of the problem-solving method, the DL algorithm is done through end-to-end problem solving, whereas in the case of ML, it breaks down into multiple sub-problems. Therefore, the DL algorithm is much better than the traditional ML algorithm for mapping the GES zone. Thus, the greatest advantage of using the DLNN algorithm is that this model is capable of building a high-level feature from a raw dataset scientifically, and is also capable of delivering forecasting results using time series data. In addition to this, DLNN consists of a different topology than the general neural network of a single hidden layer; thus, more than one hidden layer is present in this algorithm. For this reason, in various research areas, the DL algorithm has better performance than the conventional ML algorithm [30]. In the case of PSO, it is also used to conquer the problems of local optima through feature selection methods. PSO determines the quality of a dataset's features through a multi-objective fitness function [31]. As a result, the output layer of different hidden layers is optimized by the PSO algorithm to obtain more accurate predictions [32].

Therefore, the present research work has been carried out to predict GES mapping in Shirahan watershed, which is tremendously affected by water-induced gully erosion. To fulfill our research objective, we used thirteen suitable GECFs with a total of 132 gully head-cut points (each for gully and non-gully), splitting them into a 70/30 ratio for training and testing datasets. Furthermore, to creatively model the GES mapping, we used a DL as well as a conventional ML algorithm. In this study, we used DLNN, PSO, artificial neural network (ANN), and support vector machine (SVM) algorithms.

According to several literature surveys on GES mapping and the best of our knowledge, it was noticed that the DLNN model has not been used in GES assessment so far; thus, this study was carried out to investigate the potential application of the DLNN model for GES mapping. In this study, an attempt was also made to use the PSO algorithm to optimize the parameters of the deep learning model (DLNN) in the training phase and to introduce a new approach of an ensemble of PSO and DLNN in GES modeling. Not only this, but a comparison was also made between the ensemble of PSO-DLNN and conventional ANN and SVM algorithms. Thus, the application of DL and the PSO-DLNN ensemble approach for GES mapping is the novelty in this research study, as the result of this approach improved the prediction accuracy compared with any single ML algorithm. Thereafter, all of the output results were validated through sensitivity (SST), specificity (SPF), positive predictive values (PPV), negative predictive values (NPV), receiver operating characteristic-area under the curve (ROC-AUC), likelihood ratio, F-measures, and maximum probability of correct decision (MPCD) statistical analyses. Thus, the DL and PSO-DLNN ensemble methods can help to forecast and control the creation and development of gullies in Shirahan watershed, Iran.

2. Materials and Methods

2.1. Description of the Study Area

Shirahan watershed is located at a longtitude of 20° 57′ to 28° 57′ and a latitude of 51° 25′ to 51° 26′, in the central part of Hormozgan Province and to the south of Bandar-e Jask city (Figure 1). The area of the watershed is 138 km^2, the minimum height of the area is 2 m, and the maximum height is 214 m above sea level. According to statistics recorded at Jask Synoptic Station over a period of 28 years (1989–2017), precipitation in this region is very heavy, and more than 50% of it occurs in winter. According to the information of the above station, the average annual rainfall is 116.75 mm, the maximum annual rainfall is 320 mm, and the minimum is 27 mm. The climate of the region is hot and dry according to the Ambregeh method and hot/dry based on the Domarten method. Soil texture is generally silt/loam and loam. In this area, the percentage of clay increases with increasing depth. However, changes in the percentage of sand and silt do not follow a specific trend and have high fluctuations. In this area, the horizon of 75–75 cm has the highest degree of salinity (Table 1). Pictures of ditch erosion are shown in Figure 2. To study the geometric features and physical and chemical properties of the soil, 20 ditches were sampled in the study area. Studies showed further expansion of ditch erosion in salt marshes, which are located in the plain type. The general plan of ditches is compound, and their cross-sectional shape is trapezoidal. The average depth of ditches is 2.7 m; the average width is 10.3 m. Laboratory studies were used to evaluate the soil characteristics of the gullies of Shirahan watershed. Meanwhile, soil samples were taken from the soil surface to a depth of 290 cm and sent to the laboratory of Bandar Abbas Agricultural and Natural Resources Research Center for soil testing. Laboratory results showed that the soil texture in the area up to the depth under study is loose. The physical and chemical properties of the soil at 6 different soil depths are shown for the ditches studied in Table 1. Some field photographs of a gully in the present study area of Shirahan watershed are shown in Figure 2.

Table 1. Physical and chemical properties of soil in the gullies of Shirahan watershed.

Features	Soil Depth (cm)					
	0–30	30–75	75–130	130–180	18–250	250–290
PH	8.06	7.59	8.19	7.69	8.32	7.38
EC (mmhos/cm)	2.26	34.6	2.23	33.9	2.4	33.2
Na (Meq/lit)	8.82	285	8.87	285	8.8	248
Ca + Mg (Meq/lit)	13.6	64.4	13.7	62.4	13.4	63.2
SAR	3.4	50.2	3.4	50.2	3.4	50.2
Clay (%)	24	26	26	27	29	28
Silt (%)	58	30	60	32	56	30
Sand (%)	18	44	14	41	15	42
Soil texture	Silt-Loam	Loam	Silt-Loam	Loam/clay-loam	Silty-Clay-Loam	Clay-Loam

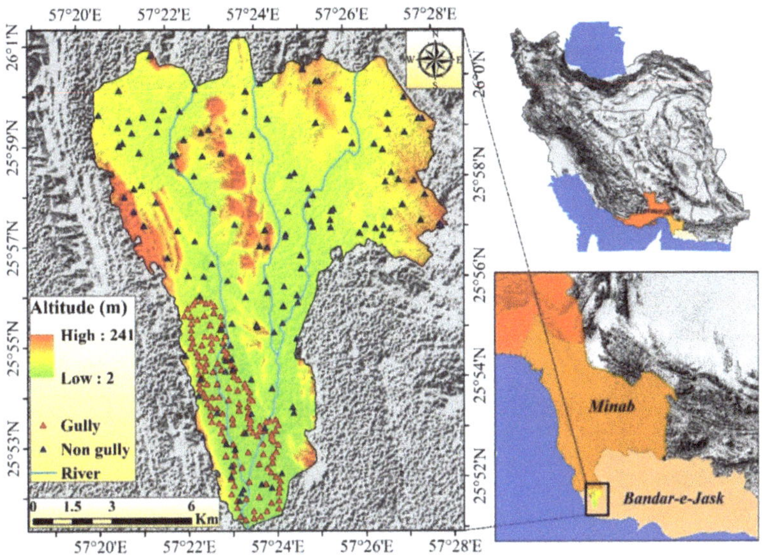

Figure 1. Location of the study area in Iran.

Figure 2. Images of gully erosion in Shirahan watershed: (**a**) head-cut gully; (**b**) gully erosion in salt land area; (**c**) measurement of gully morphometric properties.

2.2. Methodology

The methodological approach used in this research work is discussed in the following section, and the respective flowchart is presented in Figure 3.

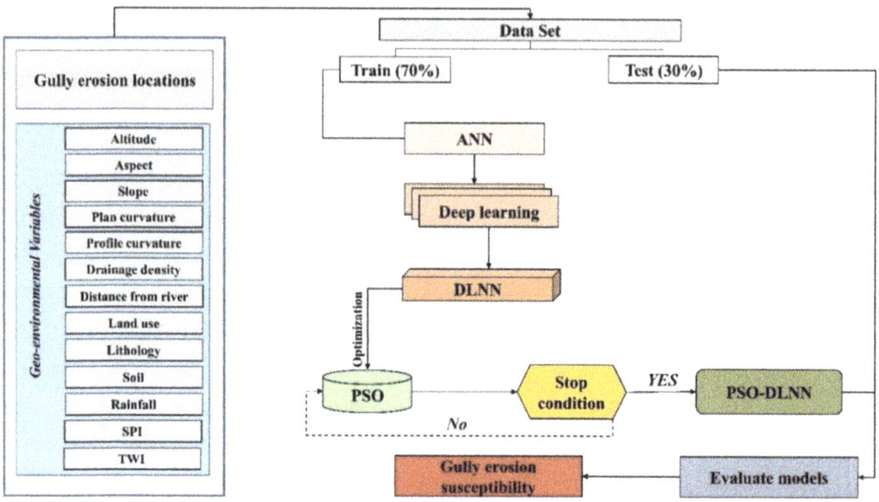

Figure 3. Methodological flowchart of particle swarm optimization (PSO)-deep learning neural network (DLNN) in gully erosion susceptibility.

Firstly, a gully erosion inventory map was prepared based on the 132 gully head-cut points (with gully and non-gully for each). These gully head-cut points were identified based on field visits and information from the Administration of Natural Resources of Hormozgan Province. Along with this, the non-gully points were randomly selected throughout the basin area with the help of the geographic information system (GIS) environment. Besides this, a total of thirteen (13) of gully erosion conditioning factors (GECFs), i.e., target variables, were considered for modeling GES based on the local topographical and climatological factors in association with several literature studies. These GECFs are altitude, aspect, slope, plan curvature, profile curvature, drainage density (DD), distance from a river, land use, lithology, soil, rainfall, stream power index (SPI), and topographic wetness index (TWI). Thereafter, multi-collinearity analysis of variance inflation factor (VIF) and tolerance (TOL) techniques were used among different GECFs to determine the linear relationship among the variables. Afterwards, modeling of GES was done by using SVM, ANN, and DLNN machine learning (ML) algorithms, and a novel ensemble of PSO-DLNN. Lastly, the results of the several GES models were validated through a ROC curve analysis to assess their accuracy.

The methodology of the present research work was carried out to solve classification problems using the aforementioned ML and DL algorithms for prediction GES mapping. Besides this, the several target variables used in this study are a combination of logical, discrete, and continuous variables. During the processing of all of these variables' data, it was recognized among the variables whether each one was a logical, discrete, or continuous one, in the SPSS 25 statistical software designed by International Business Machines (IBM), New York, USA. In this study, we also analyze affected areas of gully erosion susceptible zone, by using the presence of gully head-cut points, and we also compute the GES zones based on the gully/non-gully head-cut points along with several conditioning factors for sustainable management of the gully-affected areas.

2.3. Dataset Preparation for Spatial Modeling

In this study, a gully erosion inventory map (Figure 1) was prepared based on field visits and information from the Administration of Natural Resources of Hormozgan Province, which resulted in a total of 132 gully points. To determine the non-gully points, GIS software was used and 132 points were randomly selected. The digital elevation model (DEM) map was obtained with a pixel size of 12.5 m from the Advance Land Observatin Satellite/Phased Array type L-band Synthetic Aperture Radar (ALOSPALSAR) sensor. The topographical factors such as slope map, direction curve, plan curvature, and profile curvature were prepared based on DEM in the GIS environment. The map of the distance from a river based on the Euclidean extension was obtained in GIS software. A drainage density map was prepared using a line density extension. SAGAGIS software was used to map TWI and SPI. The soil type map of the region was obtained based on the map prepared by the Administration of Natural Resources of Hormozgan Province. The lithological map was prepared based on the geological map of 1:100,000 of the country's mapping organization. Land use maps were prepared based on Landsat satellite images and Operational Land Imager (OLI) measurement, using the maximum probability algorithm in the ENVI software environment. The precipitation map of the constituency was prepared from the statistics of 4 climatological factors in the constituency over a period of 28 years (1989–2017) and based on the inverse distance weighting (IDW) interpolation method. Details about the data sources used in this research work are presented in Table 2.

Table 2. Details about the data sources of several factors used in this study.

Parameters	Data Source	Time (Year)	Spatial Resolution/Scale
Altitude, slope, aspect, profile curvature, plan curvature, drainage density (DD), distance from river, stream power index (SPI), topographic wetness index (TWI)	ALOS PALSAR DEM (Alaska Satellite Facility)	2012	12.5 m
Rainfall	Iran Meteorological Organization (IMO) (http://www.weather.ir/)	1989 to 2017	
Lithology	Geological Survey of Iran (GSI) (http://www.gsi.ir/)	2019	1:1,000,000
Land use	Landsat OLI 8 satellite image (USGS)	2019	30 m
Soil texture	Soil and Water Research Institute (http://www.iran.swri.com)	2019	1:1,000,000

A total of 13 GECFs were selected for GES mapping in this research work, namely, altitude, aspect, slope, plan curvature, profile curvature, drainage density (DD), distance from a river, land use, lithology, soil, rainfall, stream power index (SPI), and topographic wetness index (TWI) (Figure 4a–m).

The altitude of the present study area ranges from 2 to 241 m (Figure 4a). Altitude is an important factor for the occurrence of gullies due to influences on rainfall-runoff processes, and it is largely employed in GES mapping [3]. Slope aspect indirectly affects the occurrence of gully erosion as it affects the reception of sunlight, vegetation cover, and humidity [33]. Here, the slope aspect map has nine classes, i.e., flat, N, NE, E, SE, S, SW, W, and NW (Figure 4b). Slope angle influences the pattern of runoff and infiltration rate. Therefore, depending on the slope, the erosional rate also varies from place to place, i.e., high slope areas have high erosion rates and vice versa. The slope map is shown in Figure 4c, and the value ranges from 0% to 362.74%. In a particular direction, the rate of gradient change is known as curvature, within which, plan and profile curvature generally represent the topographic characteristics of an area. The value of plan curvature ranges from −30.27 to 24.08 (Figure 4d) and profile curvature from −29.63 to 30.93 (Figure 4e). DD directly impacts occurrences of gully erosion. Horton's (1932) following equation was used to calculate DD. In this study, the DD value ranges from 0 to 2.27 km/km^2 (Figure 4f).

$$DD = \frac{\sum_{i=1}^{n} S_i}{a} \tag{1}$$

where $\sum\limits_{i=1}^{n} S_i$ is the length of drainage in km, and 'a' indicates the total area of the drainage basin in km^2.

Distance from a river also influences occurrences of gully erosion as it greatly impacts the wetting capacity of surface area and associated erosional activities. The value of the distance from a river ranges from 0 to 4680.17 m (Figure 4g). The land use type of the area is very much responsible for the occurrence of gully erosion. Bare or less vegetated areas of the land surface are highly prone to gully erosion. In this study, four types of land use were recognized, i.e., agricultural land, rangeland, rock surface, and salt land (Figure 4h). The lithological factor of an area is highly responsible for erosional activities such as the development of a gully [34]. The present study area of Shirahan watershed consists of five types of lithological unit (Figure 4i). The soil map of the study area is shown in Figure 4j, and it is classified into two categories, i.e., entisols/aridisols and badlands. Rainfall is the most important factor for the formation of a gully and its development, mainly in the arid and semi-arid areas. High-intensity rainfall with short duration is the most devastating for gullies. Here, 28 years of rainfall data have been used to prepare a rainfall map (Figure 4k), and it ranges from 125 to 175 mm. SPI indicates the stream's erosional capacity [25]. SPI value was calculated by using the following equations, and the value ranges from 0 to 2.625 in this research work (Figure 4l).

$$SPI = As \times tan\beta \quad (2)$$

where A_S represents the upslope contributing area, and β represents the slope angle.

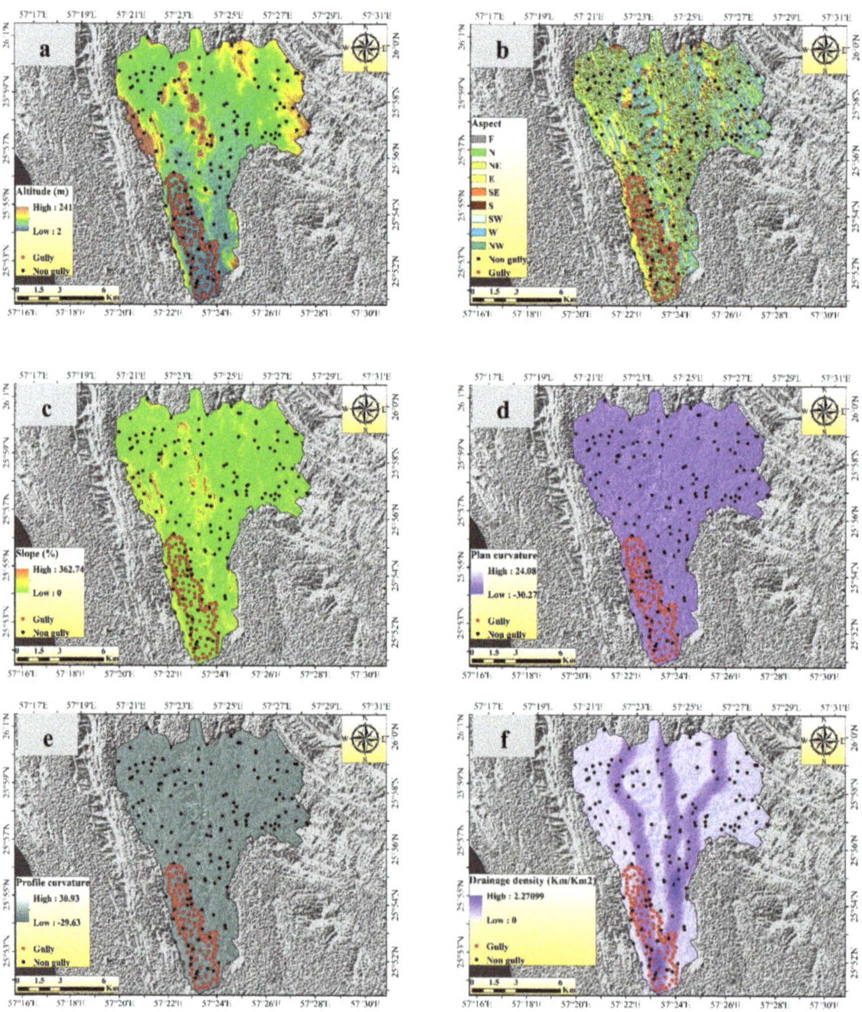

Figure 4. *Cont.*

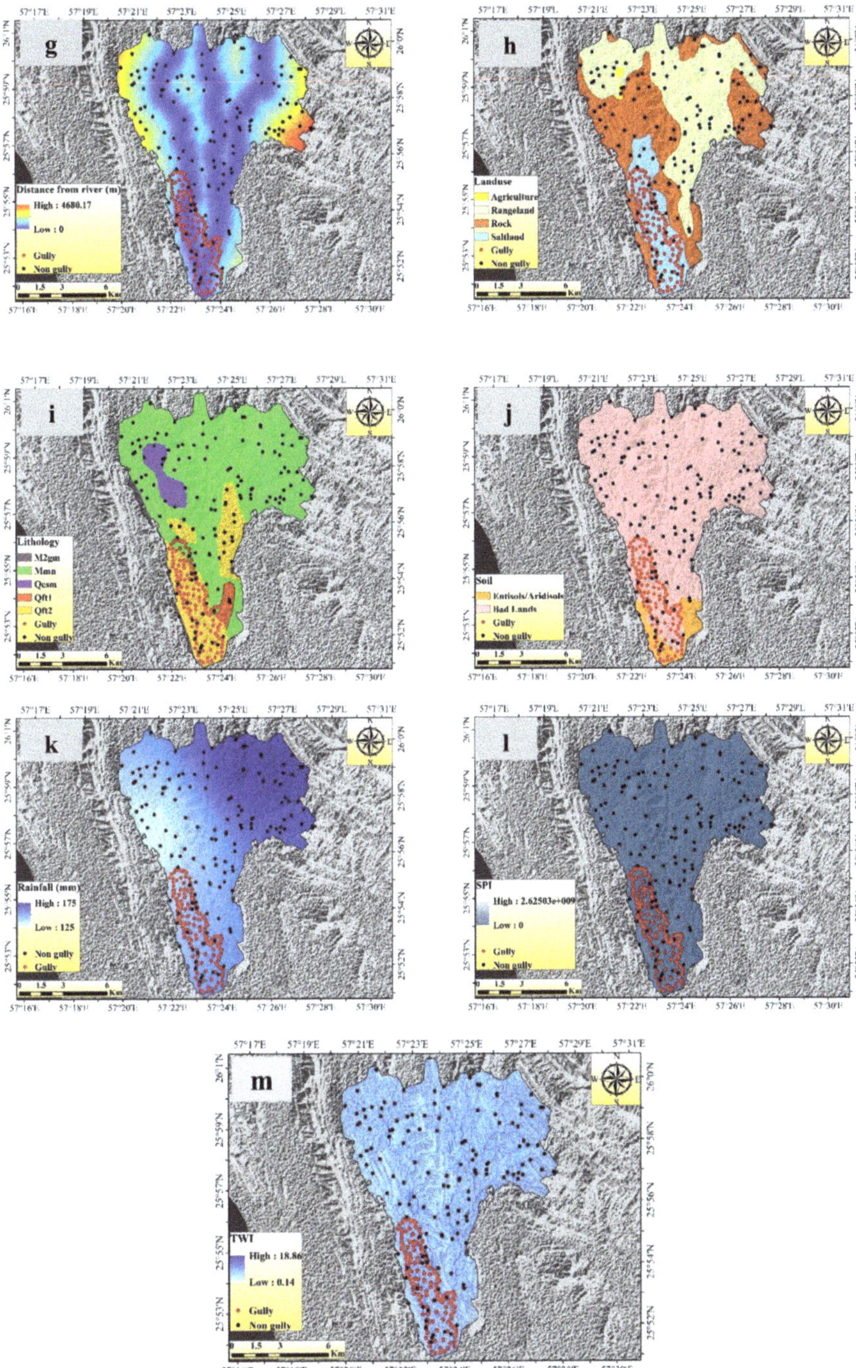

Figure 4. Gully erosion conditioning factors: (**a**) altitude, (**b**) slope, (**c**) aspect, (**d**) plan curvature, (**e**) profile curvature, (**f**) drainage density, (**g**) distance from a river, (**h**) land use, (**i**) soil, (**j**) lithology, (**k**) rainfall, (**l**) stream power index (SPI), and (**m**) topographic wetness index (TWI).

TWI determines transport capacity along with flow velocity [7], and it is an essential factor for identifying gully erosion-prone areas [35]. The following equation was used to calculate TWI value, and it ranges from 0.14 to 18.86 (Figure 4m).

$$TWI = In\left(\frac{A_s}{tan\beta}\right) \tag{3}$$

where A_s represents the area of a catchment in m^2, and β represents the gradient of the slope in radians.

2.4. Multi-Collinearity Analysis

Multi-collinearity analysis always gives the perfect outcome to evaluate the linear dependency of different geo-environmental factors in an ML model [15,36]. It is a statistical analysis and is able to find two variables of high correlation in a multiple regression study. Thus, it is very much essential to analyze the multi-collinearity of a model to obtain better results through removing the high multi-collinearity factors and minimizing the bias of the model [37]. Several researchers throughout the world have used multi-collinearity analysis in different fields such as GES mapping [21], floods [38], and landslide susceptibility mapping [39]. Multi-collinearity can be analyzed through variance inflation factor (VIF) and tolerance (TOL) [40]. As a general rule, if the TOL value is <0.10 or 0.20 and the VIF value is >5 or 10, then the result indicates high multi-collinearity among the variables [41]. The following equations were used to calculate TOL and VIF in a dataset:

$$TOL = 1 - R_j^2 \tag{4}$$

$$VIF = \frac{1}{TOL} \tag{5}$$

where R_j^2 indicates the regression value of j on other different variables in a dataset.

2.5. Machine Learning Method Used in Modeling the Gully Erosion

2.5.1. Support Vector Machine (SVM)

SVM is a very popular machine learning algorithm and was introduced by Vapnik and Chervonenkis in 1963. Several researchers throughout the world have used this machine learning classifier in the field of predicting different natural hazards such as in GES mapping [42], landslide prediction [43], flood susceptibility mapping [44], etc. SVM is implemented to solve regression analysis and multi-faceted classifier problems [45]. Vapnik [46] stated that SVM is based on the principle of structural risk minimization and statistical learning, and it is a supervised machine learning model. SVM is very much effective to reduce the error of the complexity of a linear computation and model overfitting [47]. Two types of statistically induced problems are engaged in SVM modeling. The first one is linear separating of the hyperplane by using statistical data, and the second one is converting non-linear data into linearly separable data [48]. Generally, the data processing in SVM of a non-linear relationship is done through the kernel function [49]. In addition to this, two classes can be discretely generated in SVM modeling by an optimal hyperplane, in which one class indicated above the hyperplane is assigned as 1 and the other one, located below the hyperplane, is assigned as 0, i.e., in this case gully erosion and non-gully erosion, respectively [50]. SVM has been developed for regression estimation, particularly paying attention to the solution of inverse problems. The novelty of the SVM model is that it has attempted to relocate the idea through kernel techniques for working out the inner products of unsupervised learning. Besides this, it can also be applied for singular components where the distribution of data is not well-defined. Therefore, a large class of functions can be applied for non-linearity mapping with high feature space by using this kernel trick. The hyperplane in an SVM can be calculated by using following equations:

$$Min\sum_{i=1}^{n}\varphi_i - \frac{1}{2}\sum_{i=1}^{n}\sum_{j=1}^{n}\varphi_i\varphi_j y_i y_j\left(x_i, x_j\right) \tag{6}$$

subject to

$$Min\sum_{i=1}^{n}\varphi_i y_j = 0 and 0 \leq \alpha_i \leq D \tag{7}$$

where $x = x_i$, $i = 1, 2, \ldots n$ are input variables of the vector; $y = y_i$, $j = 1, 2, \ldots n$ are output variables of the vector, and φ_i represents Lagrange multipliers.

Finally, the decision function of SVM can be classified as

$$f(x) = sgn\left(\sum_{i=j}^{n} y_i\varphi_i K\left(x_i, x_j\right) + a\right) \tag{8}$$

where a represents a bias, which indicates the linear distance of the hyperplane from the origin, $K\left(x_i, x_j\right)$ represent kernel functions, i.e., polynomial (POL) and radial basis function (RBF), and these can be expressed as follows [51]:

$$K_{POL}\left(x_i, x_j\right) = \left((x * y) + 1\right)^d \tag{9}$$

$$K_{RBF}\left(x_i, x_j\right) = e^{-y\|x - x_i\|^2} \tag{10}$$

2.5.2. Artificial Neural Network (ANN)

ANN is a popular machine learning algorithm that simulates the neural networks of a human brain and can work in a specific way [52,53]. It is used to analyze and predict non-linear statistical datasets by using different algorithms [54]. ANN has been widely used in pattern recognition and classification studies [55]. Therefore, classifications of the landscape in different ordinal areas of the GES zone are treated as a classification problem. Different types of algorithms have been used in ANN modeling; among them, multi-layer perceptron (MLP) is the most popular, based on its outcome results and frequency of use by researchers [56]. To run and analyze ANN algorithms, some basic knowledge is needed to understand the structure of input data and the relationship between the variables [57]. The ANN model with the MLP algorithm consists of three layers, namely, the input layer, hidden layer, and output layer. A schematic diagram of the feed-forward artificial neural network model is shown in Figure 5. In this research work, the input layers are training points for the erosion of the gully and the various GECFs, which have finally been connected to the output layer. Input nodes help to predict and analyze the model structure through input and hidden layers and, ultimately, to evaluate the output layer result [58,59]. This output layer gives us the GES map. The output layer consists of Boolean values of 0 and 1, in which 0 represents non-gully erosion and 1 represents gully erosion. Feed-forward of the ANN algorithm model deals with three stages, namely, feed-forward of input data, calculation, and backpropagation of related errors and their adjustments [57].

The novelty of the ANN model is that it can learn the model through a non-linear and complex relationship. Thus, the model's uniqueness is evaluated based on observation of the coherence of the network dynamics compared with the other models. It also has the ability of model generalization and can predict unseen data within the model through understanding the hidden relationship.

The ANN algorithms were elaborated using the following equations by Hagan et al. (1996):

$$net_j^l(t) = \sum_{i=o}^{p}\left(y_i^{i-1}(t) w_{ji}^l(t)\right) \tag{11}$$

The net input of the jth neuron of layer l and $i$iteration

$$y_j^l(t) = f(net_j^{(l)}(t) \tag{12}$$

$$f(net) = \frac{1}{1 + e^{(-net)}} \tag{13}$$

$$e_j(t) = c_j(t) - a_j(t) \tag{14}$$

$$\delta_j^l(t) = e_j^l(t) a_j(t) \left[1 - a_j x(t)\right] \tag{15}$$

δ factor for the jth neuron in the ith output layer

$$\delta_j^l(t) = y_j^l(t)\left[1 - y_j(t)\right] \sum \delta_j^l(t) w_{kj}^{(l+1)}(t) \tag{16}$$

δ factor for the jth neuron in the ith hidden layer

$$w_{ji}^l(t+1) = w_{ji}^l(t) + \alpha\left[w_{ji}^l(t) - w_{ji}^l(t-1)\right] + n\delta_j^{(l)}(t) y_j^{(l-1)}(t) \tag{17}$$

where α is the momentum rate and n is the learning rate.

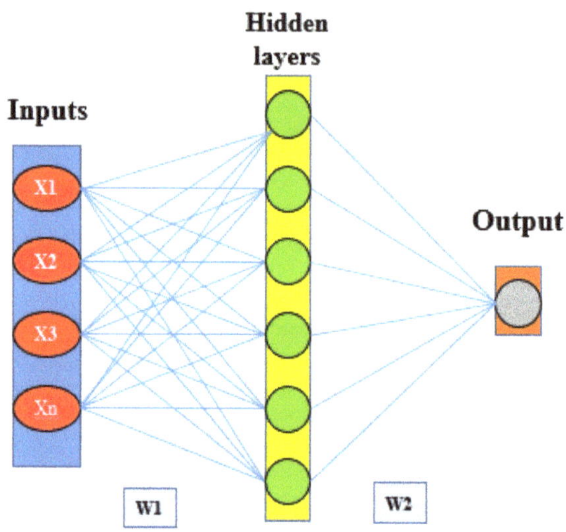

Figure 5. Schematic of feed-forward artificial neural network.

2.5.3. Deep Learning Neural Network (DLNN)

DLNN is a well-accepted machine learning model among research groups throughout the world. This ML model has a prominent advantage in appropriately constructing a high-level feature by using the raw dataset [27]. DLNN consists of three layers, i.e., an input layer, several hidden layers, and resulting in an output layer [60]. The speculative configuration of the DLNN model used for GES mapping in this research work is shown in Figure 6. The general structure of the DLNN model is to run in such a way that the input layer receives signals that are different GECFs, this information is processed and analyzed in several hidden layers, and finally, the output model's result is presented in the last layer, i.e., the output layer. The output layer has two possible labels, i.e., the first one is a negative label (non-gully erosion) and the second one is a positive label (gully erosion). These classification results are obtained from the last hidden layer and shown in the output layer [61].

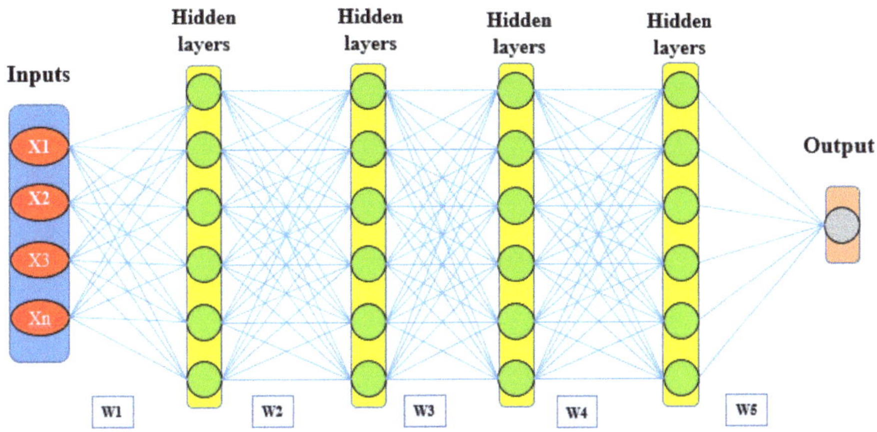

Figure 6. Schematic of deep learning neural network.

DLNN has some specific compensations over the traditional ML algorithm, and thus, in the field of prediction analysis, the use of the DLNN model has been given much more emphasis. Therefore, DLNN has showed some novel performances over the other ML models, namely, maximum utilization of unstructured data through relevant insights to understand the training dataset, being robust enough to recognize the novel data, and being able to develop additional learning models through adding more layers into the neural network system.

According to Kim (2017), the following mathematical equation is used in a DLNN machine learning model:

$$h(x) = \begin{cases} x \text{ } if \text{ } x > 0 \\ 0 \text{ } if \text{ } x \leq 0 \end{cases} = \max(0, x) \tag{18}$$

where x represents the input signal, and h indicates the activation function.

Based on the ReLU activation function, this can be as expressed as follows:

$$h'(x) = \begin{cases} 1 \text{ } if \text{ } x > 0 \\ 0 \text{ } if \text{ } x \leq 0 \end{cases} \tag{19}$$

The cost function is the difference between experiential and predicted class outputs. The loss function (L) of a cross-entropy used for pattern recognition and expressed as follows:

$$L = -\frac{1}{N_D} \sum_{n=1}^{N_D} T1n(Y) + (1 - T)1n(1 - Y) \tag{20}$$

where N_D represents the number of the training datasets, T indicates observed class outputs, and Y indicates predicted class outputs.

2.5.4. Particle Swarm Optimization (PSO)

The algorithm of PSO is a meta-heuristic and was originally developed by an American social psychologist named Kennedy [62]. In our research work, we are faced with some non-linear problems, and to find the correct solution, the PSO method was developed and widely used. The PSO algorithm was used to locate the best possible food route for bird and fish intelligence. Here, birds are the particles and try to find a solution to the problem. Particles always try to find the best possible solution to a problem through n-dimensional space, in which n represents each problem's different parameters [63]. Optimization of position and velocity is the basic principle of each particle.

Therefore, let us suppose that $x_i^t = \left(x_{i1}^t, x_{i2}^t, \ldots, x_{in}^t\right) and\ v_i^t = \left(v_{i1}^t, v_{i2}^t, \ldots, v_{in}^t\right)$ are the position and velocity of changing position designed for the *i*th particle in the *t*th iteration accordingly. The following equations are used for the *i*th particle's position and velocity in the *(t+1)*th iteration:

$$v_i^{t+1} = \omega.v_i^t + c_1.r_1.\left(p_i^t - x_i^t\right) + c_2.r_2.\left(g_i^t - x_i^t\right)\ with - v_{max} \leq v_i^{t+1} \leq v_{max} \tag{21}$$

$$x_i^{t+1} = \left(x_i^t + v_i^{t+1}\right) \tag{22}$$

where x_i^t represents the previous *i*th position; p_i^t represents the optimal found position; g_i^t represents the particle's best position; r_1 and r_2 represent random numbers of either 0 or 1; ω is weights of inertia; c_1 is a coefficient; and c_2 represents the social coefficient. There are numerous methods for particle weight assignment [64,65]; among them, standard 2011 PSO has been widely used and can be calculated by the following equation:

$$\omega = \frac{1}{2\ In\ 2} and\ c_1 = c_2 = 0.5 + In\ 2 \tag{23}$$

Therefore, it is believed that when the concentration of all particle swarms in a certain point and space has been achieved, the problem has been solved. The intelligence-based PSO algorithm has been widely used in high-efficiency swarm paralleling and optimization property. Using a multi-objective fitness function, PSO determines the quality of several features in a dataset. Finally, the ensemble structure of particle swarm optimization (PSO) and deep learning neural network (DLNN) is shown in Figure 7. Therefore, this ensemble method is the novel approach in this research study for GES mapping with high accuracy.

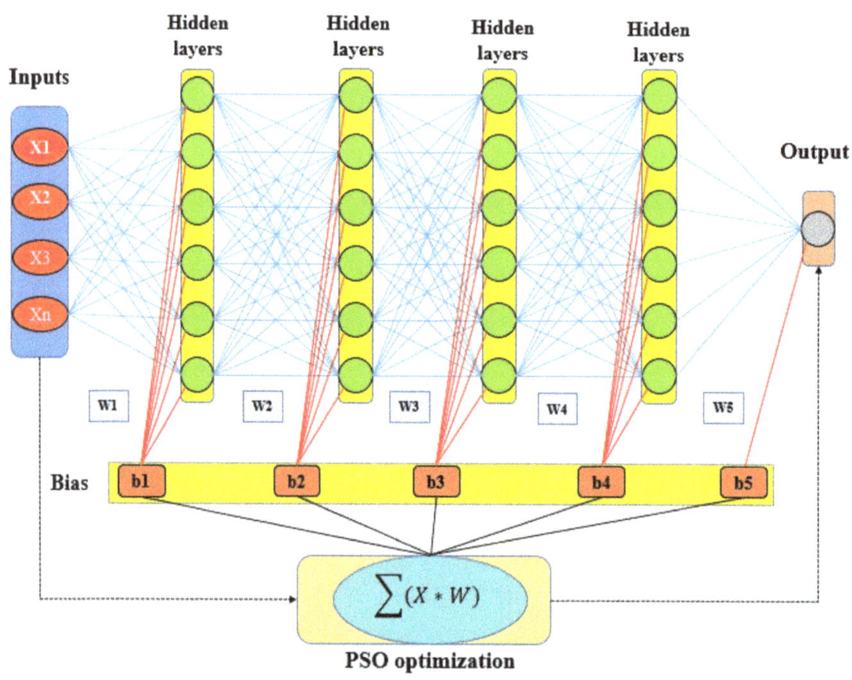

Figure 7. Schematic of ensemble particle swarm optimization and deep learning neural network.

2.6. Methods of Validation and Accuracy Assessment

GES maps were prepared based on the prediction performance of the training and validation datasets by using different machine learning models. Therefore, it is necessary to evaluate the model performance to ascertain the validity of the results. In the present research work, statistical indices along with the area under the receiver operating characteristic (AUROC) curve were used to predict the accuracy of ML and ensemble models.

2.6.1. Statistical Indices

In this study, sensitivity (SST), specificity (SPF), positive predictive values (PPV), and negative predictive values (NPV) were used to evaluate the predictive results. Four types of possible consequences were used to analyze these statistical indices, namely, true positive (TP), true negative (TN), false positive (FP), and false negative (FN). TP is when gully pixels are correctly classified as a gully, and FP is when gully pixels are incorrectly classified as a gully. On the other hand, if gully pixels are correctly or incorrectly classified as non-gully, then they are TN and FN, respectively [36]. If higher values are found among these statistical indices then the model gives better results and vice versa [22]. The following equations were used to calculate the value of these four statistical indices:

$$PPV = \frac{TP}{FP + TP} \tag{24}$$

$$NPV = \frac{TP}{TP + FN} \tag{25}$$

$$SST = \frac{TP}{TP + FN} \tag{26}$$

$$SPF = \frac{TN}{FP + TN} \tag{27}$$

2.6.2. ROC Curve

ROC curve is one of the most widely used tools for analyzing the performance validation of the ML model. ROC curve has two dimensions, i.e., events and non-events phenomena [66]. This curve is plotted on 'X' and 'Y' co-ordinates, known as sensitivity and 1-specificity, respectively, and represents true positive and false positive. The optimum value in both cases, i.e., in sensitivity (detected gullies) and specificity (detected non-gullies), is 1 [3]. The value of ROC-AUC ranges from 0.5 to 1, in which 0.5 indicates poor performance and 1 indicates very good performance. Beside this, in a proper way it can be classified into five classes, i.e., poor (0.5–0.6), moderate (0.6–0.7), good (0.7–0.8), very good (0.8–0.9), and excellent (0.9–1) [67]. The following equation was used to compute the ROC-AUC:

$$S_{AUC} = \sum_{k=1}^{n} (X_{k+1} - X_k)\left(S_k + 1 - S_{k+1} - \frac{S_k}{2}\right) \tag{28}$$

where S_{AUC} indicates area under the curve, X_k indicates 1-specificity, and S_k indicates the sensitivity of the receiver operating characteristic (ROC) curve.

Besides the above validation methods, here we also used Likelihood Ratio (LR), F-measure, and Maximum Probability of Correct Decision (MPCD) analyses to better understand the accuracy assessment of the result. In this study, the LR model is the relationship between the distribution of gully head-cut points and related GECFs. Therefore, the LR model emphasized the ratio of the probability of events and non-events phenomena of the gully occurrences. In this method, if the ratio is higher than 1, there is a high relationship among the gully erosion and associated factors. On the other hand, if the ratio is less than 1, a low relationship is found between the gully erosion and associated factors. Thus, the linear relationship of LR can be expressed as follows:

$$GESI = \sum Fr \tag{29}$$

where $GESI$ represents the gully erosion susceptibility index, and Fr represents the rating of several factors' range.

F-measure is a popular validation method in the field of classification and information retrieval communities. *F-measure* balances between precision and recall. The following equation was used to calculate the *F-measure* in this study:

$$F - measure = 2 \times TP/2 \times TP + FP + FN \tag{30}$$

In a classification performance, MPCD is a probabilistic-based measure. It is a sensitive method for recognition of class rather than just to estimate the proportion of guesses. The following equation was used to calculate the MPCD:

$$MPCD = (1 - \alpha)(1 - \beta) \tag{31}$$

where α is $\frac{FP}{FP+TN}$ and β is $\frac{FN}{FN+TP}$.

3. Results

3.1. Multi-Collinearity Analysis

Maintaining the given VIF and TOL limits, 13 gully erosion conditioning parameters were selected for gully erosion modeling. The co-linear factors (i.e., distance from a road, geomorphology, and bulk density) were excluded from this analysis. The three factors of distance from a road (TOL 0.028 and VIF 35.65), geomorphology (TOL 0.032 and VIF 31.63), and bulk density (TOL 0.022 and VIF 45.23) are associated with co-linearity problems. The range of VIF for the selected parameters is 1.06 to 3.04. In the case of TOL, the range of variation among the selected conditioning factors is 0.33 to 0.94 (Table 3). Among the 13 GECFs, altitude has the highest VIF value of 3.04 and the lowest TOL value of 0.33. On the other hand, the aspect factor has the highest TOL value of 0.94 and the lowest VIF value of 1.06. Therefore, this indicates that no multi-collinearity has been found between the thirteen conditioning factors of gully erosion used in this study.

Table 3. Multi-collinearity analysis to determine the linearity of the independent variables.

Variables	VIF	Tolerance
Altitude	3.04	0.33
Slope	1.34	0.75
Aspect	1.06	0.94
Plan curvature	1.83	0.55
Profile curvature	1.82	0.55
Distance from river	2.93	0.34
Drainage density	2.07	0.48
Rainfall	1.41	0.71
Land use	1.81	0.55
Lithology	2.07	0.48
Soil	1.11	0.90
SPI	1.58	0.63
TWI	1.94	0.52

3.2. Gully Erosion Susceptibility Modeling

In the SVM model, the very low GES areas are mainly concentrated in the eastern and northern portions of the region. The low GES areas are mainly found in the middle and western parts of the region. The moderate susceptibility areas are mainly concentrated in the middle and southern parts of the region (Figure 8a). The very high and high GES areas are mainly found in the southern portion of the watershed. The areal coverages of very low, low, moderate, high, and very high gully erosion susceptibility areas in the SVM model are 65.86 (52.08%), 28.92 (22.87%), 10.7 (8.46%), 8.0 (6.33%), and 12.97 km^2 (10.26%), respectively (Table 4).

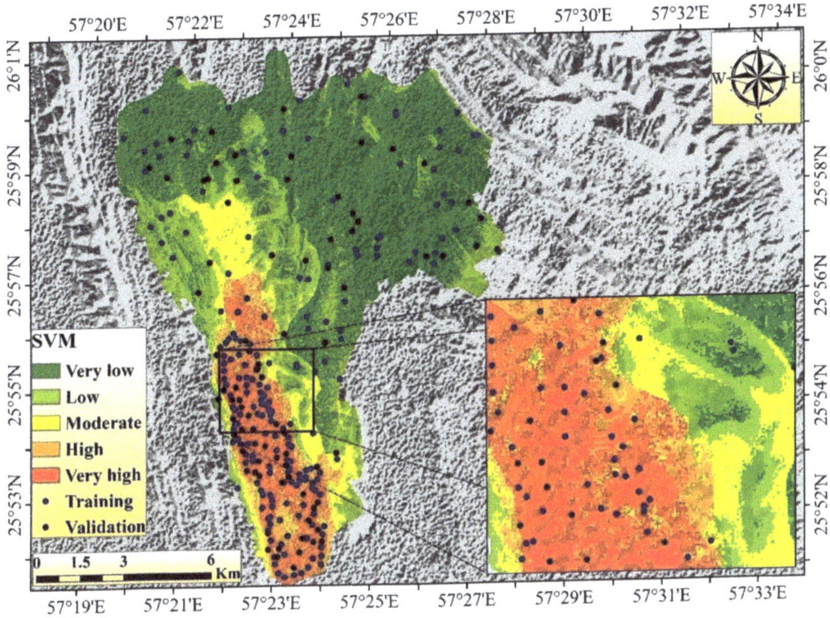

(a)

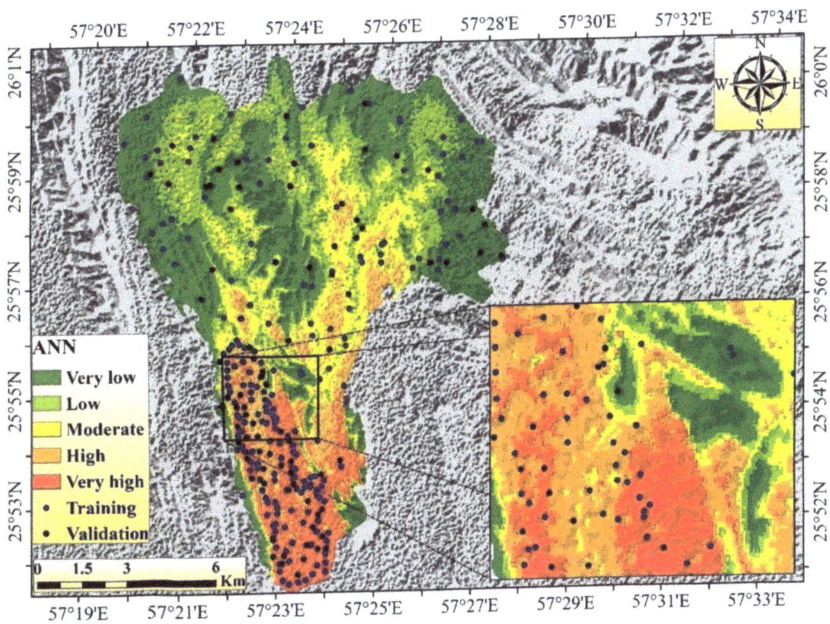

(b)

Figure 8. *Cont.*

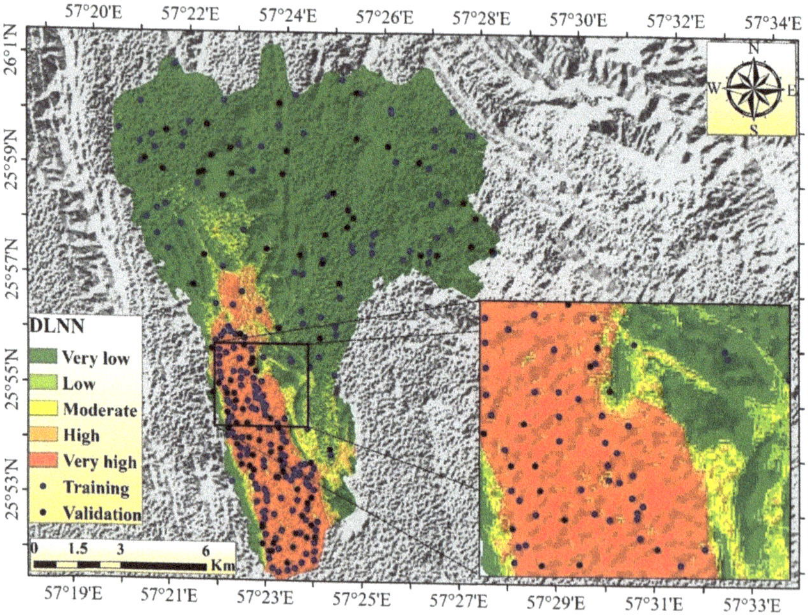

(c)

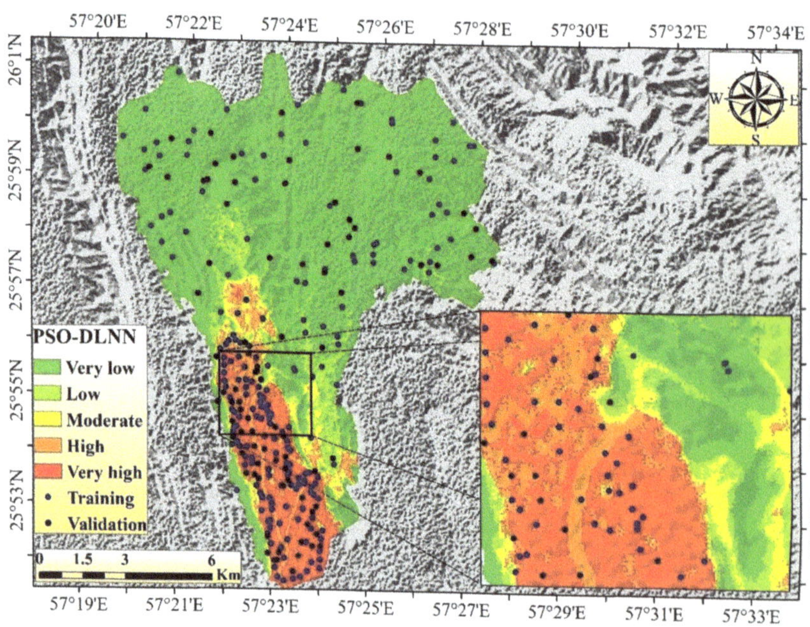

(d)

Figure 8. Head-cut gully erosion map using the four models: (**a**) SVM; (**b**) ANN; (**c**) DLNN; (**d**) PSO-DLNN.

Table 4. Areas of gully erosion susceptibility classes.

Models	Area	Susceptibility Class				
		Very Low	Low	Moderate	High	Very High
SVM	km^2	65.86	28.92	10.7	8	12.97
	%	52.08	22.87	8.46	6.33	10.26
ANN	km^2	55.76	26.85	16.85	13.48	13.51
	%	44.10	21.23	13.33	10.66	10.68
DLNN	km^2	96.34	5.85	2.73	3.17	18.36
	%	76.19	4.63	2.16	2.51	14.52
PSO-DLNN	km^2	94.58	8.15	4.03	6.31	13.38
	%	74.80	6.45	3.19	4.99	10.58

In ANN, the areal coverages for very low, low, moderate, high, and very high gully erosion susceptibility areas are 55.76 (44.10%), 26.85 (21.23%), 16.85 (13.33%), 13.48 (10.66%), and 13.51 km^2 (10.68%), respectively. According to the GES map of the ANN model, the largest portion of the area is occupied by very low (44.10%) to low (21.23%) susceptibility classes, while very high (10.68%), high (10.66%), and moderate (13.33%) susceptibility classes cover the rest of the studied region. In this model, the very high, high, and moderate susceptibility areas are mainly concentrated in the southern, middle, and eastern portions of the watershed (Figure 8b). The rest of the portion of this watershed is associated with very low to low GES zones.

In the case of DLNN, the areal coverages for very low, low, moderate, high, and very high gully erosion susceptibility zones are 96.34 (76.19%), 5.85 (4.63%), 2.73 (2.16%), 3.17 (2.51%), and 18.36 km^2 (14.52%), respectively. According to the GES map of the DLNN model, the largest portion of the area is occupied by very low (76.19%) to low (28.73%) susceptibility classes, while very high (14.52%), high (2.51%), and moderate (2.16%) susceptibility classes occupy the rest of the studied region. In this model, the very high to moderate susceptibility areas are mainly concentrated in the southern and middle portions of the watershed, and the rest of the portions are associated with very low to low susceptibility zones (Figure 8c).

In the PSO-DLNN model, the areal coverages of low, low, moderate, high, and very high gully erosion susceptibility zones are 94.58 (74.80%), 8.15 (6.45%), 4.03 (3.19%), 6.31 (4.99%), and 13.38 (10.58%) km^2, respectively. According to the GES map of the PSO-DLNN model, the major portion of the area is occupied by very low (74.80%) to low (6.45%) susceptibility classes, while very high (10.58%), high (4.99%), and moderate (3.19%) susceptibility classes cover the rest of the studied region respectively. Very high, high, and moderate gully erosion susceptibility zones mainly occupy the southern portion of the watershed, and the rest of the portions are associated with very low to low susceptibility zones (Figure 8d).

3.3. *Validation of the Models*

PSO-DLNN is the most optimal model in this analysis and is associated with maximum accuracy. The AUC value from ROC considering the testing datasets of PSO-DLNN is 0.89, which is associated with superb accuracy. The rest of the models are also associated with optimal accuracy and have similar values to the PSO-DLNN model; the AUC values from ROC of DLNN, SVM, and ANN for testing datasets are 0.87, 0.85, and 0.84, respectively (Figure 9). Apart from this, various statistical indices were considered for estimating the optimal capacity of all the models for GES modeling. The values of sensitivity in PSO-DLNN, DLNN, SVM, and ANN for training datasets are 0.98, 0.95, 0.99, and 0.99, respectively. The same values for the validation datasets in PSO-DLNN, DLNN, SVM, and ANN are 0.95, 0.90, 0.82, and 0.95, respectively. The values of specificity for the training datasets in PSO-DLNN, DLNN, SVM, and ANN are 0.85, 0.82, 0.86, and 0.87, respectively. In the case of validation datasets, the values of specificity in PSO-DLNN, DLNN, SVM, and ANN are 0.74, 0.74, 0.69, and 0.67, respectively. The values of PPV in the case of training datasets in PSO-DLNN, DLNN, SVM, and ANN are 0.87, 0.85, 0.88, and 0.89, respectively. When we consider the validation datasets, the values of PPV in PSO-DLNN, DLNN, SVM, and ANN are 0.77, 0.77, 0.71, and 0.73 (Table 5). In PSO-DLNN, DLNN, SVM, and ANN models, the values of NPV for the training datasets are 0.97, 0.94, 0.99, and 0.99, respectively. In the case of validation datasets, the values of NPV in PSO-DLNN, DLNN, SVM, and ANN are 0.94, 0.89, 0.81, and 0.93, respectively. The F-measure values in validation datasets for PSO-DLNN, DLNN, SVM, and ANN models are 0.66, 0.635, 0.63, and 0.64, respectively.

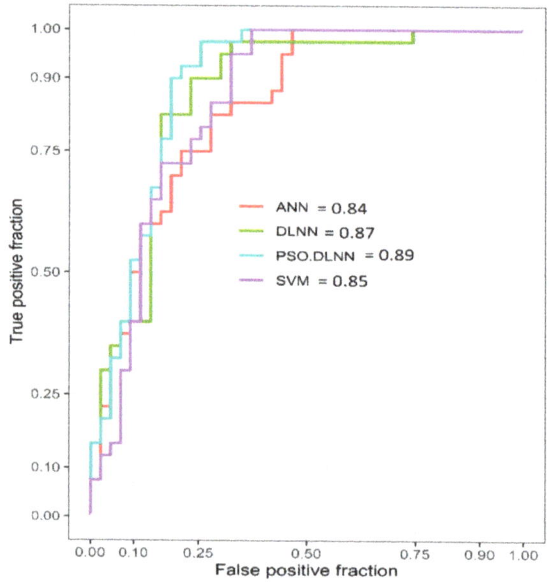

Figure 9. Receiver operating characteristic (ROC) curve analysis for four head-cut gully erosion models using the testing dataset.

Table 5. Predictive capability of guully erosion susceptibility (GES) models using training and testing datasets.

Models	Stage	Parameters					
		Sensitivity	Specificity	PPV	NPV	AUC	F-Measure
SVM	Training	0.99	0.87	0.89	0.99	0.94	0.84
	Validation	0.95	0.67	0.73	0.93	0.85	0.63
ANN	Training	0.99	0.86	0.88	0.99	0.94	0.83
	Validation	0.82	0.69	0.71	0.81	0.84	0.64
DLNN	Training	0.95	0.82	0.85	0.94	0.91	0.82
	Validation	0.90	0.74	0.77	0.89	0.87	0.65
PSO-DLNN	Training	0.98	0.85	0.87	0.97	0.93	0.84
	Validation	0.95	0.74	0.77	0.94	0.89	0.66

Details about DLNN and its associated parameters are shown in Table 6. Details about the combination of PSO and DLNN and its associated parameters are shown in Table 7. The values of population, iteration, phi, phi1, phi2, W, C1, C2, and best cost are 50, 500, 4.1, 2.05, 2.05, 0.73, 1.49, 1.49, and 0.26. The objective cost function of the PSO-DLNN model is shown in Figure 10.

Table 6. Results of optimal parameters in the DLNN model.

Parameters	Optimum
Input number of units	13
Output	2
Activation Function	ReLU
Activation	'softmax'
Function	Sigmoid
reluLeak	0.01
eta	0.8
Hidden layer unit	3-3
Iteration	200

Table 7. Parameters used in PSO algorithms in combined DLNN.

Parameters	Number
Population	50
Iteration	500
phi	4.1
phi1	2.05
Phi2	2.05
W	0.73
C1	1.49
C2	1.49
Best Cost	0.26

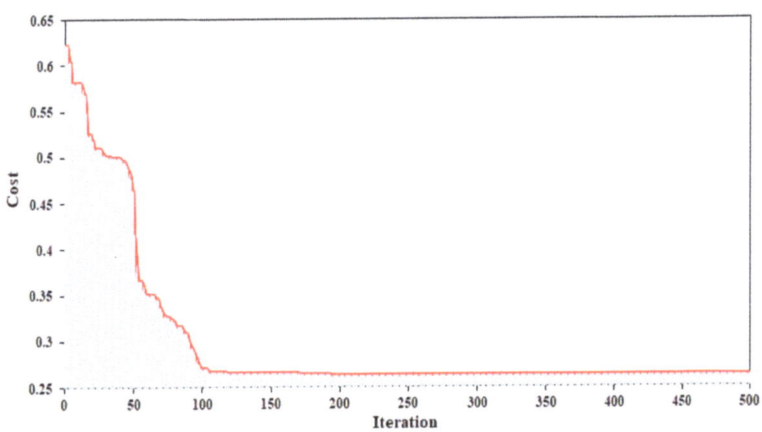

Figure 10. Convergence graph of the objective cost function (MSE) in the PSO-DLNN model.

3.4. Variable Importance

The conditioning factor for GES modeling for this region was selected considering the different kinds found in the literature. The most important parameters for the creation and development of gullies in this region are land use, altitude, lithology, rainfall, and distance from a road, etc. The relative importance of land use, altitude, lithology, rainfall, and distance from a road for the GES models are 100, 97.94, 59.51, 46.94, and 29.48, respectively. The rest of the factors (i.e., profile curvature, TWI, plan curvature, slope, soil, drainage density, SPI, and aspect) are associated with moderate to very low relative importance for GES. The relative importance of profile curvature, TWI, plan curvature, slope, soil, drainage density, SPI, and aspect for gullies are 16.22, 14.37, 11.31, 7.89, 7.1, 6.91, 5.12, and 0 (Table 8). Here, apart from the topographical and geohydrological characteristics, the impact of anthropogenic activities accelerates the rate of land degradation in the form of gullies.

Table 8. Variable importance analysis based on the PSO-DLNN model.

Variables	Importance
Altitude	97.94
Aspect	0
Slope	7.89
Plane curvature	11.31
Profile curvature	16.22
Drainage density	6.91
Distance from river	29.48
Land use	100
Lithology	59.51
Soil	7.1
Rainfall	46.94
SPI	5.12
TWI	14.37

4. Discussion

Land degradation through various forms of soil erosion can cause extensive damage, and it has an adverse impact on society and people's livelihoods throughout the world [68]. There are various forms of erosion, i.e., sheet erosion, formation of rills, formation and development of gullies and ravines, etc. [69]. Of these, the formation and development of gullies and their associated erosion is the most destructive element of land degradation worldwide [2]. Although it is a natural process of erosion, this process can be greatly accelerated by anthropogenic activities and have a serious impact on the ecosystem [70]. With this type of erosion, agricultural activities have not only affected it but have also been associated with damage to manmade infrastructure. On the one hand, this erosion is responsible for removing the top soil, but on the other hand, it is responsible for the creation and accumulation of sediment in the lower catchment area [71]. The life span of the reservoir will cause serious damage to the sediment deposition resulting from this type of erosion [72,73].

Shirahan watershed in Iran has recently faced severe gully erosion, which is responsible for large-scale erosion and is the main barrier to sustainable land management practices. Therefore, identifying vulnerable regions with the most optimal model is very useful so that appropriate soil and water conservation measures can be put in place. For this purpose, we considered SVM, ANN, DLNN, and PSO-DLNN in order to estimate the GES of this region with the maximum possible accuracy and to suggest the most suitable model. The erosion of a gully is controlled by various causal factors, and we attempted to determine the importance of these factors for gulling. Apart from the topographic and hydrogeomorphic attributes, land use is the most important variable for gully erosion, which indicates the large anthropogenic impact on the development of gullies. Other factors, such as altitude, lithology, rainfall, and the distance from a river, are very influential too on gully erosion and promote gulling. The transformation of land use is a crucial element and is responsible for large-scale erosion [74]. Alterations in land use influence landscape ecology functions, with far-reaching implications for natural ecosystems and land reclamation [75]. The character and volume of the surface runoff may change directly with the changing pattern of land use in the region. From this perspective, the nature of erosion in the form of gullies can have a significant effect on the impact of rainfall and its associated runoff characteristics in a changing environment. This type of outcome is similar to some of the findings from the studies of a number of researchers in this diversified discipline. This finding has been highlighted by many other contributions in which morphological and geological properties are assigned as the determinants of the highest possible location of GES [24,76]. Other research outcomes suggest that environmental and hydrological parameters are very significant and responsible for gulling.

All the predicted models are associated with high accuracy, but PSO-DLNN is the most optimal, with the AUC of this model being 0.89. The efficiency of all predicted models is excellent, with the AUC values for DLNN, SVM, and ANN being 0.87, 0.85, and 0.84, respectively. Apart from this, considering various statistical indices, PSO-DLNN is the best model among the models used in this study. According to the PSO-DLNN model, 18.76% of the total area is associated with a moderate to very high susceptible area of gully erosion. The southern portion of this watershed is mainly associated with higher gully occurrences. The complex geohydrological characteristics of this region are favorable for large-scale erosion in the form of gullies.

A deep learning framework is associated with higher accuracy compared with conventional ANN and SVM ML methods. This model can handle a larger number of samples and even a large amount of big data, and can estimate the results with optimum accuracy. The traditional ML algorithm is not capable of handling this large a number of samples, and the outcome from this perspective is less optimal compared to the deep learning framework. Significant progress in DLNN-dependent deep learning (DL) systems has significantly increased the consistency of machine learning for various purposes. While the standardized features of multi-layer NNs are well-established, the main advantage of DL is its structured method of self-governing the training of DLNN layer organizations. The benefits of structured data and expertise descriptions were recognized before the recent increase in interest in DLNNs. This definition is widespread in the physical sciences where the proposed method is popular for both specific theoretical structures and complicated system implementations in practice.

First, PSO produces an arbitrary solution and then discovers accurate solutions with an incremental optimum fitness attribute. This type of methodology has already been used primarily for backpropagation (BP) genetic algorithms, due to the efficiency of simple installation, fast response, and accuracy of predictions. It also demonstrated dominance in the resolution of complex applications and was initially implemented in the context of DL. The best function of the PSO algorithm is to combine various particles that are interlinked to each other to achieve an optimum position. The same technique indicates the position, velocity, and highest accuracy of each particle, which are dictated by the basic concepts used to enhance the problem. Particularly in comparison to other optimization algorithms, the advantage of the PSO algorithm is that the PSO technique usually involves a quick and important search procedure, is easy to perform, and can find the globally optimal path that is closest to the concrete ideas.

5. Conclusions

It is necessary to choose the most efficient machine learning algorithm in order to decrease the inconsistencies associated with predicting gully erosion susceptibility. The main objectives in most cases of susceptibility modeling are to identify the optimal model according to its predictive capabilities. The identification of key parameters for the formation and development of gullies is necessary to estimate the susceptibility mapping of the spatial distribution of gully erosion. Therefore, to control damage in the future, it is important to make an appropriate selection of a model to manage areas that are prone to gully degradation. The primary objective of this research was to estimate the optimal model with maximum predictive capability. For this reason, various ML algorithms, i.e., ANN, SVM, DLNN, and PSO, were considered for estimating the GES zone with optimal capacity. PSO-DLNN is the best-fitted model and is associated with the highest AUC value (0.89). Here, all the datasets were randomly partitioned with a 70/30 ratio as training and validation datasets. Topographical, hydrological, and environmental factors were most dominant and were influential factors in susceptibility modeling. The role of land use in susceptibility modeling is more significant than that of any other component. Most of the region of this watershed is associated with very low to low susceptibility zones, while 15.57% of the area is associated with a very high susceptibility zone. This study region must take appropriate planning initiatives to reduce the level of vulnerability and to protect this precious resource. In future research, it would be desirable to develop the PSO-DLNN algorithm by incorporating some new components or to develop the same algorithm with slight modifications. This would be a great contribution to the research community as well as to society. Apart from this, the selection of inappropriate parameters can reduce the efficiency of the predicted models. Thus, the selection of the most appropriate variables for susceptibility modeling is one of the important tasks for researchers.

Author Contributions: S.J. acquired the data; S.J., S.S.B., S.C.P. and A.M. conceptualized and performed the analysis; S.S.B., S.C.P., A.S., R.C. and M.S. wrote the manuscript, discussion, and analyzed the data; S.S.B. supervised and carried out funding acquisition; S.S.B. and A.M. provided technical insights, as well as edited, restructured, and professionally optimized the manuscript. All authors discussed the results and edited the manuscript. All authors have read and agreed to the published version of the manuscript.

Funding: This research was funded by the Hungarian-Mexican bilateral Scientific and Technological project, grant number 2019-2.1.11-TET-2019-00007.

Acknowledgments: We acknowledge the support of the German Research Foundation (DFG) and the Bauhaus-Universität Weimar within the Open-Access Publishing Program.

Conflicts of Interest: The authors declare no conflict of interest.

References

1. Keesstra, S.; Mol, G.; De Leeuw, J.; Okx, J.; De Cleen, M.; Visser, S.; Molenaar, C. Soil-related sustainable development goals: Four concepts to make land degradation neutrality and restoration work. *Land* **2018**, *7*, 133. [CrossRef]
2. Lal, R. Soil degradation by erosion. *L. Degrad. Dev.* **2001**, *12*, 519–539. [CrossRef]
3. Conoscenti, C.; Angileri, S.; Cappadonia, C.; Rotigliano, E.; Agnesi, V.; Märker, M. Gully erosion susceptibility assessment by means of GIS-based logistic regression: A case of Sicily (Italy). *Geomorphology* **2014**, *204*, 399–411. [CrossRef]
4. Vanwalleghem, T.; Poesen, J.; Van Den Eeckhaut, M.; Nachtergaelel, J.; Deckers, J. Reconstructing rainfall and land-use conditions leading to the development of old gullies. *Holocene* **2005**, *15*, 378–386. [CrossRef]
5. Zabihi, M.; Pourghasemi, H.R.; Motevalli, A.; Zakeri, M.A. Gully erosion modeling using GIS-based data mining techniques in northern Iran: A comparison between boosted regression tree and multivariate adaptive regression spline. In *Natural Hazards GIS-Based Spatial Modeling Using Data Mining Techniques*; Springer: Cham, Switzerland, 2019; pp. 1–26.
6. Pimentel, D.; Burgess, M. Soil erosion threatens food production. *Agriculture* **2013**, *3*, 443–463. [CrossRef]
7. Arabameri, A.; Rezaei, K.; Pourghasemi, H.R.; Lee, S.; Yamani, M. GIS-based gully erosion susceptibility mapping: A comparison among three data-driven models and AHP knowledge-based technique. *Environ. Earth Sci.* **2018**, *77*, 1–22. [CrossRef]
8. Arabameri, A.; Pradhan, B.; Rezaei, K. Gully erosion zonation mapping using integrated geographically weighted regression with certainty factor and random forest models in GIS. *J. Environ. Manag.* **2019**, *232*, 928–942. [CrossRef]
9. Vaezi, A.R.; Abbasi, M.; Bussi, G.; Keesstra, S. Modeling sediment yield in semi-arid pasture micro-catchments, NW Iran. *L. Degrad. Dev.* **2017**, *28*, 1274–1286. [CrossRef]
10. Poesen, J.; Nachtergaele, J.; Verstraeten, G.; Valentin, C. Gully erosion and environmental change: Importance and research needs. *Catena* **2003**, *50*, 91–133. [CrossRef]
11. Poesen, J. Soil erosion in the Anthropocene: Research needs. *Earth Surf. Process. Landf.* **2018**, *43*, 64–84. [CrossRef]
12. Valentin, C.; Poesen, J.; Li, Y. Gully erosion: Impacts, factors and control. *Catena* **2005**, *63*, 132–153. [CrossRef]
13. Chaplot, V. Impact of terrain attributes, parent material and soil types on gully erosion. *Geomorphology* **2013**, *186*, 1–11.
14. Angileri, S.E.; Conoscenti, C.; Hochschild, V.; Märker, M.; Rotigliano, E.; Agnesi, V. Water erosion susceptibility mapping by applying stochastic gradient treeboost to the Imera Meridionale river basin (Sicily, Italy). *Geomorphology* **2016**, *262*, 61–76. [CrossRef]
15. Saha, S.; Roy, J.; Arabameri, A.; Blaschke, T.; Tien Bui, D. Machine learning-based gully erosion susceptibility mapping: A case study of Eastern India. *Sensors* **2020**, *20*, 1313. [CrossRef] [PubMed]
16. Moradi, H.R.; Avand, M.T.; Janizadeh, S. landslide susceptibility survey using modeling methods. In *Spatial Modeling in Gis and R for Earth and Environmental Sciences*; Elsevier: Amsterdam, The Netherlands, 2019; pp. 259–276.
17. Watson, G.L.; Telesca, D.; Reid, C.E.; Pfister, G.G.; Jerrett, M. Machine learning models accurately predict ozone exposure during wildfire events. *Environ. Pollut.* **2019**, *254*, 112792. [CrossRef]

18. Arabameri, A.; Pradhan, B.; Rezaei, K.; Yamani, M.; Pourghasemi, H.R.; Lombardo, L. Spatial modelling of gully erosion using evidential belief function, logistic regression, and a new ensemble of evidential belief function—Logistic regression algorithm. *L. Degrad. Dev.* **2018**, *29*, 4035–4049. [CrossRef]
19. Dube, F.; Nhapi, I.; Murwira, A.; Gumindoga, W.; Goldin, J.; Mashauri, D.A. Potential of weight of evidence modelling for gully erosion hazard assessment in Mbire District—Zimbabwe. *J. Phys. Chem. Earth* **2014**. [CrossRef]
20. Pourghasemi, H.R.; Youse, S.; Kornejady, A.; Cerdà, A. Performance assessment of individual and ensemble data-mining techniques for gully erosion modeling. *Sci. Total Environ.* **2017**, *609*, 764–775. [CrossRef]
21. Roy, P.; Chakrabortty, R.; Chowdhuri, I.; Malik, S.; Das, B.; Pal, S.C. Development of Different Machine Learning Ensemble Classifier for Gully Erosion Susceptibility in Gandheswari Watershed of West Bengal, India. In *Machine Learning for Intelligent Decision Science*; Springer: Singapore, 2020; pp. 1–26.
22. Gayen, A.; Pourghasemi, H.R. Spatial Modeling of Gully Erosion: A New Ensemble of CART and GLM Data-Mining Algorithms. In *Spatial Modeling in GIS and R for Earth and Environmental Sciences*; Elsevier: Amsterdam, The Netherlands, 2019; pp. 653–669.
23. Yunkai, L.; Yingjie, T.; Zhiyun, O.; Lingyan, W.; Tingwu, X.; Peiling, Y.; Huanxun, Z. Analysis of soil erosion characteristics in small watersheds with particle swarm optimization, support vector machine, and artificial neuronal networks. *Environ. Earth Sci.* **2010**, *60*, 1559–1568. [CrossRef]
24. Saha, A.; Ghosh, M.; Pal, S.C. Understanding the Morphology and Development of a Rill-Gully: An Empirical Study of Khoai Badland, West Bengal, India. In *Gully Erosion Studies from India and Surrounding Regions*; Springer: Cham, Swizerland, 2020; pp. 147–161.
25. Avand, M.; Janizadeh, S.; Naghibi, S.A.; Pourghasemi, H.R.; Khosrobeigi Bozchaloei, S.; Blaschke, T. A Comparative Assessment of Random Forest and k-Nearest Neighbor Classifiers for Gully Erosion Susceptibility Mapping. *Water* **2019**, *11*, 2076. [CrossRef]
26. Shi, S.; Xu, G. Novel performance prediction model of a biofilm system treating domestic wastewater based on stacked denoising auto-encoders deep learning network. *Chem. Eng. J.* **2018**, *347*, 280–290. [CrossRef]
27. Ghorbanzadeh, O.; Blaschke, T.; Gholamnia, K.; Meena, S.R.; Tiede, D.; Aryal, J. Evaluation of different machine learning methods and deep-learning convolutional neural networks for landslide detection. *Remote Sens.* **2019**, *11*, 196. [CrossRef]
28. Schmidhuber, J. Deep learning in neural networks: An overview. *Neural Netw.* **2015**, *61*, 85–117. [CrossRef]
29. Coelho, I.M.; Coelho, V.N.; Luz, E.J.; da S. Luz, E.J.; Ochi, L.S.; Guimarães, F.G.; Rios, E. A GPU deep learning metaheuristic based model for time series forecasting. *Appl. Energy* **2017**, *201*, 412–418. [CrossRef]
30. Hong, H.; Pradhan, B.; Sameen, M.I.; Kalantar, B.; Zhu, A.; Chen, W. Improving the accuracy of landslide susceptibility model using a novel region-partitioning approach. *Landslides* **2018**, *15*, 753–772. [CrossRef]
31. Berlin, S.J.; John, M. Particle swarm optimization with deep learning for human action recognition. *Multimed. Tools Appl.* **2020**, *79*, 1–23. [CrossRef]
32. Poli, R.; Kennedy, J.; Blackwell, T. Particle swarm optimization. *Swarm Intell.* **2007**, *1*, 33–57. [CrossRef]
33. Conforti, M.; Aucelli, P.P.C.; Robustelli, G.; Scarciglia, F. Geomorphology and GIS analysis for mapping gully erosion susceptibility in the Turbolo stream catchment (Northern Calabria, Italy). *Nat. Hazards* **2011**, *56*, 881–898. [CrossRef]
34. El Maaoui, M.A.; Felfoul, M.S.; Boussema, M.R.; Snane, M.H. Sediment yield from irregularly shaped gullies located on the Fortuna lithologic formation in semi-arid area of Tunisia. *Catena* **2012**, *93*, 97–104. [CrossRef]
35. Arabameri, A.; Pradhan, B.; Rezaei, K.; Conoscenti, C. Gully erosion susceptibility mapping using GIS-based multi-criteria decision analysis techniques. *Catena* **2019**, *180*, 282–297. [CrossRef]
36. Kalantar, B.; Ueda, N.; Saeidi, V.; Ahmadi, K.; Halin, A.A.; Shabani, F. Landslide Susceptibility Mapping: Machine and Ensemble Learning Based on Remote Sensing Big Data. *Remote Sens.* **2020**, *12*, 1737. [CrossRef]
37. Wang, G.; Chen, X.; Chen, W. Spatial Prediction of Landslide Susceptibility Based on GIS and Discriminant Functions. *ISPRS Int. J. Geo-Inf.* **2020**, *9*, 144. [CrossRef]
38. Chowdhuri, I.; Pal, S.C.; Chakrabortty, R. Flood susceptibility mapping by ensemble evidential belief function and binomial logistic regression model on river basin of eastern India. *Adv. Sp. Res.* **2020**, *65*, 1466–1489. [CrossRef]
39. Youssef, A.M.; Pourghasemi, H.R. Landslide susceptibility mapping using machine learning algorithms and comparison of their performance at Abha Basin, Asir Region, Saudi Arabia. *Geosci. Front.* **2020**. [CrossRef]

40. Green, I.R.A.; Stephenson, D. Criteria for comparison of single event models. *Hydrol. Sci. J.* **1986**, *31*, 395–411. [CrossRef]
41. Kutner, M.H.; Nachtsheim, C.J.; Neter, J.; Li, W. *Applied Linear Statistical Models*; McGraw-Hill Irwin: New York, NY, USA, 2005; Volume 5.
42. Gayen, A.; Pourghasemi, H.R.; Saha, S.; Keesstra, S.; Bai, S. Gully erosion susceptibility assessment and management of hazard-prone areas in India using different machine learning algorithms. *Sci. Total Environ.* **2019**, *668*, 124–138. [CrossRef] [PubMed]
43. Hong, H.; Liu, J.; Zhu, A.-X.; Shahabi, H.; Pham, B.T.; Chen, W.; Pradhan, B.; Bui, D.T. A novel hybrid integration model using support vector machines and random subspace for weather-triggered landslide susceptibility assessment in the Wuning area (China). *Environ. Earth Sci.* **2017**, *76*, 652. [CrossRef]
44. Tehrany, M.S.; Pradhan, B.; Mansor, S.; Ahmad, N. Flood susceptibility assessment using GIS-based support vector machine model with different kernel types. *Catena* **2015**, *125*, 91–101. [CrossRef]
45. Mountrakis, G.; Im, J.; Ogole, C. Support vector machines in remote sensing: A review. *ISPRS J. Photogramm. Remote Sens.* **2011**, *66*, 247–259. [CrossRef]
46. Vapnik, V.N. *The Nature of Statistical Learning Theory*; Springer Science & Business Media: Berlin, Germany, 1995.
47. Abedini, M.; Ghasemian, B.; Shirzadi, A.; Bui, D.T. A comparative study of support vector machine and logistic model tree classifiers for shallow landslide susceptibility modeling. *Environ. Earth Sci.* **2019**, *78*, 560. [CrossRef]
48. Yao, X.; Tham, L.G.; Dai, F.C. Landslide susceptibility mapping based on support vector machine: A case study on natural slopes of Hong Kong, China. *Geomorphology* **2008**, *101*, 572–582. [CrossRef]
49. Naghibi, S.A.; Moghaddam, D.D.; Kalantar, B.; Pradhan, B.; Kisi, O. A comparative assessment of GIS-based data mining models and a novel ensemble model in groundwater well potential mapping. *J. Hydrol.* **2017**, *548*, 471–483. [CrossRef]
50. Pradhan, B. A comparative study on the predictive ability of the decision tree, support vector machine and neuro-fuzzy models in landslide susceptibility mapping using GIS. *Comput. Geosci.* **2013**, *51*, 350–365. [CrossRef]
51. Kavzoglu, T.; Colkesen, I. A kernel functions analysis for support vector machines for land cover classification. *Int. J. Appl. Earth Obs. Geoinf.* **2009**, *11*, 352–359. [CrossRef]
52. Haykin, S. *Neural Networks: A Comprehensive Foundation*, 2nd ed.; Prentice Hall Inc.: Upper Saddle River, NJ, USA, 1999.
53. Cherkassky, V.; Krasnopolsky, V.; Solomatine, D.P.; Valdes, J. Computational intelligence in earth sciences and environmental applications: Issues and challenges. *Neural Netw.* **2006**, *19*, 113–121. [CrossRef]
54. Saha, A.K.; Gupta, R.P.; Arora, M.K. GIS-based landslide hazard zonation in the Bhagirathi (Ganga) valley, Himalayas. *Int. J. Remote Sens.* **2002**, *23*, 357–369. [CrossRef]
55. Kawabata, D.; Bandibas, J. Landslide susceptibility mapping using geological data, a DEM from ASTER images and an Artificial Neural Network (ANN). *Geomorphology* **2009**, *113*, 97–109. [CrossRef]
56. Kosko, B. *Neural Networks and Fuzzy Systems: A Dynamical Systems Approach to Machine Intelligence*; Prentice-Hall Inc.: Upper Saddle River, NJ, USA, 1992.
57. Mandal, S.; Mondal, S. Machine Learning Models and Spatial Distribution of Landslide Susceptibility. In *Geoinformatics and Modelling of Landslide Susceptibility and Risk*; Springer: Cham Switzerland, 2019; pp. 165–175.
58. Falaschi, F.; Giacomelli, F.; Federici, P.R.; Puccinelli, A.; Avanzi, G.; Pochini, A.; Ribolini, A. Logistic regression versus artificial neural networks: Landslide susceptibility evaluation in a sample area of the Serchio River valley, Italy. *Nat. Hazards* **2009**, *50*, 551–569. [CrossRef]
59. Chen, W.; Pourghasemi, H.R.; Zhao, Z. A GIS-based comparative study of Dempster-Shafer, logistic regression and artificial neural network models for landslide susceptibility mapping. *Geocarto Int.* **2017**, *32*, 367–385. [CrossRef]
60. Kim, P. Matlab deep learning. *Mach. Learn. Neural Netw. Artif. Intell.* **2017**, *130*, 21.
61. Lewis, N.D.C. Deep Learning Made Easy with R: A Gentle Introduction for Data Science. Advances in Swarm Intelligence. In Proceedings of the 11th International Conference, ICSI (AusCov), Belgrade, Serbia, 14–20 July 2020.
62. Kennedy, J.; Eberhart, R.C.; Shi, Y. The particle swarm. *Swarm Intell.* **2001**, 287–325.
63. Kennedy, J.; Eberhart, R. Particle swarm optimization. In Proceedings of the ICNN'95-International Conference on Neural Networks, Perth, WA, Australia, 27 November–1 December 1995; Volume 4, pp. 1942–1948.

64. Clerc, M.; Kennedy, J. The particle swarm-explosion, stability, and convergence in a multidimensional complex space. *IEEE Trans. Evol. Comput.* **2002**, *6*, 58–73. [CrossRef]
65. Olsson, A.E. *Particle Swarm Optimization: Theory, Techniques and Applications*; Nova Science Publishers, Inc.: Hauppauge, NY, USA, 2010.
66. Frattini, P.; Crosta, G.; Carrara, A. Techniques for evaluating the performance of landslide susceptibility models. *Eng. Geol.* **2010**, *111*, 62–72. [CrossRef]
67. Yesilnacar, E.; Topal, T. Landslide susceptibility mapping: A comparison of logistic regression and neural networks methods in a medium scale study, Hendek region (Turkey). *Eng. Geol.* **2005**, *79*, 251–266. [CrossRef]
68. Biot, Y.; Blaikie, P.M.; Jackson, C.; Palmer-Jones, R. *Rethinking Research on Land Degradation in Developing Countries*; The World Bank: Washington, DC, USA, 1995.
69. Sirviö, T.; Rebeiro-Hargrave, A.; Pellikka, P. Geoinformation in gully erosion studies in the Taita Hills, SE-Kenya, preliminary results. In Proceedings of the 5th AARSE conference (African Association of Remote Sensing of the Environment), Nairobi, Kenya, 7–22 October 2004; pp. 18–21.
70. Dotterweich, M.; Rodzik, J.; Zgłobicki, W.; Schmitt, A.; Schmidtchen, G.; Bork, H.-R. High resolution gully erosion and sedimentation processes, and land use changes since the Bronze Age and future trajectories in the Kazimierz Dolny area (Nałeczów Plateau, SE-Poland). *Catena* **2012**, *95*, 50–62. [CrossRef]
71. Pal, S.C.; Chakrabortty, R. Modeling of water induced surface soil erosion and the potential risk zone prediction in a sub-tropical watershed of Eastern India. *Model. Earth Syst. Environ.* **2019**, *5*, 369–393. [CrossRef]
72. Chakrabortty, R.; Pal, S.C.; Chowdhuri, I.; Malik, S.; Das, B. Assessing the Importance of Static and Dynamic Causative Factors on Erosion Potentiality Using SWAT, EBF with Uncertainty and Plausibility, Logistic Regression and Novel Ensemble Model in a Sub-tropical Environment. *J. Indian Soc. Remote Sens.* **2020**, 1–25. [CrossRef]
73. Pal, S.C.; Chakrabortty, R. Simulating the impact of climate change on soil erosion in sub-tropical monsoon dominated watershed based on RUSLE, SCS runoff and MIROC5 climatic model. *Adv. Sp. Res.* **2019**, *64*, 352–377. [CrossRef]
74. Borrelli, P.; Robinson, D.A.; Fleischer, L.R.; Lugato, E.; Ballabio, C.; Alewell, C.; Meusburger, K.; Modugno, S.; Schütt, B.; Ferro, V.; et al. An assessment of the global impact of 21st century land use change on soil erosion. *Nat. Commun.* **2017**, *8*, 2013. [CrossRef]
75. Peng, T.; Wang, S. Effects of land use, land cover and rainfall regimes on the surface runoff and soil loss on karst slopes in southwest China. *Catena* **2012**, *90*, 53–62. [CrossRef]
76. Arabameri, A.; Pradhan, B.; Pourghasemi, H.R.; Rezaei, K.; Kerle, N. Spatial modelling of gully erosion using GIS and R programing: A comparison among three data mining algorithms. *Appl. Sci.* **2018**, *8*, 1369. [CrossRef]

sensors

MDPI

Article

Amalgamation of Geometry Structure, Meteorological and Thermophysical Parameters for Intelligent Prediction of Temperature Fields in 3D Scenes

Yuan Cao [1,2], Ligang Li [1,*], Wei Ni [1], Bo Liu [1,2], Wenbo Zhou [1,2] and Qi Xiao [1,2]

1 National Space Science Center, Key Laboratory of Electronics and Information Technology for Space System, Chinese Academy of Sciences, Beijing 100190, China; caoyuan20@mails.ucas.ac.cn (Y.C.); niwei@nssc.ac.cn (W.N.); liubo183@mails.ucas.ac.cn (B.L.); zhouwenbo20@mails.ucas.ac.cn (W.Z.); xiaoqi19@mails.ucas.ac.cn (Q.X.)

2 School of Computer Science and Technology, University of Chinese Academy of Sciences, Beijing 100049, China

* Correspondence: liligang@nssc.ac.cn; Tel.: +86-1312-152-1820

Abstract: Temperature field calculation is an important step in infrared image simulation. However, the existing solutions, such as heat conduction modelling and pre-generated lookup tables based on temperature calculation tools, are difficult to meet the requirements of high-performance simulation of infrared images based on three-dimensional scenes under multi-environmental conditions in terms of accuracy, timeliness, and flexibility. In recent years, machine learning-based temperature field prediction methods have been proposed, but these methods only consider the influence of meteorological parameters on the temperature value, while not considering the geometric structure and the thermophysical parameters of the object, which results in the low accuracy. In this paper, a multivariate temperature field prediction network based on heterogeneous data (MTPHNet) is proposed. The network fuses geometry structure, meteorological, and thermophysical parameters to predict temperature. First, a Point Cloud Feature Extraction Module and Environmental Data Mapping Module are used to extract geometric information, thermophysical, and meteorological features. The extracted features are fused by the Data Fusion Module for temperature field prediction. Experiment results show that MTPHNet significantly improves the prediction accuracy of the temperature field. Compared with the v-Support Vector Regression and the combined back-propagation neural network, the mean absolute error and root mean square error of MTPHNet are reduced by at least 23.4% and 27.7%, respectively, while the R-square is increased by at least 5.85%. MTPHNet also achieves good results in multi-target and complex target temperature field prediction tasks. These results validate the effectiveness of the proposed method.

Keywords: three-dimensional scene; temperature field; intelligent prediction; network; geometry structure; meteorological parameters; thermophysical parameters

Citation: Cao, Y.; Li, L.; Ni, W.; Liu, B.; Zhou, W.; Xiao, Q. Amalgamation of Geometry Structure, Meteorological and Thermophysical Parameters for Intelligent Prediction of Temperature Fields in 3D Scenes. *Sensors* **2022**, *22*, 2386. https://doi.org/10.3390/s22062386

Academic Editors: Moulay A. Akhloufi and Mozhdeh Shahbazi

Received: 9 February 2022
Accepted: 18 March 2022
Published: 20 March 2022

1. Introduction

Infrared imaging technology has the characteristics of high penetration, strong anti-interference, good concealment, and high precision, which can significantly compensate for visible-light imaging technology's lack of night vision capability. With the rapid development of infrared imaging technologies, infrared imaging systems have been widely applied to military, industrial, and civilian applications [1]. To develop such systems, it is essential to obtain the appropriate system parameters in advance. This requires a large number of sample images under different lighting conditions for testing and evaluation. However, owing to the complex influences of region, scenery, time-of-day and meteorological conditions, obtaining a sufficient number of samples often requires extensive re-sources and labor. Under extreme conditions, it is impossible to obtain a sufficient number of test samples. To overcome this limitation, infrared simulation has been proposed. It obtains infrared images by simulating actual infrared imaging processes by traversing three-dimensional (3D) scene construction, temperature field and radiation calculations, atmospheric radiation transmission calculation, and imaging instrument simulation. Among these, the temperature field calculation is the most important.

Research on temperature field calculations has undergone a remarkable evolution. Initially, researchers used empirical or semi-empirical models to calculate a temperature field. For instance, Jacobs [2] used a one-dimensional thermal model to calculate the temperatures of simple geometries. Biesel and Rohlfing [3] obtained an object's surface temperature by setting a series of assumptions for the heat balance equation. Curtis and Rive-ra [4] established an empirical surface temperature model that comprehensively considers the influences of time, material type, meteorological conditions, and object orientation. Balfour and Bushlin [5] established a general expression of surface temperature with respect to the sun, sky, air temperature and wind speed. However, these models are labor- and resource-intensive. Moreover, they cannot adapt to changes in details, and their accuracy is low.

To meet the requirements of accuracy, first-principle models were used for temperature field calculations. This model is based on the principle of heat transfer. The heat balance equation is established by considering various factors that affect the temperature change of the object; the temperature value is calculated by numerical calculations. For instance, Gonda et al. [6] introduced the temperature prediction model, which uses a hot node network method to calculate the temperature field distribution on the surface of an object. Sheffer and Cathcart [7] developed a thermal calculation model using a first-principle model, which considers factors, such as solar and sky radiation, mass transfer process, fluid transmission, occlusion, and multiple reflections, and can more accurately obtain the temperature change of the object. Currently, several commercial temperature field calculation software programs, such as TAITherm (https://thermoanalytics.com/taitherm, accessed on 27 February 2022) [8], Fluent (https://www.ansys.com/zh-cn/products/fluids/ansys-fluent, accessed on 27 February 2022), and Vega, which are based on first-principle models, have been developed. They realize high-precision target temperature field calculations by setting thermophysical and meteorological parameters. However, for calculating a temperature field to deter-mine a target, it is usually necessary to input several parameters, such as material, thickness, shape, atmospheric temperature, and wind speed and direction. This impedes calculations at different periods and under varying meteorological conditions, and over-whelms the current GPUs. Hence, it cannot support real-time infrared simulations.

Considering the first-principle models' calculation speed bottleneck, Hu et al. [9] proposed a scheme that uses the temperature field calculation method to generate the temperature data of a typical target scene under typical environmental conditions in advance and save it in a database lookup table. The temperature value is then obtained using database interpolation. A look-up table significantly increases the simulation speed, but it is limited in its ability to accommodate sampling resolution design and interpolation methods with numerous input meteorological parameters. Moreover, accuracy cannot be guaranteed if only the main input parameters are considered.

The temperature field calculation method proposed in this study is based on machine learning and is designed to meet real-time, high-precision and flexible infrared simulation requirements. It uses a data-driven approach to establish a mapping of parameters affecting the model's temperature field distribution to the model temperature, which essentially fits the heat-balance equation established by the first-principle model. Huang and Wu [10] proposed a similar method based on a combined back-propagation (BP) neural network to establish a relationship between model temperature and meteorological parameters. Huang et al. [11] screened meteorological parameters using the heat balance equation and used the ν-support vector regression (ν-SVR) model [12] to fit the model temperature. This meets the real-time requirements of simulations. However, contemporary machine learning models only consider the influence of meteorological parameters on temperature and ignore the influence of other factors, which affects accuracy.

This study provides a novel temperature field calculation method based on machine learning for high-precision real-time prediction of temperature field under the influence of multiple environmental variables in the real-time simulation of a 3D scene's infrared im-aging. It addresses the limitations of the contemporary models by comprehensively considering geometry structure, meteorological, and thermophysical parameters, which meets the requirements of real-time and accurate temperature field prediction. The main contributions of this study are as follows:

(1) A multivariate temperature field prediction network based on heterogeneous data (MTPHNet), which combines the characteristics of heterogeneous thermo-physical and meteorological data as 3D model parameters to predict temperature using fusion features and to improve model generalizability;
(2) To solve the problem of memory explosion when the Transformer (http://nlp.seas.harvard.edu/2018/04/03/attention.html, accessed on 27 February 2022) structure deals with 3D model thermophysical parameters, we propose the PointNet (https://github.com/charlesq34/pointnet, accessed on 27 February 2022) structure as the 3D model thermophysical feature extraction module and imitate the parameter sharing idea of a convolutional neural network to extract local and global features separately. The final fitting effect proves the effectiveness of the method;
(3) We used a multilayer perceptron (MLP) module to map the meteorological parameters to fuse the meteorological and thermophysical parameters so that the mapped features and thermophysical parameters have the same size, which is convenient for the subsequent fusion process.

The experimental results validate the effectiveness of our proposed algorithm. The remainder of this article is organized as follows: Section 2 describes our analysis process and the proposed method in detail. In Section 3, the data formats, evaluation metrics, and training methods used for training are introduced. In Section 4, corresponding experiments are designed to verify the effectiveness of this method, and the experimental results are analyzed and discussed. Section 5 draws some conclusions about our method.

2. Materials and Methods

2.1. Analysis of the Parameters That Affect the Temperature Field Distribution of the 3D Model in the Natural Environment

2.1.1. Calculation Principle

A series of heat transfer processes with different mechanisms occur between the surface of a 3D model in a natural environment and the atmospheric boundary layer. A 3D model comprises different materials; the methods and speeds of heat exchange between different materials and the external environment are different.

Figure 1 illustrates the energy interactions between an object and the external environment, which ultimately results in thermal equilibrium. For example, temperature differences between objects cause heat transfer (Figure 1A). Energy can also be transmitted directly to objects by solar radiation (Figure 1B). Atmospheric particles can also transfer energy to objects after absorbing external radiation (Figure 1C). Heat energy can be transferred from surrounding objects to the target object (Figure 1D). Fluid flow also contributes to energy transfer (Figure 1E). Lastly, energy transfer can be caused by the evaporation of water, water vapor condensation, and migration (Figure 1F).

Based on the law of the conservation of energy and the processes illustrated, the heat balance equation of an object's surface is as follows:

$$k_i \frac{\partial T}{\partial n}\bigg|_i = a_s E_{sun} + a_l E_{sky} + \sum_{j=1}^{M} Q_{rj} - \epsilon\sigma T^4 \pm Q_c \pm Q_{ec} \tag{1}$$

where $k_i \frac{\partial T}{\partial n}\Big|$ is the heat conduction of the object, $a_s E_{sun}$ denotes the ability of an object to absorb solar radiation, $a_l E_{sky}$ denotes the ability of an object to absorb radiation from the sky, $\sum_{j=1}^{M} Q_{rj}$ denotes the radiative heat transfer from other objects around, $\epsilon\sigma T^4$ denotes the self-radiation of the object, Q_c denotes convective heat transfer, and Q_{ec} denotes hidden heat.

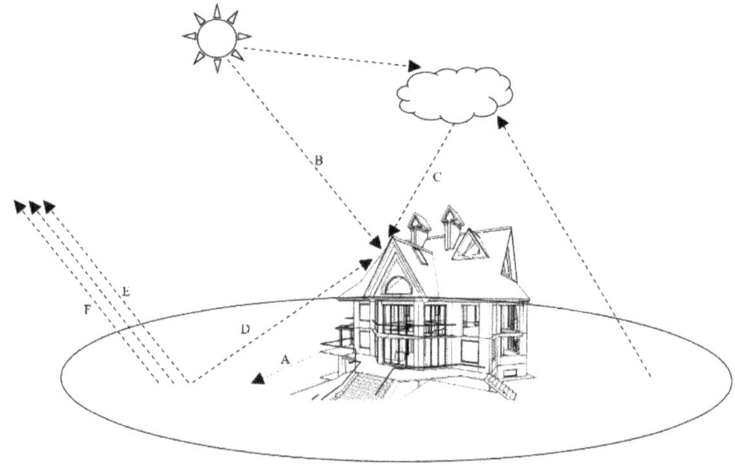

Figure 1. A 3D model of processes of energy interactions to reach thermal equilibrium under natural conditions: (A) heat transferred by temperature differences between objects; (B) energy directly transmitted to objects by solar radiation; (C) energy transferred by particles to objects after absorbing external radiation; (D) heat radiation energy transferred from surrounding objects to the target object; (E) energy transferred by fluid flow; and (F) energy transferred by water evaporation, water vapor condensation and migration.

With Equation (1), when the boundary conditions at each moment are known, the temperature field distribution at each moment can be calculated. The calculation result at the current moment is also the boundary condition at the next moment. By analyzing the above-mentioned energy transfer process, we can filter the variables that play a key role in the calculation of the temperature field distribution.

By analyzing the above-mentioned energy transfer process, we can filter out the variables that play a key role in the calculation of the temperature field distribution:

(A) Heat conduction: Owing to the collision of numerous molecules and subatomic particles, energy flows from a high-temperature object to a low-temperature object. For a temperature change caused by heat conduction, the main influencing factors are the properties of the object itself, including thermal conductivity, thickness, shape, etc.

(B) Sun radiation: Objects absorb radiant energy from the sun, which is a form of radiant heat transfer. When the object is on a clear and cloudless level surface, the formula is as follows:

$$E_{s0} = [1 - A(U^*, \beta)](0.349E_0)sin\beta + \left(\frac{1-\rho_0}{1-\rho_0\overline{\rho_g}}\right)(0.651E_0)sin\beta \quad (2)$$

where E_0 denotes the solar radiation of the entire waveband, $A(U^*, \beta)$ denotes the absorbable coefficient, which is a function of relative humidity, air temperature, and solar altitude, β denotes the solar elevation angle, $\overline{\rho_g}$ denotes the reflectivity of the ground, and ρ_0 is the Rayleigh reflectivity of the atmosphere, which is a function of the solar elevation angle.

Considering the cloudy sky, Equation (2) is modified to obtain the following formula:

$$E_{fsun} = E_{s0} \cdot CF \tag{3}$$

where CF is a function related to cloud coverage.

Therefore, the main factors influencing the temperature changes caused by solar radiation are relative humidity, air temperature, solar altitude angle, and cloud coverage. The solar altitude angle is related to the longitude, latitude, time zone, and date. In this study, it is assumed that the temperature field is calculated in a fixed scene; hence, the longitude, latitude, and time zone are invariant. Therefore, the main influencing factors of temperature changes caused by solar radiation are relative humidity, air temperature, date, and cloud coverage.

(C) Sky radiation: Atmospheric particles, such as carbon dioxide and water vapor, are present in the atmosphere. These particles absorb external radiation; thus, they have a certain temperature. Therefore, sky radiation is essentially generated by the thermal radiation of atmospheric particles, and it affects objects on the ground. The formula for sky radiation is as follows:

$$E_{sky} = \left(a + b\sqrt{e}\right)\sigma T_a^4 \tag{4}$$

where T_a is the sky temperature, which can be calculated from cloud coverage, atmospheric temperature, humidity, and altitude; a and b are related to the location and time of the measurement, and e is a function of relative humidity and atmospheric temperature.

Because altitude and location are constant in this study, the temperature changes caused by sky radiation are related to cloud coverage, atmospheric temperature, humidity, and time.

(D) Radiation from other objects: When the temperature of an object is higher than absolute zero, it spontaneously radiates energy. Therefore, when there are other objects around it, it is affected by their radiation. Hence, it is necessary to obtain the surrounding objects' temperature data.

(E) Convection heat transfer: Fluid flow further affects temperature changes. For ground objects, the main influencing factors of temperature changes caused by convective heat transfer are wind speed and direction, and air temperature.

(F) Latent heat is the energy transfer caused by the evaporation of water; and condensation and migration of water vapor. The object studied in this study does not involve heat exchange in this area.

2.1.2. Determination of the Parameters That Affect the Surface Temperature Field of the Object

Because this study focuses on the calculation of a 3D target's temperature field at a fixed altitude and location, the main meteorological parameters are date, atmospheric temperature, solar radiation, wind speed, relative humidity, cloud cover, and wind direction. The main thermophysical parameters are space coordinates, density, specific heat, conductivity, thickness, convection method, emissivity, absorptivity, and initial temperature.

2.2. Design of 3D Target Temperature Field Prediction Model Based on Heterogeneous Data Fusion

Predictive modelling of temperature fields based on machine learning is essentially a fitting of first-principle models of thermodynamics. According to the analysis in Section 2.1, this mainly includes three heat transfer processes: heat conduction, heat radiation, and heat convection.

In the first-principles model, the factors affecting the temperature of the model can be divided into two categories: the first category is meteorological parameters, which are time series data that record meteorological indicators at each moment, such as atmospheric temperature, wind speed, and direction, which characterizes the energy exchange between the object and the atmosphere, mainly reflects the heat radiation process and the heat convection process in the heat transfer process; the other is thermophysical parameters.

If the object is regarded as composed of countless particles, then the thermophysical parameters can be regarded as a kind of point cloud data, which record the emissivity, thickness, and specific heat of each particle, which characterizes the energy exchange between points in the object and mainly reflects the heat conduction process in the heat transfer process. In addition, the spatial location distribution will cause occlusion and other phenomena and will also affect the exchange of energy. Therefore, the geometric structure information will also affect the distribution of the temperature field. Although it is not a thermophysical parameter, it corresponds to each point, so we classify it as a thermophysical parameter. These two types of data are heterogeneous and determine the temperature field distribution of the 3D model.

The existing temperature field prediction model based on machine learning only considers the influence of meteorological parameters on the temperature of the target model, while ignoring the influence of thermophysical parameters, which is equivalent to considering only the thermal convection and thermal radiation models in the first principle, while ignoring heat transfer. This results in poor prediction accuracy. We introduced a Transformer [13] to solve this problem.

The Transformer is a classic work by Google. It completely abandons the traditional neural network structure and uses an attention module [14] to process data. The use of self-attention to process data, which can effectively integrate is effective for integrating heterogeneous data.

This study comprehensively considers the thermophysical and external meteorological parameters that affect the temperature of a 3D target model. The proposed MTPHNet method improves the structure of the Transformer model using meteorological parameters as the input of the encoder, and thermophysical parameters as the input of the decoder. It uses the self-attention module to fuse the two parts of data to improve the generalization ability of the model. The structure of MTPHNet is shown in Figure 2.

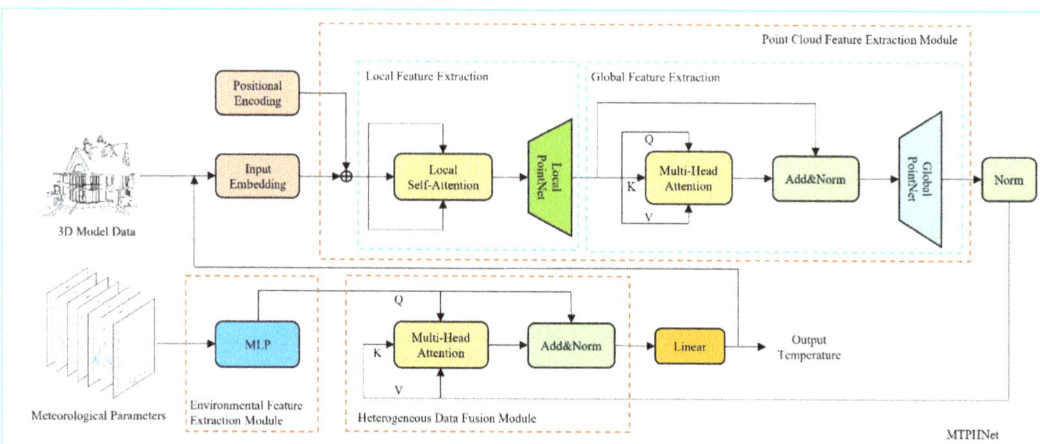

Figure 2. Structure of the multivariate temperature field prediction network based on heterogeneous data (MTPHNet). MLP: Multilayer perceptron.

The use of MTPHNet to predict the model temperature field can be expressed by Equation (5):

$$Y_{temp} = \psi_t\Big(\phi(Enc(x_{obj}), Dec(x_{env})\,)\Big) \tag{5}$$

where x_{obj} denotes the thermophysical parameters of the 3D model, such as space coordinates, thermal conductivity, and reflectivity; x_{env} denotes meteorological parameters, such as atmospheric temperature, wind speed, and direction; ϕ denotes the fusion process of thermophysical parameters and meteorological parameters to obtain fusion features; and ψ_t represents the regression prediction process, which calculates the temperature value to be predicted.

In this study, the 3D target model is represented as point cloud data. Each data point is considered an object in space and has its corresponding attribute information, such as material, thickness, and thermal conductivity. Therefore, $x_{obj} \in \mathbb{R}^{P \times A}$, where P denotes the number of points in the 3D target model, and A denotes the number of attributes corresponding to each data point. Meteorological parameters are time-related, and each moment corresponds to a set of meteorological data. Therefore, $x_{env} \in \mathbb{R}^{T \times E}$, where T denotes the duration, which is obtained by sampling at a fixed step in a period, and E denotes the number of attributes used to describe the external environment.

It is evident that thermophysical and meteorological parameters are two sets of heterogeneous data with different dimensions. However, in the natural environment, meteorological parameters act on the thermophysical parameters of the 3D model. Simultaneously, the temperature of each data point is affected by the temperature of other points around it. Therefore, the thermophysical parameters of the 3D target model affect each other. These highly coupled heterogeneous data determine the temperature of the 3D target model; therefore, this complexity cannot be handled by general data fusion methods. Thus, we use the thermophysical parameters of each point as input to the Transformer encoder. A point corresponds to a token, and the interaction between the points of the 3D target model is simulated using the encoder's calculation. Subsequently, the meteorological parameters are used as input to the Transformer's decoder for feature mapping. Finally, the two parts of the features are fused, and the fused features are regressed to calculate the predicted value of the 3D target temperature field.

2.2.1. Point-Cloud Feature Extraction Module (PCEM)

In the real environment, objects can be envisioned as a composition of countless particles, and different points have different materials and spatial positions. Different materials will often have very distinct emissivity, absorption, and scattering [15] properties, which can result in a variety of particle energy absorptions and releases. Different spatial positions will lead to phenomena, such as occlusion and shadows, resulting in uneven energy distributions. Therefore, the thermophysical parameters of the 3D object are crucial to the establishment of a temperature field. To improve the accuracy of temperature field prediction, we must extract the object's thermophysical parameters.

The thermophysical parameters include spatial coordinates, emissivity, and specific heat, which can be regarded as point cloud data with additional attributes. Temperature field prediction requires the calculation of the entire 3D target model, and each point interacts with all other points, implying that the thermophysical parameters of each are dot-produced with the thermophysical parameters of other points. The computational complexity of the original Transformer is proportional to the square of the length of the input sequence [16]; however, the number of 3D point cloud points is large, which is unsuitable for most hardware.

To solve this problem, we apply PointNet [17], a feature extraction layer for point-cloud data of the 3D target model. PointNet, proposed by Qi et al. (2017), can be directly used to process point cloud data. The model extracts features via feature mapping and maximum pooling of point cloud data and satisfactorily completes the classification task. However, because the model extracts features from single points and does not consider the relationship between points, its local feature extraction ability is weak [18]. Therefore, it is impossible to analyze complex scenes.

In this study, the 3D target model is assembled from different parts. First, we group the point clouds of the 3D target model according to the types of parts. The point cloud attributes of the same part are similar; however, the point cloud attributes of different parts are different. Subsequently, the point cloud data are organized according to the part category and each group of point cloud data are first sent to a self-attention module for calculation to obtain the relationship between points. The calculation results are sent to the PointNet for local feature extraction. The feature extraction process for the point cloud data of the 3D target model is shown in Figure 3.

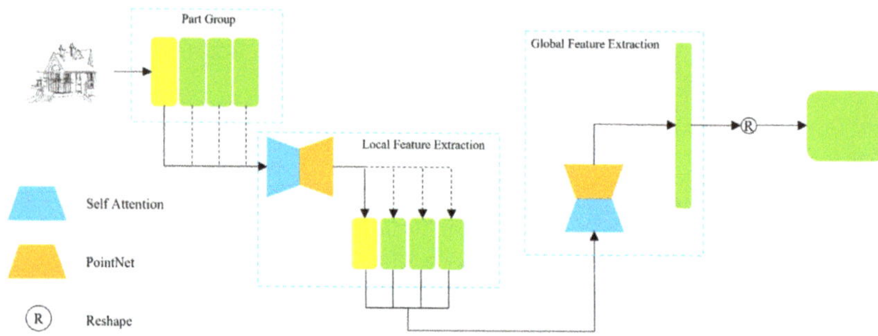

Figure 3. Flow chart of feature extraction of 3D target model.

Figure 3 indicates that we do not configure a self-attention module and PointNet, respectively, for the point cloud data of each group of parts to extract features. We rather refer to the convolution kernel of weight sharing in convolutional neural network [19] and use a unique self-attention module and PointNet which perform feature extraction on the point cloud data of different parts. Each group of parts is extracted as a local feature vector; the feature items extracted from all the parts are formed into a new sequence; and are then sent to a new self-attention module and PointNet to extract global features. This way, all features for the 3D target model are extracted.

2.2.2. Environmental Data Feature Mapping Module (EMM)

Meteorological parameters directly affect the temperature of objects. Rain reduces the surface temperature of objects, the shielding effect of clouds weakens solar radiation, and wind accelerates the heat transfer between the air and the surface of the object [20].

The thermophysical parameters of the 3D target model are mapped into a fixed-size feature block after passing through the PCEM. The thermophysical and meteorological parameters of the 3D model are heterogeneous data from different sources. To achieve the integration of heterogeneous data, we introduced a multi-layer perceptron (MLP) module to map meteorological parameters to a high-dimensional space through feature mapping and map them to a fixed size to match the feature block of the thermophysical parameters. The EMM is illustrated in Figure 4.

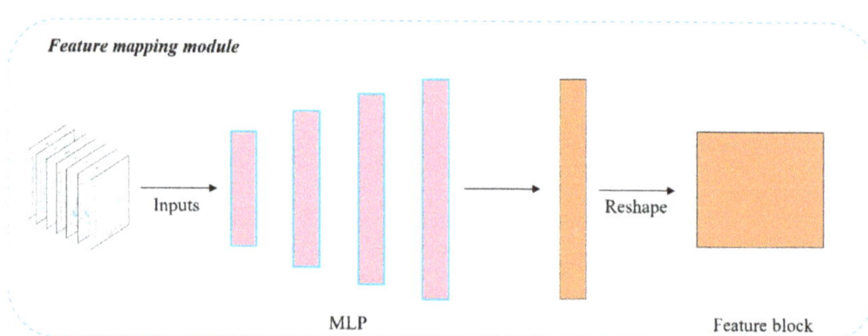

Figure 4. Schematic of environmental data features mapping module.

2.2.3. Data Fusion Module (DFM)

In the natural environment, meteorological and thermophysical parameters undergo complex physical interactions to determine the temperature field distribution of objects. In this study, we use a self-attention module to fuse the thermophysical and meteorological parameters. We use the feature block output by the encoder as the K and V of the self-attention module, and the feature block output by the decoder as the Q of the self-attention module. This process simulates the interaction between meteorological parameters and the 3D target model in the natural environment. A schematic of the integration of thermophysical and meteorological parameters is shown in Figure 5.

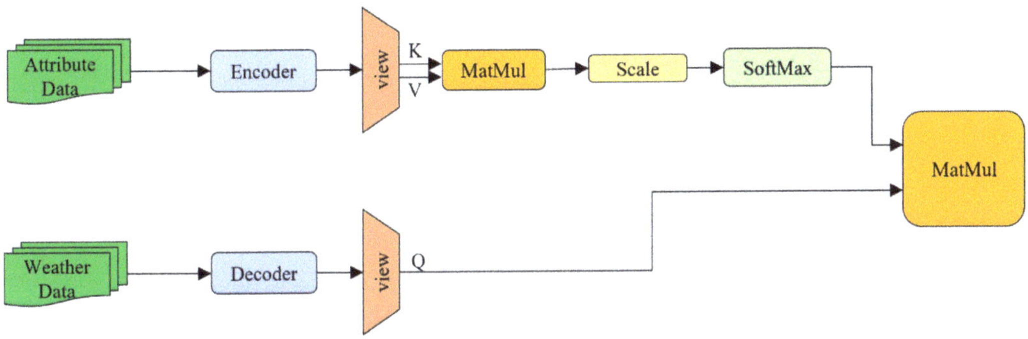

Figure 5. Schematic of the integration of internal parameter variables and external environmental parameter variables.

2.2.4. Pseudocode

Based on the above analysis, the pseudocode of MTPHNet shown in Figure 2 is summarized, and the algorithm is given in Algorithm 1.

Algorithm 1 program pseudo code of MTPHNet.

Input
x_{env}^t: meteorological parameter at the current moment.
x_{obj}^t: thermophysical parameter at the current moment.
Y_{temp}^{t-1}: the target temperature value at the last moment.

Output
Y_{temp}^t: the target temperature value at the current moment.

1 **For** $t = 1$ to t_{max}
2 Replace: $x_{encin}^t = dimension\_replace\left(x_{obj}^t, Y_{temp}^{t-1}\right)$
3 **For** $i = 1$ to P
4 $x_{encin}^t[i] = attn\_feature\left(x_{encin}^t[i], x_{encin}^t[i], x_{encin}^t[i]\right)$
5 $x_{encin}^t[i] = pointnet\_feature\left(x_{encin}^t[i]\right)$
6 **End for**
7 $x_{encout}^t = pointnet\_feature\left(attn\_feature\left(x_{encin}^t, x_{encin}^t, x_{encin}^t\right)\right)$
8 $x_{dec\_in}^t = MLP\left(x_{env}^t\right)$
9 $Y_{temp}^t = Linear\left(attn\_feature\left(x_{decin}^t, x_{encout}^t, x_{encout}^t\right)\right)$
10 **End for**

Algorithm 1 shows that t_{max} is the maximum duration of temperature field prediction, and P is the number of parts in the 3D model.

In line 2, the algorithm replaces the dimension representing temperature in x_{obj}^t with Y_{temp}^{t-1}. From lines 3 to 6, the algorithm extracts the local features of the 3D model using $attn\_feature$ and $pointnet\_feature$ for each part. In line 7, the algorithm uses $attn\_feature$ and $pointnet\_feature$ to extract the global features of the 3D model. In line 8, the algorithm uses MLP to make a feature map of x^t_{env}. Finally, the algorithm uses $attn\_feature$ to fuse $x^t_{dec\_in}$ and $x^t_{enc\_out}$, and $Linear$ to obtain the temperature value.

3. Experimental Details and Data Exploitation

3.1. Experimental Environment and Index Design

The experiment was conducted on an AMD Ryzen 7 CPU 5800H with 16 GB of RAM, NVIDIA GeForce RTX 3090 with 24 GB of memory, Python 3.7.2, and PyTorch 1.9.0 for network model training and testing.

To evaluate the effect of temperature field prediction, mean absolute error (MAE) [21], root mean square error ($RMSE$) [22], and R^2 [23] were selected as the evaluation criteria for the model quality. The calculation formulas are as follows:

$$MAE(y,\hat{y}) = \frac{1}{n}\sum_{i=1}^{n}|y_i - \hat{y}_i| \tag{6}$$

$$RMSE(y,\hat{y}) = \frac{1}{n}\sum_{i=1}^{n}\|y_i - \hat{y}_i\|_2^2 \tag{7}$$

$$R^2(y,\hat{y}) = 1 - \frac{RSS}{TSS} = \frac{ESS + 2\sum_{i=0}^{n}(y_i - \hat{y}_i)(\hat{y}_i - \overline{y})}{TSS} \tag{8}$$

$$TSS = \sum_{i=1}^{n}(y_i - \overline{y})^2 \tag{9}$$

$$RSS = \sum_{i=1}^{n}(y_i - \hat{y}_i)^2 \tag{10}$$

$$ESS = \sum_{i=1}^{n}(\hat{y}_i - \overline{y})^2 \tag{11}$$

where y denotes the true value; $\hat{y}$ denotes the predicted value; $\overline{y}$ denotes the average value of the true value; TSS is the Total sum of squares, which defines the difference between y and $\overline{y}$; RSS is the Residual sum of squares, which defines the difference between y and $\hat{y}$; and ESS is the Explained sum of squares, which defines the difference between $\hat{y}$ and $\overline{y}$.

Among the selected evaluation indicators, MAE and RMSE are used to measure the size of error between the predicted and real data; R-squared measures the quality of the fit.

3.2. Dataset

The training data used by existing temperature field prediction models based on machine learning methods were collected by instruments. These type of data are closer to reality. However, owing to the variability of natural environmental parameters and the in-stability of instruments, the data acquired by the instrument are noisy and costly.

We use our own temperature dataset constructed by ourselves, which includes the thermophysical parameters, meteorological parameters, and temperature field data of 3D objects.

3.2.1. Dataset Format

We use the thermophysical and meteorological parameters of the dataset as input to MTPHNet and the corresponding temperature data as its output to train and optimize the model parameters.

The shape of the 3D target model has an impact on the temperature field formation. Under the same environmental conditions, different shapes will cause uneven heat distribution in the 3D target model, for instance, objects in shadow will be cooler than objects in direct sunlight. Therefore, the thermophysical parameters in the dataset first need to obtain the spatial position information of the 3D target model. We built several 3D models using 3D modeling software and exported them to OBJ file format. Because OBJ file uses the face element data structure to build the 3D model and the proposed model uses the point cloud data structure, we processed the exported OBJ file and calculated the center coordinates of each face element to replace the face element. Figure 6 shows the constructed 3D model.

Figure 6. A 3D model and its corresponding point cloud data.

Figure 6 shows that each 3D object has several data points. In addition to spatial coordinates, each data point contains additional attribute information, such as material, thickness, and initial temperature. Table 1 shows the point cloud data format of the 3D target model during training.

Table 1. Point cloud data format for 3D targets.

Physical Parameters	Space Coordinates	Density	Specific Heat	Conductivity	Thickness	Convection	Emissivity	Absorptivity	Initial Temperature
Unit	(mm)	(kg/m^3)	(J/kg·K)	(W/m·K)	(mm)	Bool	/	/	°C

In addition to the 3D point cloud data, meteorological parameter data are required. For this study, we collect meteorological parameter data for four seasons. Combined with the parameters that must be collected in the analysis above, we selected date, atmospheric temperature, solar radiation, wind speed and direction, relative humidity, and cloud coverage as environmental parameter variables. Figure 7 shows the meteorological parameters related to time, and the changing trends.

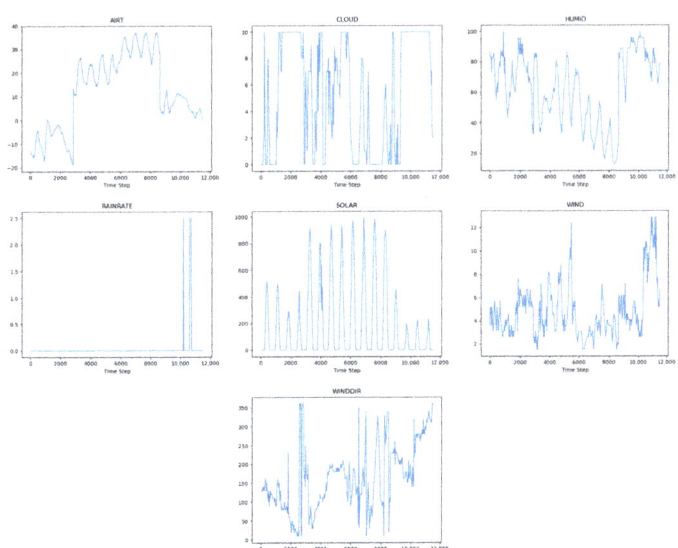

Figure 7. External meteorological parameters for temperature field prediction. Parameters from the left to right and top to bottom are atmospheric temperature, cloud cover, relative humidity, rainfall rate, solar radiation, wind speed, and direction.

According to the collected thermophysical and meteorological parameters, we use an internal temperature field calculation software to calculate the temperature field distribution of the 3D model data and add the calculation results to the dataset for training the model.

From Equation (5), $Y_{temp} \in \mathbb{R}^{P \times T}$, which means P points, and each point has T temperature values.

3.2.2. Teacher Forcing

As discussed before, the temperature of the 3D target is affected by the sun, atmosphere, and surrounding objects. It is evident that the temperature of the 3D target model at each moment is determined by the meteorological and thermophysical parameters at the current moment.

Among the features of the thermophysical parameters, one dimension of the feature represents the temperature of the point cloud data. Because the 3D model is represented by a point cloud, each point represents a distinct object. Therefore, this dimension represents the distribution of the temperature field at the current moment.

The temperature field distribution is obtained at the next moment by entering the data into MTPHNet to measure the difference between the two one time step.

Because unknown information cannot be used in the test, the calculated value is assigned to this dimension of the input data to calculate the temperature value at the next moment, after calculating the temperature value at the current moment. Figure 8 illustrates the process.

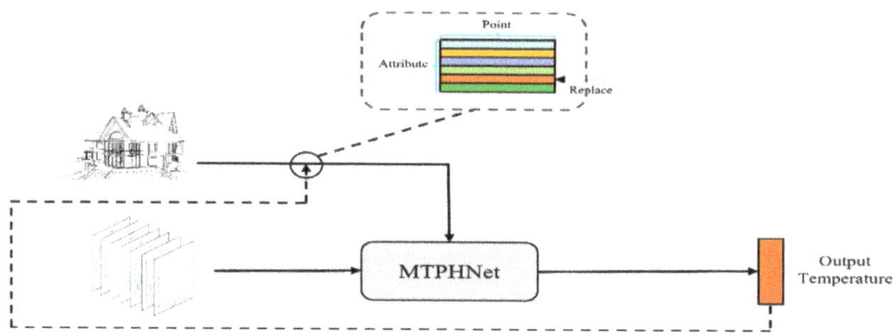

Figure 8. Temperature substitution process. The temperature value calculated by the model replaces a certain dimension of the input to simulate the temperature change of all objects in the temperature field at each moment.

During the training, the temperature value at any time is known. Therefore, there is no need to use the temperature value calculated by the model to replace the value of the dimension, which allows parallel calculations during the training.

4. Results and Discussion

4.1. Performance of the MTPHNet

To demonstrate that the MTPHNet successfully integrates an object's thermophysical and meteorological parameters can further improve the prediction of the temperature field, we used temperature field data, thermophysical parameters, and meteorological parameters as training data and compared the performance of MTPHNet with those of v-SVR and a combined BP neural network (CBPNN) model.

When training MPTHNet, the hyperparameters needed by the model included batch size, epoch, number of multi-heads, and initial learning rate. The batch size affects the degree of optimization and model speed. The size of the epoch affects the fitting effect of the model. Tuning the number of multi-heads helps the network capture richer features. The initial learning rate determines if and when the objective function converges to a local minimum. To obtain better hyperparameter values, we used Microsoft's automatic parameter tuning tool, NNI, for hyperparameter selection, which runs the code in a loop to obtain the optimal hyperparameter values. The results of the operation are shown in Figure 9:

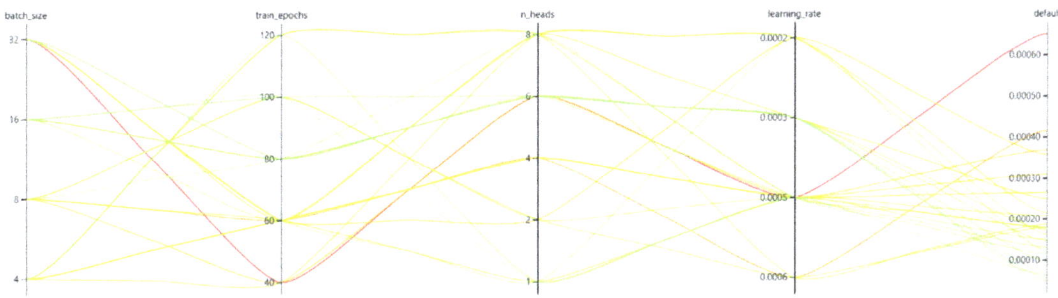

Figure 9. Results of NNI.

As can be seen in Figure 9, the batch size was set to 16; the number of epochs was 100, the number of multi-heads was 6 and the initial learning rate was 0.0003. As it is based on the Transformer structure, the MTPHNet model is large and needs a significant amount of memory. Considering computational efficiency and fitting accuracy, we selected the Huber loss function and the Adam optimizer for optimization. Dropout was used for overfitting mitigation, and the deletion ratio, p, was 0.05.

Because the thermophysical parameters of the 3D target were considered, the MTPHNet trained different 3D models with the same number of point clouds. However, v-SVR and CBPNN only consider the impact of meteorological parameters on the temperature of the 3D target and cannot simultaneously predict the temperature field of different 3D target models. For referential significance, all three models were trained with the same training set, which includes the temperature field distribution data of a single 3D model. MTPHNet was better than the other prediction models after testing on the test set. Table 2 presents the generalization performance of the models.

Table 2. Comparison of generalization performance of MTPHNet, v-SVR, and CBPNN.

Algorithm Model	MAE	RMSE	R-Squared
v-SVR	17.329	21.17	−388.6
CBPNN	2.249	3.474	0.889
MTPHNet	1.722	2.512	0.941

As shown in Table 2, the MTPHNet prediction error, was significantly lower than that of the existing temperature field prediction models. Compared with the CBPNN model, its MAE and RMSE decreased by 23.4% and 27.7%, respectively, whereas the R-squared increased by 5.85%. Figure 10 shows the prediction effects of the models.

Because the 3D model was composed of patches, in the experiment, we extracted several patches by generating random numbers to show the effect of temperature field prediction. We selected patches 96,231 and 423 for presentation. The experiments demonstrated that, although the existing temperature field prediction methods fit the temperature field on the change trend, their accuracies were insufficient. Therefore, it is necessary to combine the energy interaction mode of the 3D object in the natural environment and its meteorological and thermophysical parameters to further improve prediction accuracy.

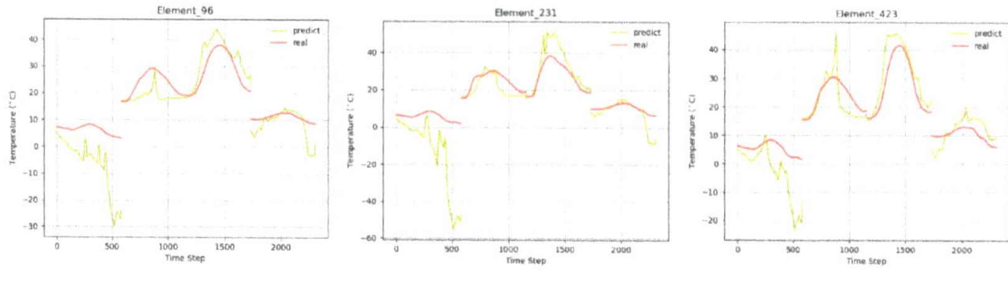

(**a**)

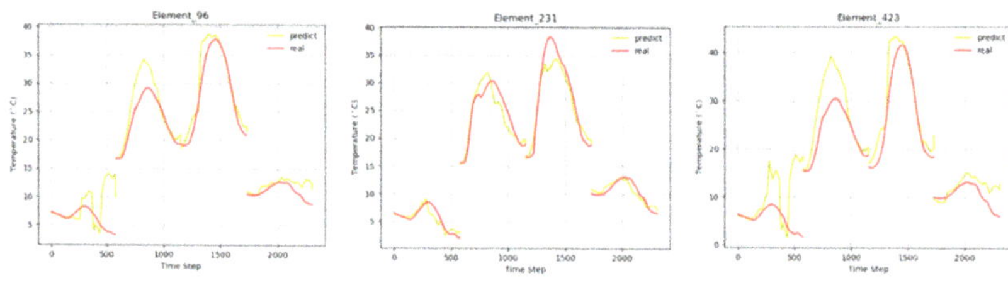

(**b**)

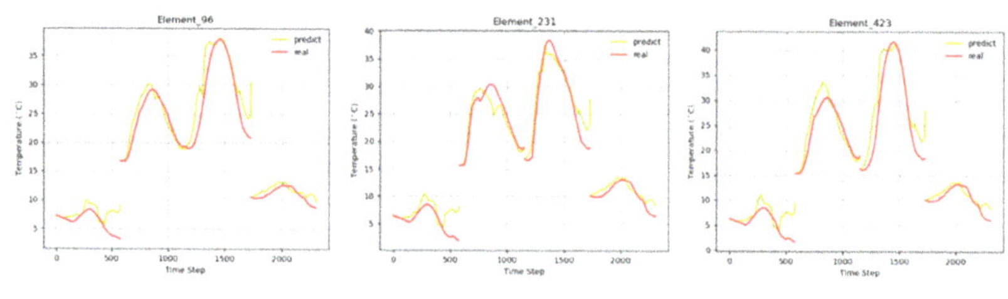

(**c**)

Figure 10. Generalization performance renderings. Prediction effect diagrams of (**a**) v-SVR; (**b**) CBPNN; and (**c**) MTPHNet.

4.2. Advantages of MTPHNet

4.2.1. Multi-Object Temperature Field Prediction

The results indicate that MTPHNet has a better fitting ability than the existing temperature field prediction models. Because MTPHNet comprehensively considers the various energy exchanges between the object and the environment and combines the thermophysical parameters of the object for training, it simultaneously trains and predicts the temperature field for different 3D targets. Existing temperature field prediction models cannot achieve this.

In this study, we summarized the 3D target temperature field data shown in Figure 5 and imported them into MTPHNet for training and verification. Table 3 presents the fitting performances.

As shown in Table 3, when the MTPHNet model was used to predict the temperature field of multiple objects, the values of its various indicators were satisfactory. The experimental results demonstrate that the thermophysical parameters of the 3D target model are significant for temperature field prediction. Figure 11 shows the effects of the multi-object temperature field prediction. Here, three materials were used for the temperature field calculation. For a convenient comparison, we selected patch 1 for presentation. The same patch shows the effect of different materials on temperature and the adaptability of MTPHNet to different temperature changes.

Table 3. MTPHNet's generalization performance for multi-object temperature field prediction.

Model	Material	MAE	RMSE	R-Square
Box	1	2.077	2.568	0.938
	2	3.953	5.664	0.877
	3	1.785	2.153	0.929
Cylinder	1	4.419	6.224	0.855
	2	2.497	3.320	0.901
	3	5.572	7.976	0.821
Sphere	1	1.910	2.329	0.918
	2	2.556	3.245	0.897
	3	4.843	6.927	0.831

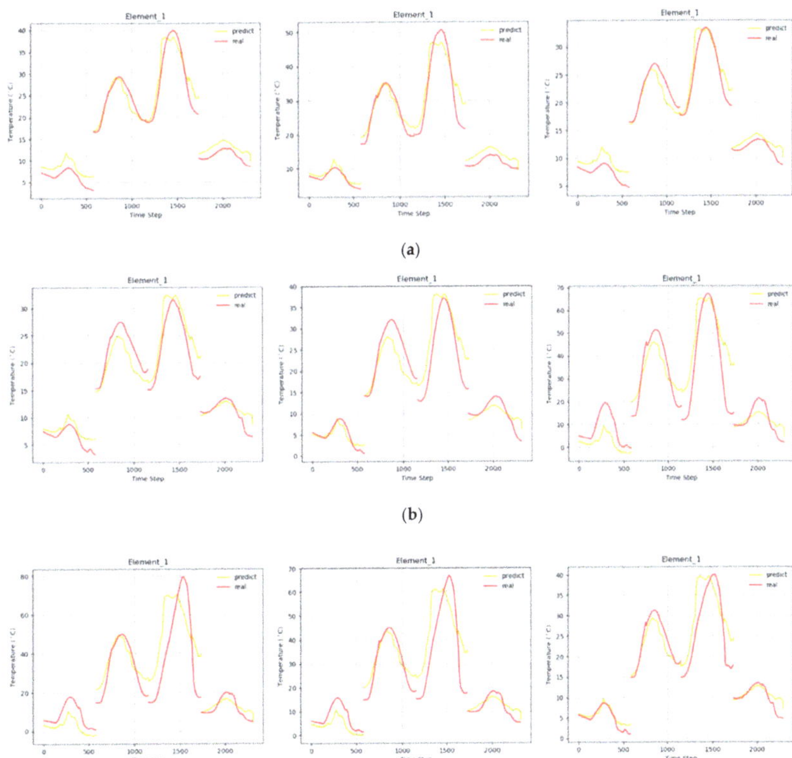

(a) (b) (c)

Figure 11. Multi-object temperature field prediction effects. Fitting of different materials of (**a**) box; (**b**) cylinder; and (**c**) sphere.

4.2.2. Prediction of Temperature Field of Complex Objects

When predicting multiple objects, this study assumed that each object had only one part; thus, the attribute data of different points are the same. In reality, however, a complex 3D object is composed of different materials, and the energy exchange between them is more complicated than that of a single material. Therefore, we chose a complex model for training and prediction. Figure 12 shows the geometry of the model.

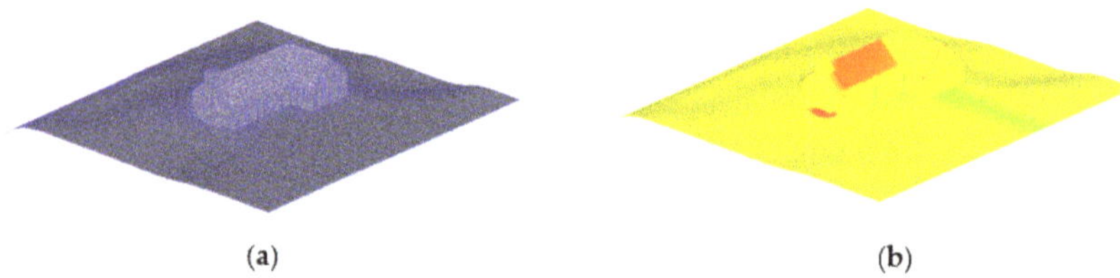

Figure 12. Complex house model with 5660 patches and 30 parts: (**a**) geometric structure; (**b**) temperature field distribution at a given moment.

Table 4 demonstrate that the model has a good generalization performance for the temperature field prediction of complex models, which further reflects the superiority of MTPHNet. Figure 13 shows the prediction effect of the temperature field of complex objects. We randomly selected patches 1032 and 4000 for presentation.

Table 4. MTPHNet's generalization performance for temperature field prediction of complex objects.

Model	MAE	RMSE	R-Square
House	2.645	3.522	0.964

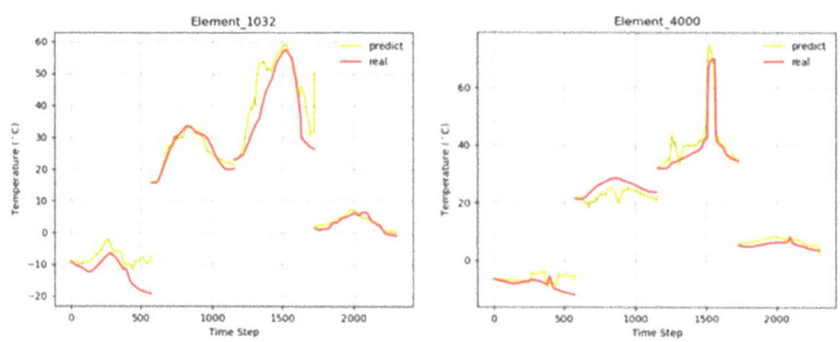

Figure 13. Prediction of the temperature field of complex objects.

4.3. Ablation Analysis

To verify the effectiveness of our proposed network, we conducted three ablation experiments to verify the performance of the main design components: environmental data feature mapping module (EMM), point cloud feature extraction module (PCEM), and data fusion module (DFM). The proposed MTPHNet is given as MTPHNet-A, and its variants for ablation are MTPHNet-B, MTPHNet-C, and MTPHNet-D. All variants were trained and validated using the same procedure described in Section 4.1. Each ablation experiment was performed three times and the results were averaged, and shown in the Table 5 and Figure 14.

Table 5. Quantitative evaluation metrics of MTPHNet and its variants. All models follow the same procedure and training environment as described in Section 4.1 and are evaluated on the same test set. The best results are shown in bold.

Model	MAE	RMSE	R-Square
MTPHNet-A (Original)	**1.722**	**2.512**	**0.941**
MTPHNet-B (no EMM)	8.734	10.362	−0.011
MTPHNet-C (no PCEM)	2.277	3.516	0.885
MTPHNet-D (no DFM)	2.303	3.431	0.89

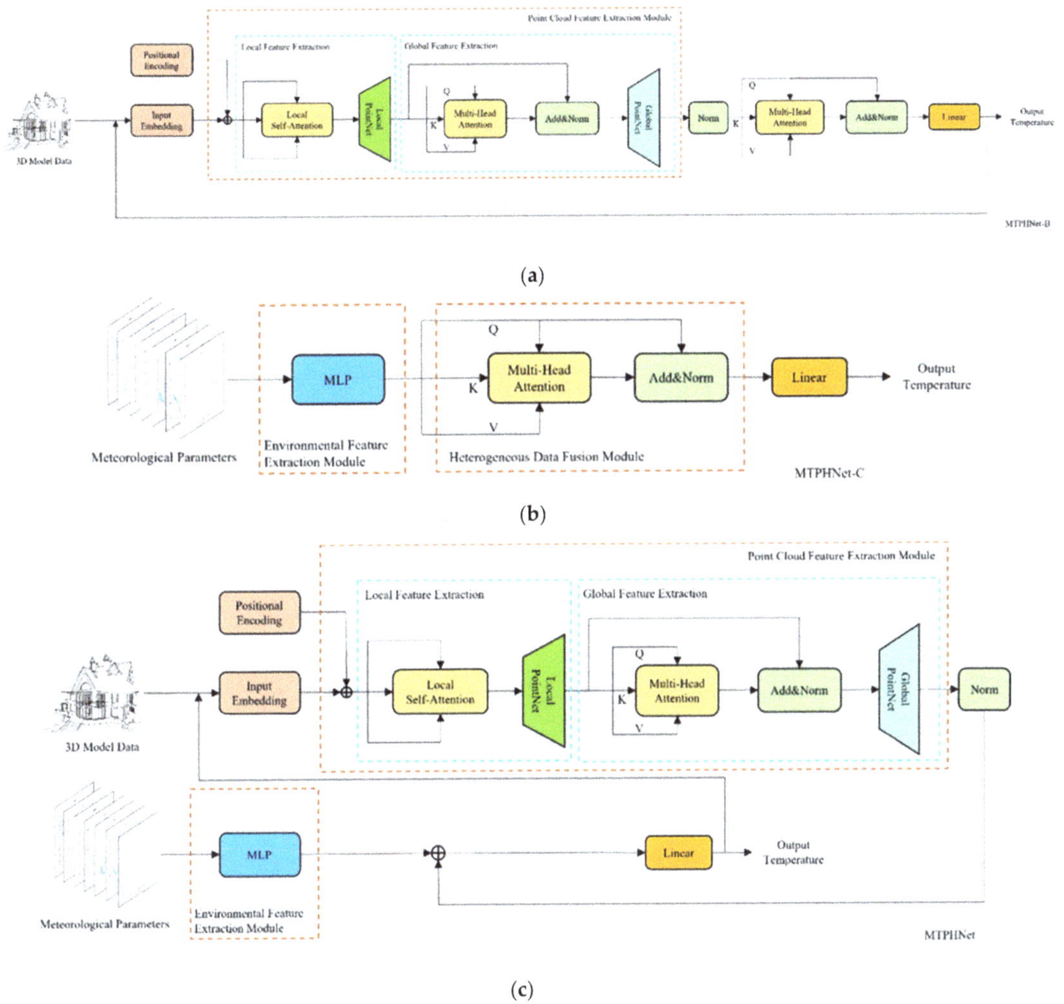

Figure 14. Multivariate temperature field prediction network based on heterogeneous data (MTPH-Net) variants for ablation experiments: (**a**) MTPHNet-B removes the environmental data feature mapping module (EMM) to study the effect of meteorological parameters on temperature field prediction; (**b**) MTPHNet-C removes the point cloud feature extraction module (PCEM) to study the effect of thermophysical parameters on temperature field prediction; and (**c**) MTPHNet-D re-places the data fusion module (DFM) with an additive fusion method to study the effect of data fusion on temperature field prediction.

4.3.1. Effectiveness Analysis of EMM

To measure the EMM's contribution, we designed a variant model without EMM, as described in Table 5: MTPHNet-B. It can be seen that the prediction effect of MPTHNet-A without EMM is extremely poor; it cannot even predict the temperature. The quantitative results show that the EMM is the core of temperature prediction.

4.3.2. Effectiveness Analysis of PCEM

We believe the use of PCEM would further improve the accuracy of temperature prediction. To substantiate it, we designed a variant without the PCEM: MTPHNet-C. In

Table 5, MTPHNet-A outperforms MTPHNet-C in all metrics. The quantitative results clearly show that PCEM improved the prediction performance.

4.3.3. Effectiveness Analysis of DFM

DFM fuses the features extracted from meteorological and thermophysical parameters, which is a crucial step. To confirm this, we designed a variant model, MTPHNet-D, which replaces the DFM with an additive fusion module. In Table 5, MTPHNet-A outperforms MTPHNet-D in all metrics, and MTPHNet-D is closer to MTPHNet-C in terms of metrics. The quantitative results show that DFM and PCEM contribute similarly to improve the prediction performance.

5. Conclusions

This study comprehensively considered the thermophysical and meteorological parameters affecting the temperature field distribution of a 3D target. Combined with temperature field distribution data, an intelligent temperature field prediction model, MTPHNet, was proposed. To fuse meteorological and thermophysical parameters, MTPHNet used PCEM to calculate the interaction between 3D target attributes and extract thermophysical features. Simultaneously, it used EMM to map meteorological parameters to meteorological features so that the mapped data and thermophysical data would be of the same size, which facilitated the subsequent data fusion. Finally, DFM fused the parts and used the results to predict the temperature. Considering PCEM's tendency of memory explosion when processing point cloud attribute data, we introduced PointNet as a feature extraction network to reduce the memory burden and divide the feature extraction process into local feature and global feature extraction activities to further streamline memory use. Compared with v-SVR and CBPNN, the MAE and RMSE of MTPHNet were reduced by at least 23.4% and 27.7%, respectively, whereas the R2 value increased by at least 5.85%. The results show that MTPHNet effectively improves model generalizability to more efficiently and accurately predict temperature fields while meeting real-time infrared simulation processing requirements. In complex object temperature field prediction tasks that simulate real environments, MTPHNet is advantageous in that it considers realistic energy interaction processes. Its MAE, RMSE, and R2 values were 2.645, 3.522, and 0.964, respectively, demonstrating the model's high adaptability to real scenes.

It should be noted that when MTPHNet performs multi-model prediction tasks, the number of point clouds of different 3D models are required to be the same, which significantly increases the difficulty of data collection. Therefore, in a future work, we plan to change the model structure so that it can be further adapted to 3D models varying numbers of point clouds.

Author Contributions: Conceptualization, Y.C. and L.L.; Data curation, Y.C., B.L., W.Z. and Q.X.; Formal analysis, L.L., W.Z. and Q.X.; Investigation, Y.C.; Methodology, Y.C., L.L. and B.L.; Project administration, L.L. and W.N.; Software, Y.C., W.N. and B.L.; Supervision, L.L. and W.N.; Validation, W.N., W.Z. and Q.X.; Visualization, Y.C. and W.Z.; Writing—Original draft, Y.C. and L.L.; Writing—Review and editing, Y.C. and L.L. All authors have read and agreed to the published version of the manuscript.

Funding: This research received no external funding.

Institutional Review Board Statement: Not applicable.

Informed Consent Statement: Not applicable.

Data Availability Statement: The data used to support the findings of this study are available from the corresponding author upon request.

Conflicts of Interest: The authors declare no conflict of interest. The founding sponsors had no role in the design of the study; in the collection, analyses, or interpretation of data; in the writing of the manuscript, and in the decision to publish the results.

References

1. Gurton, K.P.; Felton, M. Remote detection of buried land-mines and IEDs using LWIR polarimetric imaging. *Opt. Express* **2012**, *20*, 22344–22359. [CrossRef] [PubMed]
2. Jacobs, P.A.M. *Simulation of the Thermal Behaviour of an Object and Its Nearby Surroundings*; TNO-Report PH; TNO Publication: The Hague, The Netherlands, 1980.
3. Biesel, H.; Rohlfing, T. Real-Time Simulated Forward Looking Infrared (FLIR) Imagery For Training. In *Infrared Image Processing and Enhancement*; International Society for Optics and Photonics: Bellingham, WA, USA, 1987; Volume 781, pp. 71–80. [CrossRef]
4. Curtis, J.O.; Rivera, J.S. Diurnal and seasonal variation of structural element thermal signatures. In *Proceedings of the SPIE 1311, Characterization, Propagation, and Simulation of Infrared Scenes*; International Society for Optics and Photonics: Bellingham, WA, USA, 1990; pp. 136–145. [CrossRef]
5. Balfour, L.S.; Bushlin, Y. Semi-empirical model-based approach for IR scene simulation. In *Proceedings of the SPIE 3061, Infrared Technology and Applications XXIII*; International Society for Optics and Photonics: Bellingham, WA, USA, 1997; pp. 616–623. [CrossRef]
6. Gonda, T.G.; Jones, J.C.; Gerhart, G.R.; Thomas, D.J.; Martin, G.L.; Sass, D.T. PRISM Based Thermal Signature Modeling Simulation. In Proceedings of the SPIE Symposium on IR Sensors and Sensor Fusion, Orlando, FL, USA, 4–6 April 1988.
7. Sheffer, A.D.; Cathcart, J.M. Computer generated IR imagery: A first principles modeling approach. In *Proceedings of the SPIE 0933, Multispectral Image Processing and Enhancement*; International Society for Optics and Photonics: Bellingham, WA, USA, 1988; pp. 199–206. [CrossRef]
8. Schwenger, F.; Grossmann, P.; Malaplate, A. Validation of the thermal code of RadTherm-IR, IR-Workbench, and F-TOM. *SPIE Def. Secur. Sens.* **2009**, *7300*, 73000J. [CrossRef]
9. Hu, H.; Guo, C.; Hu, H. Real Time Infrared Scene Simulation System Based on Database Lookup Table Technology. *Infrared Technol.* **2013**, *6*, 329–333+344.
10. Huang, C.; Wu, X. Infrared Signature Simulation Based on the Modular Neural Network. *J. Proj. Rocket. Missiles Guid.* **2006**, *4*, 272–275.
11. Huang, C.; Wu, X.; Tong, W. Infrared image simulation based on statistical learning theory. *Int. J. Infrared. Millim. Waves* **2007**, *28*, 1143–1153.
12. Drucker, H.; Burges, C.J.; Kaufman, L.; Smola, A.; Vapnik, V. Support vector regression machines. *Adv. Neural. Inf. Process. Syst.* **1997**, *9*, 155–161.
13. Vaswani, A.; Shazeer, N.; Parmar, N.; Uszkoreit, J.; Jones, L.; Gomez, A.N.; Kaiser, Ł.; Polosukhin, I. Attention is all you need. *Adv. Neural Inf. Process. Syst.* **2017**, 5998–6008.
14. Mnih, V.; Heess, N.; Graves, A. Recurrent models of visual attention. *Adv. Neural Inf. Process. Syst.* **2014**, *2*, 2204–2212.
15. Nishijima, Y.; To, N.; Balčytis, A.; Juodkazis, S. Absorption and scattering in perfect ther-mal radiation absorber-emitter metasurfaces. *Opt. Express* **2022**, *30*, 4058–4070. [CrossRef] [PubMed]
16. Zhou, H.; Zhang, S.; Peng, J.; Zhang, S.; Li, J.; Xiong, H.; Zhang, W. Informer: Beyond efficient Transformer for long sequence time-series forecasting. In Proceedings of the AAAI Conference on Artificial Intelligence, Palo Alto, CA USA, 2–9 February 2021; Volume 35, pp. 11106–11115.
17. Qi, C.R.; Su, H.; Kaichun, M.; Guibas, L.J. Pointnet: Deep learning on point sets for 3D classification and segmenta-tion. In Proceedings of the IEEE Conference on Computer Vision and Pattern Recognition, Honolulu, HI, USA, 21–26 July 2017; pp. 77–85. [CrossRef]
18. Qi, C.R.; Yi, L.; Su, H.; Guibas, L.J. Pointnet++: Deep hi-erarchical feature learning on point sets in a metric space. *arXiv* **2017**, arXiv:1706.02413.
19. LeCun, Y.; Boser, B.; Denker, J.S.; Henderson, D.; Howard, R.E.; Hubbard, W.; Jackel, L.D. Backpropagation Applied to Handwritten Zip Code Recognition. *Neural Comput.* **1989**, *1*, 541–551. [CrossRef]
20. Chen, J.; Zhu, F.; Han, Y.; Chen, C. Fast prediction of complicated temperature field using Conditional Mul-ti-Attention Generative Adversarial Networks (CMAGAN). *Expert Syst. Appl.* **2021**, *186*, 115727. [CrossRef]
21. Willmott, C.J.; Matsuura, K. Advantages of the mean absolute error (MAE) over the root mean square error (RMSE) in assessing average model performance. *Clim. Res.* **2005**, *30*, 79–82. [CrossRef]
22. Hyndman, R.J.; Koehler, A.B. Another look at measures of forecast accuracy. *Int. J. Forecast.* **2006**, *22*, 679–688. [CrossRef]
23. Chicco, D.; Warrens, M.J.; Jurman, G. The coefficient of determination R-squared is more informative than SMAPE, MAE, MAPE, MSE and RMSE in regression analysis evaluation. *PeerJ Comput. Sci.* **2021**, *7*, e623. [CrossRef] [PubMed]

sensors

MDPI

Article

Off-Grid DOA Estimation Based on Circularly Fully Convolutional Networks (CFCN) Using Space-Frequency Pseudo-Spectrum

Wenqiong Zhang [1,2], Yiwei Huang [1,2], Jianfei Tong [1], Ming Bao [1,*] and Xiaodong Li [1,2]

1 Key Laboratory of Noise and Vibration Research, Institute of Acoustics, Chinese Academy of Sciences, Beijing 100190, China; zhangwenqiong@mail.ioa.ac.cn (W.Z.); huangyiwei@mail.ioa.ac.cn (Y.H.); tongjianfei@mail.ioa.ac.cn (J.T.); lxd@mail.ioa.ac.cn (X.L.)

2 University of Chinese Academy of Sciences, Beijing 100049, China

* Correspondence: baoming@mail.ioa.ac.cn

Abstract: Low-frequency multi-source direction-of-arrival (DOA) estimation has been challenging for micro-aperture arrays. Deep learning (DL)-based models have been introduced to this problem. Generally, existing DL-based methods formulate DOA estimation as a multi-label multi-classification problem. However, the accuracy of these methods is limited by the number of grids, and the performance is overly dependent on the training data set. In this paper, we propose an off-grid DL-based DOA estimation. The backbone is based on circularly fully convolutional networks (CFCN), trained by the data set labeled by space-frequency pseudo-spectra, and provides on-grid DOA proposals. Then, the regressor is developed to estimate the precise DOAs according to corresponding proposals and features. In this framework, spatial phase features are extracted by the circular convolution calculation. The improvement in spatial resolution is converted to increasing the dimensionality of features by rotating convolutional networks. This model ensures that the DOA estimations at different sub-bands have the same interpretation ability and effectively reduce network model parameters. The simulation and semi-anechoic chamber experiment results show that CFCN-based DOA is superior to existing methods in terms of generalization ability, resolution, and accuracy.

Keywords: off-grid; DOA estimation; circularly fully convolutional networks; space-frequency pseudo-spectrum; high resolution

Citation: Zhang, W.; Huang, Y.; Tong, J.; Bao, M.; Li, X. Off-Grid DOA Estimation Based on Circularly Fully Convolutional Networks (CFCN) Using Space-Frequency Pseudo-Spectrum. *Sensors* **2021**, *21*, 2767. https://doi.org/10.3390/s21082767

Academic Editor: Moulay A. Akhloufi

Received: 16 March 2021
Accepted: 8 April 2021
Published: 14 April 2021

1. Introduction

Direction of arrival (DOA) estimation is an important research direction in array signal processing, and it has been widely used in many military and civilian fields such as radar, communications, sonar, seismic, exploration and radio astronomy [1,2]. In many application scenarios, such as the Internet of Things and unattended ground sensor (UGS) systems [3], which focus on the remote targets, e.g., vehicles or helicopters, detection in the field, the array aperture is strictly limited so that it far exceeds the Rayleigh limit, which has made low-frequency multi-source DOA estimation complex problem for a long time.

Among traditional DOA estimation algorithms, subspace-based methods are considered to be high-resolution, such as MUSIC (multiple signal classification) [4] and ESPRIT (estimation of signal parameters via rotational invariance techniques) [5]. MUSIC-based methods employ the orthogonality of the signal subspace (steering vectors) and the noise subspace to search the spatial spectrum to achieve high-resolution. ESPRIT-based methods avoid spectrum search by the signal subspace rotation invariance properties and reduce computational complexity. When the uncertainty of the system or background noise leads to model errors, e.g., the wrong number of sources, subspace-based methods need to solve high-dimensional non-linear parameter estimation problems. Although many improved algorithms [6–9] based on MUSIC and ESPRIT have been developed to estimate the number of sources jointly, sometimes in order to solve the singular matrix of the spatial covariance, they may sacrifice the array aperture [10] and deteriorate the resolution. Later, compressed sensing (CS)-based methods [11–13] have been widely studied due to the consideration of both the super-resolution capability and the ability to detect the number of sources by exploiting spatial spectrum sparsity. Nevertheless, they suffer from a large amount of computational load caused by on-grid or off-grid search, especially in the case of wideband, which is challenging to apply to engineering implementation.

Alternatively, machine learning-based approaches, e.g., an artificial neural network (ANN), can provide a means of mapping from input features to DOA [14–18]. Moreover, ANNs consist of elementary mathematical calculations and have an advantage in computing speed compared with conventional DOA estimation algorithms. At the earliest, the radial basis function (RBF) neural network [19] is introduced to DOA estimation, which successfully learns the single source direction finding function from data sets despite the lack of resolution. Then, ref. [20] employs support vector regression (SVR) to improve the DOA resolution, while SVR is a small sample set learning method with a solid theoretical foundation and limits its application in practice. To this end, ref. [14] utilizes a multilayer perceptron (MLP) neural network to enhance the non-linear interpretation ability to the DOA mapping model. However, as the number of MLP network layers increases, the generalization performance may not necessarily be improved, which means that the accuracy of DOA estimation is insufficient.

In terms of generalization performance, deep learning has made significant progress, and the generalization performance is further improved as the training set increases. Various methods for DOA estimation based on deep learning have been proposed [21–27]. In [21], a deep neural network (DNN) was devised to perform logistic regression for each DOA to achieve high accuracy, while this method requires the known number of sources. To overcome this drawback, refs. [24,26] formulate multi-source DOA estimation as a multi-label multi-classification problem by convolutional neural networks (CNNs) and convolutional-recurrent neural networks (CRNNs), respectively, while the accuracy depends on the number of grids. Ref. [27] employs DNN to achieve rough candidate DOAs, and then takes the method of amplitude interpolation to estimate the signal directions of non-integer impinging angles. Similarly, ref. [25] first adopted CRNNs as spatial filters to obtain rough candidate sectors and then used classifiers to extract the precise directions. In [23], MLP neural networks are adopted to estimate possible sub-areas, and RBF neural networks are utilized for fine position estimation. Ref. [22] uses the multitask autoencoder for spatial filtering and then realizes spatial spectrum estimation by a fully connected multilayer neural network. From DP-based object detection models such as Faster R-CNN [28], and RefineDet++ [29] in the image field, we can learn that high-accuracy object position can be obtained from region proposal networks and regression refinement networks. Based on this inspiration, ref. [30] developes a two-stage cascaded neural network for DOA estimation, which includes a CNN and a DNN-based regressor for the discrete angular grid and the mismatch between true DOAs and discrete grids. However, the CNN-based on-grid method mainly focuses on the narrowband signal case. The input of the regressor, including the input and output of the first-stage CNN, may increase the complexity of the training data set. As can be seen from the above, high-resolution multi-source DOA estimation usually has two-stage networks for coarse search and fine search. Note that the DOA classifiers used by these algorithms map the features of all frequency sub-bands to the spatial spectrum. This means that the generalization ability of the model is sensitive to the frequency characteristics of the training data. In this way, the training data set samples are usually required to be large enough, making it difficult to exhaust all types of sources in practical applications.

This paper proposes a new off-grid DOA estimation method, which has two networks: (i) An on-grid multi-label classifier based on circularly fully convolutional networks (CFCNs) which was devised for rough grid proposals. The classifier provides the mapping from the phase map to the space-frequency pseudo-spectrum. Moreover, by performing circular convolution calculation on the sensing axis, the phase features at each sub-band can be extracted to a greater extent than linear convolution. To achieve high-resolution DOA grid proposals, the CFCN increases the feature dimension in exchange for an increase in space dimension by rotating convolutional networks; (ii) Based on these grid proposals, an off-grid regression network which was developed for precise DOA adjustment. The regressor obtains the actual deviation in the grid-gap from the features produced by the CFCN. The proposed models are trained by synthetic single-source white noise signals, which avoids the tedious and exhaustive data set. The main contributions of the proposed method are as follows:

- The circular convolution calculation enhances the phase feature acquisition ability.
- The CFCN ensures that the DOA estimations at different sub-bands have the same interpretation ability and effectively reduces network model parameters.
- The proposal DOA grids and the corresponding features provide a more feasible traversal within the grid-gap for the regression training data set and reduce the complexity of the data set.

The simulation and semi-anechoic chamber experiment results show that under the conditions of single/dual-source targets with different band-limited/signal-to-noise ratios (SNR)s, the CFCN-DOA is superior to existing methods in terms of generalization ability, resolution, and accuracy.

The rest of this paper is organized as follows: Section 2 introduces the problems of deep learning models on DOA estimation; Section 3 describes details of the off-grid DOA estimation model based on CFCNs, data set generation methods, and trained model performance comparison results with different parameters; Section 4 carries out simulation results and performance evaluation; Section 5 shows experimental verification; Section 6 discusses the experiments; and Section 7 summarized the whole work.

2. Problem Formulation

For the off-grid multi-source DOA estimation problem, we first need to determine the inputs and outputs of the model. The problem is split into two sub-problems, i.e., on-grid and off-grid problems, which are formulated as a multi-label and multi-classification problem and a regression problem, respectively. The multi-label classifier provides the DOA proposals, and the regressor produces precise DOA in the corresponding grid-gap, which fully meets the requirement of precise multi-source orientation.

In a model based on deep learning, the input is required to have sufficient information representation. For micro-aperture arrays detecting far-field sources, the signal strength features received by different sensors are not significant. Thus, DOA mainly depends on the phase difference between different sensors at different frequencies. In this paper, the phase map is chosen as the input of the model, and it is a $M \times F$ matrix $\mathbf{\Phi}_k$ at time k, where M is the number of sensors and F is the number of all sub-band signals. $\mathbf{\Phi}_k$ is denoted as

$$\mathbf{\Phi}_k = \begin{bmatrix} \phi_{k,1,1} & \cdots & \phi_{k,1,F} \\ \vdots & \ddots & \vdots \\ \phi_{k,M,1} & \cdots & \phi_{k,M,F} \end{bmatrix}, \tag{1}$$

where $\phi_{k,m,f}$ is the phase of the received signal of the m-th sensor at the f-th sub-band, which is obtained by a N point discrete Fourier transform (DFT).

The outputs of the model should be accurate DOAs of multiple sources, while it is challenging to obtain DOAs directly when the number of sources is unknown. Therefore, we formulate the model as a two-stage network, which is shown in Figure 1. At stage 1, this is an on-grid problem. Firstly, all the raw signals are transformed into the phase map by DFT. Then, all the features are extracted by the feature extraction network. Finally, the multi-label classifier provides DOA grid proposals. At stage 2, based on the features and DOA grid proposals given by the stage 1, the regressor produces precise DOAs. The network at stage 1 is first trained with the data set labeled by DOA grids, and then the regressor is trained with the data set labeled by precise DOAs in the grid-gap.

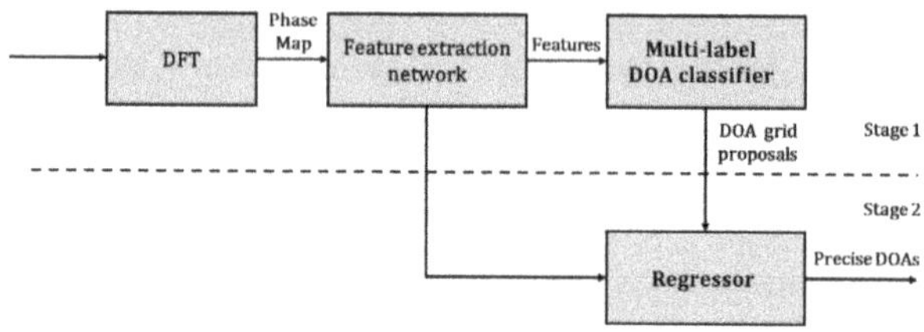

Figure 1. Block diagram of the proposed off-grid direction of arrival (DOA) estimation.

In this paper, we assume that there is only one source in a grid. The problem of the multi-label classifier at stage 1 is to obtain reliable DOA grid proposals based on the phase features of all sub-bands by using the space-frequency pseudo-spectrum. In addition, the problem of the regressor at stage 2 is to choose sufficient features as its inputs.

3. Off-Grid DOA Estimation

This section focuses on the proposed architecture, the generation of the data set, and the design of structural parameters in the network.

3.1. The Off-Grid DOA Estimation Based on Circularly Fully Convolutional Networks (CFCNs)

In the CNN convolution calculation, the linear convolution calculation used may ignore some sensor phase difference features, which are probably important information, caused by the most spaced sensors. If this phase difference is ignored, it may affect the target resolution performance. To learn the phase features of each sub-band to the greatest extent, the model needs to have the same interpretation capability for each sub-band. We develop a circularly fully convolutional network (CFCN)-based architecture consisting of two networks: the on-grid multi-label classifier and the off-grid regressor. This architecture is shown as Figure 2.

For the multi-label classifier, to ensure the independence between the spectral features of the source and space, the size of the convolution kernel is designed to be $M/2 \times 1$. When performing circular convolution calculation, the input phase matrix needs to be extended by $M/2 - 1$ length in the spatial dimension, i.e., the sensing axis. The feature extraction network has L layers, and each layer has C_l neurons/channels, where C_l can be called the feature dimension. The neurons are activated by rectified linear unit (ReLU). Since the output dimension of the fully convolutional network (FCN) is consistent with the input dimension, up-sampling is usually used to achieve dimension upgrades in a certain dimension. Up-sampling does not increase the total amount of features/information but increases data storage. In this paper, we can increase the feature dimension in exchange for increasing the space dimension by rotating the convolutional network, i.e., transposing the feature dimension and the space dimension. In the last layer of the convolution feature extraction layers, the number of neurons is consistent with DOA grids. Finally, the 1×1 convolution layer fuses the phase features of all sensors at each sub-band. To obtain the posterior probability $p(\theta_{i,f}|\mathbf{\Phi}_k)$ of DOA at the i-th space grid and the f-th sub-band at time k, the last layer is activated by the Sigmoid function, where $\theta_{i,f}$ represents the direction corresponding to the i-th DOA class at the f-th sub-band. The $I \times F$ posterior probability matrix is the space-frequency pseudo-spectrum, where I is the number of space grids and F is the number of sub-bands. For the multi-source DOA estimation, the pseudo-posterior probability of each DOA class is obtained by averaging the probabilities of all sub-bands:

$$p(\theta_i|\mathbf{\Phi}_k) = \frac{1}{F}\sum_{f=1}^{F} p(\theta_{i,f}|\mathbf{\Phi}_k). \tag{2}$$

From the pseudo-posterior probabilities, H DOA classes corresponding to the H peaks with the highest probability are selected as candidate DOAs $\gamma_h, h = 1, \cdots, H$, which are treated as DOA proposals for the regressor. In this work, we use a simple peak detection method as the DOA proposal search to verify the effectiveness of the proposed algorithm.

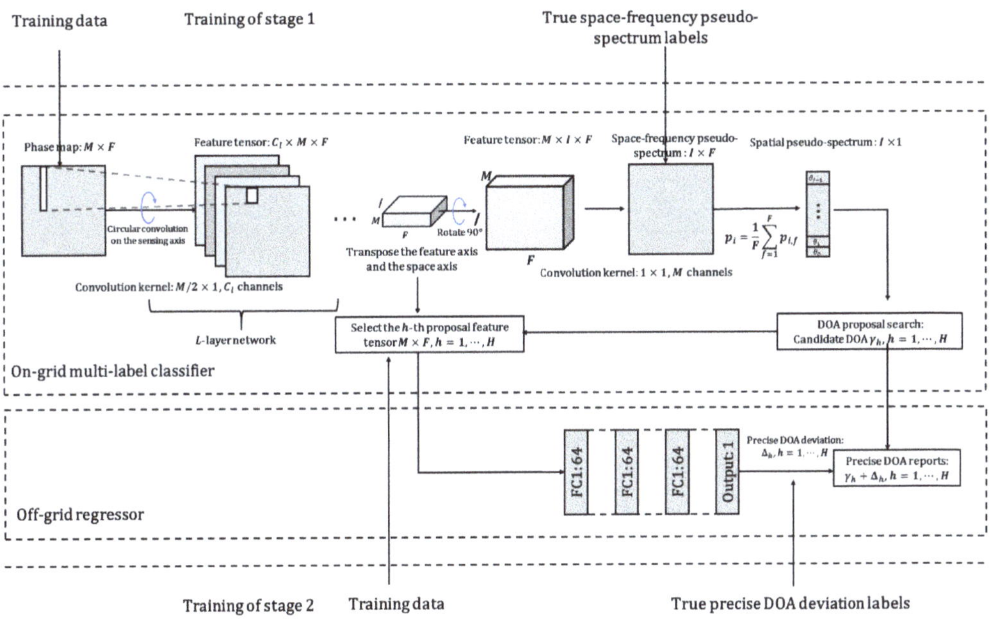

Figure 2. Proposed architecture.

Before the last 1×1 convolution layer fusing features of all sensors, the $M \times I \times F$ feature tensor R has the sufficiency of the DOA information in the grid-gap. On-grid methods cannot provide precise DOA estimation. Thus, the regression-based off-grid method is proposed. For this regressor, the accurate DOA deviation Δ_h of the DOA proposal γ_h is mapped from the h-th proposal feature $M \times F$ matrix. Therefore, the final off-grid DOA φ_h is:

$$\varphi_h = \gamma_h + \Delta_h, h = 1, \cdots, H. \tag{3}$$

In the training process, the on-grid multi-label classifier is trained first, and then the on-grid regressor is trained.

3.2. The Generation of the Data Set

For the case of long-distance, low-frequency, and low-SNRs, this article does not temporarily consider the influence of complex reverberation/multi-path. Under different targets, different source operating states, and single-source/multi-source, the signals received by the array are different. If the signals with different features at different locations are exhaustively enumerated under different SNR conditions, the data set will be extensive, and it cannot cover all combinations. In addition, when the signal frequency ranges of different targets overlap, the non-linear superposition of different sources will make the entire network challenging to train.

For the data set of the multi-label classifier, signals received by the array are generated by a single white noise source traversing different orientations and different SNRs. In this way, the features of all sub-bands vs. space orientations can be represented. The generation process is shown in Figure 3. First, the single-source white noise is transformed to the frequency domain by DFT, and the phase compensation is performed for each sub-band signal to obtain the true signal received by each sensor, where the compensation factor is the steering vector $\mathbf{A}_f(\theta_i), f = 1, \cdots, F$. The corresponding label is a $I \times F$ space-frequency pseudo-spectrum matrix, whose i-th row is equal to 1, and the other items are 0. Then, the true signals are converted into the time domain by inverse discrete Fourier transform (IDFT), mixing the sensor noise. Considering the worse noise conditions, the SNR is subject to $[-6\,\text{dB}, 60\,\text{dB}]$ uniform random distribution. The training data set referred to in this article has $I \times 4000$ samples, and the testing set has $I \times 400$ samples.

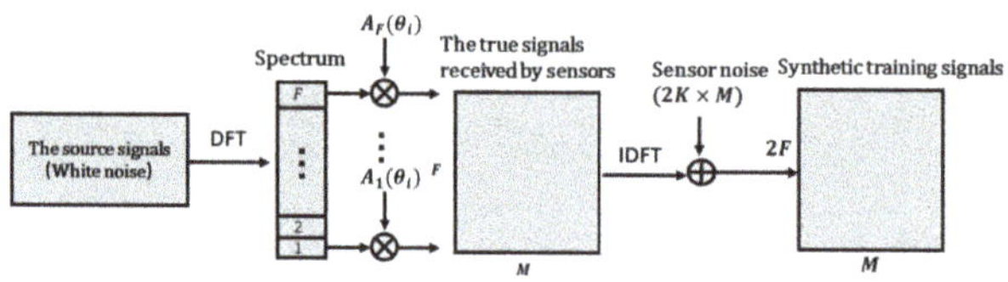

Figure 3. The generation process of data set signals.

For the data set of the regressor, the inputs are the features extracted by the multi-label classifier from the phase map, and the labels are the accurate DOA deviations in the grid-gap. If this array is symmetrical, the mapping between the DOA deviation and the feature is consistent for each grid-gap uniformly divided in space. Therefore, we can train the regressor in one grid-gap to generalize to other grid-gaps. The signals of the training set can still be traced to the original phase map input, which is generated by the process in Figure 3 according to the new DOA parameter φ_i

$$\varphi_i = \theta_i + \Delta_i, \tag{4}$$

where Δ_i is the DOA deviation at the i-th grid-gap, and it is also the label of the i-th proposal feature $M \times F$ matrix. Δ_i is subject to $[-g/2, g/2]$ uniform random distribution, where g is the DOA width of a grid-gap. Other parameter settings refer to parameters of the data set in the multi-label classifier. The numbers of samples in training and testing sets are 4000 and 400, respectively.

3.3. Training Methods and Results

The proposed deep networks are realized and trained in Pytorch on a PC with a single graphic processing unit (GPU) RTX2080Ti and an Intel i7-8700 processor. We use the stochastic gradient descent algorithm with a momentum of 0.9 to train the CFCN-based multi-label classifier and the regressor. The number of samples in each batch is set as I, binary cross-entropy is used as the loss function for the multi-label classifier, and mean-squared error (MSE) is used for the regressor. The cyclic learning rate scheduler [31] is used for training, and the learning rate range is from 10^{-6} to 10^{-1}. Xavier [32] is used for initialization. Note that this model only uses a single-source white noise data set as training, and no other multi-source band-limited signal data sets are used for training here.

In this work, we take a uniform circular array with parameters of $M = 8.70$ mm aperture as an example to compare the performance of different network structures. The number of DOA grids is $I = 72$, i.e., the width of the grid-gap is 5°. The sampling frequency is Fs = 3 kHz, the length of each snapshot is $N = 512$, and then the number of sub-bands is $F = N/2 = 256$, up to the Nyquist frequency. The influence of different convolution kernel sizes, different convolution layer widths, and different convolution layer numbers on the parameter quantity and performance of the model is discussed. The evaluation results are shown in Table 1. The structure of the model in Figure 2 is encoded for convenience. "L$*$" means the number of convolutional network layers is "$*$". "FC" means a fully connected network. "K$*$" means the size of the convolution kernel $* \times 1$. "$[\cdot] \times *$" means "$\cdot$" repeated "$*$" times. For example, CNN + FC: L7-K2- [64] $\times$ 7 + FC[[512] $\times$ 2, 72] means: the model consists of seven convolutional layers, the convolution kernel size is 2×1, each layer has 64 channels, and the fully connected network structure is $[512 \times 512 \times 72]$.

Table 1 shows the comparison of different model parameters, accuracy, and total floating point operations (FLOPs), which are obtained using an open source neural network analyzer (https://github.com/Swall0w/torchstat (accessed on 31 March 2021)). We can see that although CFCN-based methods slightly increase computational complexity, they have fewer parameters and higher accuracy than CNN + FC [24]. The main reason is that a fully connected network occupies a lot of parameters, and the phase difference features between sensors will decrease significantly as the frequency decreases. All sub-band features extracted by the CNN are input to the fully connected network, and then the fully connected network may sacrifice low phase difference features in the low-frequency range to highlight the important feature contributions in the high-frequency range in order to improve the score in training. This leads to a weak generalization ability on low-frequency signals. This inference can be verified in the simulation experiment in the next section.

Table 1. Comparison of different model parameters, accuracy, and total floating point operations (FLOPs).

The Structure of the Model	Number of Parameters (in Millions)	Accuracy (%)	Total FLOPs (Million)
CNN + FC [24]: K2-[64] × 7-FC[[512] × 2, 72]	8.48	89.4	53.67
CFCN: L4-K4-[[128] × 3, 72]	0.1689	92.74	345.66
CFCN: L4-K4-[72] × 4	0.0628	91.2	128.83
CFCN: L4-K2-[[128] × 3, 72]	0.0847	90.93	173.84
CFCN: L5-K4-[[128 × 4], 72]	0.2346	95.1	479.88
CFCN: L5-K2-[72] × 5	0.0420	94.5	86.53
CFCN: L5-K4-[72] × 5	0.0836	92.1	171.43
CFCN: L6-K4-[[128] × 5, 72]	0.3569	1.2	614.09
CFCN: L6-K4-[72] × 6	0.1044	98.5	214.02
CFCN: L6-K2-[72] × 6	0.0524	89.2	107.98
CFCN: L7-K4-[72] × 7	0.1252	1.3	256.62
CFCN: L7-K2-[72] × 7	0.0629	92.8	129.42
CFCN: L8-K2-[72] × 8	0.0733	95.5	150.86
CFCN: L9-K2-[72] × 9	0.0837	96.7	172.31
CFCN: L10-K2-[72] × 10	0.0942	97.1	193.75
CFCN: L11-K2-[72] × 11	0.1046	97.8	215.2
CFCN: L12-K2-[72] × 12	0.1151	1.4	236.64

The CFCN does not use a fully connected network to classify all features but uses a separate phase feature classification for each sub-band. In this way, it effectively reduces the parameters and ensures that each sub-band feature has an equal contribution to the DOA estimation.

When the CFCN has a 4×1 convolution kernel, the DOA estimation accuracy rate increases rapidly as the number of convolutional network layers increases. When the convolutional layer width is 72 channels, and the number of convolutional layers is increased to six layers, the accuracy rate reaches the highest 98.5%. When the width of the convolutional layer is 128, the number of convolutional layers can be up to five, and the accuracy rate is 95.1%. If the size of the convolution kernel is reduced to 2×1, the convolution width is set to 72, the number of convolution layers can be up to 11, and the accuracy rate can reach 97.8%. Therefore, a fully convolutional network with an I width of $M/2 + 2$ layers is used in the following text.

For the regressor, we utilized a four-layer fully connected network to approximate the mapping between the deviation of DOA in the grid-gap and the corresponding features. After training, the final mean absolute error (MAE) of the off-grid DOA reaches 0.782°.

3.4. Computational Complexity

In the implementation of computational processing, deep learning-based methods have been more optimized for parallel computing than traditional subspace-based algorithms such as MUSIC. Table 2 shows the average computing time of different methods on the platform of central processing unit (CPU) and GPU used by this paper. Deep learning-based DOA algorithms have better real-time performance. For the wide-band MUSIC, the total computational complexity is $O(FM^3 + FM^2I)$ [33], where $O(FM^3)$ and $O(FM^2I)$ are the complexity of eigen-decomposition and the spatial pseudo-spectrum search for F sub-bands, respectively. According to [34] and network structures involved in the article, the computational complexity of the CFCN and CNN-DOA are $O(\frac{M^3}{4}I^2F)$ and $O(M \times MF \times M/2 \times 64 \times 64 + 2 \times 64 \times F \times 512 + 2 \times 512 \times I)$, respectively. As the numbers of sensors and DOA grids increase, the proposed method has more computational complexity than CNN-DOA. If the input signals overlap 50%, i.e., the report refresh period is $256/3000 = 0.085$ s, the CFCN based on the structure L6-K4-[72] $\times$ 6 requires a processing speed of more than 2.52 GigaFLOPs (GFLOPs) per second (FLOPs/S). For the current embedded processors such as the RK3399Pro IoT device with the neural process engine reaching 2.4 TeraFLOPs/S (TFLOPs/S) [35] and NVIDIA TX2 with compute unified device architecture (CUDA) cores reaching 1.5 TFLOPs/S [36], the calculation requirements of CFCN-based methods are quite acceptable. The CFCN can even be implemented in a full-hardware system [37] for better real-time performance.

Table 2. Comparison of different model average computing time.

	CNN + FC [24]: K2-[64] × 7-FC[[512] × 2, 72]	MUSIC	CFCN: L6-K4-[72] × 6 + Regressor
CPU	6.6 ms	486 ms	9.7 ms
GPU	1.2 ms	-	2.2 ms

4. Simulation Experimental Evaluation

In this section, simulation experiments are implemented to evaluate the generalization ability and DOA estimation performance of the trained network under different conditions.

4.1. Baselines and Objective Measures

The performance of the proposed method is compared to two common algorithms: MUSIC [4] and CNN-based DOA [24]. To ensure a fair comparison, we set similar parameter settings for other methods, e.g., the DOA grid is 5°. The wideband MUSIC method averages the spatial pseudo-spectrum of all sub-bands to obtain the wideband spatial pseudo-spectrum. The H highest peak values are selected as the final DOA estimates. Two hundred Monte Carlo experiments were performed for the statistics.

For the objective evaluation, OSPA [38] (optimal sub-pattern assignment) was used as the multi-source DOA error metric:

$$\begin{cases} D_{p,c}(\mathbf{X},\mathbf{Y}) = \left[\frac{1}{|\mathbf{X}|}\left(\min_{\pi\in\Pi_{|\mathbf{X}|}} \sum_{i=1}^{|\mathbf{X}|} d_c^p(x_i,y_i) + (|\mathbf{X}|-|\mathbf{Y}|)\cdot c^p\right)\right]^{\frac{1}{p}}, & |\mathbf{X}| \le |\mathbf{Y}| \\ D_{p,c}(\mathbf{X},\mathbf{Y}) = \left[\frac{1}{|\mathbf{X}|}\min_{\pi\in\Pi_{|\mathbf{X}|}} \sum_{i=1}^{|\mathbf{X}|} d_c^p(x_i,y_i)\right]^{\frac{1}{p}}, & |\mathbf{X}| > |\mathbf{Y}| \end{cases}, \quad (5)$$

where $\mathbf{X}$ is the measured DOA set, $\mathbf{Y}$ is the true DOA set, $x_i \in \mathbf{X}$, $y_i \in \mathbf{Y}$, $|\cdot|$ is the cardinality of the set $\cdot$, p is the order of OSPA, $\Pi_{|\mathbf{X}|}$ is the set of $|\mathbf{X}|$ elements extracted and permuted and combined from $\mathbf{Y}$, and $d_c(x,y)$ is the cut-off distance:

$$d_c(x,y) = \min\{c, d(x,y)\}. \quad (6)$$

In (6), $d(x,y)$ is the difference between two angles:

$$d(x,y) = \min\{|x-y|, |360^\circ - |x-y||\}. \tag{7}$$

In this article, we set that $c = 45^\circ$, and $p = 2$. For the J Monte Carlo tests, the mean OSPA $\overline{D}_{p,c}$ is:

$$\overline{D}_{p,c} = \frac{1}{J}\sum_{j=1}^{J} D_{p,c}(x_i, y_i). \tag{8}$$

For the single-source case, the mean OSPA is the MAE. If the mean OSPA is more than 20°, the algorithm is evaluated as a failure.

4.2. Simulation Experiments

4.2.1. Simulation Settings

The sampling frequency of the input signal is Fs = 3 kHz, and the data length of each time frame is $N = 512$. To evaluate the performance of the model, we design challenging simulation conditions including different bandwidth-limited signal sources, single and dual sources, and different SNRs, where the dual sources are angularly separated by 135°, and the SNR range is $[-6\,\text{dB}, 0\,\text{dB}, 6\,\text{dB}, 12\,\text{dB}, 20\,\text{dB}]$. The specific simulation conditions are as follows:

(1) Single-source situation:

- Low frequency: 0–200 Hz;
- Full frequency: 0–1500 Hz;

(2) Dual-source situation:

- Overlapping low frequency: 0–200 Hz;
- Non-overlapping frequency: 0–200 Hz, 200–500 Hz;
- Overlapping full frequency: 0–1500 Hz;

4.2.2. Simulation Results

According to the above simulation test conditions, the simulation results of CFCN-DOA, MUSIC and CNN-DOA for the 200 Monte Carlo tests are as follows:

(i) Single-source situation:

(a) Low-frequency band-limited source cases: The true DOA is set to 182.5° in the middle of grids. Then, the on-grid-based methods cannot provide the precise DOA. Simulation evaluation results for the low-frequency case are shown in Figure 4. Figure 4a shows the mean OSPAs (MAEs) of three methods over different SNRs, and then we can see that the MAEs of CNN-DOA are all more than 20°, which means that the CNN-DOA fails. However, the CFCN-DOA has lower errors under the lower SNR conditions. Even when the SNR is 0 dB, the accuracy can still reach 17.5°, but the MUSIC cannot work at this time. As the SNR increases, the MAE of CFCN-DOA continues to decline, all lower than that of MUSIC, reaching 1.45° at 20 dB SNR. We can also see the details in space-frequency pseudo-spectra of CFCN-DOA and MUSIC at 0 dB SNR, which are plotted in Figure 4c,d, respectively. In the area of $360^\circ \times$ 0–200 Hz, the space-frequency pseudo-spectrum of MUSIC is almost flat, while the bright spots of CFCN can be identified. These characteristics can be seen more clearly from the average spatial pseudo-spectra in the area, displayed in Figure 4b. The CFCN-DOA has a higher resolution than MUSIC at a lower frequency. Note that CNN-DOA fails for low-frequency band-limited sources, and the estimated DOAs always tend towards some other unrelated points.

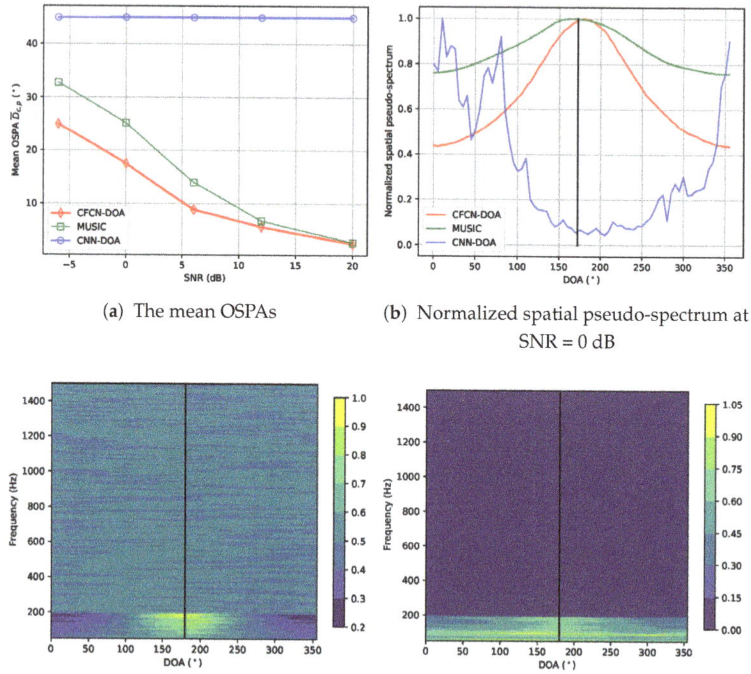

(**a**) The mean OSPAs (**b**) Normalized spatial pseudo-spectrum at SNR = 0 dB

(**c**) Space-frequency pseudo-spectrum of CFCN at SNR = 0 dB (**d**) Space-frequency pseudo-spectrum of MUSIC at SNR = 0 dB

Figure 4. The results of the simulation under the low-frequency single-source condition.

(b) Full-frequency source cases: The true DOA is also set to 182.5°. Simulation evaluation results are exhibited in Figure 5. Compared with the low-frequency band-limited case, the mean OSPAs of these methods are remarkably improved, which can be seen from Figure 5a. At this condition, CNN-DOA can work and reaches the same accuracy, i.e., 2.5°, with the MUSIC beyond 6 dB SNR, while the off-grid CFCN-DOA can further obtain higher accuracy. We can also see the superiority of CFCN-DOA from normalized space-frequency pseudo-spectra shown in Figure 5b,c. The spatial directivity of CFCN-DOA at each sub-band is more consistent than MUSIC. The spatial directivities of average spatial pseudo-spectra of MUSIC, CFCN-DOA, and CNN-DOA are enhanced in turn, which is displayed in Figure 5d. Normalizing the spatial spectrum for each sub-band, we can obtain a three-dimensional pseudo-beam pattern of CFCN, which is shown in Figure 5e. Then, the −3 dB pseudo-beam width in the frequency range 50–1500 Hz can be calculated by statistics at 20 dB SNR (See Figure 5f). The pseudo-beam width of CFCN-DOA is lower than the MUSIC at each sub-band, especially at 50 Hz CFCN-DOA reaching a pseudo-beam of less than 90° MUSIC has nearly 160° width. Since testing signals are similar to the training data set, the CNN-DOA performs very well, and the accuracy reaches the limit, i.e., the grid-gap, until the SNR is 20 dB. At this moment, the off-grid CFCN-DOA can achieve 1.45°.

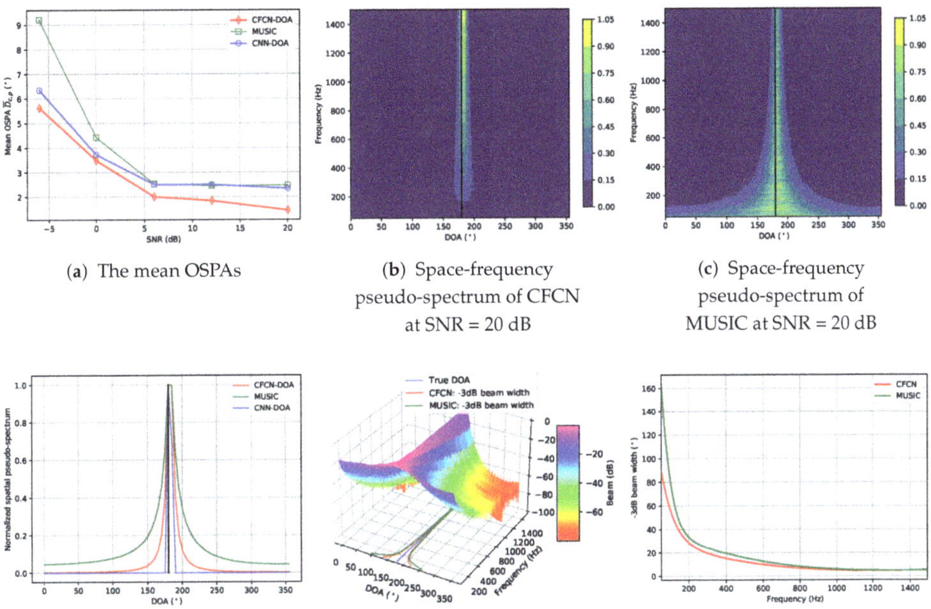

(**a**) The mean OSPAs

(**b**) Space-frequency pseudo-spectrum of CFCN at SNR = 20 dB

(**c**) Space-frequency pseudo-spectrum of MUSIC at SNR = 20 dB

(**d**) Normalized spatial pseudo-spectrum at SNR = 20 dB

(**e**) 3D display of pseudo-beam pattern at SNR = 20 dB

(**f**) Pseudo-beam width at SNR = 20 dB

Figure 5. The results of the simulation under the full-frequency single-source condition.

(ii) Dual-source situation: The true DOAs of two sources are set to 92.5° and 227.5°, respectively. Simulation results are counted in Figure 6. From Figure 6a–c, we can conclude that CNN-DOA cannot estimate multiple sources whether the frequency ranges are overlapping or non-overlapping, or the band is limited or non-limited. Although the other two methods can work in these three situations, MUSIC fails when the SNR is less than 20 dB under overlapping low-frequency, 6 dB under non-overlapping frequency, and 10 dB under overlapping full-frequency. On the contrary, CFCN-DOA is more capable of fighting low-SNR situations. The accuracy under the three conditions mentioned is 9.18°, 7.39°, and 10.48°, respectively. The space-frequency pseudo-spectra of CFCN under the three conditions mentioned are shown in Figure 6d–f respectively, and those of MUSIC are plotted in Figure 6g–i, respectively. From the space-frequency pseudo-spectra in the critical conditions of these MUSIC failures, CFCN-DOA can highlight the more spatial features of different frequencies. This also further verifies the generalization ability of the CFCN-DOA method based on learning space-frequency pseudo-spectrum.

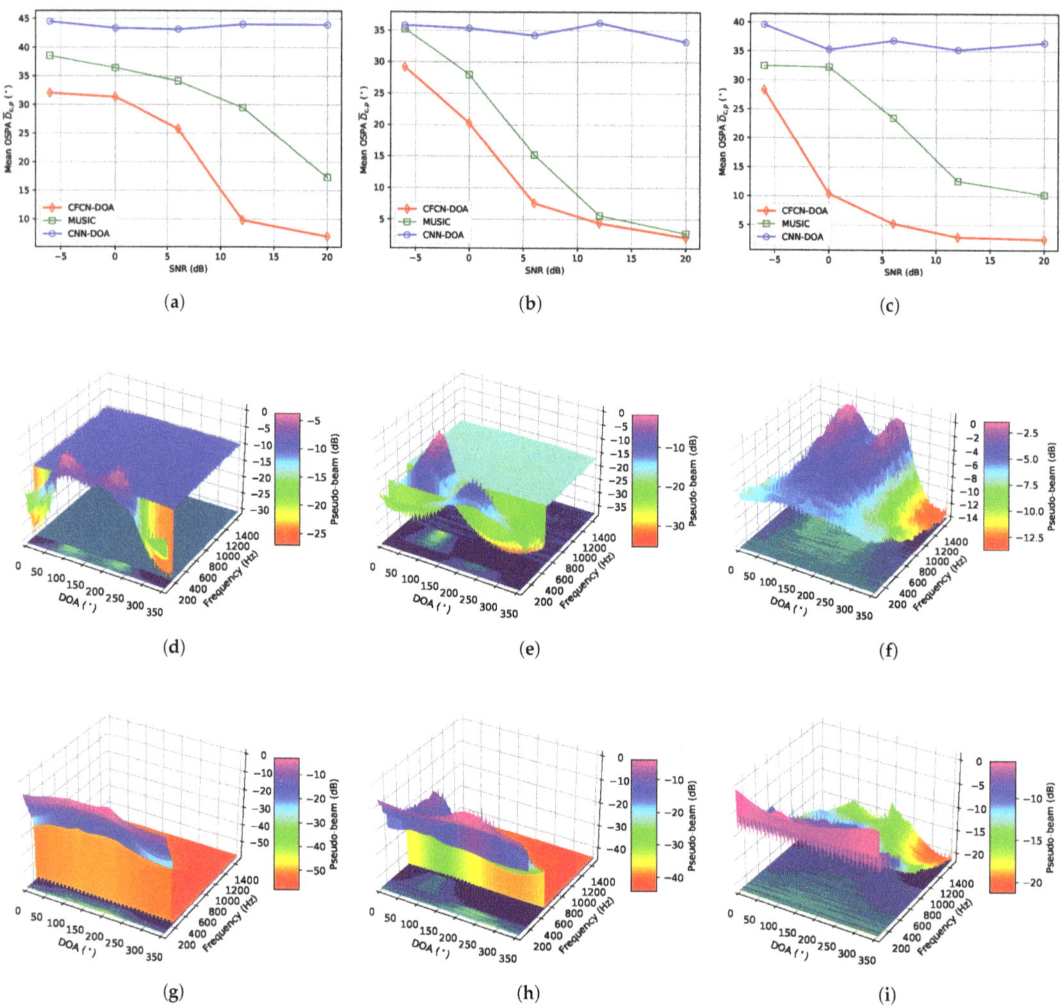

Figure 6. The results of the simulation under the dual-source condition. (**a**) The mean OSPAs under the overlapping low-frequency condition. (**b**) The mean OSPAs under the non-overlapping frequency condition. (**c**) The mean OSPAs under the overlapping full-frequency condition. (**d**) Space-frequency pseudo-spectrum of CFCN under the conditions: overlapping low-frequency and SNR = 12 dB. (**e**) Space-frequency pseudo-spectrum of CFCN under the conditions: non-overlapping frequency and SNR = 6 dB. (**f**) Space-frequency pseudo-spectrum of CFCN under the conditions: overlapping full-frequency and SNR = 0 dB. (**g**) Space-frequency pseudo-spectrum of MUSIC under the conditions: overlapping low-frequency and SNR = 12 dB. (**h**) Space-frequency pseudo-spectrum of MUSIC under the conditions: non-overlapping frequency and SNR = 6 dB. (**i**) Space-frequency pseudo-spectrum of MUSIC under the conditions: overlapping full-frequency and SNR = 0 dB.

5. Experimental Verification in Semi-Anechoic Chamber

To verify the actual generalization performance of the proposed algorithm, we implement the semi-anechoic room test. The indoor noise is less than 30 dB. The experimental scene layout is shown in Figure 7. We adopt Hivi speakers as acoustic sources and use the recorder Zoom-F8n as the acquisition equipment. The distance between the speaker and the acoustic array is 1.4 m. The angular interval of the dual sources is set to 135°. Single/dual-source tests in different frequency ranges are carried out. Then, the simulation data are replayed. The experiment time of each test is more than 40 s, and the data are reprocessed with a 512-point Hanning window and 50% overlapping. Then, the first 200 frames of data are selected for statistics. Considering the limitation of the front-end MEMS microphone frequency response parameters, the processing frequency range is set to 50–1500 Hz.

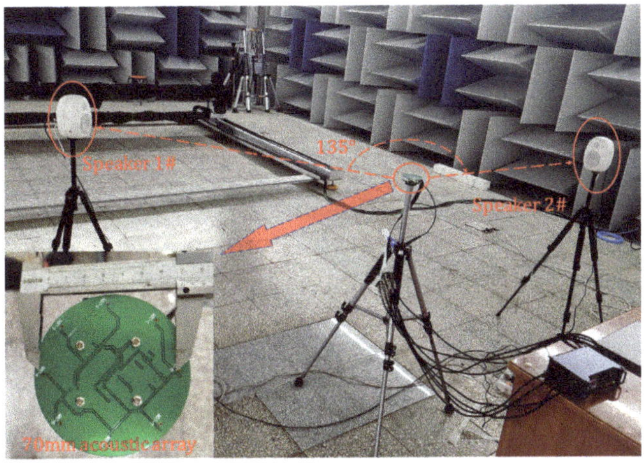

Figure 7. The settings of the semi-anechoic chamber experiment.

The results of the experiment in a semi-anechoic chamber are summarized as follows. Table 3 describes the mean OSPAs under all the conditions. Due to the limitation of the speaker frequency response, the speakers can not play pure low-frequency signals, and their harmonic components usually pollute the high-frequency parts. For instance, Figure 8 shows the frequency spectrum of a sensor receiving signal from harmonic interference under the dual-source 0–200 Hz condition.

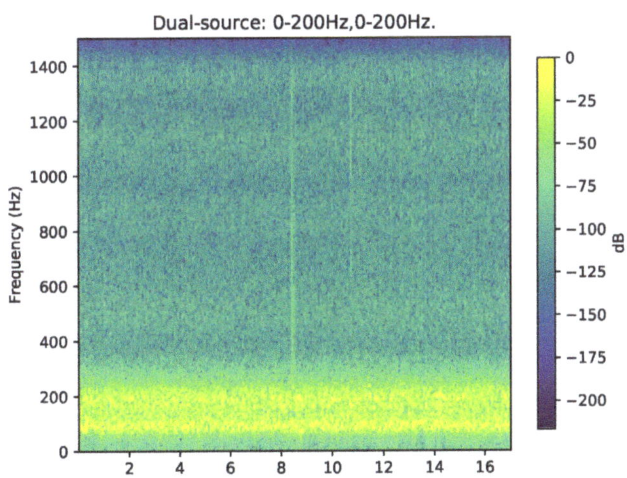

Figure 8. The frequency spectrum of a sensor receiving a signal from harmonic interference under the dual-source 0–200 Hz condition.

Table 3. Mean OSPAs for the experiment in semi-anechoic chamber (Unit: °).

Methods	Single-Source, Low-Frequency, (SNR = 81.5 dB)	Single-Source, Full-Frequency, (SNR = 89.9 dB)	Dual-Source, Overlapping Low-Frequency, (SNR = 120 dB)	Dual-Source, Non-Overlapping Low-Frequency, (SNR = 120 dB)	Dual-Source, Overlapping Full-Frequency, (SNR = 116 dB)
CFCN-DOA	2.15	0.15	1.52	1.14	3.45
MUSIC	4.64	2.76	6.84	2.56	3.75
CNN-DOA	3.42	1.23	28.89	33.04	26.75

Coupled with the quiet environment in the semi-cancellation room, CNN-DOA can give full play to its performance. For the single-source low-frequency case, CNN-DOA can reach 3.42° in such a quiet environment. However, CNN-DOA cannot deal with multi-source cases. The proposed off-grid methods have higher accuracy than MUSIC. If the frequency ranges of sources are non-overlapping, the accuracy of CFCN-DOA is up to 1.14°. Note that the error of CFCN-DOA under the dual-source overlapping full-frequency condition is larger than that under the dual-source overlapping low-frequency condition because the improved estimation performance caused by contaminated high-frequency signals for the low-frequency case is greater than the effect of frequency overlap for the full-frequency case. From Figure 9, the generalization performance is further verified by the high directivity at the space-frequency area. Note that the space-frequency pseudo-spectrum of the CFCN in Figure 9a spreads to the high-frequency range due to the weakly leaking high-frequency harmonic signals from speakers.

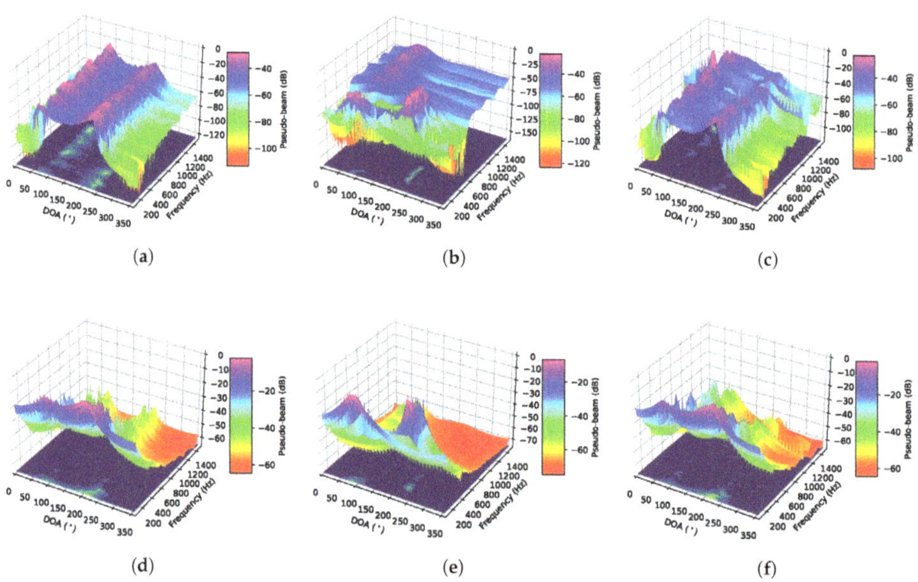

Figure 9. Space-frequency pseudo-spectra obtained by CFCN-DOA and CNN-DOA in the experiment in the semi-anechoic chamber. (**a**) CFCN under the overlapping low-frequency dual-source condition. (**b**) CFCN under the non-overlapping frequency dual-source condition. (**c**) CFCN under the overlapping full-frequency dual-source condition. (**d**) MUSIC under the overlapping low-frequency dual-source condition. (**e**) MUSIC under the non-overlapping frequency dual-source condition. (**f**) MUSIC under the overlapping full-frequency dual-source condition.

6. Discussion

From the simulation and semi-anechoic chamber experiments, the traditional CNN-based models have weak generalization ability for band-limited signals or multi-source, while the CFCN-based off-grid method can overcome these difficulties and its resolution and accuracy are better than MUSIC. There are two critical differences between these two deep learning-based methods.

(i) On the one hand, the input data of the two methods are the same, but their labels are different, i.e., the spatial pseudo-spectra as the labels of CNN-based methods and space-frequency pseudo-spectra as the labels of CFCN-based methods. The spatial features of each sub-band can be learned individually by the backbone of CFCN. In this way, other sources with limited frequency bands that do not match the features in the training data set can still be located.

(ii) On the other hand, the backbones of these two methods are different. The architecture of FCN can be split into an independent network for each sub-band, and the structure of each network is consistent. Then, CFCN can have the same interpretation capability on each sub-band. On the contrary, the features of all sub-bands are fused and mapped to the spatial pseudo-spectrum by the fully connected network in CNN-based methods. In this way, the model is more sensitive to frequency features.

Compared with traditional MUSIC-based methods, CFCN-based approaches richer nonlinear interpretation capabilities. The higher resolution and accuracy can be achieved by adjusting the network structures and related regressors. Although the CFCN has slightly high computational requirements, it is quite feasible to achieve online real-time processing for the current computing power.

7. Conclusions

In this paper, we propose an off-grid deep learning-based DOA estimation algorithm, which is based on the circularly fully convolutional network (CFCN). The backbone of this network is trained by the data set labeled by space-frequency pseudo-spectra and provides on-grid DOA proposals. Then, the regressor is developed to estimate the precise DOAs in the corresponding grid proposals. The simulation and semi-anechoic chamber experiment results show that under the conditions of single/dual sources with different band-limited/SNRs, the proposed algorithm is superior to existing methods in terms of generalization ability, resolution, and accuracy. Especially for the case of dual-source low-frequency and 12 dB SNR, CFCN can still distinguish multiple sources with an accuracy of 8.26°, while MUSIC and CNN-DOA fail at this time. Also, the -3 dB pseudo-beam width of CFCN reaches 90° at 50 Hz, which is much lower than the 160° width of MUSIC. In future work, we hope that this proposed method can be extended to the case of coherent signals.

Author Contributions: Conceptualization, W.Z., M.B. and X.L.; methodology, W.Z.; software, W.Z.; validation, Y.H. and J.T.; writing—original draft preparation, W.Z.; visualization, Y.H.; supervision, M.B.; project administration, M.B. and X.L.; funding acquisition, M.B. All authors have read and agreed to the published version of the manuscript.

Funding: This work was supported in part by the National Natural Science Foundation of China (Grant No. 11774379).

Institutional Review Board Statement: Not applicable.

Informed Consent Statement: Not applicable.

Data Availability Statement: All the simulation/experimental data and codes will be made available on request to the correspondent author's email with appropriate justification.

Conflicts of Interest: The authors declare no conflict of interest.

References

1. Krim, H.; Viberg, M. Two decades of array signal processing research: The parametric approach. *IEEE Signal Process. Mag.* **1996**, *13*, 67–94. [CrossRef]
2. Trees, H.L.V. Introduction. In *Optimum Array Processing*; John Wiley & Sons: Hoboken, NJ, USA, 2002; Chapter 1, pp. 1–16.
3. Fargeas, J.C.L.; Kabamba, P.T.; Girard, A.R. Cooperative Surveillance and Pursuit Using Unmanned Aerial Vehicles and Unattended Ground Sensors. *Sensors* **2015**, *15*, 1365–1388. [CrossRef] [PubMed]
4. Schmidt, R. Multiple emitter location and signal parameter estimation. *IEEE Trans. Antennas Propag.* **1986**, *34*, 276–280. [CrossRef]
5. Roy, R.; Kailath, T. ESPRIT-estimation of signal parameters via rotational invariance techniques. *IEEE Trans. Acoust. Speech Signal Process.* **1989**, *37*, 984–995. [CrossRef]
6. Vallet, P.; Mestre, X.; Loubaton, P. Performance Analysis of an Improved MUSIC DoA Estimator. *IEEE Trans. Signal Process.* **2015**, *63*, 6407–6422. [CrossRef]
7. Zhou, L.; Zhao, Y.; Cui, H. High resolution wideband DOA estimation based on modified MUSIC algorithm. In Proceedings of the 2008 International Conference on Information and Automation, Changsha, China, 20–23 June 2008; pp. 20–22.
8. Ying, H. Algorithm on high resolution DOA estimation under condition of unknown number of signal sources. *J. China Inst. Commun.* **2005**, *26*, 58–63.
9. Tuo, Q.; Feng, X.; Huang, J. Underwater Multi-source DOA Estimation under Condition of Unknown Number of Signals. *Comput. Simul.* **2009**, *21*, 981–983.
10. Wan, F.; Wen, J.; Liang, L. A Source Number Estimation Method Based on Improved Eigenvalue Decomposition Algorithm. In Proceedings of the 2020 IEEE 20th International Conference on Communication Technology (ICCT), Nanning, China, 28–31 October 2020; pp. 1184–1189.
11. Malioutov, D.; Cetin, M.; Willsky, A. A sparse signal reconstruction perspective for source localization with sensor arrays. *IEEE Trans. Signal Process.* **2005**, *53*, 3010–3022. [CrossRef]
12. Yang, Z.; Xie, L.; Zhang, C. Off-Grid Direction of Arrival Estimation Using Sparse Bayesian Inference. *IEEE Trans. Signal Process.* **2013**, *61*, 38–43. [CrossRef]
13. Zhang, Z.; Rao, B.D. Sparse Signal Recovery With Temporally Correlated Source Vectors Using Sparse Bayesian Learning. *IEEE J. Sel. Top. Signal Process.* **2011**, *5*, 912–926. [CrossRef]
14. Agatonovi, M.; Agatonović, M.; Stankovi, Z.; Stanković, Z.; Ov, N.D.; Doncov, N.; Sit, L.; Milovanovi, B.; Milovanović, B.; Zwick, T. Application of Artificial Neural Networks for Efficient High-Resolution 2D DOA Estimation. *Radio Eng.* **2012**, *21*, 1178–1186.
15. Agatonovic, M.; Stankovic, Z.; Milovanovic, B. High resolution two-dimensional DOA estimation using artificial neural networks. In Proceedings of the 2012 6th European Conference on Antennas and Propagation (EUCAP), Prague, Czech Republic, 26–30 March 2012; pp. 1–5.
16. Matsumoto, T.; Kuwahara, Y. 2D DOA estimation using beam steering antenna by the switched parasitic elements and RBF neural network. *Electron. Commun. Jpn. (Part I Commun.)* **2006**, *89*, 22–31. [CrossRef]
17. Agatonovic, M.; Stankovic, Z.; Milovanovic, I.; Doncov, N.S.; Sit, L.; Zwick, T.; Milovanovic, B. Efficient neural network approach for 2D doa estimation based on antenna array measurements. *Prog. Electromagn. Res.* **2013**, *137*, 741–758. [CrossRef]
18. Fonseca, N.; Coudyser, M.; Laurin, J.J.; Brault, J.J. On the Design of a Compact Neural Network-Based DOA Estimation System. *IEEE Trans. Antennas Propag.* **2010**, *58*, 357–366. [CrossRef]
19. Southall, H.; Simmers, J.; O'Donnell, T. Direction finding in phased arrays with a neural network beamformer. *IEEE Trans. Antennas Propag.* **1995**, *43*, 1369–1374. [CrossRef]
20. Pastorino, M.; Randazzo, A. A smart antenna system for direction of arrival estimation based on a support vector regression. *IEEE Trans. Antennas Propag.* **2005**, *53*, 2161–2168. [CrossRef]
21. Huang, H.; Gui, G.; Sari, H.; Adachi, F. Deep Learning for Super-Resolution DOA Estimation in Massive MIMO Systems. In Proceedings of the 2018 IEEE 88th Vehicular Technology Conference (VTC-Fall), Chicago, IL, USA, 27–30 August 2018; pp. 1–5.
22. Liu, Z.M.; Zhang, C.; Yu, P.S. Direction-of-Arrival Estimation Based on Deep Neural Networks with Robustness to Array Imperfections. *IEEE Trans. Antennas Propag.* **2018**, *66*, 7315–7327. [CrossRef]
23. Chen, X.; Wang, D.; Yin, J.; Wu, Y. A Direct Position-Determination Approach for Multiple Sources Based on Neural Network Computation. *Sensors* **2018**, *18*, 1925. [CrossRef]
24. Chakrabarty, S.; Habets, E.A.P. Multi-Speaker DOA Estimation Using Deep Convolutional Networks Trained with Noise Signals. *IEEE J. Sel. Top. Signal Process.* **2019**, *13*, 8–21. [CrossRef]
25. Yao, Y.; Lei, H.; He, W. A-CRNN-Based Method for Coherent DOA Estimation with Unknown Source Number. *Sensors* **2020**, *20*, 2296. [CrossRef]
26. Perotin, L.; Serizel, R.; Vincent, E.; Guerin, A. CRNN-Based Multiple DoA Estimation Using Acoustic Intensity Features for Ambisonics Recordings. *IEEE J. Sel. Top. Signal Process.* **2019**, *13*, 22–33. [CrossRef]
27. Liu, W. Super resolution DOA estimation based on deep neural network. *Sci. Rep.* **2020**, *10*, 19859. [CrossRef]
28. Ren, S.; He, K.; Girshick, R.; Sun, J. Faster R-CNN: Towards Real-Time Object Detection with Region Proposal Networks. *IEEE Trans. Pattern Anal. Mach. Intell.* **2017**, *39*, 1137–1149. [CrossRef]
29. Zhang, S.; Wen, L.; Lei, Z.; Li, S.Z. RefineDet++: Single-Shot Refinement Neural Network for Object Detection. *IEEE Trans. Circuits Syst. Video Technol.* **2021**, *31*, 674–687. [CrossRef]

30. Chung, H.; Seo, H.; Joo, J.; Lee, D.; Kim, S. Off-Grid DoA Estimation via Two-Stage Cascaded Neural Network. *Energies* **2021**, *14*, 228. [CrossRef]
31. Smith, L.N. Cyclical Learning Rates for Training Neural Networks. In Proceedings of the 2017 IEEE Winter Conference on Applications of Computer Vision (WACV), Santa Rosa, CA, USA, 27–29 March 2017; pp. 464–472.
32. Glorot, X.; Bengio, Y. Understanding the difficulty of training deep feedforward neural networks. In Proceedings of the Thirteenth International Conference on Artificial Intelligence and Statistics, Sardinia, Italy, 13–15 May 2010; pp. 249–256.
33. Stoeckle, C.; Munir, J.; Mezghani, A.; Nossek, J.A. DoA Estimation Performance and Computational Complexity of Subspace- and Compressed Sensing-based Methods. In Proceedings of the 19th International ITG Workshop on Smart Antennas (WSA 2015), Ilmenau, Germany, 3–5 March 2015; pp. 1–6.
34. He, K.; Sun, J. Convolutional neural networks at constrained time cost. In Proceedings of the Conference on Computer Vision and Pattern Recognition (CVPR), Boston, MA, USA, 7–12 June 2015; pp. 5353–5360.
35. Zhen, P.; Liu, B.; Cheng, Y.; Chen, H.B.; Yu, H. Fast video facial expression recognition by deeply tensor-compressed LSTM neural network on mobile device. In Proceedings of the 4th ACM/IEEE Symposium on Edge Computing, Washington, DC, USA, 7–9 November 2019; pp. 298–300.
36. Amert, T.; Otterness, N.; Yang, M.; Anderson, J.H.; Smith, F.D. GPU Scheduling on the NVIDIA TX2: Hidden Details Revealed. In Proceedings of IEEE Real-Time Systems Symposium (RTSS), Paris, France, 5–8 December 2017; pp. 104–115.
37. Avitabile, G.; Florio, A.; Coviello, G. Angle of Arrival Estimation Through a Full-Hardware Approach for Adaptive Beamforming. *IEEE Trans. Circuits Syst. II Express Briefs* **2020**, *67*, 3033–3037. [CrossRef]
38. Schuhmacher, D.; Vo, B.T.; Vo, B.N. A Consistent Metric for Performance Evaluation of Multi-Object Filters. *IEEE Trans. Signal Process.* **2008**, *56*, 3447–3457. [CrossRef]

sensors

MDPI

Article

Radar Signal Modulation Recognition Based on Sep-ResNet

Yongjiang Mao [1,2,3], Wenjuan Ren [1,2,*] and Zhanpeng Yang [1,2]

1 Key Laboratory of Network Information System Technology, Institute of Electronics, Chinese Academy of Sciences, Beijing 100190, China; myj@whu.edu.cn (Y.M.); zhanpengyang@mail.ie.ac.cn (Z.Y.)
2 Aerospace Information Research Institute, Chinese Academy of Sciences, Beijing 100190, China
3 University of Chinese Academy of Sciences, Beijing 100049, China
* Correspondence: renwj@aircas.ac.cn; Tel.: +86-134-6672-1075

Abstract: With the development of signal processing technology and the use of new radar systems, signal aliasing and electronic interference have occurred in space. The electromagnetic signals have become extremely complicated in their current applications in space, causing difficult problems in terms of accurately identifying radar-modulated signals in low signal-to-noise ratio (SNR) environments. To address this problem, in this paper, we propose an intelligent recognition method that combines time–frequency (T–F) analysis and a deep neural network to identify radar modulation signals. The T–F analysis of the complex Morlet wavelet transform (CMWT) method is used to extract the characteristics of signals and obtain the T–F images. Adaptive filtering and morphological processing are used in T–F image enhancement to reduce the interference of noise on signal characteristics. A deep neural network with the channel-separable ResNet (Sep-ResNet) is used to classify enhanced T–F images. The proposed method completes high-accuracy intelligent recognition of radar-modulated signals in a low-SNR environment. When the SNR is −10 dB, the probability of successful recognition (PSR) is 93.44%.

Keywords: radar modulation signal; time–frequency analysis; complex Morlet wavelet; image enhancement; channel-separable ResNet

Citation: Mao, Y.; Ren, W.; Yang, Z. Radar Signal Modulation Recognition Based on Sep-ResNet. *Sensors* **2021**, *21*, 7474. https://doi.org/10.3390/s21227474

Academic Editors: Moulay A. Akhloufi and Mozhdeh Shahbazi

Received: 28 September 2021
Accepted: 7 November 2021
Published: 10 November 2021

1. Introduction

Radar modulation signal (RMS) recognition is the basis of radar electronic countermeasures and electronic jamming, and is a necessary problem in electronic warfare [1]. With the use of various multi-band and full-coverage communication equipment, electronic interference and signal aliasing have appeared in space. This makes the electromagnetic environment more complicated [2,3], which brings difficulties to the recognition of RMS in low-SNR environments. Because RMS has good performance in complex spaces, it is important to improve the probability of the successful recognition of RMS in low-SNR environments.

For the traditional recognition methods of RMS, those methods based on signal characteristic parameter matching proposed by [4] and the judgment method based on the expert system proposed by [5] are used to identify RMS. Both of them are disturbed by human factors and are not stable. In [6], the principal component analysis (PCA) was used to extract radar signal features. In [7], the support vector machine (SVM) and T–F distribution images of signals were used to identify RMS. These methods, based on traditional machine learning, tend to select the characteristics of RMS manually. The methods require much a priori knowledge and struggle to meet the recognition requirements for the new radar systems and the variety of modulated signals [8]. With the development of artificial intelligence, deep learning has also been applied to the recognition of radar signals. Classification and recognition based on deep learning have many advantages. Without human assumptions and intervention about the features to be extracted, the deep neural network can effectively learn the features of the signals [9]. Deep learning can better resist the interference of noise in the extraction of signal features, thereby improving the generalization ability and accuracy in the identification of RMS [10]. In [11], the input of the classification network is a one-dimensional RMS sequence based on the time domain, frequency domain, and autocorrelation domain. By combining a convolutional neural network (CNN), a long short-term memory network (LSTM), and a deep neural network (DNN), the recognition of RMS was completed. However, this method completed the signal recognition on a one-dimensional basis. The method had no denoising processing in the case of one-dimensional signal sequences. The result was not ideal in a low-SNR environment of ≤ -10 dB. In [12], the improved AlexNet and Choi–Williams T–F distribution is used to complete the recognition of RMS based on two-dimensional images. This method did not denoise the T–F images, and the classification network was designed simply, resulting in weak anti-noise ability. This method cannot extract the features of T–F images well, and the PSR is low. In [13], the Cohen T–F transform method was used to convert the RMS to obtain T–F images, and the surface features of the image were extracted by CNN, and then the recurrent neural network was used to classify the T–F images of RMS. Because the images obtained by the Cohen transform had strong cross-terms, the characteristics of the signal experienced interference by the cross-terms, and the network could not extract the T–F image features well in low-SNR environments, resulting the PSR of the radar signal not being ideal when the SNR was -8 dB.

In this paper, in response to the difficulty in recognizing the RMS in low-SNR environments, we propose a novel method that combines the CMWT and the Sep-ResNet to accurately identify the RMS in a low-SNR environment of -13 dB. Through the enhancement T–F images by CMWT and the classification network of Sep-ResNet, the method in this paper has a strong anti-noise interference ability. This method can identify seven types of RMS, including normal signal (NS), linear frequency modulation (LFM), non-linear frequency modulation (NLFM), two-frequency shift keying (2FSK), two-phase shift keying (2PSK), four-frequency shift keying (4FSK) and four-phase shift keying (4PSK). The overall PSR of our method for seven types of RMS could reach 96.57% from 8 dB to -13 dB. In the case of SNR ≥ 2 dB, the PSR was 100%. In the case of low-SNR environments of -10 dB and -13 dB, the PSR was 93.44% and 88.24%, respectively.

Our major contributions are summarized as follows:

- The CMWT was introduced into the T–F analysis, which made it possible to avoid the interference of the T–F distribution cross-terms in the signal characteristics, and the T–F images had high T–F resolutions;
- The T–F images were denoised and enhanced through adaptive filtering and morphological methods. Effective morphological structural elements were designed to filter out noise on the T–F images and reduce the interference of noise in signal characteristics;
- By improving the residual unit structure, named Sep-ResNet, and multiple receptive fields for extracting features, as well as fusing multi-channel feature maps, the PSR was improved 2.51% in a low-SNR environment of -13 dB.

The remainder of this paper is organized as follows. Section 2 introduces the related work in the field of radar-modulation signal recognition. Section 3 introduces the recognition system framework and the proposed method, including the T–F analysis of CMWT, the T–F image-enhancing algorithm, and the improved classification network of Sep-ResNet. Section 4 shows the experimental data and results, and discusses the effectiveness of our method. Finally, Section 5 includes the conclusion of the whole work.

2. Related Work

In the past, many scholars have devoted themselves to exploring the automatic recognition system of RMS in applications. They have proposed several practicable approaches, making the system more intelligent, more robust, and less artificial. These achievements have pushed forward the development of the field of RMS recognition.

The recognition of RMS includes the extraction and classification of characteristics. In [7], the T–F analysis was used to extract signal characteristics. The SVM and auto encoder were used to classify the signal. This method introduced a slack variable to consider a non-linearly separable problem to find the best hyperplane so that the classification result had widths of the maximum margin. The method solved the problem of high-dimensional classification by selecting a suitable kernel function. The method identified RMS successfully, and the PSR was 82% in an SNR environment of −6 dB. The authors of [11] proposed a network combining CNN, LSTM, and DNN. They successfully identified six types of RMS when the SNR was from −14 to 20 dB. This method extracted the characteristics of the signal as the original one-dimensional sequence in the time domain, the fast Fourier transform sequence in the frequency domain, and the result of signal autocorrelation in the autocorrelation domain. A CNN was used to extract the surface features of the signal in different domains, and the features extracted by the CNN were used as the input of the LSTM, and the DNN was used to classify the characteristics of the signal. The length of the signal sequence extracted by this method needed to be set in advance, and the sequence length was different in the time domain, frequency domain, and autocorrelation domain. Under the preset optimal sequence length, the PSR was about 90% when the SNR was −6 dB. This method required preprocessing to obtain the optimal sequence length of a signal for a specific domain, which had limitations. Moreover, the features of one-dimensional sequence were not as rich as that of two-dimensional T–F images, and the recognition accuracy was not as high as that obtained in [14]. When signal features were extracted by the one-dimensional sequence, some feature parameters needed to be manually selected for the data. The T–F analysis method can overcome the shortcomings of the Fourier transform and reflect the signal characteristics in the two-dimensional space of T–F. By obtaining the T–F images, the order of appearance of each frequency component can be well distinguished. The T–F analysis method can adequately extract the characteristics of non-stationary signals, such as RMS. In [12], the Choi–Williams distribution (CWD) was a method of T–F analysis that was used to extract the features of a modulated signal $x(t)$. The expression is as follows:

$$C_x(t,\omega) = \frac{1}{4\pi^2}\iiint_{\infty} \phi(\tau,v)e^{-j(vt+\omega\tau)}x\left(u+\frac{\tau}{2}\right)x^*\left(u-\frac{\tau}{2}\right)e^{jvu}dvd\tau du \tag{1}$$

where t and ω are the time and frequency coordinates, respectively. $x^*(t)$ is the conjugate expression of $x(t)$. $\phi(\tau,v)$ and σ are the kernel function and filter bandwidth, respectively. In [12], the kernel function was a Gaussian kernel $\phi(\tau,v) = exp\left[-\frac{(\tau v)^2}{\sigma}\right]$ and $\sigma = 1$. The larger σ, the better aggregation of signal on the T–F images. However, a large σ will bring more serious cross-terms by $x(t)$ times $x^*(t)$. Cross-terms will reduce the quality of T–F images. In the improved AlexNet, the dropout is added to reduce overfitting, the size of the convolution kernel is modified, the receptive field of convolution is increased, and the fully connected layer is reduced to decrease the weight parameters. When the SNR is −2 dB, the recognition rate can reach 80%. In [15], the signal characteristics were extracted through the improved phase difference short-time Fourier transform (STFT). The STFT of the signal is expressed as follows:

$$STFT(t,f) = \int_{-\infty}^{+\infty} x(\tau)w(\tau-t)exp(-j2\pi f\tau)d\tau \tag{2}$$

where $w(t)$ and $x(t)$ are fixed-width window functions and signals to be analyzed, respectively. Through the continuous sliding of the $w(t)$ window, the Fourier transform was performed in the window to extract the signal characteristics. It reduced the influence of noise by increasing the order of the phase difference. However, an increase in the phase difference order would increase the complexity of the algorithm. In addition, the STFT used a fixed-width window function to extract signal characteristics, resulting in low T–F resolution. When the SNR was −6 dB, the PSR of recognition result was 93%. In [14], by improving the kernel function of CWD, the original Gaussian kernel function was changed to $\phi(\tau, v) = exp\left(-\frac{\alpha\tau^2+\beta v^2}{\sigma}\right)$, which solved the problem of $\phi(\tau, v) = 0$ when $\tau = 0$ or $v = 0$. To a certain extent, it was able to reduce the interference of the cross-terms in the T–F distribution images. Furthermore, the T–F images were denoised by 2D-Wiener filtering. This method improved the PSR to 96% when the SNR was −6 dB, and the PSR was found to be 78% when the SNR was −8 dB. Due to the existence of cross-terms, this method does not perform well in low-SNR environments. In [16], the wavelet transform T–F analysis method was used to extract the characteristics of the signal. The wavelet transform represents an improvement upon STFT; the signal to be analyzed is decomposed into a series of superpositions of wavelet functions. The wavelet functions are obtained from the mother wavelet through translation and scaling transformation. The mother wavelet is stretched at low frequencies and compressed at high frequencies, and has the characteristics of multi-scale refinement. In order to restore the characteristics of the signal on the time and frequency scale, the wavelet transform method continually approximates the signal that is to be analyzed.

A large number of RMS recognition systems have been designed, but the identification of a variety of RMS with high accuracy in low-SNR environments of ≤ -10 dB remains a challenging problem.

3. System Framework and Method

This paper designs an intelligent method to identify RMS with high accuracy in low-SNR environments. The first step is to transform the RMS into two-dimensional T–F images using CMWT. In the second step, the T–F images are grayed out, and the T–F images are denoised and enhanced through adaptive filtering and morphological processing to reduce the interference of noise on signal characteristics. In the last step, the enhanced T–F images are fed into the Sep-ResNet to train the model, and the trained model is used to accurately predict the type of RMS. The system framework is shown in Figure 1.

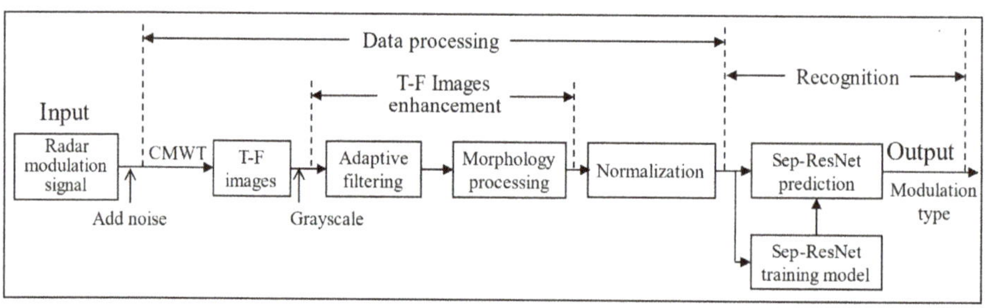

Figure 1. RMS recognition system framework.

The received RMS model is as follows:

$$x(t) = s(t) + n(t) \tag{3}$$

where $x(t)$ represents the received RMS, and t is the time variable; $s(t)$ and $n(t)$ represent the transmitted RMS and the received noise, respectively. The transmitted RMS is as follows:

$$s(t) = Arect\left(\frac{t}{T}\right)e^{-j(2\pi f_c t+\varphi(t)+\varphi_0)} \tag{4}$$

$$rect\left(\frac{t}{T}\right) = \begin{cases} 1, |t/T| \leq 1/2 \\ 0, |t/T| \geq 1/2 \end{cases} \tag{5}$$

where A represents the signal amplitude, and $A = 1$ during simulation. The $rect(\bullet)$ is the rectangular threshold function as shown in Equation (5), T is the pulse width of the signal,

f_c is the signal carrier frequency, ϕ_0 is the initial phase of the signal, and $\phi(t)$ is the different phase function, which determines the different signal modulation modes.

To simulate the received noise in the real channel of signals, Gaussian noise, white noise, narrow-band Gaussian noise, and carrier frequency random disturbance are added [17]. Gaussian noise and white noise are some of the most common noises in actual channels. The probability density function of Gaussian noise obeys a normal distribution, with a mean value of 0 and a variance of 1. The power spectral density of white noise is a constant over the entire bandwidth and obeys a uniform distribution, $S(\varpi) = N_0/2, \varpi \in (-\infty, +\infty)$. When adding white noise, it only needs to be added within the actual signal bandwidth. Often, in a real radar communication system, a band-pass filter of the target signal bandwidth is added at the receiving end. Because the communication frequency of the radar signal is high, much larger than the bandwidth of the band-pass filter, the generation of Gaussian narrow-band noise occurs. Narrow-band Gaussian noise obeys the Rayleigh distribution on the random envelope and obeys the uniform distribution on the phase. It is generally a stationary random process. Its mathematical model is $\overline{n(t)} = n_c(t)cos\overline{\varpi}_c t - n_s(t)sin\overline{\varpi}_c t$, where $\overline{n(t)}$ represents the average power of narrow-band Gaussian noise, $n_c(t)$ is the co-directional component of $\overline{n(t)}$, and $n_s(t)$ is the orthogonal component of $\overline{n(t)}$. The mean value of each component is $E\left[\overline{n(t)}\right]= E[n_c(t)]= E[n_s(t)\right] = 0$, and the variance is $\sigma^2{}_n$=1. The jitter component of the carrier frequency is set to a random number in the range of 0 to 0.05 multiplied by the carrier frequency to simulate the error of the actual transmission carrier frequency and the interference received during transmission.

The T–F analysis method can transform one-dimensional signal sequences into two-dimensional T–F images to obtain RMS characteristics. Through the image enhancement method, the enhanced images are fed to the Sep-ResNet to complete the RMS recognition.

3.1. *Complex Morlet Wavelet Transform*

Wavelet transform (WT) is a T–F analysis method. WT decomposes the original signal into a series of superpositions of wavelet functions through mother wavelet translation and scaling transformation, which solves the problem that the fixed-width window function does not change with frequency in the STFT [15]. WT does not involve the conjugate multiplication of the signal itself, which avoids the occurrence of cross-terms in [12,15]. WT can obtain high T–F resolution. The mathematical model is as follows:

$$CWT_x(a,b) = \langle x(t), \phi_{a,b}(t)\rangle = \int_{-\infty}^{+\infty} x(t)\phi^*{}_{a,b}(t)dt = \frac{1}{\sqrt{a}}\int_{-\infty}^{+\infty} x(t)\phi^*\left(\frac{t-b}{a}\right)dt \quad (6)$$

where $x(t)$ is the signal to be analyzed, and $\phi^*{}_{a,b}(t)$ is a series of wavelet functions after the mother wavelet is translated and stretched. The transformation of the mother wavelet is as follows:

$$\phi^*{}_{a,b}(t) = \frac{1}{\sqrt{a}}\phi\left(\frac{t-b}{a}\right) \quad (7)$$

where a is the scale expansion factor and $a \neq 0$. When a increases, $\phi_{a,b}(t)$ will widen and the amplitude will decrease, showing that the wavelet is caused by the compression of the amplitude and the stretching of the width, corresponding to the analysis of low-frequency signals. When a decreases, the wavelet becomes narrower and the amplitude increases, corresponding to the analysis of high-frequency signals. b is the time shift factor, which changes the center position of the wavelet. WT has adaptive capabilities. By selecting the appropriate mother wavelet (symmetry, orthogonality, and similarity), more detailed features can be obtained in the T–F resolution.

In the WT of our method, the complex Morlet function was chosen as the mother wavelet. CMWT is a complex exponential sinusoidal Gaussian wavelet, which has symmetry and non-orthogonality. CMWT has a good ability to extract the local characteristics of signals in the T–F domains and improves the resolution of T–F images. The complex Morlet mother wavelet model is given in Equation (8):

$$\phi(t) = exp\left(\frac{-t^2}{2}\right)exp(j\omega_0 t) \tag{8}$$

$$\Phi(\omega) = \sqrt{2\pi}exp\left(\frac{-(\omega-\omega_0)^2}{2}\right) \tag{9}$$

Equation (9) is the Fourier transform of complex Morlet wavelet, where ω_0 represents the center frequency. The complex Morlet mother wavelet $\phi(t)$ is divided into two parts: the real part and the imaginary part. A series of wavelet functions $\phi^*{}_{a,b}(t)$ can be obtained after the translation and scaling transformation using Equation (7). Incorporating Equation (8) into Equation (6), $CWT_x(a,b) = CWT_R + jCWT_i$. By undertaking T–F analysis, the added imaginary part of the complex Morlet wavelet can express more changeable phase information on the original signal [18]. The complex Morlet wavelet has non-orthogonality and Gaussian adjustment, which make it possible to obtain T–F images of high time and frequency resolution through a series of variable scale wavelet functions. The CMWT avoids the interference of cross-terms in signal characteristics in low-SNR environments and improves the quality of T–F images. CMWT is suitable for the T–F analysis of RMS and can obtain clear T–F images. Figure 2 shows the T–F images of seven types of radar modulation signals of CMWT without adding noise.

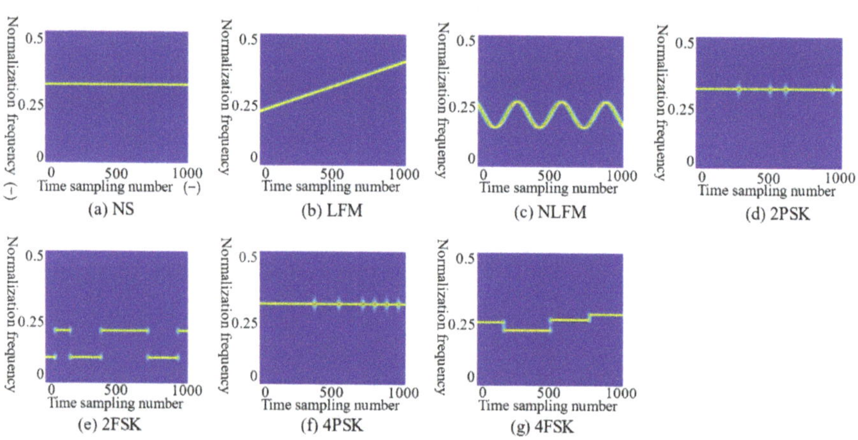

Figure 2. The T–F images of seven types of radar modulation signals. The images from **(a)**–**(g)** are the T–F images of NS, LFM, NLFM, 2PSK, 2FSK, 4PSK, and 4FSK without noise. The abscissa of the image is the number of sampling points. The ordinate is the normalized frequency, which is the signal frequency divide by the sampling frequency. According to the Nyquist sampling theorem, the sampling frequency must be greater than twice the signal frequency to avoid signal aliasing. The normalization frequency is between 0 and 0.5.

In the absence of noise, the T–F images obtained by CMWT can clearly obtain the characteristics of different RMSs. The T–F images have no cross-terms interference, the signal characteristics will not be distorted, and the T–F resolution is high. Generally, the actual received radar signal will contain a lot of noise. The SNR will seriously affect the performance of the signal characteristics on the T–F images. The SNR is defined as follows:

$$SNR_{dB} = 10log_{10}(P_s/P_n)$$

$$P_s = \frac{1}{N}\sum_{t=0}^{N-1}|s(t)|^2 \tag{10}$$

$$P_n = \frac{1}{N}\sum_{t=0}^{N-1}|n(t)|^2$$

where $s(t)$ is the modulated radar signal, $n(t)$ is the noise signal, P_s is the power of the signal, P_n is the noise power, and N is the signal length. The lower SNR, the greater the noise power, and the characteristics of signal are submerged by noise on the T–F images. Figure 3 shows T–F images of the noise with different SNR added to LFM signal by CMWT.

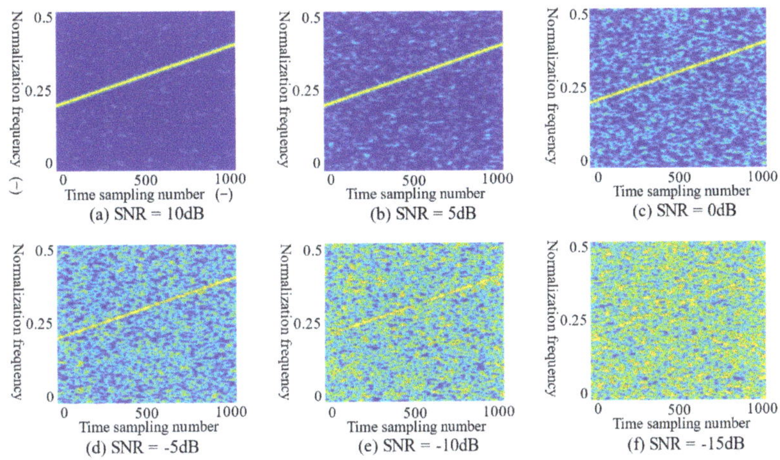

Figure 3. The T–F images of LFM signal in different SNR. The images from **(a)**–**(f)** are LFM T–F images under 10 dB, 5 dB, 0 dB, −5 dB, −10 dB and −15 dB noise, respectively.

In Figure 3, the SNR values are 10, 5, 0, −5, −10 and −15 dB, respectively. As the SNR decreases, although the characteristics of the LFM signal are still preserved, the quality of the T–F images deteriorates, and the signal characteristics are overwhelmed by noise. The difficulty of identifying the RMS will increase. Therefore, in this paper, the T–F images of the signal are properly denoised and enhanced before feeding into the CNN. T–F image enhancement can reduce the interference of noise, while better retaining the original characteristics of the signal. In addition, the enhancement algorithm improves the recognition rate of the RMS.

3.2. T–F Image Enhancement

The T–F analysis method is usually used to extract RMS characteristics and obtain T–F images. Before recognition, it is necessary to denoise and enhance the T–F images, which includes the following steps: image cropping and gray-scale, adaptive filtering [19], morphology processing [20], and normalization. Normalization involves the down-sampling of T–F images to reshape the images into a 64 × 64-pixel form. The enhancement of the T–F images will affect the Sep-ResNet extraction features and make possible the identification of the RMS. Algorithm 1 shows the enhancement algorithm for the adaptive filtering and morphological processing of T–F images.

The eroding operation is $A \odot S_1 = \{z|(S_1)_z \subseteq A\}$, and S1 is the structural element for eroding. The shape of S1 is designed to be round-like, which can better eliminate round-like noise points generated on the T–F images. The dilating operation is $A \oplus S_2 = \{z|(S_2)_z \cap A \neq \varnothing\}$, and S2 is used as the structural element of dilating, which can enhance the characteristics of the RMS on the T–F images. The Opening Operation is first eroding and then dilating, as in formula: $(A \odot S_1) \oplus S_2$. The enhancement algorithm for T–F images is shown in Algorithm 1.

Algorithm 1 Enhancement of T–F images

```
Input: Grayscale images before enhancement
Adaptive Filter:
Step A:
for origin_pixel in images:
    Initialize A1, A2, window_size = 5
    A1 = median_pixel–min_pixel, A2 = median_pixel–max_pixel
    if A1 > 0 and A2 < 0: to Step B
    else: Increase the window size
    if window_size > (max_window = 13): return median_pixel
Step B:
    Initialize B1, B2
    B1 = origin_pixel-min_pixel, B2 = origin_pixel-max_pixel
    if B1>0 and B2<0: return origin_pixel
    else: return median_pixel
end for
Morphology Processing:
Initialize the structure_element: S1, S2
for pixel in images:
    the S1 Erode pixel
    the S1 and S2 Opening Operation pixel for twice
    the S1 Erode pixel
end for
Output: Grayscale images after enhancement
```

where S1 and S2 are structural elements, as shown in Figure 4:

0	1	1	0
1	0	0	1
1	0	0	1
0	1	1	0

(a) Structure element S_1

0	0	1	1
1	1	1	1
1	1	1	1
1	1	0	0

(b) Structure element S_2

Figure 4. The morphological structural elements.

3.3. *Classification Network of the Sep-ResNet*

After the RMS is transformed by CMWT and enhanced, these T–F images were fed into the Sep-ResNet for the extraction of features and the classification of images to complete the recognition of RMS. At this point, the traditional neural networks will encounter some problems such as feature information loss, gradient vanishing, and gradient exploding as the network depth increases. It is hard to design a deeper network to extract the deep features of the images [21]. However, this paper uses the idea of residual learning, and introduces residual blocks and shortcut channels [22,23]. The classification network Sep-ResNet was designed, which solves problems of the loss of feature information, gradient vanishing, and gradient exploding. The Sep-ResNet can be designed to extract richer image features with greater depth. The Sep-ResNet structure was designed as shown in Figure 5.

In Figure 5, the Pre Conv uses three 3 × 3 convolution kernels to convolve the input image. The first one uses the convolution kernel step size s = 2 to conduct down sampling. The remaining two convolutions have the same receptive field as the original 7 × 7 convolution kernel, but the number of parameters is reduced by 45%. Furthermore, the features extracted by the smaller convolution kernel are more refined. Stage 1 includes two parts: the Down Sampling and the Residual Block. The Down Sampling part first adjusts the number of channels in Path A by using a 1 × 1 convolution kernel and then uses a kernel of s = 2 at the 3 × 3 convolution. Reference [22] uses s = 2 for convolution at the 1 × 1 convolution kernel, which will completely lose 50% of the information of the feature map. At Path B, the convolution operation with a size of 1 × 1 and s = 2 is also replaced with an average pooling with a size of 2 × 2 and s = 2. The above adjustment can ensure that the information of the feature map will not be lost when conducting the Down Sampling. The output of Down Sampling involves the width and height of the feature map being reduced by half, and the number of output channels is increased. Another part of the improved Residual Block is shown in Figure 6.

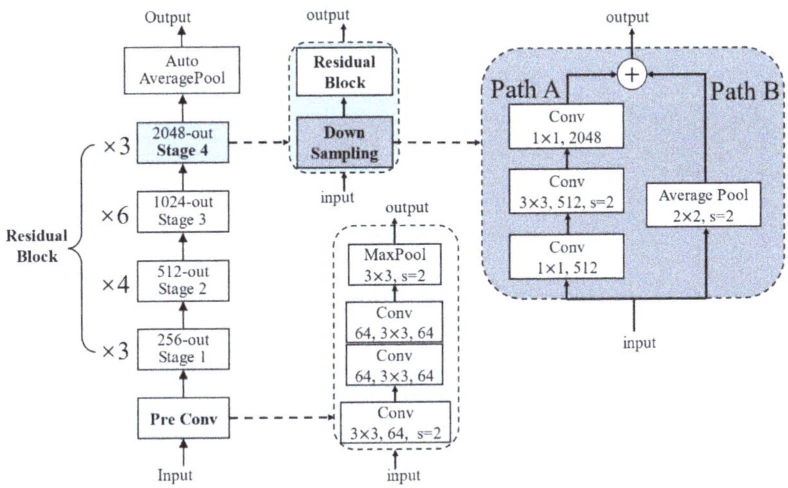

Figure 5. The Sep-ResNet structure developed in this paper.

In Figure 6, the three parameters for convolution in the rectangle represent the input channel, kernel size, and output channel. The input channel of the residual block is M feature maps. The first layer of convolutional kernel size is 1 × 1 for convolution, and all of the four out channels are obtained M/16 feature maps. The obtained feature maps use the activation function of Leaky ReLu and conduct convolution with the kernel size of 3, 5, 7 and 9, respectively. The different sizes of convolutional kernels make it possible to obtain different receptive fields and extractions of the multi-scale features. The M/16 feature maps of four channels are stacked in a concatenated manner on the channel to obtain M/4 feature maps. This allows the obtaining of fused feature maps. The obtained feature maps use kernels size is 1 × 1 for convolution and mapping to obtain M feature maps. Finally, the obtained M feature maps and the original M feature maps before convolution are correspondingly added to each channel to obtain a residual block. This method has a larger receptive field than the original residual structure, and the extracted features are more abundant. Although the increase in residual block parameters is caused by increasing the receptive field of the kernels, there is no increase in the number of convolutions. The residual block only separates the channels instead of increasing the number of channels. The residual block output of Sep-ResNet is X_{n+1}, as follows:

$$X_{n+1} = X_n + \sum_{i=1}^{4} F_i(X_n, W_n) \quad (11)$$

where $F_i(X_n, W_n)$ is the output of X_n after convolution in the i-th channel. The improved residual block can extract the features of T–F images with multiple scales and multiple receptive fields. The Sep-ResNet extracted features are more abundant in low-SNR environments, which increases the recognition accuracy of the RMS. Stage 1 repeats the residual block 3 times, and the output of Stage 1 is the input of Stage 2. Stage 2 repeats the above residual block 4 times, and the output of Stage 2 is the input of Stage 3, and so on. The structure of Sep-ResNet has a total of 53 layers of the network, including 51 convolutional layers, one auto average-pool layer, and one fully connected layer. In the Batch Normalization (BN) layer, the data of a mini batch are normalized to uniformly distributed data with a mean of 0 and a variance of 1, which can better prevent the problems of overfitting and vanishing of gradients [24]. The activation function is Leaky ReLu: $x = max\{0.01x, x\}$.The initial learning rate (LR) of the training model is LR = 0.001, and the learning rate is adjusted to LR = LR * 0.5 every 20 epochs. As the number of iterations increases, the learning rate decreases, for a total of 100 epochs. The auto average pool makes the feature maps of any width and height become size = 1 × 1 feature maps and then maps them to seven types of radar modulation signals through a full connection. The final Softmax layer maps the output probabilities of each type to between 0 and 1.

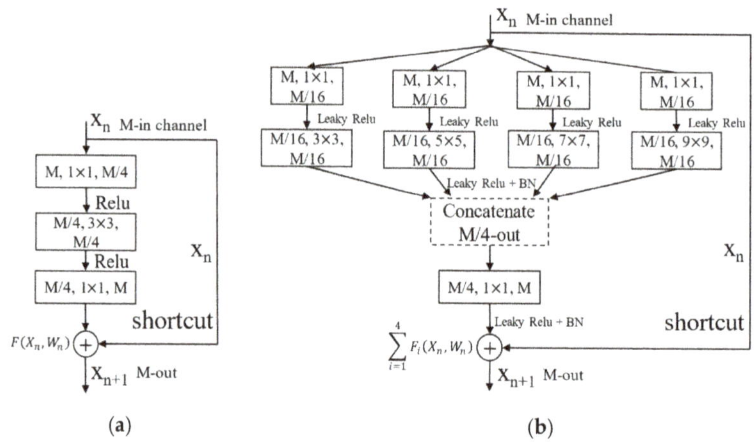

Figure 6. Comparison of the residual block. (**a**) is the original residual block of ResNet50 in [22]; (**b**) is the improved residual block in this paper. The residual block of (**b**) has multi-channel feature fusion and the multiple receptive fields.

The loss function in this experiment is cross-entropy, and label smoothing [25] is introduced to reduce the over-fitting of the model, as shown in Equation (12).

$$y' = (1 - \varepsilon) \times y + \frac{\varepsilon}{K} \tag{12}$$

$$loss = -\sum_{i=0}^{n} [y' \times logp + (1 - y') \times log(1 - p)] \tag{13}$$

where y is the original label (named the hard label) and y' is the smoothing label (named the soft label), the allowable error rate ε is 0.1, the number of categories K is 7, p is the prediction result, and *loss* is the error between the prediction result and the given truth label. The hard label only has values of 0 and 1. If there is a label error, the model will learn the features of the image in the wrong direction, resulting in poor model generalization and easy overfitting. The soft label allows a certain error tolerance rate, which can alleviate the overfitting of the model. Considering that the signal characteristics are not obvious and are severely polluted by noise when the SNR is low, the signals can easily be misclassified in low-SNR environments. The addition of label smoothing also causes the model have a certain anti-noise ability, which alleviates the problem of the loss function of cross-entropy being easily overfitted. Finally, back propagation is used to update the weight parameters of each layer to complete the training of the RMS recognition model.

4. Experimental Results and Discussion

In this part, the experimental dataset and the results are given. According to the framework of Figure 1, the method processes radar signals to obtain T–F images, and the separable Sep-ResNet channels classify the enhanced T–F images. Our method accurately identified seven radar modulation signals in low-SNR environments of ≤ -10 dB.

4.1. Experimental Dataset

The experimental environment used for the generation of the simulation signals and T–F images was MATLAB2018a. The deep learning framework for training and predicting model was Pytorch1.5.

The experimental data comprised the radar signal of seven modulation modes generated by simulation, namely NS, LFM, NLFM, 2FSK, 2PSK, 4FSK, and 4PSK. The frequency of the modulation signal was the normalized frequency, which was the signal frequency divided by the sampling frequency. Table 1 shows the specific modulation method, carrier frequency, bandwidth and Baker codes of RMS.

Table 1. The specific simulation parameters of RMS.

Signal Type	Parameter	Range
NS	Carrier frequency f_c	(180~230) MHz
LFM	f_c Bandwidth $\triangle f$	(180~230) MHz (40~60) MHz
NLFM	f_c $\triangle f$	(180~230) MHz (20~40) MHz
2PSK	f_cBarker codes Symbol width	(180~230) MHz Length = {7, 11, 13} 0.04 µs
2FSK	f_{c1}, f_{c2} Barker codes Symbol width	(180~200), (280~300) MHz {7, 11, 13} 0.04 µs
4PSK	f_c Baker codes Symbol width	(180~230) MHz {5, 7, 11, 13} 0.03 µs
4FSK	f_{c1}, f_{c2} f_{c3}, f_{c4} Baker codes Symbol width	(180~190), (210~220) MHz (240~250), (270~280) MHz {5, 7, 11, 13} 0.03 µs

Note: The pulse width for each type signal is 0.5 µs and sampling frequency is 2 GHZ.

In Table 1, the simulated RMS parameters are the dynamic range [9]. To simulate the actual received signal, the modulated signal had the range of a certain parameter, and the noise was added. The SNR of radar modulation signals was -13, -10, -7, -4, -1, 2, 5 and 8 dB, respectively. An SNR point was taken every 3 dB, for a total of eight SNR points. A total of 400 T–F images were taken for each SNR point, and each radar modulation signal contained 3200 T–F images. There were a total of 22,400 T–F images in this dataset. Overall, 60% of the dataset was used as the training set—a total of 13,440 T–F images; 40% was used as the test set—a total of 8960 T–F images.

The T–F images shown in Figure 7 could obtain the characteristics of the NLFM signal well, but the characteristics of the T–F images were gradually overwhelmed by noise as the SNR decreased. When the SNR was ≥ -1 dB, the signal characteristics were clear in the T–F images. When the SNR was ≤ -4 dB, the characteristics became vague. When the SNR was -10 dB, the noise seriously interfered with the signal characteristics. Furthermore, when the SRN was -13 dB, the characteristics were completely overwhelmed by noise.

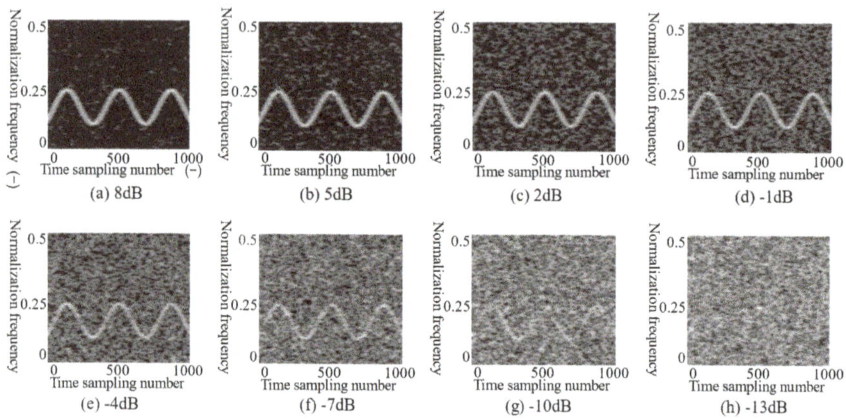

Figure 7. T–F images of the NLFM signal in different SNR from −13 dB to 8 dB in the training set. The images from (**a**)–(**h**) are NLFM grayscale T–F images under 8 dB, 5 dB, 2 dB, −1 dB, −4 dB, −7 dB, −10 dB and −13 dB noise, respectively.

4.2. Experimental Results

The T–F images of the RMS in the training set were fed into the CNN after denoising and enhancement to the train models for the recognition of the RMS.

4.2.1. Verification of the Effectiveness of CMWT

In this paper, the RMS used three T–F analysis methods, including STFT [15], CWD with an improved kernel function [14], and CMWT, to obtain the T–F images. The T–F images were fed into AlexNet, the improved AlexNet [12], and ResNet50 [22], and our Sep-ResNet for the classification of T–F images, which used an enhancement algorithm, is shown in Algorithm 1. The probability of the successful recognition (PSR) of different T–F analysis methods were compared, as shown in Figure 8:

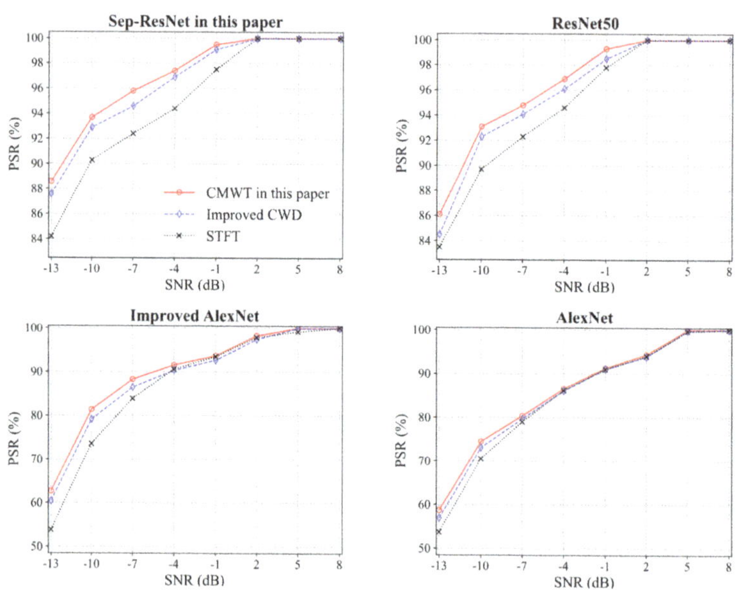

Figure 8. The PSR of three T–F analysis methods in different classification networks.

In Figure 8, the red line is the CMWT in this paper, the blue line is the improved CWD, and the black line is the STFT. The T–F analysis method of CMWT had the highest PSR of the four networks, while the CWD method with an improved kernel function had the middle PSR, and the STFT method had the lowest PSR. Because CMWT had the advantage of containing wavelets with transformable scales, there were no cross-terms and phase information in the imaginary part. It showed anti-noise ability and effectively extracted the RMS characteristics. It was found that the improved T–F analysis method can improve the PSR of the RMS. For the four CNNs, the overall PSR values of the three T–F analysis methods of the CMWT, CWD, and STFT are shown in Table 2.

Table 2. The overall PSR of RMS.

	STFT, %	CWD, %	Our CMWT, %
Our Sep-ResNet	94.59	96.05	96.57
ResNet50	94.53	95.43	95.89
Improved Alexnet	86.25	88.12	89.31
Alexnet	84.02	84.93	85.52

In Table 2, it can be seen that the T–F analysis method of the CMWT and the Sep-ResNet classification network had the highest PSR of the seven radar signals. The overall PSR of this method was 96.57%.

4.2.2. Verification of the Effectiveness of Sep-ResNet

To verify the effectiveness of the classification network of the proposed Sep-ResNet, the enhanced T–F images obtained via the CMWT were fed into the Sep-ResNet for training of the recognition model. The validation dataset was composed of 20% of the test dataset. Because the CMWT was able to better extract the RMS characteristics to obtain clear and distinguishable T–F images, the loss function successfully converged as the number of epochs increased, as shown Figure 9.

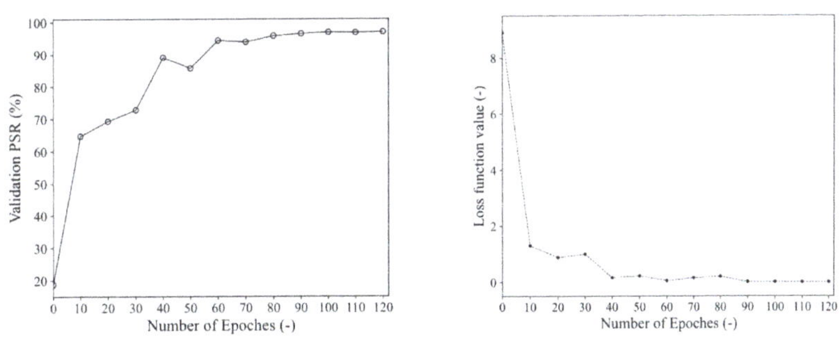

(a) (b)

Figure 9. (**a**,**b**) are the curves of the validation accuracy and loss function value, respectively, obtained during the training of Sep-ResNet. The abscissa is the number of epochs. An epoch represents all samples in the training set being trained once. The ordinate of (**b**) is the specific loss function value under the current epoch.

In the experiment, we conducted a classification comparison of AlexNet, ResNet50, VGGNet16, Inception-v3, and the backbone of U-Net models. The AlexNet structure was consistent with that reported in [12]. The residual structure of ResNet50 [22] is shown in Figure 6. VGGNet16 [26] increases the number of the output channels and uses max pooling to reduce the size of the feature map. VGGNet16 has a total of 16 convolutional layers, which is deeper than AlexNet. By decomposing large convolution filters such as 5×5 into two 3×3 filters, Inception-v3 [25] has improved performance on VGGNet. The parameters are reduced by 28%. Furthermore, Inception-v3 uses an asymmetric method to decompose the spatial convolution filter. The n × n size filter is decomposed into 1 × n and n × 1 filters to further reduce the parameters and improve performance. The main idea of U-Net is to use a feature pyramid network for feature fusion [27]. The U-Net described in this paper used a 3 × 3 size filter to convolve a T–F image of (1, 64, 64) size to obtain a feature map of (64, 32, 32) size, where the dimensions were the output channel, image width, and image height. These features were deconvolved again to obtain a feature map of size 128, 16, 16, the feature map was copied, and it was named as $m1$. Then, the convolution was continued to obtain the feature maps of size (256, 8, 8) and (512, 4, 4), which were named $m2$ and $m3$, respectively. $m3$ was upsampled to obtain feature maps of 512, 8, 8 size, and concatenated with $m2$ on the channel to obtain sizes of 768, 8, 8. Then, up-sampling allowed the feature maps of 768, 16, 16 size to be obtained, and these were added to $m1$ in order to obtain the feature maps of 896, 16, 16 size. Finally, four 3 × 3 size convolution filters and fully connected layers were mapped to seven classification nodes, corresponding to seven types of radar signals. The key to U-Net is the integration of deep and shallow features to better express image features.

During the experiment, the data processing method was consistent with the use of CMWT and the enhanced algorithm in Algorithm 1. The T–F images were fed to AlexNet, improved AlexNet, VGGNet16, Inception-v3, U-Net, ResNet50, and our Sep-ResNet, respectively. The PSR values of seven CNNs are shown in Figure 10.

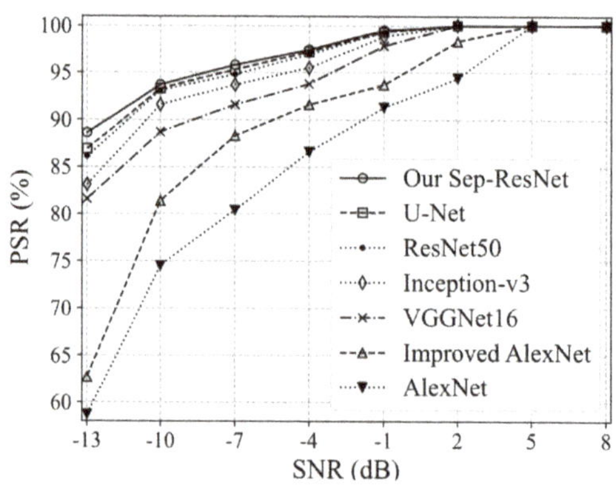

Figure 10. The PSR of seven CNNs.

The proposed Sep-ResNet had the highest PSR for the above seven radar modulation signals. Under SNR = −13 dB, the PSR of the Sep-ResNet was still 88.24%. The PSR of the AlexNet, improved AlexNet, VGGNet16, Inception-v3, ResNet50 and U-Net were 58.36%, 62.31%, 81.62%, 83.17, 85.92, and 86.81%, respectively. Because the residual network was able to design the deep enough network structure to extract T–F image features, we observed that Sep-ResNet has a multi-scale receptive field and multi-channel feature fusion, which can extract richer features of T–F images. It was found that Sep-ResNet has the best classification effect in low-SNR environments, improves the PSR, and has the anti-noise ability. Furthermore, the Sep-ResNet model showed a recognition rate of 100% for the above seven radar signals when the SNR was ≥2 dB.

Figure 11 shows the PSR of the Sep-ResNet model for seven radar modulation signals at different SNRs. Among the seven modulation signals, NLFM, 2FSK, 4FSK, and LFM had higher recognition rates. When the SNR was −13 dB, their average PSR was able to reach 93.45%, and when the SNR was ≥−4 dB, the PSR could reach 100%. The average PSR of the remaining signals—NS, 2PSK, and 4PSK—was 81.36% when the SNR was −13 dB. Their PSR was lower because the signal characteristics were relatively similar on the T–F images.

Furthermore, with the decrease in SNR, the signal characteristics were overwhelmed by noise, resulting in low recognition accuracy (see Figure 12).

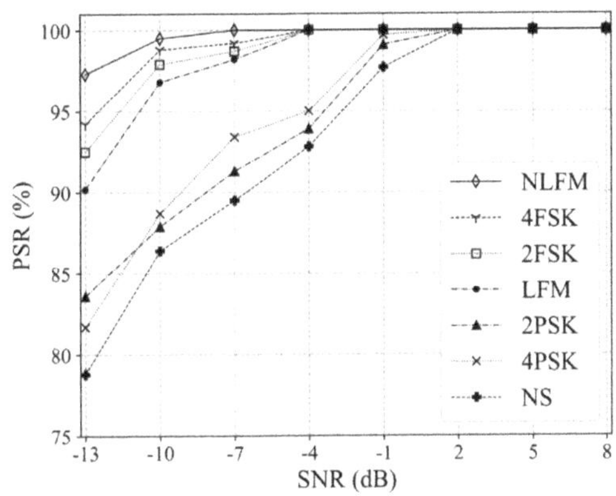

Figure 11. The PSR of seven RMSs in Sep-ResNet.

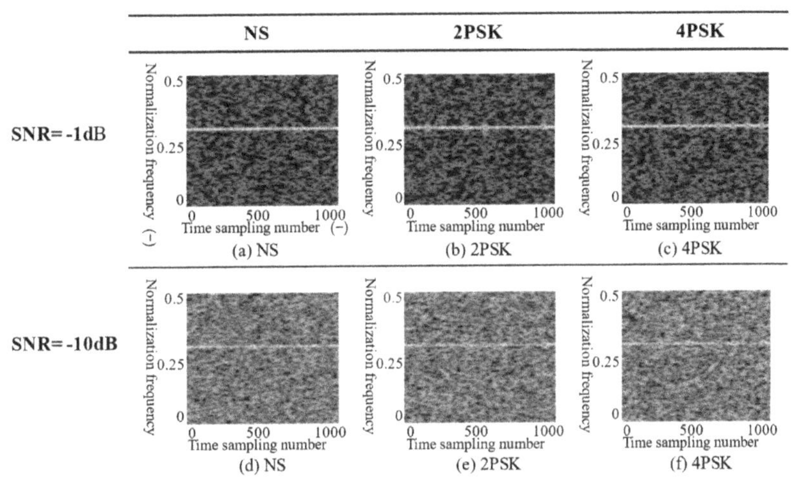

Figure 12. The T–F images of NS, 2PSK, and 4PSK when the SNR was −1 dB and −10 dB, respectively. The (**a**,**d**) are the grayscale T–F images of the signal NS under −1 dB and −10 dB noise, respectively. The (**b**,**e**) are the grayscale T–F images of the signal 2PSK under −1 dB and −10 dB noise, respectively. The (**c**,**f**) are the grayscale T–F images of the signal 4PSK under −1 dB and −10 dB noise, respectively.

The input of the confusion matrix was 100 random enhanced T–F images from the test dataset by CMWT when the SNR was −4 dB. The output on the diagonal was the recognition recall rate of each radar modulation signal. The number of identification errors was divided on the diagonal. From Table 3, we can conclude that the Sep-ResNet model had identification errors for NS, 2PSK, and 4PSK. The PSR values of NS, 2PSK, and 4PSK were 92%, 94%, and 95%, respectively. These values could be identified with 100% accuracy for the remaining four types of signals—the NS, 2PSK, and 4FSK identify errors—because their T–F images were very similar. The strong noise interfered with the characteristics of the signal, which led to errors in the Sep-ResNet classification. The LFM, NLFM, 2FSK, and 4FSK had high recognition accuracy because the features of their T–F images were distinguishable. Therefore, it can be concluded that an effective T–F analysis method is very important for the identification of radar signals. The CMWT we proposed can obtain clear T–F images, and there are no cross-terms. Our Sep-ResNet also has good classification performance.

Table 3. The confusion matrix of Sep-ResNet model for signal recognition when the SNR was −4 dB.

Input	Output						
	NS	**LFM**	**NLFM**	**2FSK**	**2PSK**	**4FSK**	**4PSK**
NS	92	0	0	0	4	0	4
LFM	0	100	0	0	0	0	0
NLFM	0	0	100	0	0	0	0
2FSK	0	0	0	100	0	0	0
2PSK	3	2	0	0	94	0	1
4FSK	0	0	0	0	0	100	0
4PSK	3	0	0	0	2	0	95

4.2.3. Verification of the Effectiveness of T–F Image Enhancement

To verify the effectiveness of the proposed enhancement algorithm for T–F images, the T–F images obtained by CMWT were enhanced using the algorithm in Algorithm 1, which also shows the image-enhancement processing results.

Figure 13b shows the grayscale T–F image of LFM signal after adaptive filtering and morphological processing. The noisy T–F images had better reductions in their noise interference, which was caused by their enhancement, and they retained the characteristics of the signal. The seven types of pre-enhanced and post-enhanced T–F images were fed into the Sep-ResNet model for recognition.

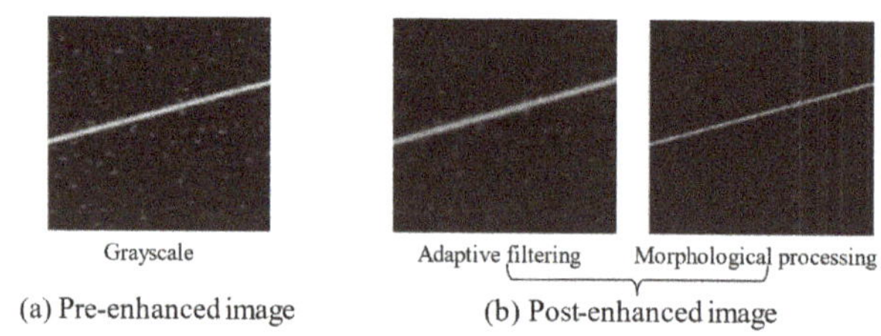

Figure 13. The LFM signal enhancement process at SNR = 8 dB. The (**a**) is the image before enhancement. The (**b**) is the image enhanced by the algorithm in this paper.

In Figure 14, it can be seen that after the T–F image enhancement algorithm was implemented, the overall PSR was improved by 2.35%, especially in the low-SNR −13, −10, −7, and −4 dB interval, by an average of 4.21%. When the SNR was −13 dB, the PSR was increased by 7.08%. The proposed enhancement algorithm could improve the PSR of radar-modulated signals in low-SNR environments. In general, a good image denoising and enhancement algorithm can reduce the interference of noise while retaining the characteristics of the signal, thereby improving the quality of T–F images and improving the system PSR of radar-modulated signals in low-SNR environments.

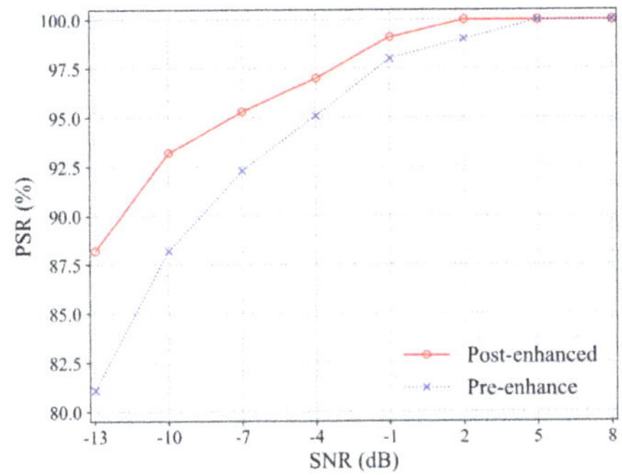

Figure 14. The PSR of the pre-enhancement and post-enhancement.

5. Conclusions

In response to the difficulty in identifying radar modulation signals in low-SNR environments, this paper proposed a method for combining the T–F analysis methods of CMWT and Sep-ResNet to intelligently identify radar modulation signals. In this paper, the T–F analysis of CMWT was used to extract the two-dimensional feature of the signal to obtain T–F images, and the images were enhanced through adaptive filtering and morphological processing. The enhanced T–F images were used as the input of Sep-ResNet for classification to intelligently and accurately achieve the recognition of radar-modulated signals in low-SNR environments. The experiments show that the T–F analysis of CMWT was better than the STFT and the improved CWD model. The classification performance of the proposed Sep-ResNet was better than AlexNet, the improved AlexNet, VGGNet16, Inception-v3, ResNet50, and the backbone of U-Net. Furthermore, the proposed enhancement algorithm was effective in filtering out the noise on T–F images. The method proposed successfully identified seven types of radar modulation signals (NS, LFM, NLFM, 2FSK, 2PSK, 4FSK, and 4PSK) in low-SNR environments. With SNR values ranging from -13 dB to 8 dB, the overall recognition rate was 96.57%, which was sufficient to effectively identify radar-modulated signals. Therefore, this method has the ability to resist noise interference, and can still maintain high PSR in low-SNR environments of ≤ -10 dB, thereby avoiding the difficulty and instability involved in the manually identification of radar modulation signals.

Author Contributions: Conceptualization, W.R., Z.Y. and Y.M.; investigation and analysis, W.R. and Z.Y.; software, Y.M.; validation, Y.M. and W.R.; writing—original draft preparation, Y.M. and W.R.; writing—review and editing, Y.M. All authors have read and agreed to the published version of the manuscript.

Funding: This research was funded by the National Natural Science Foundation of China, grant No. 61725105.

Institutional Review Board Statement: Not applicable.

Informed Consent Statement: Not applicable.

Data Availability Statement: Not applicable.

Conflicts of Interest: The authors declare no conflict of interest. We certify that the submission is original work and is not under review at any other publication. The funders had no role in the design of the study; in the collection, analyses, or interpretation of data; in the writing of the manuscript, or in the decision to publish the results.

Abbreviations

The following abbreviations were used in this manuscript:

SNR	signal-to-noise ratio
T–F	time–frequency
CMWT	complex Morlet wavelet transform
Sep-ResNet	channel-separable residual network
PSR	probability of successful recognition
RMS	radar modulation signal
WT	wavelet transform
CWD	Choi–Williams distribution
STFT	short-time Fourier transform

References

1. Latombe, G.; Granger, E.; Dilkes, F.A. Fast Learning of Grammar Production Probabilities in Radar Electronic Support. *IEEE Trans. Aerosp. Electron. Syst.* **2010**, *46*, 1262–1289. [CrossRef]
2. Gupta, M.; Hareesh, G.; Mahla, A.K. Electronic Warfare: Issues and Challenges for Emitter Classification. *Def. Sci. J.* **2011**, *61*, 228. [CrossRef]
3. Liu, Q.; Zheng, S.; Zuo, Y.; Zhang, H.; Liu, J. Electromagnetic Environment Effects and Protection of Complex Electronic Information Systems. In Proceedings of the 2020 IEEE MTT-S International Conference on Numerical Electromagnetic and Multiphysics Modeling and Optimization, Hangzhou, China, 7–9 December 2020.
4. Shen, J.; Huang, J.; Zhu, Y.; Institue, E.E. The Application of Feature Parameter Matching Method in Radar Signal Recognition. *Aerosp. Electron. Warf.* **2017**, *33*, 9–13. [CrossRef]
5. Ata'a, A.; Abdullah, S. Deinterleaving of Radar Signals and PRF Identification Algorithms. *IET Radar Sonar Navig.* **2007**, *1*, 340–347. [CrossRef]
6. Zhang, M.; Liu, L.; Diao, M. LPI Radar Waveform Recognition Based on Time-Frequency Distribution. *Sensors* **2016**, *16*, 1682. [CrossRef] [PubMed]
7. Peng, S.; Jiang, H.; Wang, H.; Alwageed, H.; Yao, Y. Modulation Classification Using Convolutional Neural Network Based Deep Learning Model. In Proceedings of the 2017 26th Wireless and Optical Communication Conference, Newark, NJ, USA, 7–8 April 2017.
8. Guo, J.; Wang, L.; Zhu, D.; Hu, C. Compact Convolutional Autoencoder for SAR Target Recognition. *IET Radar Sonar Navig.* **2020**, *14*, 967–972. [CrossRef]
9. Wu, B.; Yuan, S.; Li, P.; Jing, Z.; Huang, S.; Zhao, Y. Radar Emitter Signal Recognition Based on One-Dimensional Convolutional Neural Network with Attention Mechanism. *Sensors* **2020**, *20*, 6350. [CrossRef] [PubMed]
10. Zhang, D.; Ding, W.; Zhang, B.; Xie, C.; Li, H.; Liu, C.; Han, J. Automatic Modulation Classification Based on Deep Learning for Unmanned Aerial Vehicles. *Sensors* **2018**, *18*, 924. [CrossRef] [PubMed]
11. Wei, S.; Qu, Q.; Su, H.; Wang, M.; Shi, J.; Hao, X. Intra-Pulse Modulation Radar Signal Recognition Based on CLDN Network. *IET Radar Sonar Navig.* **2020**, *14*, 803–810. [CrossRef]
12. Yang, J.; Zhang, H. LPI Radar Signal Recognition Based on Improved AlexNet. *Mod. Electron. Technol.* **2020**, *43*, 57–60. [CrossRef]
13. Qu, Z.; Hou, C.; Hou, C.; Wang, W. Radar Signal Intra-Pulse Modulation Recognition Based on Convolutional Neural Network and Deep Q-Learning Network. *IEEE Access* **2020**, *8*, 49125–49136. [CrossRef]
14. Qu, Z.; Mao, X.; Deng, Z. Radar Signal Intra-Pulse Modulation Recognition Based on Convolutional Neural Network. *IEEE Access* **2018**, *6*, 43874–43884. [CrossRef]
15. Wang, X.; Huang, G.; Zhou, Z.; Gao, J. Radar Emitter Recognition Based on the Short Time Fourier Transform and Convolutional Neural Networks. In Proceedings of the 2017 10th International Congress on Image and Signal Processing, BioMedical Engineering and Informatics, Shanghai, China, 14–16 October 2017.
16. Li, J.; Gao, H.; Zhang, M.; Ma, Z.; Wei, L.; Zhang, C. Power Frequency Communication Signal Detection Based on Reassigned Time-Frequency Spectrogram. In Proceedings of the 2020 IEEE 20th International Conference on Communication Technology, Nanning, China, 28–31 October 2020; pp. 1213–1216.
17. Kirillov, S.N.; Lisnichuk, A.A. Analysis of Narrow-Band Interference Effect on Cognitive Radio Systems Based on Synthesized Four-Position Radio Signals. In Proceedings of the 2018 XIV International Scientific-Technical Conference on Actual Problems of Electronics Instrument Engineering (APEIE), Novosibirsk, Russia, 2–6 October 2018; pp. 50–54.
18. Zaman, S.M.K.; Marma, H.U.M.; Liang, X. Broken Rotor Bar Fault Diagnosis for Induction Motors Using Power Spectral Density and Complex Continuous Wavelet Transform Methods. In Proceedings of the 2019 IEEE Canadian Conference of Electrical and Computer Engineering, Edmonton, AB, Canada, 5–8 May 2019.
19. Sun, H.; Zhang, L.; Jin, X. An Image Denoising Method Which Combines Adaptive Median Filter with Weighting Mean Filter. In Proceedings of the 2012 International Conference on Measurement, Information and Control, Harbin, China, 18–20 May 2012; pp. 392–396.

20. Li, S.; Yu, L.; Liu, X. Algorithm of Canny Operator Edge Pre-Processing Based on Mathematical Morphology. In Proceedings of the 2020 International Conference on Computer Engineering and Application, Guangzhou, China, 18–20 March 2020; pp. 349–352.
21. Simonyan, K.; Zisserman, A. Very Deep Convolutional Networks for Large-Scale Image Recognition. *arXiv* **2015**, arXiv:1409.1556.
22. He, K.; Zhang, X.; Ren, S.; Sun, J. Deep Residual Learning for Image Recognition. In Proceedings of the 2016 IEEE Conference on Computer Vision and Pattern Recognition, Las Vegas, NV, USA, 27–30 June 2016; pp. 770–778.
23. Xie, S.; Girshick, R.; Dollár, P.; Tu, Z.; He, K. Aggregated Residual Transformations for Deep Neural Networks. In Proceedings of the 2017 IEEE Conference on Computer Vision and Pattern Recognition, Honolulu, HI, USA, 21–26 July 2017; pp. 5987–5995.
24. Santurkar, S.; Tsipras, D.; Ilyas, A.; Madry, A. How Does Batch Normalization Help Optimization? *arXiv* **2019**, arXiv:1805.11604.
25. Szegedy, C.; Vanhoucke, V.; Ioffe, S.; Shlens, J.; Wojna, Z. Rethinking the Inception Architecture for Computer Vision. In Proceedings of the 2016 IEEE Conference on Computer Vision and Pattern Recognition, Las Vegas, NV, USA, 27–30 June 2016; pp. 2818–2826.
26. Wang, L.; Guo, S.; Huang, W.; Qiao, Y. Places205-VGGNet Models for Scene Recognition. *arXiv* **2015**, arXiv:1508.01667.
27. Ronneberger, O.; Fischer, P.; Brox, T. U-Net: Convolutional Networks for Biomedical Image Segmentation. In Proceedings of the Medical Image Computing and Computer-Assisted Intervention—MICCAI 2015, Munich, Germany, 5–9 October 2015; Navab, N., Hornegger, J., Wells, W.M., Frangi, A.F., Eds.; Springer: Cham, Switzerland, 2015; pp. 234–241.

MDPI
St. Alban-Anlage 66
4052 Basel
Switzerland
Tel. +41 61 683 77 34
Fax +41 61 302 89 18
www.mdpi.com

Sensors Editorial Office
E-mail: sensors@mdpi.com
www.mdpi.com/journal/sensors

www.ingramcontent.com/pod-product-compliance
Lightning Source LLC
Chambersburg PA
CBHW061608220926
51850CB00005B/452
* 9 7 8 3 0 3 6 5 4 6 2 9 2 *